Water and Health

Prati Pal Singh · Vinod Sharma
Editors

Water and Health

Springer

Editors
Prati Pal Singh
National Institute of Pharmaceutical
Education and Research
Mohali, Punjab
India

Vinod Sharma
The National Academy of Sciences
Allahabad
India

ISBN 978-81-322-1028-3 ISBN 978-81-322-1029-0 (eBook)
DOI 10.1007/978-81-322-1029-0
Springer New Delhi Heidelberg New York Dordrecht London

Library of Congress Control Number: 2013946214

Printed on acid-free paper

Springer is part of Springer Science+Business Media (www.springer.com)

Foreword

Water has a tremendous bearing on health, development and quality of human life. Water provides a range of ecosystem services. These services provided by water are thus fundamental constituents of the economy. With each passing day, however, water availability is shrinking and water quality is deteriorating, acquiring challenging proportions.

At this critical time, the publication of the book "Water and Health" is timely and fulfils a felt need of the scientific community.

The book is a compilation of existing knowledge on water contamination and future directions to stop and reverse the current trend and conserve safe water. The book unfolds the consequences of contaminated water and opens up new knowledge on water related issues, including scientific awareness of safe water at all levels of community.

The editors Prof. Prati Pal Singh and Prof. V. P. Sharma are recognized experts on water and related issues. The editors have done admirable job in conceptualizing, collating, and editing this important compendium. The chapters in this book are written by well-established scientists in their area of expertise. Each chapter has been arranged to provide free flowing current of knowledge to benefit the reader.

The book would be of immense value to a large section of the society comprising students, scientists, educationists, public health workers, engineers, environmentalists, policy makers, planners, administrators and social workers.

The publication of the book on "Water and Health" is very timely as we embark on our 12th Five Year Plan. This takes into account the background of the United Nations Millennium Development Goals.

Prof. M. G. K. Menon, FRS
Former Minister of Science and Technology
Government of India
Former Scientific Advisor to the
Prime Minister of India, New Delhi

Prelude

Water is the driving force in nature
–Leonardo da Vinci

Water and life are inextricably intertwined, so much so, that whenever we search for life, terrestrially or extra-terrestrially, we search for water in its one or the other form. The quality and availability of water is thus bound to affect human health. Though water covers nearly 70 % of the earth's surface, yet humans can use only 0.007 % of all the water (in lakes, rivers, reservoirs and underground sources) on our planet. Our Earth has 97.5 % salt water and only 2.5 % fresh water, of which 70 % is frozen and the remaining 30 % lies in soil and underground aquifers. To meet our basic needs, we need nearly 20–50 liters of water every day. Mankind has always been in quest for pure drinking water. Ancient Sanskrit writings have documented the importance of water purification. The Egyptian inscriptions depict the methods of water treatment. Hippocrates (460–354 B. C.) emphasized the importance of water in health, and has prescribed boiling and straining (Hippocrates' sleeve) of rain water. Clean water is now regarded as the most important public health determinant. Every 20 seconds, a child dies from a water-related illness, and by 2020, 76 million people are estimated to die of water-related diseases, even if United Nations Millennium Development Goals for water are met. In India, according to the World Bank, unsafe drinking water is responsible for 21 % of communicable diseases, and the Indian economy suffers a staggering loss of 73 million working days annually due to water-related diseases. In 2011, the Human Rights Council of the United National General Assembly has adopted the resolution on "The Human Right to Safe Drinking Water and Sanitation". As it is, all these touching facts and figures prompted us to bring out this book.

The book has been divided in two parts. The first one, constituting nearly 80 % of the book, contains nineteen chapters on water and pathogens which include some of the important viral, bacterial, protozoan and helminthic diseases. Deliberately, we have chosen not to group these chapters in some particular categories *viz.* water-borne, water-washed, water-based or water-related diseases for the simple reason because in our considered opinion such a grouping of the chapters has its own limitations, and thus may not gel well with the purpose of bringing out this book. The second part contains five chapters on water and chemical contamination, wherein several issues related to water quality and pesticide toxicity are highlighted. The chapters on contamination of water by pharmaceuticals adds an entirely new dimension to the subject. Further, the chapters on water in south-west Punjab provides important insights into the water-related problems specific to this region.

Directly or indirectly, water touches us in practically each and every aspect of our physical, biological, educational, social, cultural, developmental, economic, emotional, religious and spiritual being. Among these, the relationship between water and education is very important and thus needs to be elaborated. There is need of the role of water in education and of water education. Globally, half of the children lose 443 million school days due to water-related diseases. Half of the primary schools in developing countries lack safe drinking water and toilets, and this usually results in school drop-out of girls at puberty. It is now well accepted that with every 10 % increase in women's literacy, a country's economy can grow up to 0.3 %. The availability of safe drinking water (along with improved sanitation) in schools helps in improving the attendance of children, especially girls. For any workable and sustainable solution, we need full information and data. With a good education on water-related issues, we will be better informed and it will help us in better understanding and tackling water-related problems, especially water management. Therefore, if we educate our children about water, it will help us in solving our several water-related problems.

There can be no safe drinking water without sanitation. According to the World Water Council, one billion people do not have access to safe drinking water, while 2.6 billion people lack toilets and other sanitation facilities. In this book, in view of focus and brevity, we have remained water and disease centric, but this in no way takes us away from the importance of sanitation in relation to potable water and health. In the end, let us not forget what the World Health Organization says: there is 3–34-fold economic return on every US $1 spent on water.

December 12, 2012

Prati Pal Singh
Vinod Sharma

Contents

About the Editors

Professor Prati Pal Singh, Ph.D., a renowned educationalist, microbiologist, biotechnologist and neuroimmunopharmacologist, has made important contributions to the treatment of malaria, tuberculosis, leishmaniasis and amoebiasis; he conceptualized/reported opiates as a new class of drugs. He has over 250 publications including one in *Nature Medicine*, books and international patents, and has supervised over 37 Ph.D./M. S. students. Prof. Singh is on the Editorial Board of *Journal of Neuroimmune Pharmacology, USA* and has been the *Editor-in-Chief* of *Journal of Parasitic Diseases*. Prof. Singh has been elected *F.N.A.Sc.* and *F.A.M.I.*, and awarded The Bill and Melinda Gates Foundation Global Health Award and Tulasbai Somani Award. Prof. Singh is internationally recognized, has been a co-investigator in an Indo-EU project on tuberculosis, and has traveled widely abroad for lectures/meetings. He has been Member/Chairman of committees of ICMR and other organizations.

Professor V. P. Sharma, D.Phil., D.Sc, is the Founder Director of the National Institute of Malaria Research, India. He is fellow of three science and medical academies of India and the third world academy of sciences. He has 40 years of research experience in vector biology and malaria and has published 400 scientific papers, books, video programmes etc. He has held the esteemed office of the President, The National Academy of Sciences, India. He has served International Organizations and currently he is advisor to WHO on malaria. He is the recipient of various prestigious awards of which few are: Padma Shree, WHO Darling Foundation prize, ICMR's BR Ambedkar and MOT Iyanger, Ranbaxy, FICCI, Green Scientist, OP Bhasin and Vasvik awards. Dr. Sharma is Meghnad Saha distinguished fellow of NASI at CRDT, Indian Institute of Technology, New Delhi.

Dengue: A Water-Related Mosquito-Borne Disease

1

Bhawana Jain and Amita Jain

Abstract

Dengue is a common vector-borne disease of tropical and subtropical areas, affecting about 50 million people each year and exposing more than 2.5 billion people at risk. Dengue virus (DENV), responsible for the disease, belongs to family Flaviviridae, genus Flavivirus and has four distinguished serotypes (DENV 1, DENV 2, DENV 3, and DENV 4). The main vector of the virus is *Aedes aegypti* mosquito. Water logging increases mosquito density and an increase in vector-borne diseases (such as dengue, malaria, West Nile fever) through the expansion in the number and range of vector habitats. Infection with any of the DENV serotypes may be asymptomatic in the majority of cases or may result in a wide spectrum of clinical symptoms, ranging from a mild flu-like syndrome (known as dengue fever [DF]) to the most severe forms of the disease; Dengue Hemorrhagic Fever/Dengue Shock Syndrome (DHF/DSS). Recently in 2008, WHO has revised its classification and divides the clinical illness to non-severe dengue, dengue ± warning signs, and severe dengue. Pathogenesis of severe dengue disease is very complex and not properly elucidated. Timely diagnosis of the disease is the key to control it. Routine diagnosis largely depends upon NS1 antigen, IgM, and genome detection by different methods. Confirmatory diagnosis is by cell culture method, the most important of which is C6/36 mosquito cell line or intrathoracic inoculation of mosquito. Till date, no effective vaccine is available in the market. Different approaches are being used to develop various formulations of vaccine, but are of not much use. Alternative vaccine strategies in the form of small interfering RNA (siRNA) and micro RNA (miRNA) seem to be

B. Jain · A. Jain (✉)
Department of Microbiology, Virology Laboratory,
King George's Medical University,
Lucknow 226001, India
e-mail: amita602002@yahoo.com

B. Jain
e-mail: drbhawana_kgmu@hotmail.com

P. P. Singh and V. Sharma (eds.), *Water and Health*,
DOI: 10.1007/978-81-322-1029-0_1, © Springer India 2014

quite promising in present era. Though these technologies are in very primitive phase, various studies are going on all over the world to bring them to reality.

Keywords
Dengue · Dengue virus · Dengue fever · Dengue hemorrhagic fever · Dengue shock syndrome

Introduction

Water logging increases mosquito density and an increase in vector-borne diseases (such as dengue, malaria, West Nile fever) through the expansion in the number and range of vector habitats. Standing water caused by heavy rainfall or overflow of rivers can act as breeding sites for mosquitoes and therefore enhance the potential for exposure of the population. Dengue is a mosquito-borne infection that in recent decades has reached endemic proportions in India and became a major international public health concern. Dengue is found in tropical and subtropical regions around the world, predominantly in urban and semi-urban areas.

History

The name for dengue disease comes from the Swahili language, meaning a disease caused by an evil spirit. Reports of symptoms consistent with dengue date back over two millennia (Gubler 2006). Although first reports of major epidemics of an illness thought to possibly be dengue occurred on three continents (Asia, Africa, and North America) in 1779 and 1780 (Rush 1789; Hirsch 1883; Pepper 1941; Howe 1977), reports of illnesses clinically compatible with dengue fever occurred even earlier. The earliest record found to date is in a Chinese encyclopedia of disease symptoms and remedies (Nobuchi 1979). The disease was called water poison by the Chinese and was thought to be somehow connected with flying insects associated with water. Thus, dengue or a very similar illness had a wide geographic distribution before the 18th century, when the first known pandemic of dengue-like illness began.

The Dengue Virus

Dengue virus (DENV) is positive-sense RNA viruses belonging to the mosquito-borne genus *Flavivirus* and has four circulating serotypes (DENV-1–4). The mosquito-borne flaviviruses have evolved into two major groups distinguished by their clinical presentation in humans and their ecology (Refer to Table 1.1) (Sabin 1959; Gaunt et al. 2001; Kramer and Ebel 2003). The Encephalitic viruses are zoonotic viruses with birds as the natural vertebrate host and primarily Culex species mosquitoes as vectors. The other group is more viscerotropic and may cause hemorrhagic fever. These viruses have a forest cycle with lower primates as vertebrate hosts and Aedes mosquitoes as their principal vectors. DENVs have fully adapted to humans in the urban environment and no longer require the forest cycle for maintenance (Gubler 2002).

Dengue virion has host cell–derived lipid bilayer and three viral structural proteins: capsid, envelope and membrane protein, and the RNA genome. Dengue virus has a single RNA molecule of positive polarity. Genome is surrounded by an approximately icosahedral or isometric nucleocapsid about 30 nm in diameter, composed of viral capsid protein (C), which is a small basic dimer-forming protein also found to co-localize the nucleus of infected cells (Netsawang et al. 2010). This nucleocapsid is covered by a host cell–derived lipid envelope about 10 nm deep. The complete virion is about 50 nm in diameter (Fig. 1.1).

Table 1.1 Mosquito-borne flaviviruses

Mosquito-borne flavivirus	Clinical presentation	Vector
A. *Neurotropic*		
1. Japanese encephalitis virus (JEV)	Encephalitis	Culex
2. West Nile virus (WNV)	Encephalitis	Culex
3. Murray Valley encephalitis virus (MVEV)	Encephalitis	Culex
4. St. Louis encephalitis virus (SLEV)	Encephalitis	Culex
B. Viscerotropic		
1. Yellow fever virus (YFV)	Hemorrhagic	Aedes
2. Dengue virus (DENV)	Hemorrhagic	Aedes

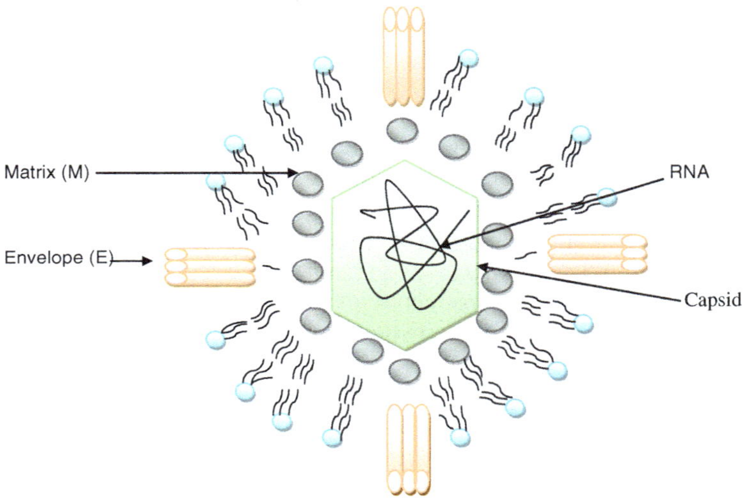

Fig. 1.1 Line diagram of dengue virus structure

In immature virions, found inside the infected cells, the envelope glycoprotein (E) is arranged in trimers that give spiky appearance to virion surface and the precursor of membrane protein (preM) prevents premature fusogenic activity of the envelope protein. During virion maturation, the prepart of the M protein is cleaved and the envelope protein is rearranged from trimers to head-to-tail dimers that lie flat against the membrane giving it smooth appearance (Kuhn et al. 2002; Zhang et al. 2004).

The genomic RNA is infectious and generates the production of virions when transported into suitable host cells. It is approximately 11 kb in length, and it is flanked by conserved untranslated regions (UTR) which form secondary structures that mediate genome circularization and have important functions in genome replication (Chambers et al. 1990). The UTRs are well conserved in sequence and structure. The dengue virus 5′ UTR is approximately 100 bp in length and has a type I cap (m7GpppAm) at the 5′ end. It does not have a poly A tail at the 3′ end (Hahn et al. 1988). In addition to the structural proteins, seven virus-encoded non-structural (NS) proteins are also detected in dengue virus (NS1, NS2A, NS2B, NS3, NS4A, NS4B, and NS5) (Fig. 1.2). The functions of all NS-proteins are not well characterized; however, they play significant role in viral replication and polyprotein processing (Bollati et al. 2009).

The NS1 protein is secreted from infected cells which can be measured in the blood of patients infected with dengue virus. It is used as an early diagnostic marker. NS1 protein interacts with host immune system and also helps in viral replication. NS2A and NS2B are small membrane-associated proteins. NS2B functions

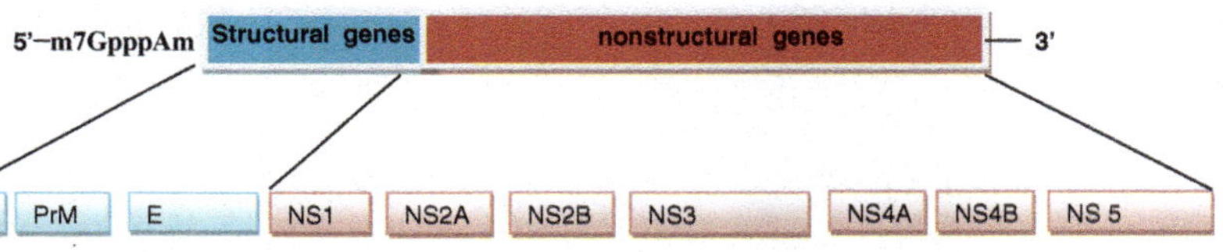

Fig. 1.2 Dengue virus RNA protein-coding regions and genome organization

as co-factor for multifunctional NS3 protein, which has trypsin-like serine protease and helicase activities (Assenberg et al. 2009). The NS4A and NS4B are small membrane-associated proteins; NS4A has been associated with membrane alterations (Miller et al. 2007), and NS4B has been associated with membrane structures involved in the replication. The NS5 protein is the viral RNA–dependent RNA polymerase mediating the genome replication (Yap et al. 2007). Of the non-structural proteins, NS1 is known to raise antibody responses. It is also known to evoke T-cell responses, in addition to NS3, NS4B, and NS5 (Rothman 2004).

Distinct genotypes or lineages (viruses highly related in nucleotide sequence) have been identified within each serotype, highlighting the extensive genetic variability of the dengue serotypes. Purifying selection appears to be a dominant theme in dengue viral evolution, however, such that only viruses that are "fit" for both human and vector are maintained. Among them, "Asian" genotypes of DEN-2 and DEN-3 are frequently associated with severe disease accompanying secondary dengue infections (Lanciotti et al. 1994; Leitmeyer 1999; Messer 2003). Intrahost viral diversity (quasispecies) has also been described in human hosts.

The Vector

Transmission of dengue virus by *A. aegypti* was first demonstrated in 1906 (Bancroft 1906). This mosquito is a tropical and subtropical species widely distributed around the world. *A. aegypti* is one of the most efficient vectors for arboviruses because it is highly anthropophilic, frequently bites several times before completing oogenesis, and thrives in close proximity to humans. Due to lower temperatures at high altitude, *A. aegypti* is relatively uncommon above 1,000 m. The immature stages are found in water-filled habitats, mostly in artificial containers closely associated with human dwellings and often indoors. Studies suggest that most female *A. aegypti* may spend their lifetime in or around the houses where they emerge as adults. This means that people, rather than mosquitoes, rapidly move the virus within and between communities. Dengue outbreaks have also been attributed to *Aedes albopictus*, *Aedes polynesiensis*, and several species of the *Aedes scutellaris* complex (WHO 2009).

In recent decades, *A. albopictus* has spread from Asia to Africa, the America, and Europe, notably aided by the international trade in used tyres in which eggs are deposited when they contain rainwater. The eggs can remain viable for many months in the absence of water.

The life cycle of a mosquito includes four separate stages: egg, larva, pupa, and adult; the first three stages require an aqueous environment (Fig. 1.3). The duration of the developmental stages depend on environmental temperature and availability of food at the larval stage. For *A. aegypti,* it takes roughly 8–10 days at room temperature to reach the adult stage (Gubler and editor 1997).

Transmission Cycles

Two transmission cycles are known to exist; one being the human to human epidemic transmission cycle occurring in urban environment and the other involving non-human primates (monkeys) and jungle mosquito (sylvatic cycle) (Fig. 1.4).

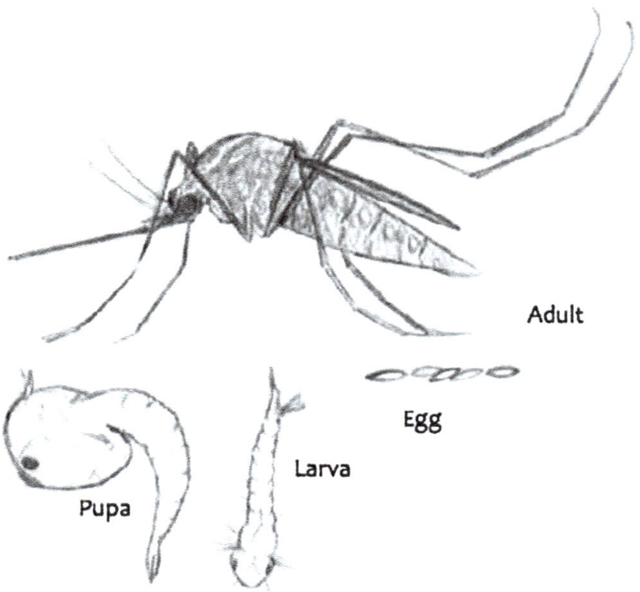

Fig. 1.3 Mosquito life cycle

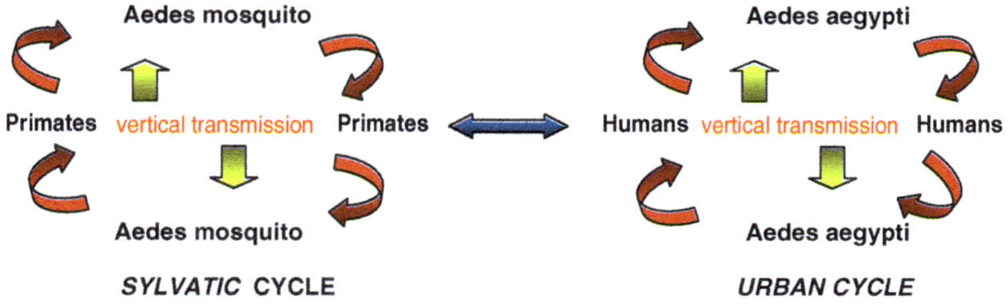

Fig. 1.4 Transmission cycles

In urban transmission cycle, humans are the main amplifying host of the virus. Dengue virus circulating in the blood of viremic humans is ingested by female mosquitoes during feeding. The virus then infects the mosquito mid-gut and subsequently spreads systemically over a period of 8–12 days. After this extrinsic incubation period, the virus can be transmitted to other humans during subsequent probing or feeding. The extrinsic incubation period is influenced in part by environmental conditions, especially ambient temperature. Thereafter, the mosquito remains infective for the rest of its life. Vertical transmission (transovarial transmission) of dengue virus has been demonstrated in the laboratory but rarely in the field. The significance of vertical transmission for maintenance of the virus is not well understood. Sylvatic dengue strains in some parts of Africa and Asia may also lead to human infection, causing mild illness. Several factors can influence the dynamics of virus transmission—including environmental and climate factors, host–pathogen interactions, and population immunological factors. Climate directly influences the biology of the vectors and thereby their abundance and distribution; it is consequently an important determinant of vector-borne disease epidemics (WHO 2009).

Dengue Disease

Dengue virus infection in humans causes a spectrum of illness ranging from in apparent or mild febrile illness to severe and fatal hemorrhagic disease (Anonymous 1986). Earlier, the WHO scheme classified symptomatic dengue virus infections into three categories: undifferentiated fever, dengue fever, and DHF/DSS. In September 2008, the global expert meeting on Dengue classification in Geneva recommended a revised case classification into dengue, dengue with warning signs, and severe dengue (WHO 2009).

Infection with any of the four serotypes causes a similar clinical presentation that may vary in severity, depending on a number of risk factors. The incubation period varies from 3 to 14 days (average, 4–7 days) (Sabin 1952). In areas where dengue is endemic, the illness is often clinically non-specific, especially in children, with symptoms of a viral syndrome that has a variety of local names. Important risk factors influencing the proportion of patients who have severe disease during epidemic transmission include the strain and serotype of the infecting virus and the immune status, age, and genetic background of the human host (Anonymous 1986; Barnes and Rosen 1974).

Dengue Fever

Dengue fever is clinically defined as an acute febrile illness with two or more manifestations (headache, retro-orbital pain, myalgia, arthralgia, rash, hemorrhagic manifestations, or leukopenia) and occurrence at the same location and time as other confirmed cases of dengue fever. The initial temperature may rise to 102–105 °F, and fever may last for 2–7 days. The fever may drop after a few days, only to rebound 12–24 h later (saddleback). A relative bradycardia may be noted despite the fever. The conjunctivae may be injected, and the pharynx may be inflamed. Lymphadenopathy is common. Rash is variable but occurs in up to 50 % of patients as either early or late eruptions. Facial flushing or erythematous mottling may occur coincident with or slightly before onset of fever and disappears 1–2 days after onset of symptoms. A second rash, varying in form from scarlatiniform to maculopapular, may appear between days 2 and 6 of illness. The rash usually begins on the trunk and spreads to the face and extremities. In some cases, an intense erythematous pattern with islands of normal skin is observed. The average duration of the second rash is 2–3 days. Toward the end of the febrile phase of illness or after the temperature falls to or below normal, petechiae may appear; these may be scattered or confluent. Intense pruritus followed by desquamation on the palms of the hands and soles of the feet may occur.

Hemorrhagic manifestations in dengue fever patients are not uncommon and range from mild to severe. Skin hemorrhages, including petechiae and purpura, are the most common, along with gum bleeding, epistaxis, menorrhagia, and gastrointestinal (GI) hemorrhage. Hematuria occurs infrequently, and jaundice is rare. Clinical laboratory findings associated with dengue fever include a neutropenia followed by a lymphocytosis, often marked by atypical lymphocytes. Liver enzyme levels in the serum may be elevated; the elevation is usually mild, but in some patients, alanine aminotransferase and aspartate aminotransferase levels reach 500–1,000 U/L. In one epidemic of DEN-4, 54 % of confirmed patients with data reported on liver enzymes had elevated levels (Dietz et al. 1996). Thrombocytopenia is also common in dengue fever; in the above epidemic, 34 % of patients with confirmed dengue fever who were tested had platelet counts of less than 100,000/mm^3 (Dietz et al. 1996). Dengue fever is generally self-limiting and is rarely fatal. The acute phase of illness lasts for 3–7 days, but the convalescent phase may be prolonged for weeks and may be associated with weakness and depression, especially in adults. No permanent sequelae are known to be associated with this infection.

Dengue Hemorrhagic Fever

A case must meet all four of the following criteria to be defined as DHF:

- fever or history of fever lasting 2–7 days;
- a hemorrhagic tendency shown by a positive tourniquet test or spontaneous bleeding;

- thrombocytopenia (platelet count 10^9/L or less); and
- evidence of plasma leakage shown either by hemoconcentration with substantial changes in serial measurements of packed-cell volume, or by the development of pleural effusions or ascites, or both.

DHF is further classified into four severity grades according to World Health Organization criteria:

Grade I—Fever accompanied by non-specific constitutional symptoms with a positive tourniquet test as the only hemorrhagic manifestation

Grade II—Same as grade I, except with spontaneous hemorrhagic manifestations

Grade III—Circulatory failure manifested by rapid, weak pulse with narrowing of the pulse pressure (<20 mmHg) or hypotension

Grade IV—Profound shock with undetectable blood pressure and pulse

Dengue Shock Syndrome (DSS)

The term dengue shock syndrome (DSS) refers to DHF grades III and IV, in which shock is present as well as all four DHF-defining criteria. Moderate shock, identified by narrowing of the pulse pressure or hypotension for age, is present in grade III DHF, whereas profound shock with no detectable pulse or blood pressure is present in grade IV DHF by using a set of clinical and/or laboratory parameters, one sees a clear-cut difference between patients with severe dengue and those with non-severe dengue.

Newer classification

For practical reasons, it was desirable to split the large group of patients with non-severe dengue into two subgroups—patients with warning signs and those without them (Refer to Table 1.2).

Pathogenesis

The pathogenesis of DHF and DSS is still controversial. Lack of suitable animal model for dengue virus study hampers the actual understanding of pathogenesis. Multiple theories are there to explain the pathogenic changes that occur in DHF and DSS. The most commonly accepted is known as the secondary infection or immune enhancement hypothesis (Halstead 1970, 1988) (Fig. 1.5). This hypothesis implies that patients experiencing a second infection with a heterologous dengue virus serotype have a significantly higher risk for developing DHF and DSS (Halstead 1988). Various in vitro studies and autopsy findings have come up with the suggestions that three organ systems are

Table 1.2 Criteria for diagnosing dengue (with or without warning signs) and severe dengue (new classification by WHO in 2008)

Dengue ± Warning signs		Severe dengue
Probable dengue	*Warning signs*	*Severe plasma leakage*
Live in/travel to dengue endemic area Fever and 2 of the following criteria 1. Nausea, vomiting 2. Rash 3. Aches and pains 4. Tourniquet test positive 5. Leukopenia 6. Any warning sign	1. Abdominal pain or tenderness 2. Persistent vomiting 3. Clinical fluid accumulation 4. Mucosal bleed 5. Lethargy, restlessness 6. Liver enlargement >2 cm 7. Laboratory: increase HCT concurrent with rapid decrease in platelet count	Leading to • Shock (DSS) • Fluid accumulation with respiratory distress *Severe bleeding* As evaluated by clinician *Severe organ involvement* • Liver: ALT or AST ≥1,000 units • CNS: impaired consciousness • Heart and other organs

HCT hematocrit, *DSS* dengue shock syndrome, *ALT* alanine transferase, *AST* aspartate transferase, *CNS* central nervous system

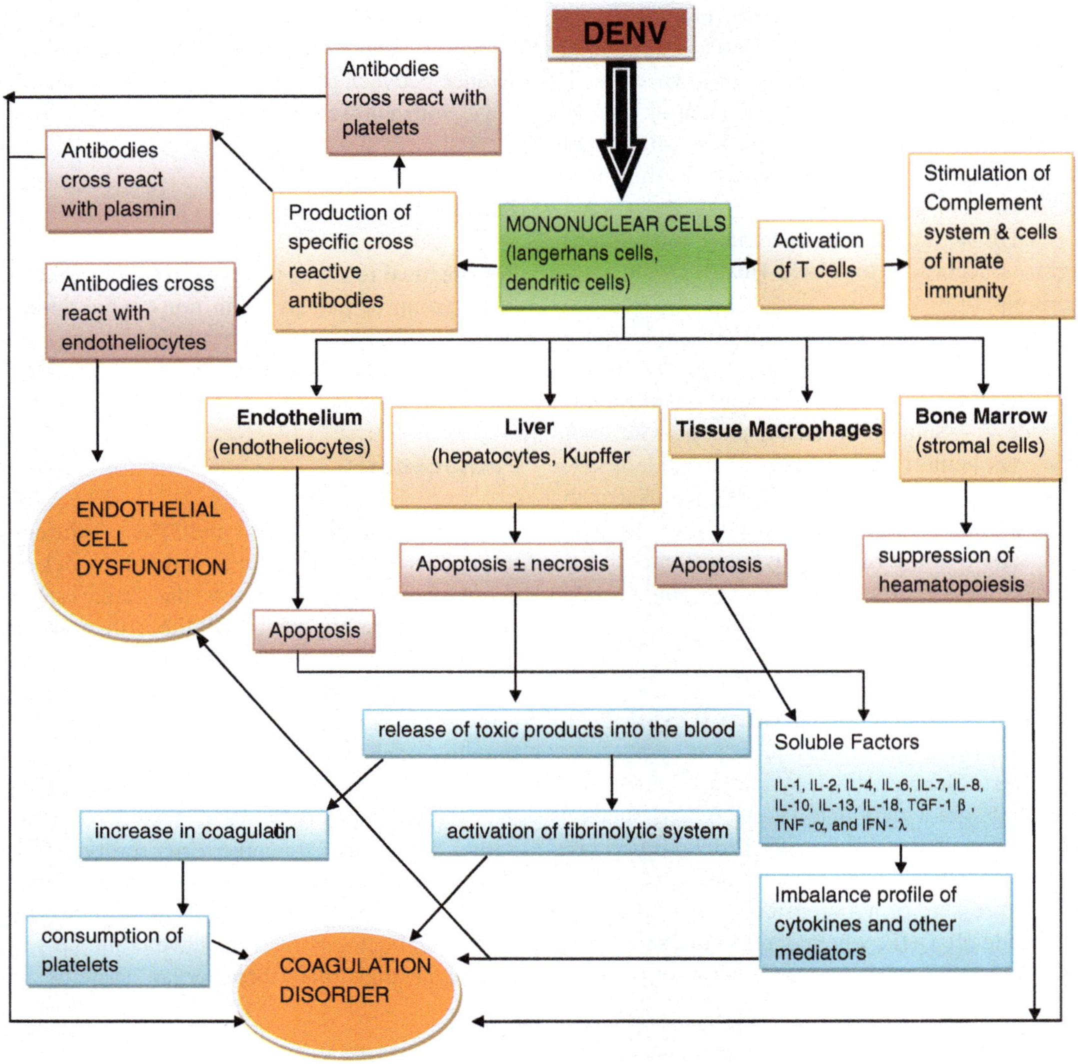

Fig. 1.5 Flow diagram depicting complex pathogenesis of severe dengue (adapted from Martina et al 2009)

basically involved in DHF/DSS pathogenesis: the immune system, the liver, and the endothelial lining. Through mosquito bite, virus enters in blood stream, with spillover in the epidermis and dermis, resulting in infection of immature Langerhans cells (epidermal dendritic cells) and keratinocytes (Limon-Flores et al. 2005). Infected cells then migrate from site of infection to lymph nodes, where monocytes and macrophages are recruited, which become targets of infection. Consequently, infection is amplified and virus is disseminated through the lymphatic system. As a result of this primary viremia, several cells of the mononuclear lineage, including liver macrophages (de Macedo et al. 2006), are infected.

Preexisting heterologous dengue antibody recognizes the infecting virus and forms an antigen–antibody complex, which is then bound to and internalized by immunoglobulin Fc receptors on the cell membrane of leukocytes, especially macrophages. Because the antibody is heterologous, the virus is not neutralized and is free to replicate once inside the macrophage.

Thus, it is hypothesized that prior infection, through a process known as antibody-dependent enhancement (ADE), enhances the infection and replication of dengue virus in cells of the mononuclear cell lineage (Halstead and Rourke 1977a, b; Halstead 1988). It is thought that these cells produce and secrete vasoactive mediators in response to dengue infection, which causes increased vascular permeability leading to hypovolemia and shock.

High viral load and activation of T cells result in "storm" of inflammatory cytokines and other mediators leading to increased plasma leakage characteristic of DHF/DSS. Higher plasma levels of IL-1, IL-2, IL-4, IL-6, IL-7, IL-8, IL-10, IL-13, IL-18, TGF-1α, TNF-β, and IFN-λ have been found in patients with severe DENV infections, in particular in patients with DSS (Bozza et al. 2008).

Epidemiology

Dengue is the most rapidly spreading mosquito-borne viral disease in the world, estimating about 50–100 million infections per year. In the last 50 years, incidence has increased 30-fold with increasing geographic expansion to new countries and, in the present decade, from urban to rural settings. The factors responsible for the dramatic increase in epidemic dengue as a global public health problem in the past 17 years are complex and not fully understood. However, the resurgence appears to be closely associated with demographic and societal changes over the past 50 years (Gubler 1987, 1998). Two major factors have been the unprecedented global population growth and the associated unplanned and uncontrolled urbanization, especially in tropical developing countries. The substandard housing, crowding, and deterioration in water, sewer, and waste management systems associated with unplanned urbanization have created ideal conditions for increased transmission of mosquito-borne diseases in tropical urban centers. A third major factor has been the lack of effective mosquito control in areas where dengue is endemic (Gubler 1987, 1989, 1997, 1998). A fourth factor responsible for the global emergence of dengue and DHF is increased air travel, which provides the ideal mechanism for the transport of dengue and other urban pathogens between population centers of the world (Gubler 1987, 1998). A fifth factor that has contributed to the epidemic dengue has been the decay in public health infrastructures in most countries in the past 30 years.

Dengue infection is endemic in India for over two centuries as a benign and self-limited disease. In recent years, the disease has changed its manifestation and presents in more severe form like DHF/DSS and with increasing frequency of outbreaks. India has experienced several outbreaks of dengue virus infection with different serotypes (Refer to Table 1.3). In our laboratory, we have detected prevalence of DENV 1, 2, and 3 in the year 2008 and 2009 while more of DENV 1 and 3 in 2010 (Pandey et al 2012).

Laboratory Diagnosis

Laboratory diagnostic methods for diagnosing dengue virus infection may involve detection of the virus, viral nucleic acid, antigens or antibodies, or a combination of these techniques. After the onset of illness, the virus can be detected in serum, plasma, circulating blood cells, and other tissues for 4–5 days. During the early stages of the disease, virus isolation, nucleic acid, or antigen detection can be used to diagnose the infection. At the end of the acute phase of infection, serology is the method of choice for diagnosis.

According to WHO, following criteria should be followed for dengue diagnosis:

Highly Suggestive

One of the following:

1. IgM + in a single serum sample
2. IgG + in a single serum sample with a HI titer of 1,280 or greater

Table 1.3 Epidemiology of circulating serotypes of dengue in last two decades in India

Years	Site in India	Virus serotype	References
1990	Calcutta	DV-3	Bhattacharjee et al. (1993)
1993	Jammu	DV-2	Padbidri et al.)(1996)
1996	Haryana	DV-2	Kumar et al. (2008)
1996	Delhi	DV-2	Broor et al. (1997), Dar et al. (1999)
1996	Lucknow (Uttar Pradesh)	DV-2	Agarwal et al. (1999)
1997	Delhi	DV-1	Kurukumbi et al. (2001)
2001	Gwalior (Madhya Pradesh)	DV-2	Parida et al. (2002)
2003	Gwalior (Madhya Pradesh)	DV-3	Dash et al. (2005)
2003	Delhi	DV-1, DV-2, DV-3, DV-4	Dar et al. (2006)
2004	Gwalior (Madhya Pradesh)	DV-3	Dash et al. (2006)
2005	Delhi	DV-3	Gupta et al. (2006)
2006	Delhi	DV-1, DV-3	Kukreti et al. (2008)
2006	Delhi	Concurrent infection with dengue serotypes	Bharaj et al. (2008)
2008	Lucknow	DV-1, DV-2, DV-3	Pandey et al. (2012)
2009	Lucknow	DV-1, DV-2, DV-3	Pandey et al. (2012)
2010	Lucknow	DV-1 and DV-3	

DV dengue virus

Confirmed

One of the following:

1. PCR+
2. Virus culture+
3. IgM seroconversion in paired sera
4. IgG seroconversion in paired sera or fourfold IgG titer increase in paired sera

1. *Serological tests*

Five basic serologic tests have been routinely used for diagnosis of dengue infection:

- Hemagglutination-inhibition (HI),
- Complement fixation (CF),
- Neutralization test (NT),
- Immunoglobulin M (IgM) capture enzyme-linked immunosorbent assay (MAC-ELISA),
- Indirect immunoglobulin G ELISA

2. *Virus isolation*

Virus isolation is not commonly used in routine diagnostics, but it constitutes definitive proof of dengue virus infection. Dengue virus can be isolated from patients' samples by inoculating suckling mice via intracranial route, using mosquito cell lines (*A. albopictus* C6/36 and *Aedes pseudoscutellaris* AP61) and inoculating live mosquitoes via intrathoracic route. Currently, the most widely used method of dengue virus culture from patient sera is to use cultured mosquito cells (Gubler et al. 1984). To identify dengue virus, cell cultures must be screened for specific evidence of infection by an antigen detection immunofluorescence assay using serotype-specific monoclonal antibodies and flavivirus group-reactive or dengue complex-reactive monoclonal antibodies.

3. *Antigen detection*

The dengue virus antigen NS1 is detectable in patients' blood (first 5 days of illness) prior to appearance of antidengue antibodies; therefore, it is suitable for early diagnosis (Young et al. 2000). Commercially available assays for NS1 are in EIA and immunochromatographic test formats; however, EIA format is shown to be more sensitive in various studies. Fluorescent antibody, immunoperoxidase, and avidin-biotin enzyme assays allow detection of dengue virus antigen in acetone-fixed leukocytes and in snap-frozen or formalin-fixed tissues collected at autopsy.

4. *Dengue virus genome detection*

Dengue virus genome detection enables specific detection at early stages of the disease when serological diagnosis is not very reliable. The viremic phase of the disease coincides with fever, duration ranging from 2–10 days (Guzman and Kouri 2002). A positive test result ensures diagnosis but a negative result does not rule out the possibility of infection as individual variation in viremic level and timing may occur. Unlike the serological methods, detection of viral genome is vulnerable to contamination and needs technical expertise and specific laboratory facilities; however, it provides precise diagnosis from single early phase sample and serotyping is possible for epidemiological follow-up. Different types of reverse transcription polymerase chain reaction (RT-PCR) are done for dengue virus RNA detection.

Prevention and Control

Prevention and control of dengue and DHF has become more urgent. Unfortunately, tools available to prevent dengue infection are very limited. There is no vaccine currently available, and options for mosquito control are the only left hope presently.

Activities to control transmission target *A. aegypti* (the main vector) in the habitats of its immature and adult stages in the household and immediate vicinity, as well as other settings where human–vector contact occurs (e.g., schools, hospitals and workplaces).

A. aegypti proliferates in many purposely filled household containers such as those used for domestic water storage and for decorative plants, as well as in a multiplicity of rain-filled habitats—including used tyres, discarded food and beverage containers, blocked gutters, and buildings under construction. Typically, these mosquitoes do not fly far, the majority remaining within 100 m of where they emerged. They feed almost entirely on humans, mainly during daylight hours, and both indoors and outdoors.

The habitats are eliminated by preventing access by mosquitoes to these containers or by frequently emptying and cleaning them, by removing the developing stages using insecticides or biological control agents, by killing the adult mosquitoes using insecticides, or by combinations of these methods.

Insecticides

Larvicides and adulticides are the two insecticides to be used for vector control. Larvicides should be considered as complementary to environmental management and—except in emergencies—should be restricted to containers that cannot otherwise be eliminated or managed.

WHO recommends following larvicides for non-potable water:

1. Organophosphates
   - Pirimiphos-methyl in 1 mg/L of active ingredient (a.i.)
   - Temephos in 1 mg/L of active ingredient (a.i.)
2. Insect growth regulators
   - Diflubenzuron in 0.02–0.25 mg/L of active ingredient (a.i.)
   - RS-methoprene in 1 mg/L of active ingredient (a.i.)
   - Novaluron in 0.01–0.05 mg/L of active ingredient (a.i.)
   - Pyriproxyfen in 0.01 mg/L of active ingredient (a.i.)

For treatment of drinking water, temephos and methoprene can be applied at dosages of up to 1 mg of active ingredient (a.i.)/L (1 ppm); pyriproxyfen can be applied at dosages up to 0.01 mg a.i. /L (0.01 ppm); and *Bti* at 1–5 mg/L.

Adulticides recommended by WHO for spraying and fogging are as follows:

1. Organophosphates
   - Fenitrothion
   - Malathion
   - Pirimiphos-methyl
2. Pyrethroid
   - Bioresmethrin

- Cyfluthrin
- Cypermethrin
- Cyphenothrin
- Deltamethrin
- D-Phenothrin
- Etofenprox
- Permethrin
- Resmethrin

Space sprays can be applied either as thermal fogs at 10–50 l/ha or as ultra-low-volume applications of undiluted or slightly diluted technical-grade insecticide in the form of a cold aerosol of droplets of controlled size (15–25 μm) at a rate of 0.5–2.0 l/ha. Portable or vehicle-mounted thermal or cold-fog generators can be used for ground application.

Biological Control

Biological control is based on the introduction of organisms that prey upon, parasitize, compete with, or otherwise reduce populations of the target species. Against *Aedes* vectors of dengue, only certain species of larvivorous fish and predatory copepods (Copepoda: Cyclopoida)—small freshwater crustaceans—have proved effective in operational contexts in specific container habitats, and even then seldom on a large scale.

Vaccine Development

Currently, no effective vaccine is available for dengue virus infection. The World Health Organization designated the development of a tetravalent dengue vaccine a priority for the most cost-effective approach to dengue prevention (Brandt 1988, 1990). With the support of the World Health Organization, considerable progress in developing a vaccine for dengue and DHF has been made in recent years (Bhamarapravati et al. 1987; Bhamarapravati and Yoksan 1989, 1997). Promising candidate attenuated vaccine viruses have been developed and have been evaluated in phase I and II trials in Thailand as monovalent, bivalent, trivalent, and tetravalent formulations (Bhamarapravati 1997). Promising progress in the development of alternative vaccine strategies using new molecular technology has also been made in recent years. Recent approaches include the use of inactivated whole-virion vaccines (Chambers 1997), synthetic peptides (Roehrig et al. 1992a, b), subunit vaccines (Mason et al. 1989; Deubel et al. 1991), vector expression recombinant live vector systems (Chambers et al. 1997; Mason et al. 1991), infectious cDNA clone-derived vaccines (Bray and Lai 1991; Chen et al. 1995; Kapoor et al. 1995) and naked DNA (Kochel et al. 1997).

Alternative Vaccine Strategy

In present scenario, there seems an urgent requirement of alternative approach to control dengue disease. RNAi (RNA interfering) is a technique which is used to silence gene with the help of small RNAs. Small RNA (sRNA) regulatory pathways (SRRPs) control gene expression through a variety of mechanisms (Fire et al. 1998). Components of the microRNA, small interfering (siRNA), and PIWI RNA pathways, three major SRRPs, are present in mosquitoes (Campbell et al. 2008a). In each of these pathways, gene expression is regulated in the cleavage and degradation of mRNAs. Cellular processes as diverse as development, antiviral defense, and maintenance of the germ line are controlled by these mechanisms (Campbell et al. 2008b; Sanchez-Vargas et al. 2009).

References

Agarwal R, Kapoor S, Nagar R, Misra A, Tandon R, Mathur A et al (1999) Clinical study of the patients with dengue hemorrhagic fever during the epidemic of 1996 at Lucknow, India. SE Asian J Trop Med Public Health 30:735–740

Anonymous (1986) Dengue hemorrhagic fever, diagnosis, treatment and control. World health organization, Switzerland

Assenberg R, Mastrangelo E, Walter TS, Verma A, Milani M, Owens RJ et al (2009) Crystal structure of a novel conformational state of the flavivirus NS3

protein: implications for polyprotein processing and viral replication. J Virol 83(24):12895–12906
Bancroft TL (1906) On the aetiology of dengue fever. Aust Med Gaz 25:17–18
Barnes WJS, Rosen L (1974) Fatal hemorrhagic disease and shock associated with primary dengue infection on a pacific island. Am J Trop Med Hyg 23:495–506
Bhamarapravati N (1997) Live attenuated tetravalent dengue vaccine. In: Gubler DJ, Kuno G (eds) Dengue and dengue hemorrhagic fever. CAB International, London, pp 367–378
Bhamarapravati N, Yoksan S (1989) Study of bivalent dengue vaccine in volunteers. Lancet 1077
Bhamarapravati N, Yoksan S, Chayaniyayothian T, Angsubhakorn S, Bunyaratvej A (1987) Immunization with a live attenuated dengue-2 virus candidate vaccine (16681-PDK 53): clinical, immunological and biological responses in adult volunteers. Bull WHO 65:185–195
Bharaj P, Chahar HS, Pandey A, Diddi K, Dar L, Guleria R, Kabra SK, Broor S (2008) Concurrent infections by all four dengue virus serotypes during an outbreak of dengue in 2006 in Delhi. India Virol J 2008(5):1–4
Bhattacharjee N, Mukherjee KK, Chakravarti SK, Mukherjee MK, De PN, Sengupta M, Banik GB, Bhowmick P, Sinha SK, Chakraborty MS (1993) Dengue haemorrhagic fever (DHF) outbreak in Calcutta—1990. J Commun Dis 25:10–14
Bollati M, Alvarez K, Assenberg R, Baronti C, Canard B, Cook S et al (2009) Structure and functionality in flavivirus NS-proteins: prospectives for drug design. Antiviral Res
Bozza FA, Cruz OG, Zagne SM, Azeredo EL, Nogueira RM, Assis EF, Bozza PT, Kubelka CF (2008) Multiplex cytokine profile from dengue patients: MIP-1beta and IFN-gamma as predictive factors for severity. BMC Infect Dis 8:86
Brandt WE (1988) Current approaches to the development of dengue vaccines and related aspects of the molecular biology of flaviviruses. J Infect Dis 157:1105–1111
Brandt WE (1990) Development of dengue and Japanese encephalitis vaccines. J Infect Dis 162:577–583
Bray M, Lai C-J (1991) Construction of intertypic chimeric dengue viruses by substitution of structural protein genes. Proc Natl Acad Sci USA 88:10342–10346
Broor S, Dar L, Sengupta S, Chakaraborty M, Wali JP, Biswas A, Kabra SK, Jain Y, Seth P (1997) Recent dengue epidemic in Delhi, India. In: Saluzzo JE, Dodet B (ed) Factors in the emergence of arbovirus diseases, Elsevier, Paris, pp 123–127
Campbell CL, Black WCT, Hess AM, Foy BD (2008a) Comparative genomics of small RNA regulatory pathway components in vector mosquitoes. BMC Genomics 9(1):425
Campbell CL, Keene KM, Brackney DE, Olson KE, Blair CD, Wilusz J, Foy BD (2008b) *Aedes aegypti* uses RNA interference in defense against Sindbis virus infection. BMC Microbiol 8:47
Chambers TJ, Hahn CS, Galler R, Rice CM (1990) Flavivirus genome organization, expression and replication. Annul Rev Microbiol 44:649–688
Chambers TJ, Tsai TF, Pervikov Y, Monath TP (1997) Vaccine development against dengue and Japanese encephalitis: report of a world health organization meeting. Vaccine 15:1494–1502
Chen W, Kawano H, Men R, Clark D, Lai C-J (1995) Construction of intertypic chimeric dengue viruses exhibiting type 3 antigenicity and neurovirulence for mice. J Virol 69:5186–5190
Dar L, Broor S, Sengupta S, Xess I, Seth P (1999) The first major outbreak of dengue hemorrhagic fever in Delhi. India Emerg Infect Dis 5:589–590
Dar L, Gupta E, Narang P, Broor S (2006) Co-circulation of dengue serotypes 1,2,3 and 4 during the 2003 outbreak in Delhi. India Emerg Infect Dis 12:352–353
Dash PK, Saxena P, Abhyankar A, Bhargava R, Jana AM (2005) Emergence of dengue virus type-3 in northern India. SE Asian J Trop Med Public Health 36:370–377
Dash PK, Parida MM, Saxena P, Abhyankar A, Singh CP, Tewari KN, Jana AM, Sekhar K, Rao PV (2006) Reemergence of dengue virus type-3 (subtype-III) in India: implications for increased incidence of DHF & DSS. Virol J 6:3–55
de Macedo FC, Nicol AF, Cooper LD, Yearsley M, Pires AR, Nuovo GJ (2006) Histologic, viral, and molecular correlates of dengue fever infection of the liver using highly sensitive immunohistochemistry. Diagn Mol Pathol 15:223–228
Deubel V, Bordier M, Megret F, Gentry MK, Schlesinger JJ, Girard M (1991) Processing, secretion, and immunoreactivity of carboxy terminally truncated dengue-2 virus envelope proteins expressed in insect cells by recombinant baculoviruses. Virology 180:442–447
Dietz V, Gubler DJ, Ortiz S, Kuno G, Casta-Velez A, Sather GE, Gomez I, Vergne E (1996) The 1986 dengue and dengue hemorrhagic fever epidemic in Puerto Rico: epidemiologic and clinical observations. PR Health Sci J 15:201–210
Fire A, Xu S, Montgomery MK, Kostas SA, Driver SE, Mello CC (1998) Potent and specific genetic interference by double-stranded RNA in Caenorhabditis elegans. Nature 391(6669):806–811
Gaunt M, Sall A, de Lamballerie X et al (2001) Phylogenetic relationships of flaviviruses correlate with their epidemiology, disease association and biogeography. J Gen Virol 82:1867–1876
Gubler DJ (1987) Dengue and dengue hemorrhagic fever in the Americas. PR Health Sci J 6:107–111
Gubler DJ (1989) *Aedes aegypti* and *Aedes aegypti*-borne disease control in the 1990s: top down or bottom up. Am J Trop Med Hyg 40:571–578
Gubler DJ (1997) Dengue and dengue hemorrhagic fever: its history and resurgence as a global public health problem. In: Gubler DJ, Kuno G (eds) Dengue and dengue hemorrhagic fever. CAB International, London, pp 1–22

Gubler DJ (1998) The global pandemic of dengue/dengue haemorrhagic fever: current status and prospects for the future. Ann Acad Med Singap 27:227–234

Gubler DJ (2002) Epidemic dengue/dengue hemorrhagic fever as a public health, social and economic problem in the 21st century. Trends Microbiol 10:100–103

Gubler DJ (2006). Dengue/dengue haemorrhagic fever: history and current status. Novartis Found Symp 277, 3–16; discussion 16–22, 71–13, 251–253

Gubler DJ, editor KG (1997) Dengue and dengue hemorrhagic fever, CAB International, UK

Gubler DJ, Kuno G, Sather GE, Velez M, Oliver A (1984) Mosquito cell cultures and specific monoclonal antibodies in surveillance for dengue viruses. Am J Trop Med Hyg 33:158–165

Gupta E, Dar L, Kapoor G, Broor S (2006) The changing epidemiology of dengue in Delhi. India Virol J 3:92

Guzman MG, Kouri G (2002) Dengue: an update. Lancet Infect Dis 2(1):33–42

Hahn YS, Galler R, Hunkapillar T, Dalrymple JM, Strauss JH, Strauss EG (1988) Nucleotide sequence of dengue 2 RNA and comparison of the encoded proteins with those of other flaviviruses. Virology 162:167–180

Halstead SB (1970) Observations related to pathogenesis of dengue hemorrhagic fever. VI. Hypotheses and discussion. Yale J Biol Med 42:350–362

Halstead SB (1988) Pathogenesis of dengue: challenges to molecular biology. Science 239:476–481

Halstead SB, O'Rourke EJ (1977a) Antibody-enhanced dengue virus infection in primate leukocytes. Nature (London) 265:739–741

Halstead SB, O'Rourke EJ (1977b) Dengue viruses and mononuclear phagocytes. I. Infection enhancement by non-neutralizing antibody. J Exp Med 146:210–217

Hirsch A (1883) Dengue, a comparatively new disease: its symptoms. In: Handbook of geographical and historical pathology, vol 1. Sydenham Society, London, pp 55–81

Howe GM (1977) World geography of human diseases. Academic Press, Inc, New York

Kapoor M, Zhang L, Mohan PM, Padmanabhan R (1995) Synthesis and characterization of an infectious dengue virus type-2 RNA genome (New Guinea C strain). Gene 162:175–180

Kochel T, Wu S-J, Raviprakash K, Hobart P, Hoffman SL, Hayes CG (1997) Inoculation of plasmids expressing the dengue-2 envelope gene elicit neutralizing antibodies in mice. Vaccine 15:547–552

Kramer LD, Ebel GD (2003) Dynamics of flavivirus infection in mosquitoes. Adv Virus Res 60:187–232

Kuhn RJ, Zhang W, Rossmann MG, Pletnev SV, Corver J, Lenches E et al (2002) Structure of dengue virus: implications for flavivirus organization, maturation, and fusion. Cell 108(5):717–725

Kukreti H, Chaudhary A, Rautela RS, Anand R, Mittal V, Chhabra M, Bhattacharya D, Lal S, Rai A (2008) Emergence of an independent lineage of dengue virus type 1 (DENV-1) and its co-circulation with predominant DENV-3 during the 2006 dengue fever outbreak in Delhi. Int J Infect Dis 12(5):542–549

Kumar R, Tripathi S, Tambe JJ, Arora V, Nag VL (2008) Dengue encephalopathy in children in Northern India: clinical features and comparison with non dengue. J Neurol Sci 269:41–48

Kurukumbi M, Wali JP, Broor S, Aggarwal P, Seth P, Handa R, Dhar L, Vajapayee M (2001) Seroepidemiology and active surveillance of dengue fever/dengue haemorrhagic fever in Delhi. Indian J Med Sci 55:149–156

Lanciotti RS et al (1994) Molecular evolution and epidemiology of dengue-3 viruses. J Gen Virol 75(Pt 1):65–75

Leitmeyer KC (1999) Dengue virus structural differences that correlate with pathogenesis. J Virol 73(6):4738–4747

Limon-Flores AY, Perez-Tapia M, Estrada-Garcia I, Vaughan G, Escobar-Gutierrez A, Calderon-Amador J, Herrera-Rodriguez SE, Brizuela- Garcia A, Heras-Chavarria M, Flores-Langarica A, Cedillo-Barron L, Flores-Romo L (2005) Dengue virus inoculation to human skin explants: an effective approach to assess in situ the early infection and the effects on cutaneous dendritic cells. Int J Exp Pathol 86:323–334

Martina BEE, Koraka P, Osterhaus ADME (2009) Dengue virus pathogenesis: an integrated view. Clin Microbiol Rev 22(4):564–581

Mason PW, Dalrymple JM, Gentry MK, McCown JM, Hoke CH, Burke DS, Fournier MJ, Mason TL (1989) Molecular characterization of a neutralizing domain of the Japanese encephalitis virus structural glycoprotein. J Gen Virol 70:2037–2049

Mason PW, Pincus S, Fournier MJ, Mason TL, Shope RE, Paoletti E (1991) Japanese encephalitis virus-vaccinia recombinants produce particulate forms of the structural membrane proteins and induce high levels of protection against lethal JEV infection. Virology 180:294–305

Messer WB (2003) Emergence and global spread of a dengue serotype 3, subtype III virus. Emerg Infect Dis 9(7):800–809

Miller S, Kastner S, Krijnse-Locker J, Buhler S, Bartenschlager R (2007) The non-structural protein 4A of dengue virus is an integral membrane protein inducing membrane alterations in a 2 K regulated manner. J Biol Chem 282(12):8873–8882

Netsawang J, Noisakran S, Puttikhunt C, Kasinrerk W, Wongwiwat W, Malasit P et al (2010) Nuclear localization of dengue virus capsid protein is required for DAXX interaction and apoptosis. Virus Res 147(2):275–283

Nobuchi H (1979) The symptoms of a dengue-like illness recorded in a Chinese medical encyclopedia. Kanpo Rinsho 26:422–425 (In Japanese)

Padbidri VS, Thakre JP, Risbud AR, Joshi GD, Singh A, Mavale MS, Gupta VP (1996) An outbreak of dengue hemorrhagic fever in Jammu. Indian J Virol 12:83–87

Pandey N, Nagar R, Gupta S, Omprakash, Khan D, Singh DD, Mishra G, Prakash S, Singh KP, Singh M, Jain A (2012) Trend of dengue virus infection at Lucknow, north India (2008–2010): a hospital based study. Indian J Med Res 136(5):862–867

Parida MM, Dash PK, Upadhyay C, Saxena P, Jana AM (2002) Serological and virological investigation of an outbreak of dengue fever in Gwalior, India. Indian J Med Res 116:248–254

Pepper OHP (1941) A note on David Bylon and dengue. Ann Med Hist 3rd Ser 3:363–368

Roehrig JT, Johnson AH, Hunt AR, Beaty BJ, Mathews JH (1992a) Enhancement of the antibody response to flavivirus B-cell epitopes by using homologous or heterologous T-cell epitopes. J Virol 66:3385–3390

Roehrig JT, Mathews JH, Risi PA, Brubaker JR, Hunt AR (1992b) Mapping of biologically active helper T-cell epitopes on the flavivirus envelope glycoprotein. In: Brown F, Chanock RM, Ginsberg H, Lerner RA (eds) Vaccines 92. Cold Spring Harbor Laboratory, Cold Spring Harbor, pp 277–281

Rothman AL (2004) Dengue: defining protective versus pathologic immunity. J Clin Invest 113(7):946–951

Rush AB (1789) An account of the bilious remitting fever, as it appeared in Philadelphia in the summer and autumn of the year 1780. Medical enquiries and observations, Prichard and Hall, Philadelphia, pp 104–117

Sabin AB (1952) Research on dengue during World War II. Am J Trop Med Hyg 1:30–50

Sabin A (1959) Survey of knowledge and problems in field of arthropod-borne virus infections. Arch Gesamte Virusforsch 9:1–10

Sanchez-Vargas I, Scott JC, Poole-Smith BK, Franz AW, Barbosa-Solomieu V, Wilusz J, Olson KE, Blair CD (2009) Dengue virus type 2 infections of *Aedes aegypti* are modulated by the mosquito's RNA interference pathway. PLoS Pathog 5(2):e1000299

WHO (2009) Dengue guidelines for diagnosis, treatment, prevention and control. World Health Organization, Geneva

Yap TL, Xu T, Chen YL, Malet H, Egloff MP, Canard B et al (2007) Crystal structure of the dengue virus RNA dependent RNA polymerase catalytic domain at 1.85 angstrom resolution. J Virol 81(9):4753–4765

Young PR, Hilditch PA, Bletchly C, Halloran W (2000) An antigen capture enzyme linked immunosorbent assay reveals high levels of the dengue virus protein NS1 in the sera of infected patients. J Clin Micribiol 38(3):1053–1057

Zhang Y, Zhang W, Ogata S, Clements D, Strauss JH, Baker TS et al (2004) Conformational changes of the flavivirus E glycoprotein. Structure 12(9):1607–1618

Water, Environmental Surveillance and Molecular Epidemiology of Poliovirus in India

2

Tapan N. Dhole and Animesh Chatterjee

Abstract

Poliomyelitis has appeared in epidemic form, become endemic on a global scale and has been reduced to near elimination, all within the span of documented medical history. Environmental surveillance of poliovirus (PV) means monitoring of PV transmission in human populations by examining environmental specimens supposedly contaminated by human faeces. The rationale for surveillance is based on the fact that PV-infected individuals, whether presenting with disease symptoms or not, shed large amounts of PV in the faeces for several weeks. As the morbidity: infection ratio of PV infection is very low, this fact contributes to the sensitivity of PV surveillance, which under optimal conditions can be better than that of the standard acute flaccid paralysis (AFP) surveillance. The World Health Organization (WHO) has included environmental surveillance of PV in the new Strategic Plan of the Global Polio Eradication Initiative for years 2010–2012 to be increasingly used in PV surveillance, supplementing AFP surveillance.

Keywords

Epidemiology · Polio · Polio vaccine virus · Public health · Surveillance

Introduction

Clean water is essential for life. Throughout the world, millions of people do not have access to microbiologically safe water for drinking, cooking and other essential purposes. It is estimated that one-sixth of the world's population do not have access to improved sources of drinking water (WHO 2010). Enteric viruses are the important causative agents of human diseases that transit easily to water environments due to varied human activity. They are usually present in insufficiently treated drinking water, groundwater, rivers and seas. Impurities from human households are a main source of water contamination. Enteroviruses may cause a wide variety of pathological symptoms and enteroviral infections

T. N. Dhole (✉) · A. Chatterjee
Department of Microbiology, Sanjay Gandhi Postgraduate Institute of Medical Sciences, Raebareli Road, Lucknow 226 014, India
e-mail: tapandhole@yahoo.co.in

P. P. Singh and V. Sharma (eds.), *Water and Health*,
DOI: 10.1007/978-81-322-1029-0_2, © Springer India 2014

that affect especially young children. Enteroviral epidemics are predominantly waterborne; therefore, water contamination poses an absolute threat to human health (Nwachuku and Gerba 2006).

Success of the Global Polio Eradication Initiative

A multidecade effort of Global Polio Eradication Initiative (GPEI) launched by WHO in 1988 has reduced the global polio incidence by >99 % and intensified its efforts to eradicate the circulation of wild poliovirus (WPV) in the four endemic countries by the end of 2012. India, which was among the world's four endemic countries besides being Pakistan, Afghanistan and Nigeria, responsible for the transmission of WPV, has experienced a year without reporting a single case for the first time. The last case of wild poliovirus type 1 (WPV1) was reported in the state of West Bengal on 13 January 2011. The states of Uttar Pradesh and Bihar in northern region have been the focus against polio because of the high population density and poor sanitation and have complicated the efforts to break the transmission cycle.

With 741 cases of polio reported in the year 2009 to one in 2011, India has seen a massive drop in the number of polio cases. Since November 2009, no cases of WPV1 have been reported in the states of Uttar Pradesh and Bihar, both of which were polio endemic and no cases of WPV3 have been reported since April 2010. But seven cases of vaccine-derived polio were seen in India in 2011. By contrast, Pakistan and Afghanistan have reported three and nine cases, so far in 2011, respectively, and Nigeria still has all the three PV strains in circulation.

It is because of the massive efforts of Indian health authorities that have made possible India free of polio for 1 year. In 2011, more than 900 million doses of oral PV were administered to Indian children and more than 172 million children were vaccinated twice on national immunization days (NIDs) every year. About 120 million children were vaccinated during an additional seven sub-national immunization days.

The WHO's executive board declared the completion of its polio eradication efforts a programmatic emergency for public health and urged the handful of countries where polio still exists to declare a national public health emergency and also called for certification-standard surveillances to identify the emergence of circulating vaccine-derived PVs, and adequate funding to interrupt wild PV transmission globally, which they believe can be achieved by the end of 2013 (Kew 2012; Kaura and Abraham 2012).

Primary Strategies for Achieving this Goal

1. **High Routine Immunization**: To immunize every child aged <1 year with at least 3 doses of oral poliovirus vaccine (OPV). Paralytic polio can be caused by any of 3 closely related strains (serotypes) of PV. Trivalent OPV (OPV3) provides immunity against all 3 types. Three routine OPV doses should be received by infants at ages 6, 10 and 14 weeks.
2. **National Immunization Days**: Conduct Pulse Polio Immunization (PPI) programme by providing additional OPV doses to every child aged <5 years at intervals of 4–6 weeks. The aim of NIDs/PPI is to "flood" the community with OPV within a very short period of time, thereby interrupting transmission of virus throughout the community. Intensification of the PPI programme is accomplished by the addition of extra-immunization rounds, adding a house-to-house "search and vaccinate" component in addition to providing vaccine at a fixed post.
3. **Surveillance of AFP**: To identify all reservoirs of wild PV transmission. This includes AFP case investigation and laboratory investigation of stool specimens collected from AFP cases, which are tested for PVs in specialized laboratories.

4. **"Mopping-up" Immunization**: When poliovirus transmission has been reduced to well-defined and focal geographical areas, intensive house-to-house, child-to-child immunization campaigns are conducted over a period of days to break the final chains of virus transmission.

Cross-Border Polio Spread: A Threat to India

India has been polio free for over a year. But the big danger now is a cross-border threat. In 2011, there were 198 cases in Pakistan and 80 cases in Afghanistan. Given the porous border, this increases India's chances of a cross-border polio spread. Last year, China reported a polio outbreak after a gap of 10 years and the virus had spread from Pakistan. Therefore, if a polio-free country becomes re-infected, the virus can spread like wildfire. This is precisely why the WHO has declared polio eradication in the Pakistan region as an "emergency". Hence, the government had made a policy that every Pakistani traveller, irrespective of age, or of vaccination status, must take a dose of OPV before travelling to India. Accordingly, India recently had set up vaccination booths at Chakdabagh (Poonch) and at Kaman (Baramulla) of Jammu and Kashmir, at Munabao railway station in Rajasthan's Barmer district and at Wagah border and Attari railway station in Punjab to administer polio drops to all children below 5 years, coming in from Pakistan. This was done to control the biggest threat of import of the virus.

Poliomyelitis: The Disease

Poliomyelitis, or polio, is a life-threatening acute paralytic disease caused by PV, a member of the genus *Enterovirus* in the family Picornaviridae (Hovi et al. 2004). PVs are transmitted from person to person following excretion in faeces and pharyngeal secretions, mainly via the hand-to-hand-to-mouth route. Because the PV receptor is only expressed on cells of humans and a few sub-human primate species, there are no known extra-human reservoirs (Racaniello 2006). Following infection, the virus replicates in the gastrointestinal tract and may cause viremia (Sabin 1956). Occasionally, the virus then invades the central nervous system and destroys lower motor neurons, causing a clinically distinctive flaccid paralysis without permanent sensory loss (Nathanson 2008). Like other RNA viruses, PVs exist as mixtures of microvariants, called quasi-species (Mulders et al. 1999). This is caused by the error-prone, virus-encoded RNA polymerase, which lacks proof-reading activity, resulting in a rapid accumulation of mutations upon replication (Mulders et al. 1999; Hovi et al. 2004). An additional mode of generating divergence between PVs and other enteroviruses (EVs) is their ability to recombine with other serotypes (intertypic recombinants) or with another genome of the same serotype (intratypic recombinants) (Mulders et al. 1999; Hovi et al. 2004). During replication in humans and upon transmission between hosts, some of the mutations are enriched, which has resulted in numerous genetic lineages within each serotype of PV that cocirculate worldwide (Mulders et al. 1999).

To date, there are three PV serotypes, designated type 1, type 2 and type 3, which were originally distinguished from the other EVs by neutralization with serotype-specific antisera and the propensity to cause paralytic illness (Georgopoulou et al. 2000) (Fig. 2.1).

Epidemiology

The disease of poliomyelitis has a long history. The first example may even have been more than 3,000 years ago. An Egyptian stele dating from the 18th Egyptian dynasty (1580–1350 BCE) shows a priest with a deformity of his leg characteristic of the flaccid paralysis typical of poliomyelitis. The first known clinical description of poliomyelitis is attributed to Michael Underwood, a British physician, who in 1789 reported observing an illness which appeared to

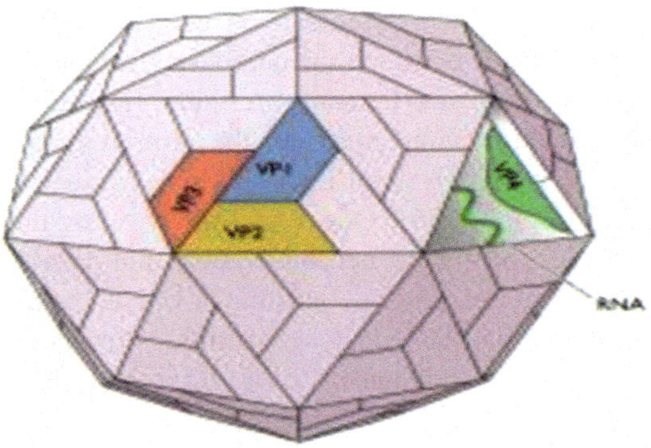

Fig. 2.1 Poliovirus serotypes. Three immunologically distinct types of PV, PV1, PV2 and PV3 have been recognized. Strains which cause severe paralysis are known as wild poliovirus (PV1 wild, PV2 wild and PV3 wild). Sabin PV1, PV2 and PV3 strains are used as efficient vaccine known as vaccine-derived poliovirus. Adapted from http://www.wenliang.myweb.uga.edu

target primarily children and left those afflicted with residual debility of the lower extremities. Initial outbreaks in Europe were documented in the early nineteenth century, and outbreaks in the United States were first reported in 1843. However, it was not until the early twentieth century that the number of paralytic poliomyelitis cases reached epidemic proportions (De Jesus 2007). The polio is in circulation in India since ages, especially in Uttar Pradesh and Bihar because of poor sanitation and high population density. If one looks at the picture of different villages of Uttar Pradesh and Bihar, old paralytic cases of AFP can easily be found. This indicates that the poliovirus has survived and circulated in these areas for years. This circulating virus has made victim of polio cases time to time, depending on the immunity in general population.

In May 1988, during the World Health Assembly, Minister of Health of all member states of the WHO voted to launch global goal to eradicate polio. As a result of this, GPEI started and estimated that global incidence of polio decreased by more than 99 % with three WHO regions (Americas, Western Pacific and Europe) being certified polio-free (Sutter et al. 2001). Intensive polio eradication programme in the South-East Asia Region (SEAR), with the use of tOPV, led to the substantial decrease in the number of polio cases. By 2001, PV circulation in India was limited primarily to northern states of Uttar Pradesh and Bihar, with 268 cases reported nationwide (Mukherji et al. 2005). However, a major resurgence occurred in 2002 with 1,600 cases nationwide (Mukherji et al. 2005; Sathyamala et al. 2005), of which majority of cases, that is, 1,363 (85 %) were from Uttar Pradesh and Bihar only (Sathyamala et al. 2005). This resurgence was attributed to the decline in OPV coverage in critical areas with vaccination coverage of children in 15 % houses in some districts (Mukherji et al. 2005). Thus, a large number of children were missed in areas with high population density, resulting in a very large birth cohort of susceptible individuals in areas of poor sanitation.

Molecular Epidemiology of Wild Poliovirus Circulation in India

The role of molecular surveillance in eradication initiatives of poliomyelitis has proved to be an extremely powerful tool for assessing the transmission pathways, monitoring quality of the national immunization programme, assessing vaccine coverage and monitoring the success of eradication strategies. The poliovirus has three serotypes PV1, PV2 and PV3, and sufficient genetic clusters (John et al. 2011) have been observed in each serotype. The genotype and cluster, and lineage and sub-lineage are the operational taxonomic units for molecular surveillance. For all practical purposes, VP1-906 nt has been sequenced and compared (Martin 2011). It is essential to know the indigenous baseline genotype/cluster/lineage circulating in different states of India and to monitor the changes with accelerated efforts for elimination of these lineages. Currently, the poliovirus wild-type 1 has three clusters with multiple lineages circulating in different parts of India, while wild-type 3 has four clusters with few lineages. P2 wild poliovirus has already been eradicated from India (Barrett 2009). Surveillance for poliomyelitis is a dynamic process, and continuous

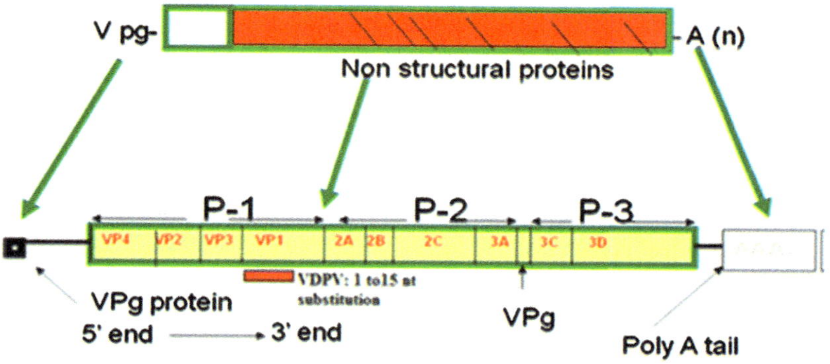

Fig. 2.2 Structure of poliovirus. Poliovirus genome consists of a single molecule of single-stranded RNA, *7,500 nucleotides long. It acts like mRNA in infected cell and is translated into a single large polypeptide. This polypeptide is cleaved by virus-coded enzymes into capsid proteins (VP1, VP2, VP3 and VP4) and non-structural proteins which include proteases and RNA-dependent RNA polymerase

monitoring of appearance and disappearance of different lineages in any given region is imperative. The most effective and useful strategy of strain surveillance is crossing over from endemic to relatively non-endemic area. Depending on the effective and optimal immunization coverage, circulation of indigenous lineage may persist or disappear. The close similarity of sequences within each lineage indicates good quality of surveillance. Some of the wild PV strains may disappear without surfacing for more than 2 years (silent transmission) or may appear as importation from other countries. It is a barometer of current status of polio eradication which helps in refining the strategies to achieve the goal in shortest possible time (Fig. 2.2).

Poliovirus Vaccines

Protective immunity against poliomyelitis is conferred through immunization or natural PV infection. Immunity is PV serotype specific (Ghendon and Robertson 1994). Protection against infection is associated with both circulating antibodies in the blood and secretory antibodies in the gut and upper respiratory tract, which prevent the spread of PV to the CNS (Ghendon and Robertson 1994; Wood et al. 2000; CDC 2002). PV is the only EV for which a vaccine is available (Zaoutis and Klein 1998). In 1955, the first successful vaccine against poliomyelitis was developed by Jonas Salk, known as inactivated poliovirus vaccine or IPV (CDC 2002). In 1963, Albert Sabin developed an oral, live attenuated poliovirus vaccine (OPV) that contained all three PV serotypes (Wood et al. 2000; CDC 2002). After its introduction, the OPV was adopted rapidly worldwide as the vaccine of choice (CDC 2002). Virologists were of the opinion that Salk's vaccine (IPV) could not provide long-lasting protection and that this could only be achieved with the Sabin's live attenuated version, which had sufficient immunogenicity to provide protection (Blume and Geesink 2000). However, none of these vaccines (IPV and OPV) can provide 100 % immunity against infection or re-infection with PV (Wood et al. 2000; Zaoutis and Klein 1998; Blume and Geesink 2000).

Inactivated Poliovirus Vaccine

The IPV is very effective in inducing circulating antibodies in the blood, thus preventing PV in the gut from entering and replicating in the CNS (Wood et al. 2000). The use of IPV in several Northern European countries (Denmark, Finland, Sweden, The Netherlands) succeeded in effectively eliminating wild-type PV circulation (WHO 2003a, b). The IPV provides protection for many years after a complete series; however, this duration of immunity is not certain (Wood et al. 2000; CDC 2002). Although IPV is very effective in inducing circulating antibodies against PV for individual protection, it is less effective than OPV in inducing mucosal immunity at replication sites in the gastrointestinal tract (Wood et al. 2000). The IPV stimulates production of serum antibodies in the bloodstream, which cannot prevent the PV from initially multiplying in the intestine (Pelczar et al. 1993). However, these serum antibodies do prevent PV in the bloodstream from reaching the spinal cord and causing paralysis (Pelczar et al. 1993). The OPV produces long-lasting mucosal immunity by stimulating the formation of secretory IgA antibodies in the intestine and also serum antibodies in the bloodstream (Pelczar et al. 1993). Thus, mucosal immunity restricts viral replication following exposure to PV and is important for community protection (Wood et al. 2000). The intestinal secretory antibodies can prevent the primary intestinal infection by neutralizing the infectivity of virulent PV strains that a person may encounter later (Pelczar et al. 1993).

Oral Poliovirus Vaccine

The trivalent OPV (tOPV) contains live attenuated strains of all three serotypes of PV. These attenuated PV strains replicate in the human gut, inducing mucosal immunity that inhibits replication of the virus in the gastrointestinal tract (Wood and Thorley 2003; CDC 2002). A single dose of OPV produces immunity to all three PV vaccine strains in about 50 % of the recipients, and three doses of OPV will produce immunity in 95 % of the recipients (CDC 2002). The OPV has been found to confer longer-lasting immunity, so that repeated boosters are not necessary and act quickly, immunity being achieved in a matter of days (Wood and Thorley 2003; Blume and Geesink 2000).

Repeated OPV not only potentiates the immune response of the child but may produce immune paralysis. The immunogenicity of OPV varies region to region; the seroconversion rate and protection in children of Bihar and Uttar Pradesh in India are below the Indian standard (Vashishtha 2009; Paul 2007). The herd effect of OPV is also quite low in developing countries like India (Paul 2007). Vaccine viruses are less infectious than their wild counterparts, and spread to non-immune children (contact immunization) is another reason. Both of these factors are weak in children of Uttar Pradesh and Bihar. Therefore, virtually every child must be vaccinated with repeated doses of OPV to ensure personal protection, but it is difficult to achieve where primary immunization is weak. If the immunization rates fall after achieving high level of immunity in the polio-free community, risk of large outbreaks increases rapidly among the growing cohorts of non-immune children. The routine immunization with OPV will no longer outweigh the burden of diseases either due to paralysis caused by OPV (vaccine-associated paralytic polio), or due to outbreak caused by circulating vaccine-derived poliomyelitis (cVDPV) (Heymann et al. 2006a, b). The mucosal immunity induced by OPV in India varies by location, serotype and vaccine formulation (Grassly et al. 2009). The present discrepancies in tOPV versus IPV are debatable based on scientific merit and demerit, and advantage versus disadvantage of their use in two highly populated states of the country (Blume 2005). The tOPV/monovalent OPV (mOPV) has reduced the circulation of wild PV from major part of the country without making significant impact on their circulation in 107 sub-districts of Uttar Pradesh and Bihar. Therefore, the success of tOPV cannot be ignored, but its small amount inherent problem of the vaccine needs to be addressed. It is difficult to maintain the high level of immunity in the community with tOPV because of seroconversion, GI immunity and sustaining the high level of antibody for protection. The polio eradication stands for eradication of wild as well as vaccine strain from the community. Therefore, prolonged use of tOPV will invite innumerable problems like frequent importation of wild strains, occurrence of circulating cVDPV's and vaccine-associated paralytic poliomyelitis (VAPP) and circulation of Sabin strains in the community (Modlin 2010; Heymann et al. 2006a, b; Estivariz et al. 2011).

The frequent importation has been observed from polio endemic countries to relative non-endemic countries as well as interstate within India. In India, frequent importation has been observed to neighbouring areas like Nepal to Bihar, Bihar to Bangladesh, Uttar Pradesh to Mumbai and other nearby states (Andrus et al. 2001). The importation of virus has been observed mainly in those countries from where the OPV vaccine coverage is low in general population after eradication.

The effective use of the OPV by many countries involved in the global PEI has nearly achieved elimination of wild-type PV circulation. However, maintenance of high immunization coverage is crucial to protect against imported wild-type PVs and to prevent person-to-person transmission of OPV-derived viruses (Buttinelli et al. 2003). It is important that all countries maintain a high-quality AFP surveillance system and that a global strategy is developed for the cessation of OPV immunization after global certification of polio eradication (Buttinelli et al. 2003) (Fig. 2.3).

Vaccine-Derived Polioviruses

A variety of OPV-derived viruses can be isolated from OPV recipients and their contacts (WHO 2004). The extent of sequence divergence of the VP1 capsid gene from Sabin PV strains can be used as a "molecular clock" to estimate the duration of PV replication (WHO 2004). A constant rate of accumulation of synonymous nucleotide substitutions is assumed to exist, and for the PV, genome rates of approximately 1–2 % change per year have been proposed (Kew et al. 1998). All clinical and environmental PV isolates that are related to OPV strains are VDPVs (WHO 2004). Derivatives of Sabin OPV strains have been classified into two broad categories for programmatic reasons: "OPV-like viruses" and "vaccine-derived polioviruses (VDPVs)" (WHO 2004).

The vast majorities of vaccine-related isolates are "OPV-like" and have close sequence relationships (>99 % VP1 sequence identity) to the original OPV strains (WHO 2004). Immunologically normal OPV recipients are known to excrete PVs for an average of 3–4 weeks. Short excretion periods and high population immunity normally limit the person-to-person spread of these OPV-like viruses (WHO 2004). Rare "VDPV" isolates show <99 % VP1 sequence identity to the parental Sabin PV vaccine strains,

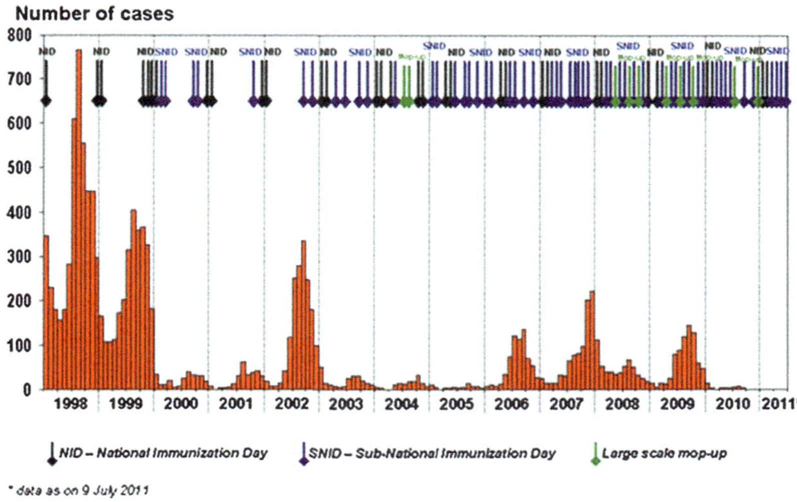

Fig. 2.3 Monthly incidence of wild poliovirus cases in India with national, sub-national and large-scale immunization. Adapted from National Polio Surveillance Project (NPSP), India

and the extent of their genetic changes indicates prolonged replication. Up to date, two categories of VDPV isolates have been identified: immunodeficient VDPVs (iVDPVs) and circulating VDPVs (cVDPVs) (WHO 2004).

Immunodeficient Vaccine-Derived Polioviruses

The potential for prolonged replication of PV vaccine stains in patients with B-cell immunodeficiencies has been recognized for many years (WHO 2004). The first iVDPV isolates to be characterized with modern molecular techniques were from patients with defects in antibody production (generally CVID or X-linked agammaglobulinemia) (Kew et al. 1998; Sutter and Prevots 1994; Yang et al. 2003). Some iVDPV isolates are highly divergent (−90 % VP1 sequence identity to the parental Sabin PV vaccine strain), suggesting that the chronic PV infections had persisted for 10 years or more (Kew et al. 1998; Sutter and Prevots 1994; Yang et al. 2003). Eighteen chronic iVDPV excretors were detected worldwide through the end of 2002, although this number may be an underestimate in the absence of systematic screening of immunodeficient patients (WHO 2004). So far, all reports of persistent iVDPV infections have been from countries with high or intermediate levels of development, where the rates of OPV coverage are high and where the survival times of immunodeficient patients may be extended by their access to appropriate clinical management (Yang et al. 2003). Currently, there is no clear evidence of spread of iVDPV from immunodeficient patients to the wider community (Yang et al. 2003; WHO 2004) (Fig. 2.4).

Circulating Vaccine-Derived Polioviruses

In regions of low OPV coverage, a VDPV may result from transmission of Sabin PV vaccine strains from one immunized individual to another and accumulation of sufficient mutations to adopt wild-type PV characteristics of neurovirulence and transmissibility (Wood and Thorley 2003). A VDPV may cause an outbreak of poliomyelitis, and if there is evidence of person-to-person transmission, based on epidemiological and phylogenetic studies, it is defined as a circulating VDPV (cVDPV) (Yang et al. 2003; Wood and Thorley 2003). According to scientific reports, any PV that is circulating will eventually recombine with another related EV and that recombination is an indicator of circulation rather than a step in the increased ability to transmit from person to person (WHO 2004). All cVDPVs but none of the iVDPVs described in scientific reports thus far appear to be recombinants with EVs closely related to PVs (Yang et al. 2003; Kew et al. 2004). The possible role of recombination in the phenotypic reversion of OPV is unclear. Recombination with EVs appears to be an indicator of circulation, as the cVDPVs in Hispaniola and Egypt had participated in successive rounds of recombination during the outbreaks (Yang et al. 2003; Kew et al. 2004).

Several outbreaks of poliomyelitis due to cVDPV have been documented (Wood and Thorley 2003). A type 2 vaccine-related PV circulated in Belarus following local cessation of OPV use from 1963 to 1966 (Kew et al. 2004). *An* outbreak of type 3 poliomyelitis in Poland in 1968 was associated with PV strains derived from the USOL-D-bac vaccine (Martin et al. 2000). In Egypt between 1983 and 1993, 32 cases of paralytic disease from a cVDPV type 2 were reported, including many retrospective cases (CDC 2001). Polio cases attributed to cVDPV type 1 have been found in Haiti, the Philippines and the Dominican Republic during 2000 and 2001 (Kew et al. 2004; Wood and Thorley 2003). The small cluster of cases marked the first polio outbreak in the Western Hemisphere in more than 9 years (WHO 2000). There have been 19 reports of AFP in the Dominican Republic and one in Haiti (WHO 2000; Dove 2001). Though AFP can be caused by conditions other than polio, laboratory tests confirmed that a cVDPV type 1 was involved in recent outbreaks (Dove 2001). In Hispaniola (the

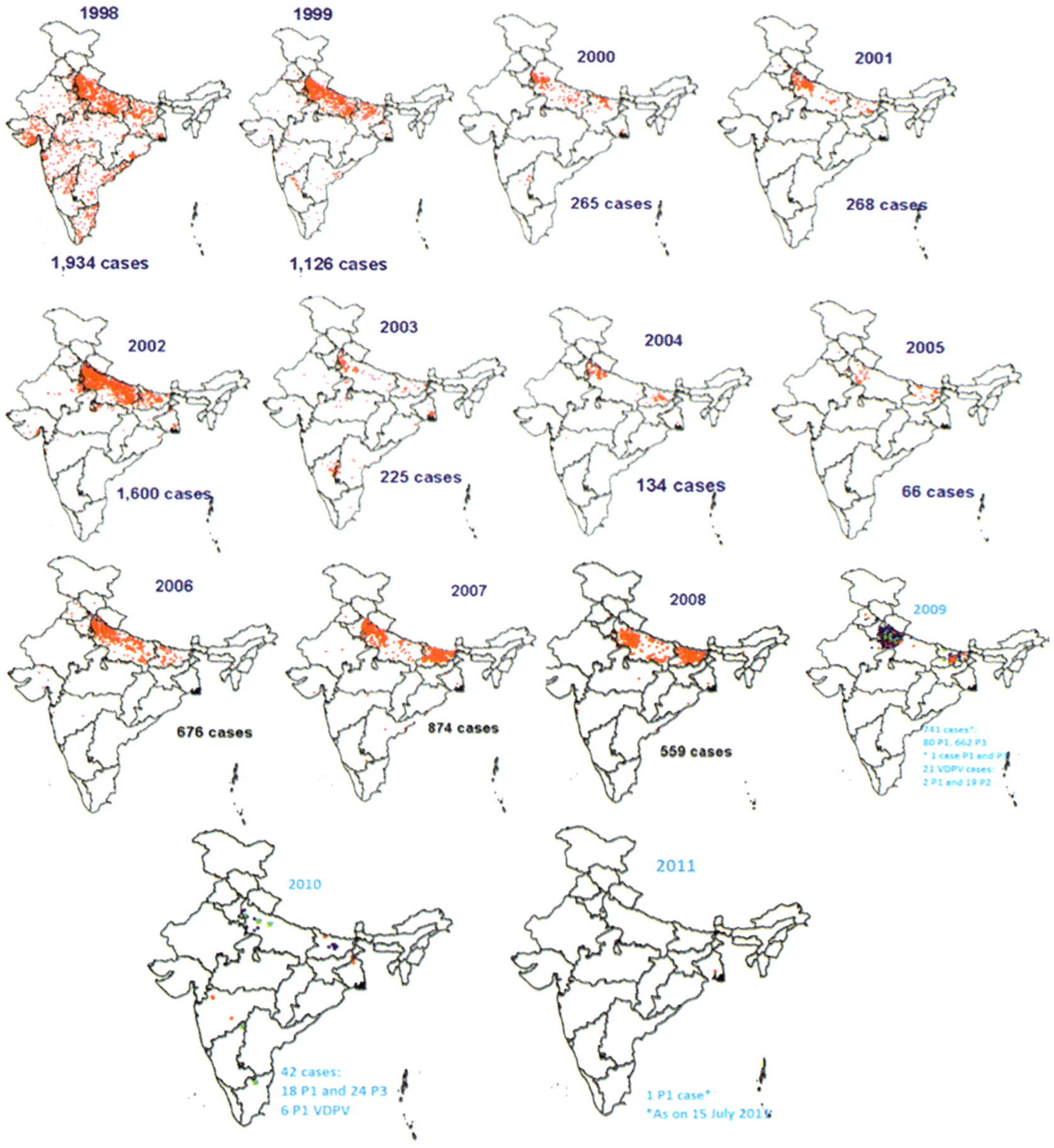

Fig. 2.4 Geographical distribution of wild poliovirus cases in India (1998–2011). Adapted from National Polio Surveillance Project (NPSP), India

Dominican Republic and Haiti) and the Philippines, the cVDPV had undergone recombination with NPEVs. This has been the first reliable report that a VDPV strain reverted to a virulent form and spread contagiously. The virus in these episodes showed more than 2 % genetic sequence difference from the parent Sabin PV vaccine strain (VP1 region of the genome) and probably circulated for more than 2 years before being detected (Kew et al. 2004). The outbreaks began when a VDPV infected inadequately vaccinated individuals, leading to the spread of the pathogenic virus (Dove 2001). Outbreaks of a similar kind have occurred more recently in the Philippines and Madagascar (WHO 2002; Rousset et al. 2003). In Madagascar, five cases of AFP associated with cVDPV type 2 were reported and partial genomic sequencing indicated that two of the PV strains had been circulating for approximately 1 and 2.5 years, respectively (Rousset et al. 2003). Mass vaccination campaigns with OPV interrupted circulation of cVDPVs in Hispaniola and have been underway in the Philippines and Madagascar (Kew et al. 2004). A common factor to all cVDPV outbreaks has been low population immunity, consistent with low OPV coverage and the apparent absence of circulating indigenous wild-type PV of the same serotype. Other risk factors are typical for wild-type PV circulation and include crowding, high birth rates, poor hygiene and sanitation and tropical climate (Kew et al. 2004).

Environmental Surveillance of Poliovirus Circulation

Environmental surveillance has been used successfully in monitoring enteric virus circulation and assessing the extent or duration of epidemic PV circulation in specific populations (WHO 2002, 2004). The rationale for environmental surveillance is based on the characteristic PV excretion pattern. Infected individuals excrete PVs in faeces for periods up to several weeks, whether or not they are symptomatic, and therefore, large numbers of PVs may remain infectious in the environment for varying lengths of time, depending on the immediate conditions (WHO 2002, 2004). Wild-type PVs and cVDPVs have been detected in the environment even in the absence of reported cases of AFP, which is of major concern, since these PVs might be transmitted and continue to circulate in a non-immune population after the cessation of polio vaccination (Friedrich 2000). A study conducted by Divizia et al. (1999) confirmed the environmental circulation in Albania of recombinant PV strains (Sabin-like PV type 2/wild PV type 1), sustained by a massive immunization effort and by the presence in the environment of a PV type 1, isolated from a river 2 months before the first case of symptomatic AFP. An unusual highly diverged derivative of the Sabin PV type 2 strain was isolated from environmental samples during routine screening for wild-type PV in Israel (Shulman et al. 2000). The extensive genetic divergence of the isolate from its parental Sabin PV type 2 vaccine strain suggested that the virus had replicated in one or more individuals for approximately 6 years (Shulman et al. 2000). According to other studies, VDPVs (with 1.4 % nucleotide divergence from the vaccine strain) were isolated from sewage and river water in Japan within 3 months following OPV vaccination, and several of these VDPV type 1 and 3 isolates showed increased neurovirulence (Yoshida et al. 2002; Horie et al. 2002). More recently, two Sabin-like PVs were found by environmental surveillance 8 and 11 months after any OPV vaccine was used in New Zealand and showed 99.8 % as well as 99.9 % homology with Sabin PV type 2 vaccine strain in the VP1 region (WHO 2003a, b). This suggested that these PVs could have been excreted by recently vaccinated children (1 or 2 months) visiting from a country using OPV (WHO 2003a, b). Furthermore, a highly evolved VDPV type 3 strain harbouring a 13 % sequence drift from Sabin PV type 3 vaccine strain has been isolated from sewage in Estonia (Blomqvist et al. 2004). Research has shown that PV isolates in the environment are genetically and epidemiologically related to those circulating in the community (Divizia et al. 1999; Shulman et al. 2000). Thus, the properties of PV isolates from sewage and river water would reflect those of PVs excreted from humans after OPV immunization, and for susceptible individuals, VDPVs have the potential to be the causative agents of poliomyelitis (Yoshida et al. 2002). However, it is difficult to address the risk of infection from the environment, since there is little chance that individuals come into direct contact with raw sewage. In contrast, access to river water or any other water source (used by the community for domestic purposes) is easy, and therefore, susceptible individuals should be regarded as at greater risk of infection from such water sources (Yoshida et al. 2002). Nonetheless, although it is possible to eliminate wild-type PV from the human community and environment, it will be difficult to eradicate poliomyelitis completely as long as OPV is not replaced by IPV (Yoshida et al. 2002). It is evident that environmental surveillance is still epidemiologically important, because the results of virus surveillance retrospectively reflect the properties of virus circulating in the community and it assesses the potential risk of infection from the environment as well as food (Divizia et al. 1999; Yoshida et al. 2002). The examination of composite human faecal samples through environmental surveillance links PV isolates from unknown individuals to populations served by the wastewater system (WHO 2004). In addition, environmental surveillance provides valuable information, particularly in urban populations where AFP surveillance is absent and where persistent PV circulation or reintroduction is suspected (WHO 2004).

Future Strategies

Eradication of polio is a public health programme of unprecedented magnitude, complexity and cost. Tremendous progress has been made in the global fight against poliovirus.

India has invested heavily to reach where we are today. There are innumerable risks associated with the future of polio eradication. Protecting the investment requires the successful completion of both wild and vaccine virus eradication and its sustenance. Plans to design a new vaccine policy for the future especially to systematically introduce IPV and to safely discontinue OPV after eradication of WPVs have to be made urgently. There must be a smooth gradual transition from the current pre-eradication vaccine strategies to post-eradication vaccine policy, to chalk out a clear strategy on how to deal with the issues like OPV cessation plans, global synchronization versus regional/national synchronization, duration of AFP surveillance, tackling of future outbreaks of both wild and vaccine viruses, role of IPV in controlling future outbreaks of cVDPVs and development of safe and affordable IPV.

References

Andrus JK, Thapa AB, Withana N et al (2001) A new paradigm for international disease control: lessons learned from polio eradication in Southeast Asia. Am J Public Health 91:146–150

Barrett S (2009) Polio eradication: strengthening the weakest links. Health Aff (Millwood) 28:1079–1090

Blomqvist S, Savolainen C, Laine P, Hirttio P, Lamminsalo E, Pentilla E et al (2004) Characterisation of a highly evolved vaccine derived poliovirus type 3 isolated from sewage in Estonia. J Virol 78:4876–4883

Blume S, Geesink I (2000) A brief history of polio vaccines. Science 288:1593–1594

Blume SS (2005) Lock in, the state and vaccine development: lessons from the history of the polio vaccines. Res Policy 34:159–173

Buttinelli G, Donati V, Fiore S et al (2003) Nucleotide variation in sabin type 2 poliovirus from an immunodeficient patient with poliomyelitis. J Gen Virol 82:79–91

Centers for Disease Control and Prevention (2001) Circulation of a type 2 vaccine-derived poliovirus-Egypt, 1982–1993. Morb Mortal Wkly Rep 50(41–42):51

Centers for Disease control and Prevention (2002) Progress towards global eradication of poliomyelitis, 2001. Morb Mortal Wkly Rep 51:253–256

De Jesus NH (2007) Epidemics to eradication: the modern history of poliomyelitis. Virol J 4:70

Divizia M, Palombi L, Buonomo E, Donia D, Ruscio V, Equestre M et al (1999) Genomic characterization of human and environmental polioviruses isolated in Albania. Appl Environ Microbiol 65:3534–3539

Dove A (2001) Paralytic poliomyelitis returns to the Western Hemisphere. PICO News, Caribbean Outbreak

Estı'variz CF, Molna'r Z, Venczel L et al (2011) Paralytic poliomyelitis associated with sabin monovalent and bivalent oral polio vaccines in Hungary. Am J Epidemiol 174:316–325

Friedrich F (2000) Genomic modifications in oral poliovirus vaccine strains after multiplication in humans and implications for the eradication of poliovirus. Acta Virol 44:109–117

Georgopoulou A, Markoulatos P, Spyrou N, Vamvakopoulos NC (2000) Improved genotyping vaccine and wild-type poliovirus strains by restriction fragment length polymorphism analysis: clinical diagnostic implications. J Clin Microbiol 38:4337–4342

Ghendon Y, Robertson SE (1994) Interrupting the transmission of wild polioviruses with vaccines: immunological considerations. Bull World Health Organ 72:973–983

Grassly NC, Jafari H, Bahl S et al (2009) Mucosal immunity after vaccination with monovalent and trivalent oral poliovirus vaccine in India. J Infect Dis 200:794–801

Heymann DL, Sutter RW, Aylward RB (2006a) A vision of a world without polio: the OPV cessation strategy. Biologicals 34(2):75–79

Heymann DL, Sutter RW, Aylward RB (2006b) Polio eradication: interrupting transmission, towards a polio-free world. Future Virol 1:181–188

Horie H, Yoshida H, Matsuura K, Miyazawa M, Wakabayashi K, Nomoto A, Hashizume S (2002) Isolation of vaccine-derived type 1 polioviruses displaying similar properties to virulent wild strain Mahoney from sewage in Japan. J Med Virol 68:445–451

Hovi T, Lindholm N, Savolainen C, Stenvik M, Burns C (2004) Evolution of wild type 1 poliovirus in two healthy siblings excreting the virus over a period of 6 months. J Gen Virol 85:369–377

John T, Jain H, Ravishankar K et al (2011) Monovalent type 1 oral poliovirus vaccine among infants in India: report of two randomized double-blind controlled clinical trials. Vaccine 5:5793–5801

Kaura G, Abraham T (2012) Polio eradication: a complex end game. BMJ 344:e2398

Kew OM, Sutter RW, Nottay BK et al (1998) Prolonged replication of a type 1 vaccine-derived poliovirus in an immunodeficient patient. J Clin Microbiol 36:3893–2899

Kew OM, Wright PF, Agol VI (2004) Circulating vaccine-derived polioviruses: current state of knowledge. Bull World Health Organ 82:16–23

Martin J, Ferguson GL, Wood DJ, Minor PD (2000) The vaccine origin of the 1968 epidemic of type 3 poliomyelitis in Poland. Virology 278:42–49

Martin J (2011) Detection and characterization of polioviruses. Methods Mol Biol (Cliffton) 665:233

Modlin JF (2010) The bumpy road to polio eradication. N Engl J Med 362:2346–2349

Mukherji WCS, Jindal LCAK, Singh BZ et al (2005) Polio eradication in India: myth or reality. Med J Arm Forces Ind 61:364–366

Mulders MN, Reimerink JHJ, Stenvik M, Alaeddinoglu I, van der Avoort HGAM et al (1999) A sabin vaccine-derived field isolate of poliovirus type 1 displaying aberrant phenotypic and genetic features, including a deletion in antigenic site. J Gen Virol 80:907–916

Nathanson N (2008) The pathogenesis of poliomyelitis: what we don't know. Adv Virus Res 71:1–50

Nwachuku N, Gerba CP (2006) Health risks of enteric viral infections in children. Rev Environ Contam Toxicol 186:1–56

Kew O (2012) Reaching the last one per cent: progress and challenges in global polio eradication. Curr Opin Virol 2:188–198

Paul Y (2007) Role of genetic factors in polio eradication: new challenge for policy makers. Vaccine 25:8365–8371

Pelczar MJ Jr, Chan ECS, Krieg NR (1993) Microbiology: concepts and applications. McGraw-Hill Inc., New York 698

Racaniello VR (2006) One hundred years of poliovirus pathogenesis. Virology 344:9–16

Rousset D, Rakoto-Andrianarivelo M, Razafindratsimandresy R et al (2003) Recombinant vaccine-derived poliovirus in Madagascar. Emerg Infect Dis 9:885–887

Sabin AB (1956) Pathogenesis of poliomyelitis; reappraisal in the light of new data. Science 123:1151–1157

Sathyamala C, Mittal O, Dasgupta R et al (2005) Polio eradication initiative in India: deconstructing the GPEI. Int J Health Serv 2:361–383

Shulman LM, Manor J, Handsher R, Delpeyroux F, MacDonough MJ, Halmut T et al (2000) Molecular and antigenic characterisation of a highly evolved derivative of the type 2 oral polio vaccine strain isolated from sewage in Israel. J Clin Microbiol 38:3729–3734

Sutter RW, Prevots DR (1994) Vaccine-associated paralytic poliomyelitis among immunodeficient persons. Infect Med 11:426–438

Sutter RW, Tangermann RH, Aylward RB et al (2001) Poliomyelitis eradication: progress, challenges for the end game, and preparation for the post-eradication era. Infect Dis Clin North Am 15:41–64

Vashishtha VM (2009) Polio eradication in India: need for caution. Indian J Pediatr 76:757

Wood DJ, Sutter RW, Dowdle WR (2000) Stopping poliovirus vaccination after eradication: issues and challenges. Bull World Health Organ 78:347–357

Wood N, Thorley B (2003) Viewpoint towards global poliomyelitis eradication: the successes and challenges for a developed country. J Paediatr Child Health 39:647–650

World Health Organization and UNICEF (2010) Progress on sanitation and drinking water: 2010 Update

World Health Organization (2000) Outbreak news. Wkly Epidemiol Rec 75:397–399

World Health Organization (2002) Manual for the virological investigation of poliomyelitis. World Health Organization (WHO/EPI/GEN/02.1), Geneva

World Health Organization (2003a) OPV virus circulation and evolution investigated in New Zealand. Polio Lab Netw Quaterly Update 9:2–3

World Health Organization (2003b) Progress towards the global eradication of poliomyletis, 2002. Wkly Epidemiol Rec 78:138–144

World Health Organization (2004) Manual for the virological investigation of poliomyletis. World Health Organization (WHO/EPI/GEN/04), Geneva

Yang CF, Naguib T, Yang SJ et al (2003) Circulation of endemic type 2 vaccine-derived poliovirus in Egypt from 1983–1993. J Virol 77:8366–8377

Yoshida H, Horie H, Matsuura K, Kitamura T, Hashizumi S, Miyamura T (2002) Prevalence of vaccine-derived polioviruses in the environment. J Gen Virol 83:1107–1111

Zaoutis T, Klein JD (1998) Enterovirus infections. Pediatr Rev 19:183–191

Hepatitis A and E in Potable Water: A Threat to Health

3

Prem Shankar, Jyotsna Mishra, and Sarman Singh

Abstract

Water is a vital asset, required to maintain the basic life process of living being. Therefore, clean water is imperative as waterborne diseases still pose a major critical risk factor in drinking water quality. The WHO has recommended the quality standards of drinking water to avoid water-borne infections. The contamination of drinking water may be through physical, chemical, and biological means. The pathogen concerned with biological way includes many types of bacteria, viruses, helminthes, and protozoa which vary generally in size, structure, and composition with different disease manifestation. The contamination of drinking water especially with viruses is a global issue not only for undeveloped and developing countries, but also for developed ones. Enterically transmitted water-borne hepatitis is recognized as a major public health problem in many developing countries. Hepatitis A virus (HAV) and Hepatitis E virus (HEV) are reported to be the most common cause of infectious hepatitis epidemic outbreaks transmitted through water, especially in developing countries. Though both viruses generally lead to self-limiting symptomatic disease, fulminant hepatic failure with fatal outcome occurs in a small proportion of patients. The control and prevention of enterically transmitted viral hepatitis remains a major public health challenge. The present chapter is intended to give a brief account of the various aspects of hepatitis A and hepatitis E, i.e. virulence, pathogenesis, clinical aspects, diagnosis, immunology, epidemiology and its prevention.

P. Shankar · J. Mishra · S. Singh (✉)
Division of Clinical Microbiology, Department of Laboratory Medicine, All India Institute of Medical Sciences, Ansari Nagar, New Delhi, India
e-mail: sarman.singh@gmail.com

P. P. Singh and V. Sharma (eds.), *Water and Health*,
DOI: 10.1007/978-81-322-1029-0_3, © Springer India 2014

Keywords
Drinking water · Health · Hepatitis A virus · Hepatitis E virus · Waterborne disease

Introduction

Water is basic and foremost requirement for living being. The earth has around 70 % of total water and 98 % water of which is salty and remaining 2 % water is suitable for drinking purpose, most of which present in glaciers. The water is essential for life as it have many biological, physical, and chemical properties by which one can continue their healthy life. Unfortunately, very few people in developing countries have access to clean water, and around 2.4 billion people still have no access to basic sanitation (WHO 2000). The lack in proper management of wastewater leads to paucity of drinking water. The lack in availability of drinking water, free from any contaminant, is major source of various gastrointestinal diseases ultimately increasing the sanitation-related diseases throughout the world (Carr 2002). As per the WHO, diarrheal infection accounts for 4.1 % of overall worldwide burden of infection and is accountable for deaths of 1.8 million individuals per year. It is expected that 88 % of this burden is caused by water supply, hygiene, and sanitation concerned to children in developing countries. People become infected when any material contaminated by feces enters to their mouth and generally it can take place through drinking of contaminated water. This has been the source of several remarkable outbreaks of fecal-oral enteric diseases such as cholera and typhoid. However, there are many other ways by which fecal material may reach up to the mouth, for instance by the hands or contaminated food. The food which propagated, washed, and prepared in contaminated water leads to foodborne illness. The germs present in feces can cause diseases by even slight contact and transfer. One gram of human or animal waste (feces) comprises approximately 10 lakhs viruses, 1 lakh bacteria, 1,000 parasite cysts, and 100 parasite eggs (UNICEF 2008). The drinking water is associated with many of the infectious diseases if it is contaminated with pathogenic bacteria, viruses, and protozoa. The majority of enteric waterborne pathogens are introduced into potable water supplies by animal or human feces but they can also exist naturally in water as native aquatic microorganisms. Approximately, each day the adult individual should drink about eight glasses of water. If it is not pure or it is contaminated, then it will lead to more deadly disease to human. Every bodily function like immune response against infection, brain function, joint lubrication, cellular communication depends upon the availability of free water at cellular stage; it shows how essential is water for human health and life. Therefore, amount of safe water free from any type of contaminant should reach to human body in order to maintain their biological process.

Water and its Biological Role on Human Health

Directly or indirectly water has an important role for human body system to regulate and maintain vital functioning of body and its intricate systems. These include digestion and absorption of food and nutrients, body temperature, and blood circulation maintenance, and it carries nutrients and oxygen to cells and as detoxifying agent. Consequently, its limpidness is utmost imperative to avoid waterborne diseases. If potable water is treated inefficiently, then it may cause health risk due to presence of enteric and adenoviruses to individuals on consumption (WHO 1997). Waterborne diseases are any illness caused by consumption of water polluted by human or animal feces, which contain pathogenic microorganisms. These may include virus, bacteria, or protozoans. The contamination of viruses into water is more dangerous than bacteria because a single virus particle is sufficient to pose health risk whereas the bacterial cells need 100–1 million numbers of cells to cause infection (Girones and Puig 1994).

Microbes in Drinking Water: A Menace to Health

Contaminated water accounts for about 80 % of the diseases in developing countries, and in these regions, potable water is arranged from unhygienic or impure sources and it is further contaminated during transportation and storage at household containers or tanks. Nowadays, it is becoming the important concern about unsafe water and health impacts on individuals. The presence of enteric virus in public water supplies through sewage or its effluent is the key source of contamination. There are various other microorganisms like bacteria, protozoans, and viruses are conscientious for waterborne illness. Most of outbreaks occurred after monsoon rains, heavy flooding, and contamination of well water and massive uptake of unprocessed sewage into urban city water management plants. Among other viruses, here, we will confine to fecal-oral enteric viruses specially HEV and HAV.

Hepatitis Viruses

The term hepatitis is derived from hepar^{G} denotes liver, and suffix-*itis* means inflammation, It means inflammation of liver. Various illnesses and conditions can be a source of inflammation to liver, for example alcohol, drugs, chemicals, immune-disorders, and many viruses also, for example, enteric viruses like HAV and HEV.

There are numbers of hepatitis viruses; they have been designated as type A, B, C, D, E, F (not confirmed), and G; there is no meaning these are just alphabetical notation. Among these, only hepatitis A and E are associated with drinking water contaminated with fecal origin. In developing countries, 5–20 % population exhibit augmented HEV IgG titer with increasing age; however, the pathogenesis of HEV infection still remains confusing. (Kuniholm et al. 2009). Hepatitis E virus (HEV) is transmitted by fecal-oral route, often through polluted or sewage contaminated water, and is responsible for large numbers of waterborne epidemics of acute hepatitis in endemic regions and for sporadic autochthonous cases in non-endemic regions (Pavio et al. 2008). It is previously called as NANB (enterically transmitted non-A, non-B hepatitis) and they are spread via fecal-oral route (Khuroo 1980). HEV is waterborne disease, and Harrison in the year 1999 (Harrison 1999) and Aggarwal et al. in year 2000 (Aggarwal and Krawczynski 2000) suggested it as a zoonosis because numerous non-human primates, sheep, pigs, goats, cows, and rodents are at risk to infection. In developing and developed countries such as Asia, Middle East, and northern Africa, hepatitis E is chief cause of acute liver infection in adult population (Perez Gracia and Rodriguez-Iglesias 2003). Hepatitis E virus causes liver inflammatory disease; it is usually self-limiting in immune competent person (Purcell and Emerson 2008). But this can lead to acute hepatitis E which is responsible for mortality and morbidity in adults. However, mortality rate of HEV is 10 times more than hepatitis A virus. The comparison between hepatitis A and E virus is summarized (Table 3.1).

History

The description of hepatitis epidemics has been documented since ancient time but Hippocrates in year 460–375 BC described clinical feature of viral hepatitis. However, only in 1940s it became evident that at least two types of hepatitis existed. At that time, no diagnostic procedures were available and disease was diagnosed on clinical and epidemiological grounds. In the early 1960s with discovery of first hepatitis B marker (Australia antigen), a probable differential diagnosis between this and other types of hepatitis could be made. In the early 1970s, viral particles were visualized by electron microscopy in feces of patients with hepatitis A (Feistone et al. 1973).

Table 3.1 Comparisons between hepatitis A and E virus

Features	Hepatitis A virus	Hepatitis E virus
Notation	HAV	HEV
Family	*Picornaviridae*	Similar to Calicivirus
Viral structure (nm)	~27	30–35
Genome size (nucleotide)	7,500 bp	7,500 bp
Envelope	No	No
Genetic material	RNA	RNA
Strand type	Single strand	Single strand
Strand polarity	Positive sense	Positive sense
Nature	Infectious hepatitis	Non- infectious
Incubation episode (weeks)	2–6	2–9
Onset nature	Generally acute	Generally acute
Mode of spread		
Fecal or oral	Generally	Generally
Parenteral	Rare	No
Other	Food or waterborne	Waterborne in developing countries
Sequelae carrier	No	No
Chronic hepatitis	No case reported	No case reported
Mortality	0.1–0.2 %	20 %
Immune response		
Homologous	Yes	Yes
Heterologous	No	No

The first isolation of HAV particle was done in the year 1973 by Purcell in suspension of fecal sample of infected person and he suggested that human being as the sole reservoir for this virus. Since the use of accurate serologic investigations in the 1980s, the epidemiology, clinical manifestations, and usual history of hepatitis A have become evident. The HEV was first visualized in the year 1983 and its genome was cloned and characterized in 1991 (Tam et al. 1991). HEV was primarily recognized in India in 1955 and has because been documented as main reason of acute hepatitis in young adults all over to Asia, the northern and western Africa, and Middle East. In the year 1991, Hatami reported first epidemic of HEV infection in Kermanshah (capital city of Kermanshah territory, situated 525 km from Tehran in the western division of Iran and about 120 km from the border of Iraq) (Hatami 1991). The indication for existence of HEV was initially provided in the year 1980 (Arankelle et al. 1994).

Transmission

The transmission of the HAV and HEV is mainly through following routes:

Fecal-Oral Transmission of Contaminated Water

Hepatitis A is an enteric infection spread by infected excreta. The elevated concentrations of viral particle are shed in stools of patients from 3 to 10 days prior to onset of illness and until 1–2 weeks after onset of jaundice. The persistent of HAV in fecal excretion remains longer in children and in immune-compromised persons (up to 4–5 months after infection). Commonly, HEV infection is transferred via the drinking water contaminated with fecal. Fecal-oral transmission of HEV occurred due to fecal contamination of drinking water and infection occurs through transfer of ingestion of fecal-contaminated food or water.

Person-to-Person Transmission

The HAV disease is usually acquired via fecal-oral route either by person-to-person contact or ingestion of polluted food or water, as well as sexually via analingus (Corey and Holmes 1980). Whereas, transmission of the HEV infection from person to person is unusual.

Bloodborne Transmission

The transfusion-transmitted transmission (TTT) is observed infrequently since viremia endures for a short period about 7–10 days, ceasing with commencement of jaundice and the development of specific antibody reaction (Krugman et al. 1959) though lengthy period of viremia have been confirmed in patient with HAV (Mannucci et al. 1994) and in animal model studies (Asher et al. 1995). Transmitted from the hepatitis virion tainted blood products. However, there is no indication of sexual transmission or for transmission by transfusion.

Primate to Human (Zoonotic) Transmission

Naturally chimpanzees infected by HAV and experimentally transmitted to other non-human primates too such as monkey tamarins, squirrel-sized New World monkeys in genus Saguinus of family Callitrichidae. This is transmitted from animals to humans via exposure with infectious body fluids of infected animals. This type of transmission is very frequent in hyper-endemic region. There is possibility of zoonotic transfer (Tei et al. 2003) of virus through non-human primates, cow, goats, pig, rodents, and sheep because these are susceptible to infection.

Vertical Transmission

Hepatitis A can be transmitted by parenteral route but very rarely by blood and blood products because of short period of HAV viremia during acute infection. The transmission of HAV to infant occurs via intrapartum exposure to contaminated maternal blood or feces (Tanaka et al. 1995). Persons in psychiatric institutions or day care centers, health care providers, military personnel, and men who have sex with men, especially when practicing anal intercourse, are at higher risk of infection (Hollinger and Ticehurst 1996). The hepatitis A virus disease is accompanied and represents chiefly as illness of children (Chadha et al. 1999a, b).

However, the spread of hepatitis E is mainly through poor sanitation quality of the livelihood. The majority of cases have been observed in rainy seasons or after floods (Uchida 1992). Hepatitis E virus is also commonly transmitted from infected mothers to their babies with significant perinatal morbidity and mortality (Singh et al. 2003). The infection transmits rapidly in adult population as compared to the children's and risk becomes more for patients of developed countries because of lack in immunity that travels to endemic regions. Eli Schwartz et al. in the year 2010 studied HEV infection in travelers (Schwartz et al. 1999).

Reservoir of the Virus

Humans are only reservoir of hepatitis A virus; it is transmitted from feces of infected patients, either by person-to-person contact or by consumption of contaminated food or potable water. If vaccination is done appropriately at extensive level, then virus could theoretically be eradicated. Among other hepatitis viruses, HEV is merely a single virus that has animal origin and it was first genetically identified and isolated from pig in 1997 from USA (Meng et al. 1997). Recently, in the year 2001, HEV strain of avian origin has been recognized and characterized from chicken having hepatitis splenomegaly (HS) syndrome (Haqshenas et al. 2001). Moreover, HEV strain also has been documented from mongooses, rabbits, dear, and rats (Nakamura et al. 2006; Johne et al. 2009; Reuter et al. 2009; Tomiyama et al. 2009; Zhao et al. 2009). Rapid increase in the data shows that the HEV is disease of zoonotic origin with swine as an animal reservoir. The strains isolated from pig and human genetically differ greatly closer and impractical to differentiate (Lu et al. 2006). As a result, zoonotic transmission of HEV exhibits a fundamental public health distress more than food security and zoonotic perilous (Meng 2009).

Epidemiology

Hepatitis A infection occurs worldwide intermittently or in epidemic outbreaks. There is an expected 1.4 million cases annually (Viral Hepatitis Prevention Board 1997). The high-income regions have very low HAV endemicity levels with incidence of 1.5 per 100,000 in industrialized countries, for example, United States or Germany (Wasley et al. 2005) as compared to developing countries (parts of Africa, Asia, Central and South America) where it may reach up to 150 per 100,000 per year (Fig. 3.1a: Adapted from the viral hepatitis slide set published by the US Centers of Disease Control and Prevention, Atlanta, GA, USA, at http://wwwnc.cdc.gov/travel/yellowbook/2012/chapter-3-infectious-diseases-related-to-travel/hepatitis-a.htm). The epidemiological pattern of HAV infection varies with hygienic and sanitary conditions in the area; the highest prevalence of infection occurs in regions with the lowest socioeconomic levels (Bell 2002), whereas epidemiology of HEV is found quite higher in developing and developed countries like Asia, Africa and Central America, and Middle East, respectively (Gupta and Smetana 1957; Belabbes et al. 1985; Arankalle et al. 1988; Velazquez et al. 1990; Tsega et al. 1991). In India only, more than 2, 00,000 sporadic cases of hepatitis E are thought to occur yearly (Scharschmidt 1995), and as per the center for disease control (CDC), currently 40 % of acute viral hepatitis infection in India has been attributed to this virus. The largest recorded hepatitis E outbreak resulting in significant morbidity occurred in China between the year 1986 and 1988 and was considered highest involving 100,000 persons (Zhuang 1992). In year 2004, Matsubayashi and Khuroo et al. have reported blood-transfused HEV epidemics in endemic area (Matsubayashi et al. 2004; Khuroo et al. 2004). The infection of hepatitis E virus was first reported in year 1955 in an outbreak occurred in New Delhi, India (Gupta 1957). The maximum numbers of outbreaks has been reported by means of fecal-oral route of contaminated water (Belabbes et al. 1985; Naik et al. 1992) and epidemiology prototype is similar to those of HAV virus (Fig. 3.1 b: Adapted from the viral hepatitis slide set published by the US Centers of Disease Control and Prevention, Atlanta, GA, USA, at http://wwwnc.cdc.gov/travel/yellowbook/2012/chapter-3-infectious-diseases-related-to-travel/hepatitis-e.htm). In developed countries, modern revolution and industrialization is responsible for sporadic cases of HEV epidemics (Miyamura 2011).

Incidence of Hepatitis A and E with Age Group

The viral hepatitis infection is strappingly depends on age, and older children or adult shows symptomatic disease whereas young one shows asymptomatic infection. The frequency of seroprevalence of HAV and HEV antibodies varies as per the age group. The analytical features comprises gastrointestinal and flu-like symptoms, jaundice with recovery period of several weeks (Cuthbert 2001) with increase in age risk of death may occur with acute liver failure (Ciocca 2000). In developing countries like India, infection often occurs in children, who are likely to be asymptomatic and with lifelong immunity. The age-specific prevalence of antibodies to HAV can be used to define several patterns of infection including disease rates and predominant transmission. HAV is universally acquired by age of 5 in both India and Egypt. An overview of age-specific seroprevalence profile of HAV antibodies in Indian population has been compiled in Table 3.2. Though, such areas have low reported disease rates and rare disease outbreaks but may have high disease rates owing to high level of circulation virus. Developing countries with translational economies such as southern and eastern Europe and some regions in Middle East are reported as intermediate endemicity with reductions in exposure to the HAV

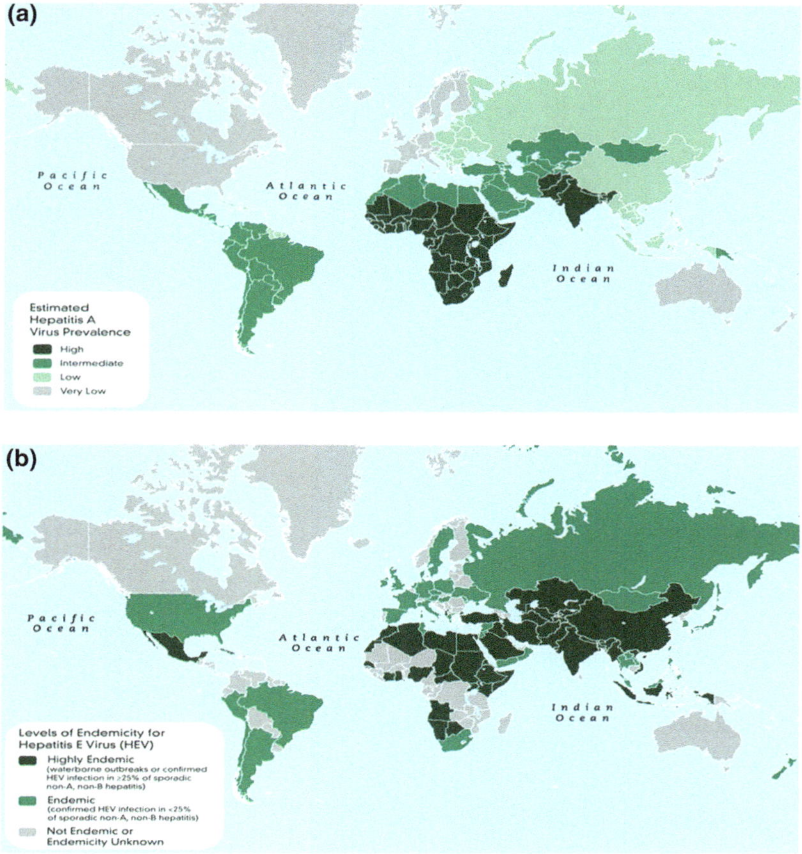

Fig. 3.1 a and b: The epidemiological pattern of HAV and HEV infection

in childhood. However, high level of HAV virus circulation via food and water transmission results in high frequency of disease in older children, adolescents, and young adults. In developed countries, Europe and Asia, the endemicity of HAV infection is transitional to low and incidence of anti-HAV varies commonly. Low exposure of virus results in less cases of hepatitis A infection; however, population remains susceptible throughout adulthood. The first retrospectively documented epidemic of viral hepatitis was in year 1955 from Delhi, and many recurrent epidemics of hepatitis E have been observed in Asia and Africa Table 3.3. The prototype of seroprevalence with age and sex is imperative for policy maker and health system organization, NGOs in order to lessen outbreaks by adopting appropriate preventive measures such as monitoring, immunization policies, and other means against waterborne enteric hepatitis.

Molecular Structure

The hepatitis A virus is 27 nm, positive-sense ssRNA, 7.5 kb in length, non-enveloped, icosahedral virus of the heparnavirus genus from Picornaviridae family (Fig. 3.2a: Electron micrographic view of hepatitis A virus Credit for HAV image: Centers for Disease Control and Prevention's Public Health Image Library (PHIL), with identification number #2739). It is separated into 3 fragments: a 5′ untranslated region (742 nucleotides); a single long ORF that synthesized a 2,227 amino acid polypeptide chain (6,681 nucleotides); and a 3′ non-coding region (63 nucleotides). The 5′ untranslated

Table 3.2 Seroprevalence of hepatitis A virus antibody in India

Geographic region	Number	Prevalence in (%)	Age groups	Reference
North/Delhi	25	100	Neonates	Tandon et al. (1984)
	73	68	Till 5 year	
	52	90	5–10 year	
North/Delhi	36	58.3	Newborn	Mittal et al. (1998)
	166	50.6	<6 month–24 month	
	91	76.9	2–5 year	
	67	79.1	>5–10 year	
West/Pune	1982 1992	1982 1992		Arankalle et al. (1995)
	0 9	0 67	<0.6 year	
	11 11	18 27	0.7–1.5 year	
	186 242	93.7 97.9	1.6– >15 year	
North/Delhi	170	35.8	1–8 year	Thakur et al. (1998)
	181	59	9–12 year	
	86	77	13–16 year	
North/Delhi	93	60.2	11–20 year	Das et al. (1998)
	302	64.2	≥21 year	
West/Mumbai	75	50.7	0–3 year	Dhawan et al. (1998)
	143	73	4–10 year	
	66	83.3	11–15 year	
	386	84.4	≥16 year	
South/Kerala	89	4.5	1–5 year	Mathews et al. (1998)
	94	8.6	6–10 year	
	186	48.3	≥11–15 year	
North/Lucknow	28	68	0–5 month	Aggarwal et al. (1999)
	22	91	6–10 month	
	23	96	11–18 month	
West/Pune	38	94.7	0–3 month	Chaddha et al. (1999a)
	30	56.7	4–6 month	
	365	28.2	7–48 month	
	66	90.9	49–72 month	
South/Hyderabad	21	14	0–5 month	Joshi et al. (2000)
	49	59.1	6–20 month	
	20	95	≥21 month	
North/Delhi	30	60	0–0.5 year	Dutta et al. (2000)
	90	21	0.5–2 year	
	90	66	>2–5 year	
	210	86.6	>5–12 year	
North/Delhi	500	71.2	19–45 year	Das et al. (2000)
North/Delhi	91	5.1	18–21 year	Jindal et al. (2002)
North/Delhi	206	86	4–7	Batra et al. (2002)
	574	91	8–11	
	644	97	12–18	
South/Chennai	19	31.6	0–2 year	Mohanavlli et al. (2003)
	77	83.1	>2–6 year	
	86	94.1	>6–12 year	

Table 3.3 Seroprevalence of hepatitis E virus antibody in India

Geographic region	Number	% Prevalence	Age groups	Reference
North/Kashmir	40	5	<14	Khuroo et al. (1994)
West/Pune	30	7	1.6–3	Arankalle et al. (1995)
	39	8	4–5	
	103	5	6–10	
	1,064	12	11–15	
	109	40	16–5	
	91	38	25–35	
	79	39	36–45	
	87	40	>45	
North/Lucknow	28	64	0–5	Aggarwal et al. (1997)
	22	59	6–10	
	25	64	11–18	
	20	50	Adults	Das et al. (2000)
North/Urban Delhi	500	35.6	General population	Das et al. (2000)
North/Rural Delhi	1,005	23.8	6 months–10 year	Mathur et al. (2001)
North/Urban Delhi	1,065	28.7	6 months–10 year	
South/Chennai	19	5.3	0–2	Mohanavalli et al. (2003)
	22	9.0	2–4	
	55	7.3	4–6	
	38	7.9	6–8	
	18	16.7	8–10	
	33	9.0	10–12	
South/Vellore	200	0.54	1–5	Daniel et al. (2004)
	200	1	6–15	
	100	8	16–40	
	100	13	>40	

regions are most conserved sequence of the genome and it comprises an internal ribosome entry site (IRES). The ORF encodes a poly protein organized into 3 functional regions: P1, P2, and P3. P1 is secondarily cleaved into four capsid proteins, VP1 to VP4, whereas P2 and P3 encode non-structural proteins of the viral structure (Jameel 1999; Chandra et al. 2008). The HEV have diameter of 27–34 nm, containing non-segmented positive-sense ssRNA virus with non-enveloped structure of 7,500 nucleotides in length. (Fig. 3.2 b: Electron micrographic view of HEV obtained from the CDC PHIL. Image credit: CDC/(PHIL #5605). It was first seen in year 1983 (Balayan et al. 1983) and later on molecularly cloned in year 1990 (Reyes et al. 1990). The sedimentation coefficient of virion particle is 183 S and buoyant density of native virion particles, isolated from infected patient stools, and is between 1.39 and 1.40 g/cm3 in cesium chloride (Li et al. 1997a, b). It consists 5′ and 3′ UTR (untranslated region) with three open reading frames, that is, ORF 1 (~5 kb), ORF 2 (~2 kb), and ORF 3 (smallest one) (Tam et al. 1991). That synthesizes non-structural polypeptides chain, a capsid protein, non-structural phosphoproteins, and 3′ untranslated region. The virion life cycle and its pathogenesis are unclear as no competent cell culture system is available. The mechanism of pathogenesis is not well explained due to limitations of culture systems. The infection is confirmed by occurrence of anti-HEV antibodies in host system. A conserved sequence of 58

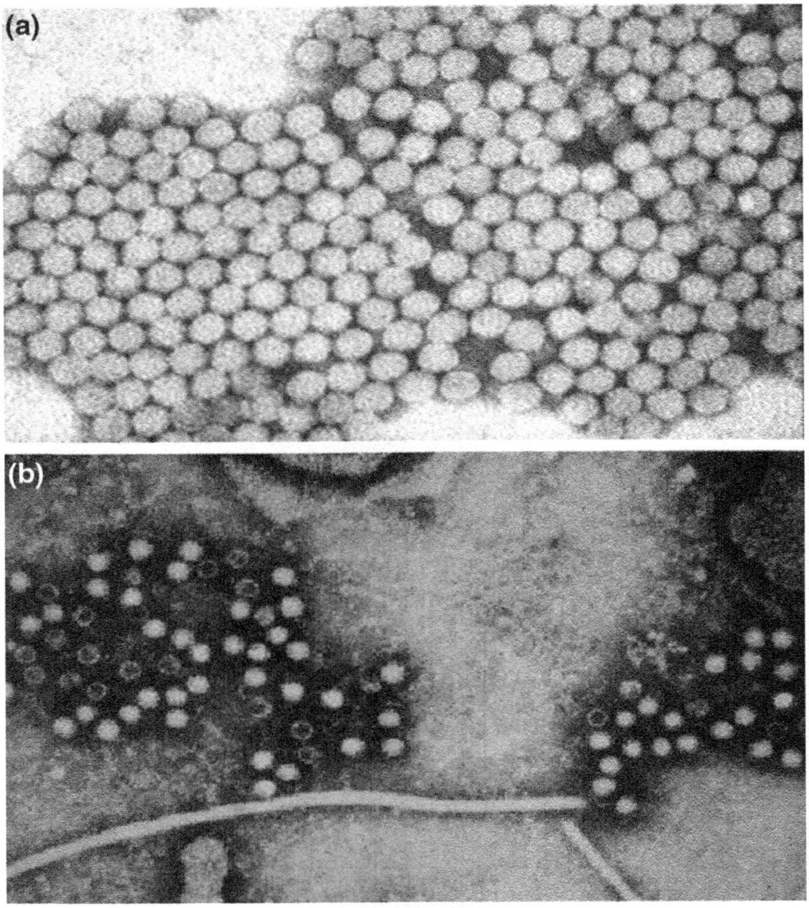

Fig. 3.2 a and b: Electron micrographic view of HAV and HEV

nucleotides from ORF-1 makes stem-loop structure with hairpin (Tam et al. 1996) and is essential for HEV genomic replication (Purdy et al. 1993a, b).

Life Cycles: Generalized Overview

The generalized life cycle for hepatitis A and E virus is as follows.

Virus Attachment and Penetration

It is the first and foremost step in initiation of viral life cycle. The attachment of virus particles is facilitated by cellular receptor of hepatocytes. Then, it is penetrated within the cells where further process takes place.

Uncoating of Viral RNA Material

The uncoating of viral nucleic acid takes place in order to release ssRNA into cytosol.

Translation and Post-translational Processing

The RNA genome of virus translated by a process involving the internal ribosome entry and resulting in production of viral polypeptides. The newly formed polypeptide goes into post-translational modification and segregated into structural and non-structural peptides.

Viral Assembling or Packaging

The assembly of structural, non-structural, and other component for virus particle synthesis starts in cytoplasm of host cell which is then assembled as complete infectious virion particle.

Maturation and Release

The newly packed virion particles released out of the cell.

Taxonomy and Genotypes

There are four human genotypes of hepatitis A virus have been identified; although considerable biological differences have not been observed (Lemon et al. 1992), it has five genotypes.

Genotype 1, genotype II, genotype III, genotype IV, genotype V, genotype VI in Asia-Africa, United States, Mexico, China and Europe, respectively (Wang et al. 1990; Tsarev et al. 1999; Schlauder et al. 1999). Regardless of important genetic inconsistency, a single serotype of HAV has been described. Sequence difference of a 168-nucleotide fragment adjoining VP1/2A junction has primarily defined seven genotypes to vary by at least 15 % and subtypes that differ by 7–7.5 % (Robertson et al. 1992). Genotypes I, II, and III separated into subtypes A and B, which are competent to transmit a disease to humans, though current literatures have reclassified previous genotype VII in the genotype II clad as subgenotype IIB (Lu et al. 2004). Data on genotype circulation represented that genotypes I and III are mainly ubiquitous genotypes isolated from humans. Subtype IA appears to be conscientious for mainstream of the hepatitis A cases globally, while subtype IB viruses have been chiefly found in Mediterranean province (Costa-Mattioli et al. 2003; Lu et al. 2004), although they have been also reported elsewhere too (Nainan et al. 2006). The Subgenotype IIIA is widespread in Central Asia (Costa-Mattioli 2003). Among HAV genotypes, subgenotypes IIA (formerly known as genotype II) and genotypes IIB (formerly genotype VII) are seldom reported.

On the basis of nucleotide sequencing pattern, HEV is member of *Hepevirus* genus, and that genus has not so far assigned to a family (Panda et al. 2007). However, it is classified as member of genus *hepevirus* under *hepeviridae* family. The putative non-structural polypeptide of HEV is highly similar to Rubella and plant furoviruses (Koonin et al. 1992). Phylogenetically, HEV shows highest, but limited, match with Rubella virus: *Togaviridae* family and with necrotic yellow vein virus of sugar beet: *Furovirus* genus, *Togaviridae* family. The capping enzymes of HEV have same properties as those of alpha viruses like super group (Ropp et al. 2000). The nucleotide sequencing classified the HEV into four major genotypes, namely genotype 1, 2, 3, and 4 (Schlauder and Mushahwar 2001). Geographically, distribution of genotypes differ from place to place, type 1 genotype is found mainly in developing countries such as Asia and Africa and are consider as epidemic strain, while majority of type 2 found in Mexico and Africa. However, genotype 3 is scattered all over the world and has been isolated from sporadic cases of acute HEV infection. Genotype 4 is found mostly in Asian continents and consisting strains from domestic pig and human origin (Meng et al. 1998; Schlauder and Mushahwar 2001; Lu et al. 2006; Lorenzo et al. 2007; Okamoto 2007). The genotypes auxiliary classifies into subgenotypes, genotype 1 have 5 subtypes, that is, 1A, 1B, 1C, 1D, and 1E. The genotype 2 have subtypes 2A and 2B while genotype 3 have highest subtypes, that is, 10 and represented as 3A, 3B, 3C, 3D, 3E, 3F, 3G, 3H, 3I, and 3J. Only seven subtypes are found for genotype 4, that is, 4A, 4B, 4C, 4D, 4E, 4F, and 4G (Lu et al. 2006). The typing of genotypes helps in epidemiological outbreaks of viruses as per geographical distributions (Norder et al. 2004). The HEV genotype 1 and 2 are

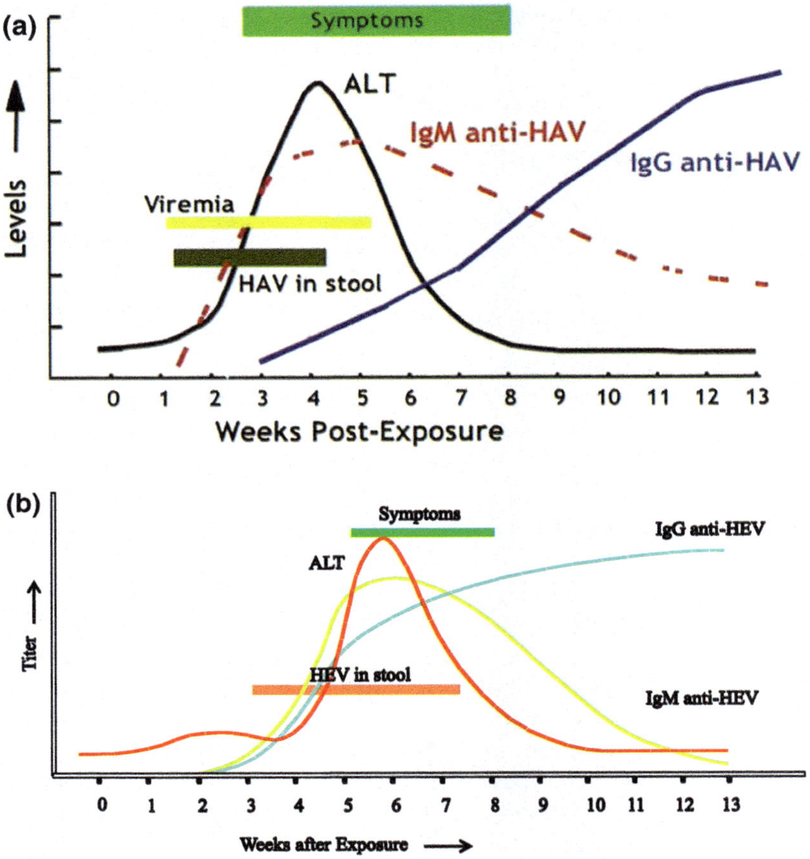

Fig. 3.3 **a** and **b**: Course of antibodies in HAV and HEV infection

mammalian strains and these are mainly associated to waterborne epidemics in endemic parts (Lu et al. 2006).

Immunological and Virological Response Against Hepatitis A and E Infection

The duration of viremia, viruses in bloodstream, for HEV is between 14 to 28 days in majority of the patients with representation with clinical symptoms (Ruan et al. 1997). Nanda et al. from Delhi have reported the presence of hepatitis RNA in serum for 112 days after the inception of clinical signs (Nanda et al. 1995). The fecal shedding or viremia of HEV has been observed in patients before the HEV infection. HEV subsequent to development of immunoglobulin M and immunoglobulin G antibodies indicates that factors other than acute-phase immunoglobulin may be essential for viral clearance (Clayson et al. 1995). Presence of virus in stool has been shown to begin up to 9 days earlier to the icteric phase of infection (Tomar 1998). Usually, presence of virus in stool lasts up to 14 days after the beginning of illness, but it has been reported to prolong until the 49th day of disease (Nanda et al. 1995) (Fig. 3.3a). The antibodies IgG and IgM are detected after the infection and antibodies concentration becomes higher between 15–30 days post-infection (Bryan et al. 1994). After infection, the anti-HEV-IgM concentration appears to decline in three months, in early phase of improvement (Kwo et al. 1997). Antibody IgG can be observed from 1–15 years subsequent to the HEV disease (Dawson et al. 1992). As per Bryan et al. in 1994, the antibody titer of IgG anti-HEV shows crest value after approximately one month and it decrease rapidly (Favorov et al. 1996). The majority of studies point out that IgG anti-HEV titers increases roughly 4 weeks after illness and then decrease speedily (Favorov et al. 1996). Mast et al. in 1998 showed that the detection limit may vary with techniques used, and recombinant protein-based assays are much more sensitive than protein-based immune assays (Mast et al. 1998). The level of circulating antibodies greatly decreases with age (De Bruyn and Song 1998). Though it is not well understood that why seroprevalence of anti-HEV and attack rates are lesser in older waterborne outbreaks, at present, the understanding between immunity and persistent of anti-HEV IgG is incomplete. The recorded epidemics of Kanpur take place in poorer social economic inhabitants which was 4–5 times as compared to privileged income group. These discoveries show that wherever poorer income group have been exposed and they got some defensive immunity against hepatitis E illness (Shrestha 1991). Therefore, for small inoculums, lower level of anti-HEV IgG could be the enough to control clinical disease, but it will not be true with heavy exposure. Therefore, infection might be happen to both types of group either low level of immune protection or people who are naives to infection previously. The resistance against disease is depended on the age group and this is discovered during 1955 Kanpur outbreak as the children exposed in that epidemic, maintained strong defensive immunity as compared to adult people disease.

Clinically, HEV and HAV infections are almost indistinguishable (Chwartz and Galun 1994). The disease may range from asymptomatic infection to mild hepatitis to subacute liver failure (Dalton et al. 2008). Among pregnant women with viral hepatitis, the rates of HEV infection have usually been higher than among non-pregnant patients (Singh et al. 2001). The illness appears in two phases: the first prodromal (pre-icteric) phase of 1–10 days with symptoms like, abdominal pain, tenderness, nausea, vomiting, and fever. The second (icteric) phase (15–40 days) marked by jaundice and dark urine is followed by viremia, liver enzyme elevations, antibody seroconversion, and clearing of the virus. Sometimes symptoms like arthral, diarrhea, pruritus, and an urticarial rash also occurs (Purdy and Krawczynski 1994). The disease is self-limited, without any chronic or carrier state after infection (Roberts and Whitlock 1992), and the majority of patients recover completely (Thomas et al. 1997).

The incubation episode of hepatitis E illness ranges from 2–8 weeks from the time of exposure which is marked by peak viremia (Labrique et al. 1999; Fig. 3.3b). The HEV RNA can be detected in blood and stool instantaneously prior to the beginning of clinical symptoms. Liver enzymes such as aspartate and alanine aminotransferase, alkaline phosphatase, y-glutamyl transpeptidase, and bilirubin rise peaking at about 6 weeks after infection and remain elevated for 20–90 days (Fletcher 1993) and falling to normal levels by week 10. After onset of the clinical symptoms, within few days to weeks, HEV RNA is cleared from the blood; however, the virus shedding in stool continues for another 2 weeks (Clayson et al. 1995). In many patients, liver biopsy specimens have shown cholestatic changes, including intracanalicular bile stasis and a gland-like transformation of parenchymal cells. Histopathological changes gradually resolve over 3–6 months (Teshale et al. 2010).

As in hepatitis A, specific IgM and IgG immune responses occur early in the disease and are detected soon after infection, with highest antibody titers at 15–30 days (Bryan et al. 1994) The immunoglobulin M (IgM) serology is often used to identify acute cases; however, it is not always detectable (Lin et al. 2000) and false-positive results occur (Takahashi et al. 2005). Unlike specific IgM titers which decline within 3 months after infection (Lin et al. 2000), IgG persists for a year and remains been detected for at least 12 years after acute infection (Chadha et al. 1999a, b).

Occurrence of Waterborne Hepatitis with Other Conditions

The management and surveillance of hepatitis A and E virus in developing countries becomes more pathetic when it is associated with other autoimmune and biological conditions. The growing population in developing country, India, needs special assessment and research to reduce waterborne infection and their effect on people's health.

Hepatitis A and E Infection with AIDS

Generally, HAV do not generate higher cruel complications in HIV-infected patients, but it may disrupt anti-retro-viral treatment (ART) (Lutwick 1999) resulting into future consequences. *Ida* et al. in year 2002 have showed that HIV-infected individual had a notably elevated HAV viral load and a considerably prolonged interval of HAV viremia with achievable viral shedding, compared to non-HIV-infected patients (Ida et al. 2002). This is likely due to result in prolonged duration of HAV transmission in a community. It is recommended that HAV vaccination should be provided before the initiation of HIV infection. If CD_4 counts become >300 cells/mm^3, then administration of HAV vaccine is preferred to maximize the antibody response to HAV vaccine.

The infection of HEV in AIDS patients is self-limiting and is responsible for liver inflammation. Constant HEV illness may build up in immune-suppressed patients; these individuals serve as a source of virus carrier for a long time. Elizabeth et al. assumed that HEV may be the root of chronic hepatitis in liver graft recipients (Haagsma et al. 2008).

Hepatitis A and E Infection in Organ Transplantation

Patients who underwent liver transplantation for HAV-allied liver dysfunction may be at elevated threat of reoccurrence of hepatitis A illness, chiefly if they necessitate anti-rejection graft treatment. Ordinary monitoring of anti-HAV antibodies-M and HAV-RNA during premature post-transplant period in HAV associated liver transplant recipients may distinguish reinfection from rigorous cellular refusal (Park et al. 2010).

The constant infection of HEV takes place in grafted patient along with HIV infection and can lead to liver cirrhosis and graft rejection (Pischke and SandWedemeyer 2010). Kumar et al. have explained additional clinical conditions associated with HEV infection in organ-transplanted people with immune-compromised conditions (Kamar et al. 2008a, b). It progresses to status of cirrhosis in some of these patients (Kamar et al. 2008a, b). These conditions are associated with immunosuppressive treatment and virus persistent during anti-hepatitis E immune prophylaxis (Kamar et al. 2008a, b).

Hepatitis A and E Infection During Pregnancy

The infection of hepatitis A during pregnancy is general basis of acute hepatitis although it is infrequently reported. HAV infection during pregnancy linked with maternal difficulty and preterm labor (Elinav et al. 2006). It is better to vaccinate and diagnose, by serologically, pregnant woman during this period especially to those woman residing in endemic areas. Distinct rise in liver enzymes during pregnancy period may cause pregnancy associated liver dysfunction such as pre-eclampsia; elevated liver enzymes, hemolysis, low platelets (HELLP syndrome); and severe fatty liver of pregnancy. In pregnant females in third trimester with viral hepatitis, the rate of HEV infection is reportedly between 40-57% (Singh et al. 2001). In addition to high attack rate, Hepatitis E infection have also been associated with a high mortality rate among pregnant women and reaches to 20 % in endemic regions (northern and central India, Pakistan) (Khuroo et al. 1995; Hussaini et al. 1997). Khuroo et al. in year 1981 reported a high prevalence of fulminant hepatitis in pregnant women during an epidemic of NANB type (Khuroo et al. 1981). Bhatia et al.

in year 2008 revealed that there is no difference in HEV infection pattern between pregnant and non-pregnant woman and it is just considered as inadequate predictive reason (Bhatia et al. 2008). Thus, prior to decide pregnancy as a foremost factor, there are other significant factors also which cannot be underestimated in defining the pathogenesis of HEV illness and these includes type of causative strain, genotype or subtype, patient immunological condition, viral load, and other coinfection during the course of HEV infection (Renou et al. 2008). According to Andersson et al. (2008), there is no case of HEV infection has been reported in pregnant woman with genotype 3 strains (Andersson et al. 2008).

Diagnosis

The diagnosis of HAV can be established by clinical characteristics followed by laboratory investigation such as biochemical assays of liver, recognition of HAV particles or virus constituent, detection of precise antibodies. The cell culture techniques do not respond well therefore their use as a diagnostic approach is very less.

Clinical Symptoms

The human being infected with HAV is attended by jaundice. The incidence of ailment depends on patient age and frequently occurs in child with anicteric circumstances. The typical ill people represent fatigue, malaise, loss of appetite, vomiting nausea, usual flu-like symptoms, generally smoker losses interest in smoking.

Hepatitis A infection has four clinical phases, although these do not occur in all patients. The first stage is an incubation period of 15–50 days (mean 28–30 days). This period is asymptomatic; however, infected individual may actively shed virus in stool. The second stage is a pre-icteric period of several days to weeks and may proceed to onset of jaundice. This prodromal phase is characterized by imprecise symptoms followed by gastrointestinal complications. Other symptoms may include myalgia (muscle pain), arthralgia (joint pain), cough, pharyngitis, diarrhea, constipation, pruritus (itchiness), and urticaria (hives). Dark urine caused by elevated bilirubin levels usually occurs prior to onset of jaundice. In third stage, characteristic yellowing of skin and eyes appear and most symptoms subside, although clinical signs such as hepatomegaly and hepatic tenderness are found in about half of the patients which usually resolve within few weeks. The final stage is a convalescent period during which the patient recovers.

Isolation and Identification of Organism

HAV has been cultured in several cell lines of human and non-human origins, mutually with primary and secondary African green monkey kidney cells (Daemer et al. 1981) and fetal rhesus monkey kidney cells (Flehmig 1980). The HAV is complex to grow and do not generate cytopathic consequences in cell lines even though unusual techniques like radio-immuno focus assay (RIFA) are essential to find infectious foci of HAV in predetermined cells. As a result, culture method is not well considered for HAV isolation. The discovery of virological distinctiveness in culture system is considered excellent for HEV identification. There are various in vitro culture systems being used for HEV culture including human lung, kidney, or liver cells (*2BS, A549, Hep-G2)*. Krawczynski et al. has done detection of virus-associated antigen in experimentally infected cynomolgus macaques (Krawczynski and Bradley 1989). Currently, there is no reliable technique available for HEV cell culture.

Biochemical Test

Subsequent to affirmation of clinical manifestation, level of aminotransferase (AST) enzyme, which is considered as sensitive indicator in serum, is evaluated (equal or more to 500 IU/

ml); bilirubin level becomes higher than 10 mg/ml, and immunoglobulin M (IgM) residue measurable during the phase. Antibodies in urine and saliva can be detected along with direct and indirect bilirubin, alanine aminotransferase (ALT), AST and alkaline phosphatase.

Acute hepatitis E diagnosis is made by biochemical assessment of liver function including laboratory evaluation of: urine bilirubin and urobilinogen, total and direct serum bilirubin, ALT, AST, alkaline phosphatase, prothrombin time, total protein, albumin, IgG, IgA, IgM and complete blood count. However HEV infection is usually confirmed by the demonstration of IgM anti-HEV in the serum by ELISA (Purcell 1996).

Microscopic Techniques

Currently, numerous methods have been developed to enhance microscopic sensitivity and specificity for hepatitis A and E diagnosis. Immune electron microscopy was the first generation test to be used by investigators. The original manifestation of hepatitis A virus in fecal samples was accomplished by immune electron microscopy (IEM; Sánchez et al. 2007). It is better than other methods as it do not involve radioactive-labeled compound. Virus like particle (VLP) in acute phase of infection can be seen in immune electron microscope (Purcell and Ticehurst 1988). Though, it has drawback because sometimes majority of the clinical samples do not have adequate VLPs in order for immune electron microscopy to be a useful tool for clinical or epidemiologic surveys. The application of IEM has been broadly used previously in detection of HAV (Feinstone et al. 1973). But it is relatively less used for routine diagnosis due to labor intensive use. Immune fluorescence microscope (IFE) techniques are use to identify semi-quantitative antibodies against HEV illness, and Krawczynski and Bradley in 1989 have used this to determine anti-HEV semi-qualitatively (Krawczynski and Bradley 1989). Although this technique is more difficult and expensive and therefore not being used for diagnostic purpose normally.

Serological Methods

Now serological diagnosis has become easy and there are various serology-based kits are available commercially and validated to detect specific antibodies specific to HEV or HAV. The diagnosis by serological methods is back dated since 1980, and the presence of HAV-specific IgM antibodies in blood confirms HAV infection (Stapleton 1995). It shows present infection whereas detection of IgG antibodies in blood represents past illness and indicates that individual is immune to further infection. The appearance of anti-HEV IgM in patient serum demonstrates HEV infection. Bruce et al. in the year 2002 developed a quantitative enzyme immunoassay (EIA) for antibody to HEV by means of shortened HEV capsid polypeptide expressed in baculo virus system to improve seroepidemiology, to add on hepatitis E identification, and to facilitate vaccine evaluations. They validate and reproduce indirect EIA to find antibody to HEV infection quantitatively. Presently, three commercial EIA tests are available for detection of anti-HEV antibodies. The First of two tests are from company Gene labs and they designed the test by using 4 short recombinant polypeptides corresponding to 3′ end of the ORF-2 (42 amino acids) and ORF-3 (33 amino acids of Burmese genotype 3) and Mexican genotype 2 strains for immunoglobulin G and immunoglobulin M antibodies detection. The third based on detection of anti-HEV IgG antibodies, manufactured by Abbott, specific to recombinant protein (123 amino acids) of complete ORF-3 of genotype 1 strain. The sensitivity immunoassays against particular HEV antibodies can be amplified by multifold by incorporating the antigens of diverse genotypes more precisely N and C terminals of the open reading frame 2 and 3 (Wang et al. 2001).

Molecular Methods

As compared to other serology and culture methods, molecular methods are highly sensitive and specific to detect HAV antigen in clinical samples. The application of restriction fragment length polymorphism (RFLP) (Goswami et al. 1997), single-strand conformational polymorphism (SSCP) (Calder et al. 2003), southern blotting (Calder et al. 2003), nucleic acid sequencing–based amplification (NASBA) (Jean et al. 2001), nucleic acid hybridization (NAH) (Zhou et al. 1991), reverse transcription-PCR (RTPCR) (Cromeans et al. 1997), antigen capture RT-PCR (AC-RT-PCR) (Jansen et al. 1990) has been used extensively by various researchers.

Real-time PCR, qualitative and quantitative, considered as good enough to detect viral nucleic acid in serum and feces (Espy et al. 2006) but has been used in fecal-contaminated water (Pina et al. 2000; Grimm and Fout 2002). These methods are more sensitive and specific and can detect recent infection too. However, this method includes need of skilled manpower and an appropriate method of viral RNA extractions with removal of inhibitory substances in order to detect virion nucleic acid.

Immunization to Protect Waterborne Hepatitis for Concern Public

The development in cell culture research laid down the birth of new and effective vaccination methods for HAV infection with production of safe and licensed vaccine (Werzberger et al. 1992; Innis et al. 1994). Around 2, 70,000 infection of HAV anticipated yearly in USA before vaccination against HAV (Calder et al. 2003). In 1996, USA Government Advisory Committee on Immunization Practices (ACIP) recommended HAV vaccine merely to explicit high-risk groups, comprising children residing in communities with prominent incidence of hepatitis illness, gay persons, travelers, and laboratory staff contacting HAV-infected patients and person prone to HAV disease (Koff 2007). At present, there are various highly immunogenic vaccines are available which can be consider good choice to provide immunity against HAV infection (Hammitt et al. 2008). There are various studies which are currently being going on for development of vaccine against HEV (Tsarev et al. 1997; Xing et al. 1999; Yarbough 1999; Zhang et al. 2001a, b; Purcell and Emerson 2001; Im et al. 2001). Although presently there is no vaccine available commercially against HEV infection, among them, a recombinant capsid antigen in baculovirus vector expressed in insect cells is considered an effective candidate for vaccine. The ORF2 polypeptide has been used to elicit immune system of individual against HEV infection (Purdy et al. 1993a, b). The immune system can be evoked by use of vaccine construct prepared by ORF-1 and ORF-2 of HEV because it is most important portion of genome responsible for antigenic properties (Li et al. 1997a, b; Purdy et al. 1993a, b; Li et al. 2000; Im et al. 2001; Zhang et al. 2001a, b; Ruan et al. 2003). Additionally, ORF-2-based vaccine has been developed and qualified phase-II trial successfully in Nepal (Shrestha et al. 2007) but Feng-Cai et al. in 2009 have used recombinant capsid protein of 50 KD size successfully completed phase-III clinical trial in Xiamen University, China (Zhang et al. 2009).

Management and Surveillance of Waterborne Contamination to Protect Hepatitis A and E Infection

As mode of the transmission is primarily via fecal origin, it takes place by intake of infected foods or water. Thus, appropriate preparation should be undertaken for endemic areas where occurrence of infection is more. The hepatitis A and E infection is self-limiting and in most of cases, it is cured by itself without damage/squeal just by proper monitoring. Symptoms resolve in 3 months in 85 % of infections, whereas it completely recovers after 6 months (Koff 1992) liver grafting needs in fulminant hepatitis patients. The general solution is related to environmental and personal cleanliness and hygiene to avoid fecal-oral transmission of pathogens. Since there is lack of effective and successful antiviral treatment for hepatitis infections, surveillance and monitoring of waterborne contamination is leftover as a significant point for their control. The better cure for infection is preventive strategies to ensure sanitation conditions especially to endemic regions. The frequency of infection can be reduced greatly by providing good quality of drinking water or food items. There should be a divided system of sewages to dispose drainage systems this will reduce possibility of cross-contamination with water pipelines. Further, drinking water can be chlorinated or boiled to kill virus present in water (Velázquez et al. 1990).

References

Aggarwal R, Krawczynski K (2000) Hepatitis E: an overview and recent advances in clinical and laboratory research. J. Gastroenterol. Hepatol. 15:9–20

Aggarwal R, Naik SR (1997) Epidemiology of hepatitis E: past, present and future. Trop. Gastroenterol. 18:49–56

Aggarwal R, Naik S, Yachha SK, Naik SR (1999) Seroprevalence of antibodies to hepatitis A virus among children in Northern India. Indian Pediatr. 36:1248–1250

Andersson MI, Hughes J, Gordon FH, Ijaz S, Donati M (2008) Of pigs and pregnancy. Lancet 372:1192

Arankelle VA, Chadha MS, Tsarev SA, Emerson SU, Risbud AR, Banerjee K (1994) Seroepidemiology of water borne hepatitis in India and evidence for a third enterically transmitted hepatitis agent. Proc. Natl. Acad. Sci. U.S.A. 91:3428–3432

Arankalle VA, Chada MS, Mehendale SM, Banerjee K (1988) Outbreak of enterically transmitted non-A, non-B hepatitis among school children. Lancet 2:1199

Arankalle VA, Tsarev SA, Chadha MS, Alling DW, Emerson SU, Banerjee K et al (1995) Age-specific prevalence of antibodies to hepatitis A and E viruses in Pune, India, 1982 and 1992. J. Infect. Dis. 171:447–450

Asher LV, Binn LN, Mensing TL (1995) Pathogenesis of hepatitis A in orally inoculated owl monkey (Aotus trivirgatus). J. Med. Virol. 47:260–268

Balayan MS, Andjaparidze AG, Savinskaya SS (1983) Evidence for a virus in non-A, non-B hepatitis transmitted via the fecal-oral route. Intervirology 20:23–31

Batra Y, Bhatkal B, Ojha B, Kaur K, Saraya A, Panda SK et al (2002) Vaccination against hepatitis A virus may not be required for school children in northern India: results of a seroepidemiological survey. Bull. World Health Organ. 80:728–731

Bell BP (2002) Global epidemiology of hepatitis A: implications for control strategies. In: Margolis HS, Alter MJ, Liang JT, Dienstag JL (eds) Viral hepatitis and liver disease. International Medical Press, London, pp 9–14

Belabbes EH, Bouguermouh A, Benatallah A, Illoul G (1985) Epidemic non-A, non-B viral hepatitis in Algeria: strong evidence for its spreading by water. J. Med. Virol. 16:257–263

Bhatia V, Singhal A, Panda SK, Acharya SK (2008) A 20-year single-center experience with acute liver failure during pregnancy: is the prognosis really worse? Hepatology 48:1577–1585

Bryan JP, Tsarev SA, Iqbal M (1994) Epidemic hepatitis E in Pakistan: patterns of serologic response and evidence that antibody to hepatitis E virus protects against disease. J. Infect. Dis. 170:517–521

Calder L, Simmons G, Thornley C, Taylor P, Pritchard K, Greening G et al (2003) An outbreak of hepatitis A associated with consumption of raw blueberries. Epidemiol. Infect. 131:745–751

Carr R (2002) Health aspects. Publication on health aspects of sanitation. http://www.sanicon.net/titles/topicintro.php/topic/Id=2 12 Sept 2003

Chadha MS, Chtambar SD, Shaikh NJ, Arankalle VA (1999a) Exposure of Indian children to hepatitis A virus and vaccination age. Indian J Med Res 9:11–15

Chadha MS, Walimbe AM, Arankalle VA (1999b) Retrospective serological analysis of hepatitis E patients: a long-term follow-up study. J Viral Hepatitis 6:457–461

Chandra V, Taneja S, Kalia M, Jameel S (2008) Molecular biology and pathogenesis of hepatitis E virus. J. Biosci. 33:451–464

Chwartz E, Galun E (1994) Prior hepatitis A virus infection enhances acute hepatitis E virus infection. Med. Hypotheses 42:198–202

Ciocca M (2000) Clinical course and consequences of hepatitis A infection. Vaccine 18:S71–S74

Clayson ET, Myint KS, Snitbhan R (1995) Viremia, fecal shedding and IgM and IgG responses in patients with hepatitis E. J. Infect. Dis. 172:927–933

Corey L, Holmes KK (1980) Sexual transmission of hepatitis A in homosexual men: incidence and mechanism. N. Engl. J. Med. 302:435–438

Costa-Mattioli M, Napoli ADi, Ferre V, Billaudel S, Perez-Bercoff R and Cristina J (2003) Genetic variability of hepatitis A virus. J Gen Virol 84:3191–3201

Cromeans TL, Nainan OV, Margolis HS (1997) Detection of hepatitis A virus RNA in oyster meat. Appl. Environ. Microbiol. 63:2460–2463

Cuthbert JA (2001) Hepatitis A: old and new. Clin. Microbiol. Rev. 14:38–58

Daemer RJ, Feinstone SM, Gust ID, Purcell RH (1981) Propagation of human hepatitis A virus in African green monkey kidney cell culture: primary isolation and serial passage. Infect. Immun. 32:388–393

Dalton HR, Bendall R, Ijaz S, Banks M (2008) Hepatitis E: an emerging infection in developed countries. Lancet Infect Dis 8:698–709

Daniel HD, Warier A, Abraham P, Sridharan G (2004) Age-wise exposure rates to hepatitis e virus in a southern Indian patient population without liver disease. Am. J. Trop. Med. Hyg. 71:675–678

Das K, Agarwal A, Andrew R, Frösner GG, Kar P (2000a) Role of hepatitis E and other hepatotropic virus in aetiology of sporadic acute viral hepatitis: a hospital based study from urban Delhi. Eur. J. Epidemiol. 16:937–940

Das K, Kar P, Chakraborty A, Gupta S, Das BC (1998) Is vaccination program against hepatitis A needed in India? Indian J. Gastroenterol. 17:158

Das K, Jain A, Gupta S, Kapoor S, Gupta RK, Chakravorty A et al (2000b) The changing epidemiological patterns of hepatitis A in the urban population of India: emergence of a trend similar to the European countries. Eur. J. Epidemiol. 16:507–510

Dawson GJ, Mushahwar IK, Chau KH (1992) Detection of long-lasting antibody to hepatitis E virus in a US traveller to Pakistan. Lancet 340:426–427

De Bruyn G, Song E (1998) Seroepidemiology of hepatotropic viral infections in Amp Pipal. Nepal Trop Doct 28:173–174

Dhawan PS, Shah SS, Alvares JF, Kher A, Shankaran, Kandoth PW et al (1998) Seroprevalence of hepatitis A virus in Mumbai and immunogenicity and safety of hepatitis A vaccine. Indian J. Gastroenterol. 17:16–8

Dutta AK, Aggarwal A, Kapoor AK, Ray GN, Batra S (2000) Seroepidemiology of hepatitis A in children in Delhi. Indian J. Pediatr. 67:77–79

Elinav E, Ben-Dov IZ, Shapira Y, Daudi N, Adler R, Shouval D et al (2006) Acute hepatitis A infection in pregnancy is associated with high rates of gestational complications and preterm labor. Gastroenterology 130:1129–1134

Espy MJ, Uhl JR, Sloan LM, Buckwalter SP, Jones MF, Vetter EA et al (2006) Real-time PCR in clinical microbiology: applications for routine laboratory testing. Clin. Microbiol. Rev. 19:165–256

Favorov MO, Khudyakov YE, Mast EE, Yashina TL, Shapiro CN, Khudyakova NS et al (1996) IgM and IgG antibodies to hepatitis E virus (HEV) detected by an enzyme immunoassay based on an HEV-specific artificial recombinant mosaic protein. J. Med. Virol. 50:50–58

Feinstone SM, Kapikian AZ, Purceli RH (1973) Hepatitis A: detection by immune electron microscopy of a virus like antigen associated with acute illness. Science 182:1026–1028

Flehmig B (1980) Hepatitis A-virus in cell culture. I. Propagation of different hepatitis A-virus isolates in a fetal rhesus monkey kidney cell line (FrhK-4). Med. Microbiol. Immunol. 168:239–248

Fletcher J (1993) A traveller returning from Nepal with hepatitis E. Med. J. Aust. 159:563

Girones R, Puig M (1994) Detection of adenovirus and enterovirus by PCR amplification in polluted waters. Water science and technology, fields virology. Lippincott-Raven, Philadelphia, pp 5–6

Goswami BB, Burkhardt W III, Cebula TA (1997) Identification of genetic variants of hepatitis A virus. J. Virol. Methods 65:95–103

Grimm AC, Fout GS (2002) Development of molecular method to identify hepatitis E virus in water. J. Virol. Methods 101:175–188

Gupta DN, Smetana HF (1957) The histopathology of viral hepatitis as seen in the Delhi epidemic. Indian J. Med. Res. 45:101

Haagsma EB, van den Berg AP, Porte RJ, Benne CA, Vennema H, Reimerink JH et al (2008) Chronic hepatitis E virus infection in liver transplant recipients. Liver Transpl. 14:547–553

Harrison TJ (1999) Hepatitis E virus an update. Liver 19:171–176

Haqshenas G, Shivaprasad HL, Woolcock PR, Read DH, Meng XJ (2001) Genetic identification and characterization of a novel virus related to human hepatitis E virus from chickens with hepatitis-splenomegaly syndrome in the United States. J. Gen. Virol. 82:2449–2462

Hatami H (1991) Epidemic report of hepatitis E in Kermanshah. Nabz J 9:23–31

Hollinger FB, Ticehurst JR (1996) Hepatitis A virus. In: Fields BN, Knipe DM, Howley PM (eds) Fields virology, 3rd edn. Lippincott-Raven, Philadelphia, pp 735–782

Hammitt LL, Bulkow L, Hennessy TW, Zanis C, Snowball M, Williams JL et al (2008) Persistence of antibody to hepatitis A virus 10 years after vaccination among children and adults. J. Infect. Dis. 198:1776–1782

Hussaini SH, Skidmore SJ, Richardson P, Sherratt LM, Cooper BT, O'Grady JG (1997) Severe hepatitis E infection during pregnancy. J. Viral Hepatitis 4:51–54

Ida S, Tachikawa N, Nakajima A, Daikoku M, Yano M, Kikuchi Y et al (2002) Influence of human immunodeficiency virus type I infection on acute hepatitis A infection. Clin. Infect. Dis. 34:379–385

Im SW, Zhang JZ, Zhuang H, Che XY, Zhu WF, Xu GM et al (2001) A bacterially expressed peptide prevents experimental infection of primates by the hepatitis E virus. Vaccine 19:3726–3732

Innis BL, Snitbhan R, Kunasol P, Laorakpongse T, Poopatanakool W, Kozik CA et al (1994) Protection against hepatitis A by an inactivated vaccine. J. Am. Med. Assoc. 271:1328–1334

Jameel S (1999) Molecular biology and pathogenesis of hepatitis E virus. Expert Rev. Mol. Med. 6:1–16

Jansen RW, Siegl G, Lemon SM (1990) Molecular epidemiology of human hepatitis A virus defined by an antigen-capture polymerase chain reaction method. Proc. Natl. Acad. Sci. U.S.A. 87:2867–2871

Jean J, Blais B, Darveau A, Fliss I (2001) Detection of hepatitis A virus by the nucleic acid sequence-based amplification technique and comparison with reverse transcription-PCR. Appl. Environ. Microbiol. 67:5593–5600

Jindal M, Rana SS, Gupta RK, Das K, Kar P (2002) Serological study of hepatitis A virus infection amongst the students of a medical college in Delhi & evaluation of the need of vaccination. Indian J. Med. Res. 115:1–4

Johne R, Plenge-Bonig A, Hess M, Ulrich RG, Reetz J, Schielke A (2009) Detection of a novel hepatitis E-like virus in faeces of wild rats using a nested broad-spectrum RT-PCR. J. Gen. Virol. 91:750–758

Joshi N, Nagarjuna KYR, Kumar A (2000) Age related seroprevalence of antibodies to hepatitis A virus in Hyderabad, India. Trop. Gastroenterol. 21:63–65

Kamar N, Selves J, Mansuy JM, Ouezzani L, Péron JM, Guitard J et al (2008a) Hepatitis E virus and chronic hepatitis in organ-transplant recipients. N. Engl. J. Med. 358:811–817

Kamar N, Mansuy JM, Cointault O, Selves J, Abravanel F, Danjoux M et al (2008b) Hepatitis E virus-related cirrhosis in kidney- and kidney-pancreas-transplant recipients. Am. J. Transplant. 8:1744–1748

Khuroo MS (1980) Study of an epidemic of non-A non-B hepatitis: possibility of another human hepatitis virus distinct from post transfusion non-A non-B type. Am. J. Med. 68:818

Khuroo MS, Kamili S, Yattoo GN (2004) Hepatitis E virus infection may be transmitted through blood transfusions in an endemic area. J. Gastroenterol. Hepatol. 19:778–784

Khuroo MS, Kamili S, Jameel S (1995) Vertical transmission of hepatitis E virus. Lancet 345:1025–1026

Khuroo MS, Rustgi VK, Dawson GJ, Mushahwar IK, Yattoo GN, Kamili S et al (1994) Spectrum of hepatitis E virus infection in India. J. Med. Virol. 43:281–286

Khuroo MS, Teli MR, Skidmore S, Sofi MA, Khuroo MI (1981) Incidence and severity of viral hepatitis in pregnancy. Am. J. Med. 70:252–255

Koff RS (2007) Review article: vaccination and viral hepatitis–current status and future prospects. Aliment. Pharmacol. Ther. 26:1285–1292

Koff RS (1992) Clinical manifestations and diagnosis of hepatitis A virus infection. Vaccine 10:S15–S17

Koonin EV, Gorbalenya AE, Purdy MA, Rozanov MN, Reyes GR, Bradley DW (1992) Computer assisted assignment of functional domains in the nonstructural polyprotein of hepatitis E virus: delineation of an additional group of positive-strand RNA plant and animal viruses. Proc. Natl. Acad. Sci. U.S.A. 89:8259–8263

Krawczynski K, Bradley DW (1989) Enterically-transmitted non-A, non-B hepatitis: identification of virus associated antigen in experimentally infected cynomolgus macaques. J. Infect. Dis. 159:1042–1049

Krugman S, Ward R, Giles JP (1959) Infectious hepatitis: detection of virus during the incubation period and in clinically in apparent infection. N. Engl. J. Med. 261:729

Kuniholm MH, Purcell RH, McQuillan GM, Engle RE, Wasley A, Nelson KE (2009) Epidemiology of hepatitis E in the United States: results from the third national health and nutrition examination survey 1988–1994. J. Infect. Dis. 200:48–56

Kwo PY, Schlauder GG, Carpenter HAl (1997) Acute hepatitis E by a new isolate acquired in the United States. Mayo Clin. Proc. 72:1133–1136

Labrique AB, Thomas DL, Stoszek SK, Nelson KE (1999) Hepatitis E: an emerging infectious disease. Epidemiol. Rev. 21:162–179

Lemon SM, Jansen RW, Brown EA (1992) Genetic, antigenic and biological differences between strains of hepatitis A virus. Vaccine 10:S40–S44

Lin CC, Wu JC, Chang TT, Chang WY, Yu ML, Tam AW et al (2000) Diagnostic value of immunoglobulin G (IgG) and IgM anti-hepatitis E virus (HEV) tests based on HEV RNA in an area where hepatitis E is not endemic. J. Clin. Microbiol. 38:3915–3918

Li F, Torresi J, Locarnini SA, Zhuang H, Zhu W, Guo X et al (1997a) Amino-terminal epitopes are exposed when full-length open reading frame 2 of hepatitis E virus is expressed in Escherichia coli, but carboxy-terminal epitopes are masked. J. Med. Virol. 52:289–300

Li F, Riddell MA, Seow HF, Takeda N, Miyamura T, Anderson DA (2000) Recombinant subunit ORF2.1 antigen and induction of antibody against immunodominant epitopes in the hepatitis E virus capsid protein. J. Med. Virol. 60:379–386

Li TC, Yamakawa Y, Suzuki K, Tatsumi M, Razak MA, Uchida T (1997b) Expression and self assembly of empty virus-like particles of hepatitis E virus. J. Virol. 71:7207–7213

Lorenzo FR, Tsatsralt-Od B, Ganbat S, Takahashi M, Okamoto H (2007) Analysis of the full-length genome of hepatitis E virus isolates obtained from farm pigs in Mongolia. J. Med. Virol. 79:1128–1137

Lu L, Li C, Hagedorn CH (2006) Phylogenetic analysis of global hepatitis E virus sequences: genetic diversity, subtypes and zoonosis. Rev. Med. Virol. 16:5–36

Lu L, ChingKZ Paulade VS, Nakano T, Siegl G, Weitz M et al (2004) Characterization of the complete genomic sequence of genotype II hepatitis A virus (CF53/Berne isolate). J. Gen. Virol. 85:2943–2952

Lutwick L (1999) Clinical interactions between human immunodeficiency virus and the human hepatitis viruses. Infect Dis Clin Pract 8:9–20

Mannucci PM, Gdovin S, Gringeri A (1994) Transmission of hepatitis a to patient with hemophilia by factor VIII concentrates treated with organic solvent and detergent to inactivate viruses. Ann. Intern. Med. 120:1–7

Mast EE, Alter MJ, Holland PV (1998) Evaluation of assays for antibody to hepatitis E virus by a serum panel. Hepatitis E virus antibody serum panel evaluation group. Hepatology 27:857–861

Mathew P, Bobba R, Zacharias P (1998) Hepatitis A seroprevalence in Kerala. Indian J. Gastroenterol. 17:S71–S72

Mathur P, Arora NK, Panda SK, Kapoor SK, Jailkhani BL, Irshad M (2001) Sero-epidemiology of hepatitis E virus (HEV) in urban and rural children of North India. Indian Pediatr. 38:461–475

Matsubayashi K, Nagaoka Y, Sakata H (2004) Transfusion-transmitted hepatitis E caused by apparently indigenous hepatitis E virus strain in Hokkaido. Jpn Transfus 44:934–940

Meng XJ, Purcell RH, Halbur PG, Lehman JR, Webb DM, Tsareva TS (1997) A novel virus in swine is closely related to the human hepatitis E virus. Proc. Natl. Acad. Sci. U.S.A. 94:9860–9865

Meng XJ (2009) Hepatitis E virus: animal reservoirs and zoonotic risk. Vet. Microbiol. 140:256–265

Meng XJ, Halbur PG, Shapiro MS, Govindarajan S, Bruna JD, Mushahwar IK et al (1998) Genetic and experimental evidence for cross-species infection by swine hepatitis E virus. J. Virol. 72:9714–9721

Mittal SK, Rastogi A, Kumar N, Talukdar B, Kar P (1998) Seroprevalence of hepatitis A in children - Implications for hepatitis A vaccine. Trop. Gastroenterol. 19:120–121

Miyamura T (2011) Hepatitis E virus infection in developed countries. Virus Res. 161:40–46

Mohanavalli B, Dhevahi E, Menon T, Malathi S, Thyagarajan SP (2003) Prevalence of antibodies to hepatitis A and hepatitis E virus in urban school children in Chennai. Indian Pediatr. 40:328–331

Naik SR, Aggarwal R, Salunke PN, Mehrotra NN (1992) A large waterborne viral hepatitis E epidemic in Kanpur, India. Bull. World Health Organ. 70:597–604

Nainan OV, Xia G, Vaughan G, Margolis HS (2006) Diagnosis of hepatitis A virus infection: a molecular approach. Clin. Microbiol. Rev. 19:63–79

Nakamura M, Takahashi K, Taira K, Taira M, Ohno A, Sakugawa H et al (2006) Hepatitis E virus infection in wild mongooses of Okinawa, Japan: demonstration of anti-HEV antibodies and a full-genome nucleotide sequence. Hepatol Res 34:137–140

Nanda SK, Ansari IH, Acharya SK (1995) Protracted viremia during acute sporadic hepatitis E virus infection. Gastroenterology 108:225–230

Norder H, Courouce AM, Coursaget P, Echevarria JM, Lee SD, Mushahwar IK et al (2004) Genetic diversity of hepatitis B virus strains derived worldwide: genotypes, subgenotypes and HBsAg subtypes. Intervirology 47:289–309

Okamoto H (2007) Genetic variability and evolution of hepatitis E virus. Virus Res. 127:216–228

Panda SK, Thakral D, Rehman S (2007) Hepatitis E virus. Rev. Med. Virol. 17:151–180

Park GC, Hwang S, Yu YD, Park PJ, Choi YI, Song GW et al (2010) Intractable recurrent hepatitis A virus infection requiring repeated liver transplantation: a case report. Transpl. Proc. 42:4658–4660

Pavio N, Renou C, Di Liberto G, Boutrouille A, Eloit M (2008) Hepatitis E: a curious zoonosis. Front Biosci 13:7172–7183

Perez Gracia MT, Rodriguez-Iglesias M (2003) Hepatitis E virus: current status. Med. Clin. 121:787–792

Pischke Sand, Wedemeyer H (2010) Chronic hepatitis e in liver transplant recipients: a significant clinical problem? Minerva Gastroenterol. Dietol. 56:121–128

Pina S, Buti M, Cotrina M, Piella J, Girones R (2000) HEV identified in serum from humans with acute hepatitis and sewage of animal origin in Spain. J. Hepatol. 33:826–833

Purcell RH (1996) Hepatitis E virus. In: Fields BN, Knipe DM, Howley PM (ed) Fields virology Lippincott-Raven, Philadelphia, pp 2831–2843

Purcell RH, Emerson SU (2001) Hepatitis E virus. In: Knipe DM, Howley PM (eds) Fields virology. Lippincot and Wilkins, Philadelphia, pp 3051–3052

Purcell RH, Ticehurst JR (1988) Enterically transmitted non-A, Non-B hepatitis: epidemiology and clinical characteristics. In: Zuckerman AJ (ed) In viral hepatitis and liver disease Alan R. Liss, Inc., New York. pp 131–137

Purcell RH, Emerson SU (2008) Hepatitis E: an emerging awareness of an old disease. J. Hepatol. 48:494–503

Purdy MA, McCaustland KA, Krawczynski K, Spelbring J, Reyes GR, Bradley DW (1993a) Preliminary evidence that a trpE-HEV fusion protein protects cynomolgus macaques against challenge with wild-type hepatitis E virus (HEV). J. Med. Virol. 41:90–104

Purdy MA, Krawczynski K (1994) Hepatitis E. Gastroenterol. Clin. North Am. 23:537–546

Purdy M, Tam A, Huang C, Yarbough P, Reyes G (1993b) Hepatitis E virus: a non-enveloped member of the 'alpha-like' RNA virus supergroup. Semin Virol 4:319–326

Renou C, Pariente A, Nicand E, Pavio N (2008) Pathogenesis of hepatitis E in pregnancy. Liver Int 28:1465–1466

Reuter G, Fodor D, Forgach P, Katai A, Szucs G (2009) Characterization and zoonotic potential of endemic hepatitis E virus (HEV) strains in humans and animals in Hungary. J. Clin. Virol. 44:277–281

Reyes GR, Purdy MA, Kim JP (1990) Isolation of a cDNA from the virus responsible for enterically transmitted non-A, non-B hepatitis. Science 247:1335–1339

Robertson BH, Jansen RW, Khanna B, Totsuka A, Nainan OV, Siegl G et al (1992) Genetic relatedness of hepatitis A virus strains recovered from different geographical regions. J. Gen. Virol. 73:1365–1377

Roberts JK, Whitlock RT (1992) Hepatitis E in a traveler to Bangladesh. Ann. Intern. Med. 117:93

Ropp SL, Tam AW, Beames B, Purdy M, Frey TK (2000) Expression of the hepatitis E virus ORF1. Arch. Virol. 145:1321–1337

Ruan B, Ma Y, Zhuang H (1997) A study on viremia in patients with hepatitis E (In Chinese). Chung Hua Yu Fang I Hsueh Tsa Chih 31:352

Ruan B, Zumbika E, Wang SY, Chen Y, Ma YL, Chen YG et al (2003) Cloning and expression of cDNAs from hepatitis E virus structural gene. Hepatobiliary Pancreat Dis Int 2:387–390

Sánchez G, Bosch A, Pintó RM (2007) Hepatitis A virus detection in food: current and future prospect. Lett Appl Microbiol 45:1–5

Schlauder GG, Desai SM, Zanetti AR, Tassopoulos NC, Mushahwar IK (1999) Novel hepatitis E virus (HEV) isolates from Europe: evidence for additional genotypes of HEV. J Med Virol 57:243–251

Schlauder GG, Mushahwar IK (2001) Genetic heterogeneity of hepatitis E virus. J Med Virol 65:282–292

Scharschmidt BF (1995) Hepatitis E: a virus in waiting. Lancet 346:519–520

Schwartz E, Jenks NP, Damme PV, Galun E (1999) Hepatitis E virus infection in travelers. Clin Infect Dis 29:1312–1314

Shrestha SM (1991) Enteric non-A, non-B hepatitis in Nepal: clinical and epidemiological observations. In: Shikata T, Purcell RH, Uchida T (eds) Viral hepatitis C, D, and E: proceedings of the international meeting on non-A, non-B hepatitis, Tokyo, 27–30 September, 1989. Elsevier Science Publishers BV, Amsterdam, pp 265–275

Shrestha MP, Scott RM, Joshi DM, Mammen MP Jr, Thapa GB, Thapa N et al (2007) Safety and efficacy of a recombinant hepatitis E vaccine. N Engl J Med 356:895–903

Singh S, Mohanty A, Joshi YK, Dwivedi SN, Deka D (2001) Outcome of hepatitis E virus infection in Indian pregnant women admitted to a tertiary care hospital. Indian J. Med. Res. 113:35–39

Singh S, Mohanty A, Joshi YK, Deka D, Mohanty S, Panda SK (2003) Mother-to-child transmission of hepatitis E virus infection. Indian J. Pediatr. 70:37–39

Stapleton JT (1995) Host immune response to hepatitis A virus. J Infect Dis 171:S9–S14

Tam AW, Smith MM, Guerra ME, Huang CC, Bradley DW, Fry KE et al (1991) Hepatitis E virus (HEV): molecular cloning and sequencing of the full-length viral genome. Virology 185:120–131

Tam AW, White R, Reed E, Short M, Zhang Y, Fuerst TR et al (1996) In vitro propagation and production of hepatitis E virus from in vivo-infected primary macaque hepatocytes. Virology 215:1–9

Takahashi M, Kusakai S, Mizuo H, Suzuki K, Fujimura K, Masuko K et al (2005) Simultaneous detection of immunoglobulin A (IgA) and IgM antibodies against hepatitis E virus (HEV) is highly specific for diagnosis of acute HEV infection. J Clin Microbiol 43:49–56

Tanaka I, Shima M, Kubota Y (1995) Vertical transmission of hepatitis A virus. Lancet 345:397

Tandon BN, Gandhi BM, Joshi YK (1984) Aetiological spectrum of viral hepatitis and prevalence of markers of hepatitis A and B virus infection in North India. Bull. World Health Organ 62:67–73

Tei S, Kitajima N, Takahashi K, Mishiro S (2003) Zoonotic transmission of Hepatitis E virus from deer to human beings. Lancet 362:371–373

Teshale EH, Hu DJ, Holmberg SD (2010) The two faces of hepatitis E virus. Clin Infect Dis 51:328–334

Thakur V, Guptan RC, Bagga S, Sood B, Sarin SK (1998) Prevalence of hepatitis A antibodies in children in India. Indian J. Gastroenterol. 17:S67

Thomas DL, Yarbough PO, Vlahov D, Tsarev SA, Nelson KE, Saah AJ et al (1997) Seroreactivity to hepatitis E virus in areas where the disease is not endemic. J Clin Microbiol 35:1244–1247

Tomiyama D, Inoue E, Osawa Y, Okazaki K (2009) Serological evidence of infection with hepatitis E virus among wild Yezo-deer, Cervus nippon yesoensis, in Hokkaido. Jpn J Hepatol 16:524–528

Tomar BS (1998) Hepatitis E in India (In Chinese). Chung Hua Min Kuo Hsiao Erh Ko I Hsueh Hui Tsa Chih. 39:150–156

Tsega E, Krawczynski K, Hansson BG (1991) Outbreak of acute hepatitis E virus infection among military personnel in northern Ethiopia. J Med Virol 34:232–236

Tsarev SA, Binn LN, Gomatos PJ (1999) Phylogenetic analysis of hepatitis E virus isolates from Egypt. J Med Virol 57:68–74

Tsarev SA, Tsareva TS, Emerson SU, Govindarajan S, Shapiro M, Gerin JL et al (1997) Recombinant vaccine against hepatitis E: dose response and protection against heterologous challenge. Vaccine 15:1834–1838

Uchida T (1992) Hepatitis E: review. Gastroenterologia Japonica 27:687–696

UNICEF (2008) International year of sanitation 2008. http://www.unicef.org/india/wes_3950.htm

Velázquez O, Stetler HC, Avila C, Ornelas G, Alvarez C, Hadler SC et al (1990) Epidemic transmission of enterically transmitted non-A, non-B hepatitis in Mexico, 1986–1987. J Am Med Assoc 263:3281–3285

Viral Hepatitis Prevention Board (1997) Viral hepatitis. News from the VHPB meeting in St. Julians, Malta 6:1

Wang Y, Ling R, Erker JC (1990) A divergent genotype of hepatitis E virus in Chinese patients with acute hepatitis. J Gen Virol 80:169–177

Wang Y, Zhang H, Li Z, Gu W, Lan H, Hao W et al (2001) Detection of sporadic cases of hepatitis E Virus (HEV) infection in china using immunoassays based on recombinant open reading frame 2 and 3 polypeptides from HEV genotype 4. J Clin Microbiol 39:4370–4379

Wasley A, Miller JT, Finelli L (2005) Surveillance for acute viral hepatitis United States. Centers for disease control and prevention. Morbidity and mortality weekly report surveillance summaries 56(3):1–24 16 Mar 2007

Werzberger A, Mensch B, Kuter B, Brown L, Lewis J, Sitrin R et al (1992) A controlled trial of a formalin-inactivated hepatitis A vaccine in healthy children. N Engl J Med 327:453–457

World Health Organization (2000) Global water supply and sanitation assessment: a snapshot of water supply and sanitation worldwide at the turn of the millennium. Retrieved 12 Sept 2003

World Health Organization (1997) Microbial aspects. Drinking water guidelines, vol 3: surveillance and control of community supplies. Geneva p 23, 4

Xing L, Kato K, Li T, Takeda N, Miyamura T, Hammar L et al (1999) Recombinant hepatitis E capsid protein self-assembles into a dual-domain T = 1 particle presenting native virus epitopes. Virology 265:35–45

Yarbough PO (1999) Hepatitis E virus. Advances in HEV biology and HEV vaccine approaches. Intervirology 42:179–184

Zhang M, Emerson SU, Nguyen H, Engle RE, Govindarajan S, Gertin JL et al (2001a) Immunogenicity and protective efficacy of a vaccine prepared from 53 kDa truncated hepatitis E virus capsid protein expressed in insect cells. Vaccine 20:853–857

Zhang JZ, Ng MH, Xia NS, Lau SH, Che XY, Chau TN et al (2001b) Conformational antigenic determinants generated by interactions between a bacterially expressed recombinant peptide of the hepatitis E virus structural protein. J Med Virol 64:125–132

Zhang J, Liu CB, Li RC, Li YM, Zheng YJ, Li YP et al (2009) Randomized-controlled phase II clinical trial of a bacterially expressed recombinant hepatitis E vaccine. Vaccine 27:1869–1874

Zhao C, Ma Z, Harrison TJ, Feng R, Zhang C, Qiao Z et al (2009) A novel genotype of hepatitis E virus prevalent among farmed rabbits in China. J Mol Virol 81:1371–1379

Zhou YJ, Estes MK, Jiang X, Metcalf TG (1991) Concentration and detection of hepatitis A virus and rotavirus from shellfish by hybridization tests. Appl Environ Microbiol 57:2963–2968

Zhuang H (1992) Hepatitis E and strategies for its control. Viral Hepatitis in China: problems and control strategies. Monogr Virol 19:126

Waterborne Viral Gastroenteritis: An Introduction to Common Agents

4

Parul Jain and Amita Jain

Abstract

Acute gastroenteritis is among the most common illnesses of human beings, and its associated morbidity and mortality are greatest among those at the extremes of age; children and elderly. During the 1970s, several viruses were associated with this syndrome, which are now known to be caused mainly by viruses belonging to four distinct families—rotaviruses, caliciviruses, astroviruses, and adenoviruses. Other viruses, such as the toroviruses, picobirnaviruses, coronavirus, and enterovirus 22, may play a role as well. Transmission by food or water has been documented for astroviruses, caliciviruses, rotaviruses, and norovirus. In developing countries, gastroenteritis is a common cause of death in children <5 years, while deaths from diarrhea are less common, much illness leads to hospitalization or doctor visits. Laboratory confirmation of waterborne illness is based on demonstration of virus particles or antigen in stool, detection of viral nucleic acid in stool, or demonstration of a rise in specific antibody to the virus. Newer methods for syndrome surveillance of acute viral gastroenteritis are being developed like multiplex real-time reverse transcriptase PCRs. Application of these more sensitive methods to detect and characterize individual agents is just beginning, but has already opened up new avenues to reassess their disease burden, examine their molecular epidemiology, and consider new directions for their prevention and control through vaccination, improvements in water quality, and sanitary practices.

P. Jain · A. Jain (✉)
Department of Microbiology, Virology Laboratory,
King George's Medical University,
Lucknow 226 001, India
e-mail: amita602002@yahoo.com

P. Jain
e-mail: parulkgmu@yahoo.com

P. P. Singh and V. Sharma (eds.), *Water and Health*,
DOI: 10.1007/978-81-322-1029-0_4, © Springer India 2014

Keywords
Adenoviruses • Astroviruses Caliciviruses • Rotaviruses • Viral gastroenteritis • Waterborne infections

Introduction

Viral gastroenteritis resulting from exposure to contaminated drinking and recreational waters has been reported worldwide. Conservative estimates put diarrhea in the top five causes of deaths worldwide with 3.5 to 5 million deaths every year and most occurring in young children in nonindustrialized countries (Guerrant et al. 1990). According to the World Health Organization (WHO), diarrheal diseases account for 4.1 % of the total DALY global burden of disease of which 88 % is attributable to unsafe water supply, sanitation, and hygiene. Several waterborne outbreaks of viral gastroenteritis have been reported throughout the world, which may be of particular importance since these outbreaks have the potential to involve large numbers of people and wide geographic areas and, perhaps, to introduce new variants to an area. Here, we have briefly described the viral causes of diarrhea.

Waterborne Outbreaks of Viral Gastroenteritis

Waterborne viral gastroenteritis is defined as an illness that occurs after consumption or use of water intended for drinking. Contamination of underground water may occur as a result of leakage from septic tanks, flooding after heavy rainfalls, and back siphoning through a cross-connection between a well and a septic tank. Runoff water after heavy rainfall may contaminate surface water sources (Hedberg and Osterholm 1993). Human fecal wastes contain the viruses, and their numbers and distribution in sewage-polluted waters depend both on the burden of viral disease in the population and on the availability of municipal sewage treatment processing like filtration or chlorination of water supply.

Several outbreaks of viral gastroenteritis have been reported all around the world, associated with consumption of contaminated drinking water (Kukkula et al. 1997; Hafliger 2000) /ice (Cannon et al. 1991) /sea food like molluscan shellfish (Dowell et al. 1995; Rippey 1994).

History of Viral Gastroenteritis

Till 1970s, the etiology of gastroenteritis remained unknown in most of the cases. In 1972, Kapikian et al. (1972) first identified Norwalk virus, by immunoelectron microscopy, in stools of gastroenteritis cases after an outbreak that had occurred at an elementary school in Norwalk, Ohio, in October 1968. In 1973, rotavirus was identified in the duodenal mucosa of children with diarrhea by Bishop et al. using electron microscopy. Later, in 1975, astroviruses were described by Madeley and Cosgrove in the stools of infants hospitalized with diarrhea and in outbreaks of gastroenteritis in newborn nurseries (Madeley and Cosgrove 1975). Adenoviruses were first isolated from civilians and army recruits who had respiratory disease (Jones et al. 2005); the enteric adenoviruses were later implicated as a cause of acute diarrhea in children.

Etiology

Acute gastroenteritis is caused by a number of different agents, including bacteria, viruses, and parasites. Noroviruses (NoVs) and rotaviruses (RVs) of group A are the leading causes of viral gastroenteritis. Sapoviruses (SaVs), astroviruses (AsVs), and enteric adenoviruses (AdVs) are other important causes. Candidate agents for whom etiologic relationship has not yet been confirmed are coronaviruses, picobirnaviruses, pestiviruses, toroviruses, and echovirus type 22.

Table 4.1 Features of causally confirmed human gastroenteritis viruses

Virus	Family	Morphology	Genome	Clinical Severity	Cultivation possible
Rotavirus (Group A)	Reoviridae	100 nm, triple-shelled, wheel-like capsid	Double-stranded segmented RNA	Major cause of severe dehydrating diarrhea in children <5 years of age	Yes
Norovirus	Caliciviridae	27–40 nm, round virus	Positive-sense single-stranded RNA	Epidemics of diarrhea and vomiting in all ages	No
Sapovirus	Caliciviridae	Star of David appearance	Positive-sense single-stranded RNA	Gastroenteritis in children <5 years of age	No
Enteric Adenovirus (Group F)	Adenoviridae	90 nm, icosahedral	Double-stranded DNA	Endemic diarrhea in children <5 years of age	Yes, fastidious
Astrovirus	Astroviridae	28–30 nm with characteristic star on surface	Positive-sense single-stranded RNA	Endemic diarrhea in children, outbreaks possible, role in HIV-related diarrhea	Yes

The morphologic and clinical features of causally confirmed gastroenteritis viruses are highlighted in Table 4.1.

Epidemiology

The incidence of bacterial and parasitic gastroenteritis is more in developing countries as compared to the developed world. This reflects difference in hygiene and sanitation and is therefore subject to control by public health measures. On the other hand, the agents of viral gastroenteritis that have other modes of transmission as well account for a greater proportion of gastroenteritis episodes in the industrialized world.

Endemic childhood gastroenteritis is caused mainly by rotavirus and, to a lesser extent, astrovirus, enteric adenovirus, and sapoviruses. Epidemic gastroenteritis is primarily caused by noroviruses though outbreaks due to rotavirus, astrovirus, and sapovirus have also been reported. The distribution of various viruses among cases of gastroenteritis is given in Table 4.2.

Brief Description of Agents of Viral Gastroenteritis

Rotavirus

Structure: The mature virions are about 100 nm in diameter, are nonenveloped, and possess a triple-layered icosahedral protein capsid composed of an outer layer, an intermediate layer, and an inner core layer. The capsid contains all enzymes for mRNA production. The genome consists of 11 segments of double-stranded RNA, each segment codes for at least one protein. The segmented genome allows genetic re-assortment during dual infections of cells or susceptible hosts, producing reassortants of mixed parentage which is a source of viral diversity (Estates and Kapikian 2007) (Fig. 4.1).

The viral proteins, structural and nonstructural: Rotavirus has six structural proteins (VP1 to VP4, VP6, and VP7) and six nonstructural proteins (NSP1 to NSP6). The innermost protein shell consists of VP2, arranged in an icosahedral lattice. The VP1 polymerase and VP3 capping

Table 4.2 Distribution of common viruses causing human gastroenteritis

	Brandt et al1983	Nguyen 2007	Li 2009
Place of study	Washington	Vietnam	Hong Kong
Time period of study	1974–1982	2002–2003	January–August 2008
Population	Pediatric	Pediatric	Pediatric
Norovirus	_[a]	5.5 %	23 %
Rotavirus	34.5%	67.4 %	28 %
Enteric Adenovirus	4.7%	3.2 %	3 %
Astrovirus	_[a]	0.6 %	0.6 %
Sapovirus	_[a]	0.8 %	0
Mixed viruses	40.1%	5.5 %	0.4 %
Unexplained viruses	19.1%		

[a] These viruses were not known till 1982

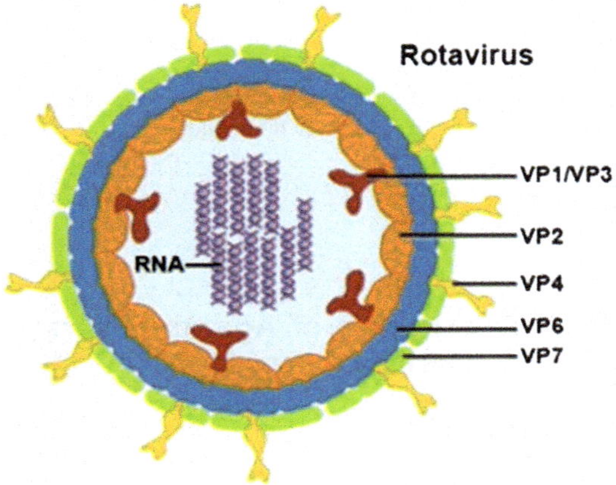

Fig. 4.1 Morphological structure of rotavirus showing location of various proteins

enzyme are anchored inside the VP2 shell. The middle layer consists of VP6, which forms thick trimeric pillars in an icosahedral lattice. VP6 is the target of the most abundant antibodies elicited by rotavirus infection, though this antigen is not exposed on the viral surface. The outermost layer consists of a coat glycoprotein VP7 and VP4 spikes, which protrude from the virion. (Estates and Kapikian 2007). The details of the proteins are shown in Table 4.3.

Classification: There are seven major groups of rotavirus (A through G); group A and, to a lesser extent, group B and group C are primarily responsible for human gastroenteritis. Group specificity of rotaviruses is determined by the internal structural protein, viral protein 6 (VP6) which is the target of commercial immunoassays (Tang et al. 1997). Within each group, rotaviruses are classified into serotypes on the basis of two outer capsid proteins, VP7 (G-protein/glycoprotein) and VP4 (P-protein/protease-sensitive protein) which induce neutralizing antibodies, on the basis of which a binary serotyping system has been developed. Till date, 15 VP7 or G serotypes have been identified. VP4 is a protease-cleaved protein, and serotypes determined by this protein are termed P serotypes. Due to a lack of readily available typing serum or monoclonal antibodies to different VP4 types, 26 different genotypes of VP4 have been identified based on sequence analysis of VP4. These genotypes correlate well with known serotypes, so the genotypes are mentioned in brackets (e.g., P1A[8]) (Estates and Kapikian 2007).

Replication cycle: Many aspects of replication cycle of rotavirus remain uncertain. The entire cycle occurs within the cytoplasm. The intestinal trypsin enzyme cleaves the rotavirus spike protein VP4 into an amino-terminal fragment, VP8, and a carboxy-terminal fragment, VP5. The VP8 fragment contains a hemagglutination domain, and the VP5 fragment contains a membrane interaction domain. Rotavirus enters the cells using several cell surface receptors. When intracellular calcium concentrations fall, rotavirus virions uncoat, forming double-layered subviral particle (DLP). DLPs are transcriptionally active particles composed of genome, VP1, VP3, and the inner two protein layers, which are assembled in viral factories

Table 4.3 Characteristics of rotavirus structural and nonstructural proteins

Genome segment	Protein product	Mol. Wt. of protein (A^0)	No. of molecules per virion	Location in virus particles	Function
1	VP1	125	<25	Core	RNA-dependent RNA polymerase; part of minimal replication complex
2	VP2	94	120	Core	Inner capsid structural protein; sequence nonspecific RNA binding activity; part of minimal replication complex
3	VP3	88	<25	Core	Guanylyltransferase; methyltransferase; part of virion transcription complex
4	VP4	86.7	120	Outer capsid spike	Dimers form outer capsid spike; P-type neutralization antigen, hemagglutination, cell attachment protein, involved in virulence; cleavage by trypsin into VP5 and VP8 enhances infectivity
5	NSP1	58.6	0	Nonstructural	Associates with cytoskeleton; extensive sequence diversity between strains; two conserved cystein-rich zinc finger motifs; RNA binding
6	VP6	44.8	780	Middle capsid	Major virion protein; middle capsid structural protein; homotrimeric structure; subgroup antigen; hydrophobic
7	NSP3	34.6	0	Nonstructural	Homodimer, specifically binds 3′ end of rotavirus mRNA; binds eIF4G1; involved in translational regulation
8	NSP2	36.7	0	Nonstructural	Nonspecific ssRNA binding; accumulates in viroplasm; involved in viroplasm formation; NT-Pase activity, homomultimer; binds NSP5 and VP1
9	VP7	37.4	780	Outer capsid	G-type neutralization antigen, N-linked high-mannose glycosylation and trimming
10	NSP4	20.2	0	Nonstructural	Enterotoxin, receptor for budding of DLP through ER membrane; N-linked high-mannose glycosylation; uncleaved signal sequence, rough ER transmembrane glycoprotein; modulates intracellular Ca^{2+}
11	NSP5	21.7	0	Nonstructural	Interacts with NSP2 and NSP6; homomultimerizes; autocatalytic kinase activity; binds single-stranded RNA
	NSP6	12	0	Nonstructural	Product of second out of frame ORF; interacts with NSP5; localizes to viroplasm

Table adapted from Estes and Kapikian 2007, Feild's virology, 5th ed.

termed *viroplasms*. The dsRNA transcribes mRNA, the newly transcribed rotavirus mRNA enters two pools, mRNA that remains sequestered within viroplasms is replicated to produce negative-strand RNA which remains associated with the positive strand, while mRNA that escapes viroplasms is translated. The dsRNA segments formed within nascent subviral particles mature by budding through the membrane of the endoplasmic reticulum, when particles acquire their outer capsid proteins (Mendez et al. 1999) (Fig. 4.2).

Epidemiology: Worldwide, nearly all children are infected with rotavirus by 3–5 years of age. In developing countries, the incidence of rotavirus diarrhea has varied between 0.07 and

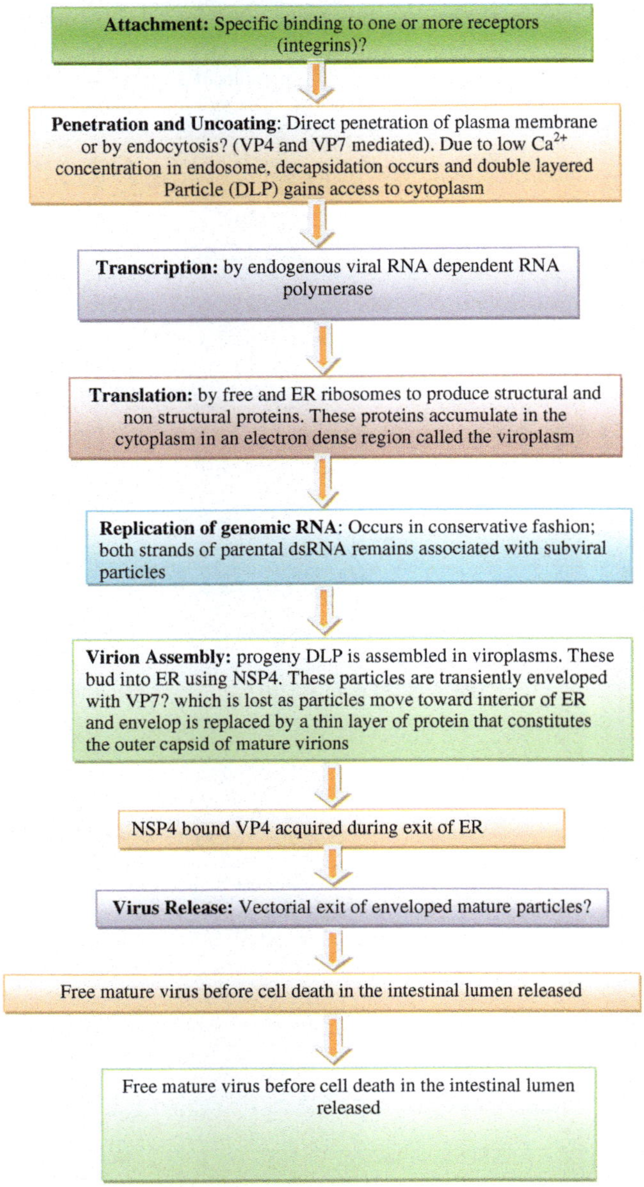

Fig. 4.2 Representation of rotavirus replication cycle in intestinal polarized cells

0.8 episodes per child per year (Bresee et al. 1999). The incidence is similar in developed countries, implying that this infection cannot be effectively controlled by improvement in sanitation and hygiene. Incidence of disease peaks among children 4 to 23 months of age, but illness in adults is also known. Neonatal infections are often asymptomatic because of protection from maternal immunity, while infections after 3 months of age are more likely to be symptomatic. Reinfections of rotavirus are common, but disease severity decreases with each repeat infection. Therefore, depending on the epidemiologic setting, the proportion of children with rotavirus diarrhea varies, for example, a median of 8 % of all cases of diarrhea in the community, 28 % of outpatient visits, and 34 % of hospitalizations are accounted for by rotavirus infections (Bresee et al. 1999).

The incidence of rotavirus serotypes has varied by geographic location and time. Gene segments that encode the G- and P-proteins can segregate independently. This gives rise to strains with at least 42 different P–G serotype combinations. Of the different serotypes of group A rotavirus identified, only 5 types (G1 through G4 and G9) are common; recently, G8 strains have emerged as important strains in parts of Africa. The four predominant P–G combinations, P[8]G1, P[4]G2, P[8]G3, and P[8]G4, comprise 88.5 % of the strains worldwide (Santos and Hoshino 2005). Group B rotaviruses have been implicated in several large outbreaks of severe gastroenteritis among adults in China since 1982 (Hung 1988). It has also been found in adults and few children in Bangladesh (Sanekata 2003) and recently in western India (Kelkar and Zade 2004; Chitambar et al. 2011) but not in other parts of the world. Group C rotaviruses have been associated with a small proportion of pediatric cases in several countries worldwide (Oseto et al. 1993). Asian Rotavirus Surveillance Network (ARSN) performed the first phase of data collection in nine countries and regions over a 2-year period during 2001–2003, when it was known that rotavirus was responsible for overall 45 % diarrhea admissions in the region. G1 was the commonest strain in Hong Kong (49 %) and Vietnam (47 %), G9 in Korea (39 %), Taiwan (37 %), and Thailand (55 %), while G3 was the commonest in China (67 %) (Nelson et al. 2008).

In temperate climates, rotavirus disease occurs predominantly during the cooler fall and winter months, while in tropical areas, the infection occurs in an endemic fashion throughout the year.

Transmission: Rotaviruses are usually transmitted by the faeco-oral route, but speculation continues whether these are also transmitted by the respiratory route. The rapid spread of rotaviruses can be attributed to its ability to survive on various surfaces under different conditions (Wilde et al. 1992). Rotaviruses have been detected in both treated and untreated sewage water. The contaminated water may be an important source for group B rotavirus outbreaks though its role in transmission of group A viruses is unknown (Estates and Kapikian 2007).

Incubation period of rotavirus diarrhea has been estimated to be <48 h though it varies from 1 to 4 days. Duration of virus excretion ranges from 4 to 57 days. In 43 % of children, viral excretion stops in 10 days of onset of diarrhea, and in 70 % of children, it stops within 20 days, while in 30 % of children, virus may be detected in stool for 25–57 days.

Pathogenesis: In immunocompetent children, rotavirus infection is restricted to mature enterocytes in the villous epithelium of the proximal small intestine, which are ultimately destroyed.

Osmotic diarrhea results from malabsorption of carbohydrates due to reduction in absorptive villous epithelium as well as due to decreased turnover of microvillar membrane disaccharidases. Also, a role of NSP4 protein that functions as an enterotoxin and alters epithelial cell function and permeability has been described. Additionally, it is suggested that rotavirus induces fluid and electrolyte secretion by activation of enteric nervous system in the intestinal wall (Gilger et al. 1992) Fig. 4.3.

Clinical manifestations: Rotavirus gastroenteritis is more frequently associated with dehydration than is gastroenteritis by other pathogens. The clinical spectrum of rotavirus infection varies from subclinical illness to severe gastroenteritis and potentially fatal dehydration. The illness has an abrupt onset, with vomiting lasting 1–2 days, followed by loose, watery diarrhea that lasts for approximately 5 days, and the stools rarely contain red or white cells. Up to one-third of patients may have a temperature of >39 °C.

Respiratory and neurologic (Carlson et al. 1978) symptoms have been reported in children with rotavirus infection, but causal associations have not been proven. Diabetes mellitus, intussusceptions, sudden infant death syndrome, and necrotizing enterocolitis have also been reported with rotavirus infection.

In children with primary or acquired immunodeficiency (particularly bone marrow and liver transplant recipients), rotavirus can cause severe and prolonged disease with prolonged viral excretion and rarely can replicate extraintestinally.

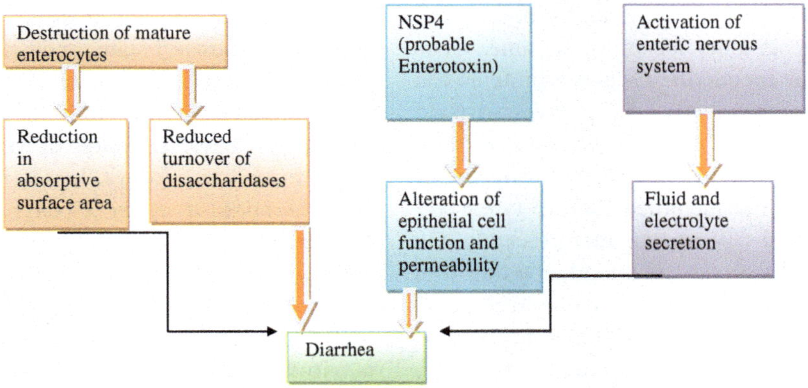

Fig. 4.3 Pathogenesis of rotavirus gastroenteritis

Immunity: Rotavirus-specific secretory IgA antibodies in the intestine appear only for a short duration during acute infection. These are correlated with a temporary protection against rotavirus disease, and therefore, reinfections with rotaviruses are common. During the acute phase, these antibodies are directed principally against VP2 and VP6 which broaden to include other proteins during the convalescent phase. Serum IgM antibodies also appear during the acute phase which is later replaced by IgG and, to a lesser extent, by IgA antibodies.

Laboratory diagnosis:

1) Virus Isolation: Human rotaviruses can be isolated in cell culture, although this is a slow and laborious process. The cell culture medium must be supplemented with proteases like trypsin or pancreatin. Cell lines in which rotaviruses can be isolated include primary simian kidney cells, intestinally derived cell lines like human colon adenocarcinoma cell line $CaCo_2$, and MA104.
2) Direct examination by electron microscope: It is now not used routinely but can be of use for detecting nongroup A rotaviruses and mixed infections with other enteric viruses.
3) Antigen detection: Rotavirus immunoassays are predominantly based on ELISA or latex agglutination principles and use either monoclonal or polyclonal antibodies. The commercial ELISA kits can have a sensitivity and specificity of up to 98 and 100 %, respectively. Monoclonal G- and P-serotype-specific antibodies to rotaviruses are also commercially available in ELISA format to type the most frequent human rotavirus strains.
4) Nucleic Acid Detection:

a. *Identification of dsRNA by RNA-PAGE*: Viral RNA from a specimen can be directly subjected to polyacrylamide gel electrophoresis and then differentiated on the basis of characteristic patterns of bands by silver staining, referred to as electropherotypes. This is of limited potential today, but appearance of an unusual electropherotype could denote a novel strain or a nongroup A virus.

b. *Dot Blot Assays*: Viral RNA is extracted from patient specimen, spotted onto nitrocellulose membrane, and hybridized with radiolabeled cDNA probes.

c. *PCR*: It is the most sensitive method for detecting group A, B, and C rotaviruses. RT-PCR can be done directly from stool samples and can also be used for P genotyping.

5) Serologic Assays: ELISA can be used to detect rotavirus-specific antibodies in stool and serum. Serum IgA detection is the most sensitive and rapid test for diagnosis of primary infection. A fourfold rise in convalescent-phase serum and stool samples IgA and IgG titers over acute-phase samples is also a reliable indicator of acute infection.

Treatment: Rotavirus disease spontaneously resolves within 1–2 weeks, but to prevent dehydration, appropriate therapy should be instituted early. In children who can take oral fluids, standard hydration solution recommended by the World Health Organization is the therapy of choice. If oral rehydration fails, if the patient is severely dehydrated or patient cannot tolerate oral fluids because of vomiting, intravenous fluids must be administered. Therapeutic role of probiotics, bismuth salicylate, and enkephalinase inhibitors like racecadotril is at present considered to be at an experimental stage. Antibiotics and antimotility agents should be avoided. Early refeeding after rehydration of children is recommended. In chronic symptomatic rotavirus infection in immunocompromised children, orally administered immunoglobulins or colostrums might be considered.

Prevention:

- Effective disinfection of contaminated material. A spray composed of 0.1 % o-phenylphenol and 79 % ethanol is an effective surface disinfectant for rotaviruses.
- Careful hand washing.
- Vaccine: The objective of rotavirus vaccine is to prevent moderate-to-severe disease but not mild disease associated with rotavirus, to duplicate the degree of protection that follows natural infection and to decrease the number of hospitalizations with dehydration, especially in resource-poor countries, where mortality associated with rotavirus is high. The different vaccines available/under development for rotaviruses are as under:

1. Vaccines based on animal rotaviruses: Considered to be naturally attenuated for humans, the Jennerian vaccines. Three monovalent animal rotavirus vaccines, two bovine rotavirus strains, RIT 4237 (P6[1]G6) and WC3 (P7[5]G6), and a simian (rhesus) rotavirus reassortant vaccine (RRV) strain (P[3]G3) have been studied. Variable efficacy was seen in field trials with these vaccines. China introduced live attenuated oral LLR vaccine that was derived from a lamb strain of rotavirus, the efficacy of which is not known (Dennehy 2008).
2. Human–rhesus RRV (RotaShield): This was the first multivalent live oral reassortant vaccine (a rhesus rotavirus tetravalent [RRV-TV] vaccine), which contained a mixture of four most common G serotypes, G1 to G4: three rhesus–human monoreassortant strains containing the VP7 genes of human serotypes G1, G2, and G4 strains were substituted for the VP7 gene of the parent RRV, and the fourth strain comprised serotype G3 of rhesus RRV (Kapikian 1996). This vaccine was licensed for use in the United States but was then withdrawn from the market because it was causally linked with intussusceptions with an estimated frequency of 1 in 10,000 vaccinated infants.
3. Currently licensed vaccines: Currently, two vaccines have been licensed for human use. First is a pentavalent human-bovine (WC3) reassortant live attenuated oral vaccine (RotaTeq) developed by Merck Research Co. Another is a live attenuated human rotavirus vaccine, strain 89-12, that was originally developed from P1A[8]G1 by tissue culture passage of a wild-type human rotavirus isolate. GlaxoSmithKline Biologicals further modified the vaccine by cloning and tissue culture passaging of the parent 89-12 vaccine strain to develop the resulting vaccine, RIX4414 (Rotarix). Salient features of these two vaccines are shown in Table 4.4.
4. Candidate Vaccines: The National Institute of Allergy and Infectious Diseases (NIAID) has developed a tetravalent human-bovine rotavirus reassortant vaccine candidate, which incorporates four reassortants with VP7 specificity for G1, G2, G3, and G4 human serotype and 10 genes from the bovine rotavirus UK strain (P[7]G6). This vaccine showed safety and efficacy in a clinical trial at Finland (Midthun et al. 1985).

Live, orally administered neonatal human rotavirus strains have been explored as vaccine candidates because they appear to be naturally attenuated. Two naturally occurring human-bovine rotavirus reassortants obtained from asymptomatically infected newborns in Delhi (116E) and Bangalore (I321) have undergone clinical trials. The former has 10 human rotavirus genes and a single gene segment encoding VP4 derived from a bovine rotavirus, whereas the latter has two human rotavirus nonstructural genes and nine bovine rotavirus genes (Glass

Table 4.4 Features of currently licensed vaccines

Vaccine	Parent strain and genotype	Formulation	Dose	Side effects	Protective efficacy
Rota Teq	Bovine rotavirus strain WC3, P7[5]G6	5 reassortants; 4 reassortants with the VP7 specificity for serotype 1, 2, 3, or 4, and 1 reassortant with the VP4 of P1A[8] specificity from the human rotavirus parent strain with the remainder of the genes from the WC3 bovine rotavirus parent	3 oral doses at 2, 4, and 6 months of age	Intussusceptions not seen. Diarrhea and vomiting may occur	84.7 % against severe gastroenteritis and hospitalization
Rotarix	Human rotavirus strain 89-12, G1 P1A[8]	No reassortants; attenuation by passage of human rotavirus 89-12 strain in tissue culture	2 oral doses at 2 and 4 months of age	Intussusceptions not seen. Diarrhea and vomiting may occur	89 % against all rotavirus disease

2005). A human neonatal P[6]G3 strain, RV3, was evaluated as an oral vaccine in 3-month-old infants and was found to be safe and well tolerated (Barnes et al. 2002).

Norwalk and Related Human Caliciviruses

History: An outbreak of acute gastroenteritis occurred among students and teachers in a school in Norwalk, Ohio, in 1968, with a primary attack rate of 50 % and secondary attack rate of 32 % (Adler and Zickl 1969). A similar illness occurred when organism-free filtrates of stools collected from affected individuals were given to human volunteers. But the agent could not be grown in cell culture and organ culture (Blacklow et al. 1972). In 1972, Kapikian et al. (1972) by using immunoelectron microscopy (IEM) identified the Norwalk virus in a fecal filtrate used to induce illness in human volunteers which were seen as 27 nm viral particles. Also, convalescent-phase serum was collected from a volunteer who became ill following inoculation with the fecal filtrate. Using an antigen–antibody reaction, the virus particles were precipitated and the complexes were visualized with an electron microscope.

All these data were compiled, and Norwalk virus (NV) was proposed to be the etiologic agent of the Norwalk, Ohio, outbreak of gastroenteritis. Other small, round-structured viruses (SRSVs) that were morphologically similar to NV were later associated with outbreaks of gastroenteritis (Dolin 1982; Thornhill 1977).

Classification: Initially, the acute gastroenteritis viruses were classified on the basis of morphology. For example, NV was the prototype of a group of agents initially called small round-structured viruses (SRSVs). Later, based on molecular biology, most SRSVs were placed in the family Caliciviridae. Family Caliciviridae is now divided into four genera—Norovirus, Sapovirus, Lagovirus, and Vesivirus. Within each genus, species have been defined based on genetic relatedness. For species Norwalk virus (genus Norovirus) and species Sapporo virus (genus Sapovirus), based on relatedness of VP1, capsid protein genogroups have been proposed which are further classified into genotypes (Atmar and Estes 2001) (Table 4.5). The noroviruses have been subdivided into at least five genogroups designated GI–GV (Lew et al. 1994), most common strains implicated in human disease fall into genogroups GI and GII (Patel et al. 2009). The GI genogroup includes the Norwalk virus, whereas the Snow Mountain and Hawaii viruses belong to genogroup GII. Individual strains are designated by Arabic numeral after the genogroup designation: for example, GI.1 (Norwalk) or GII.1 (Hawaii).

Table 4.5 Genogroups and genotypes of Caliciviridae

Genus	Species	Genogroups	Genotypes
Norovirus (NoV)	Norwalk virus (NV)	I	8
		II	19
		III	3
		IV	1
		V	1
Sapovirus (SaV)	Sapporo virus (SV)	I	3
		II	3
		III	1
		IV	1
		V	1

Structure: The Calicivirus virions exhibit icosahedral symmetry. The virions have a single major capsid protein folding into 90 dimers to form a shell domain from which capsomeres protrude that are arch like. These arches are arranged in a way that 32 cup-shaped depressions are visible by electron microscopy, from where the caliciviruses derive their name (Latin: calyx means cup). The Norwalk virus, prototype of calicivirus is nonenveloped, small (27–40 nm), and round virus (Prasad and Rothnagel 1994).

Caliciviruses have a linear, single-stranded, positive-sense RNA genome (ranging from 7.3 to 8.5 kb in length) that is organized into two or more ORFs, depending on the genus. All caliciviruses encode a relatively small ORF near the 3′ end that encodes the minor structural protein, VP2. The nonstructural proteins are encoded toward the 5′ region in ORF one and the structural proteins toward the 3′ end of genome. The norovirus and vesivirus have separate ORFs (ORF2) for VP, whereas in sapovirus and lagovirus, the first ORF codes for the nonstructural proteins as well as the capsid protein, which is found in-frame at the end of the nonstructural proteins (Fig. 4.4).

Viral proteins: *Structural*: Caliciviruses have three structural proteins, the features of which are listed in Table 4.6. *Nonstructural*: Caliciviruses derive their mature nonstructural proteins by proteolytic cleavage of a large polyprotein encoded in ORF1 that has the viral RNA polymerase (NS7), helicase (NS3), and protease functions (NS6) (Jiang et al. 1993). The norovirus and vesivirus have five cleavage sites to produce six mature proteins (NS1 to NS6), while lagovirus and sapovirus have six cleavage sites to produce seven mature proteins (NS1 to NS7). The function of rest of the nonstructural proteins is unknown. Replication cycle is not fully known. Proposed replication cycle is shown in Fig 4.5.

Epidemiology: Noroviruses and sapoviruses have a worldwide distribution. Infections occur round the year, although the incidence tends to

Fig. 4.4 Genome of Norwalk virus

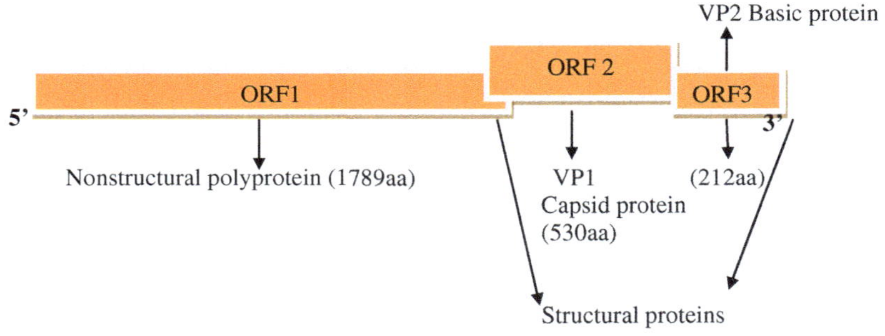

Table 4.6 Structural proteins of Norwalk virus

Protein Product	Mol. Wt. of protein (d)	No. of molecules per virion	Location in Virus particles	Function
VP1	60,000	180	Capsid	Major structural protein, determines antigenic phenotype and interactions of virus with host cells
VP2	12,000–29,000	1–2	Capsid	Minor structural protein, function unknown, may play a role in particle maturation and stability
VPg	–	1–2	Capsid	Minor component, likely functions as a nonstructural protein during replication

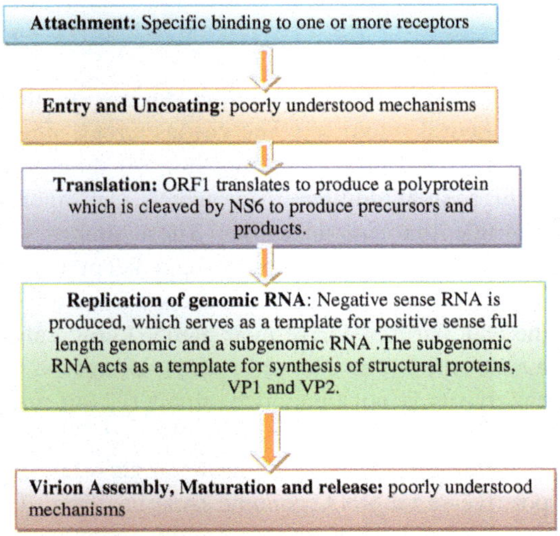

Fig. 4.5 Representation of proposed replication cycle of caliciviruses

peak during cold weather months in temperate climates. Noroviruses have been associated with infection in all age groups, although most gastroenteritis outbreaks involve mainly older school-aged children and adults. In pediatric populations, importance of noroviruses may be second to that of rotaviruses. Sapoviruses on the other hand are primarily associated with pediatric gastroenteritis, which rarely needs hospitalization. A few outbreaks have been documented due to sapoviruses. Noroviruses are the single most important cause of gastroenteritis outbreaks and account for more than 90 % of outbreaks of viral gastroenteritis and for about 50 % of all-cause outbreaks worldwide (Patel et al. 2009). Epidemics occur round the year, in all age groups and in a variety of settings like schools, camps, nursing homes, swimming facilities, and restaurants (Table 4.7).

Transmission of calicivirus occurs predominantly by the faeco-oral route, though vomitus (Chadwick and McCann 1999) has also been implicated as a vehicle of transmission. The predominant modes of spread are person-to-person contact and food- or waterborne spread. Aerosolization and contamination of surfaces by infected individuals can also lead to exposure.

Incubation periods are generally 24–48 h, though it ranges from 18 to 72 h. Virus shedding in stools is maximal over the first 24–48 h after illness. Virus can be rarely detected beyond 72 h after the onset of vomiting or diarrhea by immunoelectron microscopy. However, virus can be detected for up to 3 weeks after resolution of illness using sensitive enzyme-linked immunosorbent assay techniques (Graham et al. 1994) or PCR (Okhuysen 1995). Virus may also be shed by asymptomatically infected individuals.

Table 4.7 Epidemiology of Norwalk virus outbreaks

	Vega et al. (2011)	Lopman et al. (2003)	Werber et al. (2009)
Study period	March 2009–January 2011	1995–2000	24 August–7 September 2008
Place of study	20 state and local health laboratories, United States	10 surveillance systems in Europe network	Podgorica, Europe
Incidence of Norwalk gastroenteritis (%)	72	>85	60
Major serotype isolated	G2	Not Known	G2 & G1
New variants isolated	G2.4 New Orleans	Not Known	None isolated

GII.4 cluster is the predominant virus detected throughout the world. But major shifts in the predominant circulating strain can occur. Mixed infections of noroviruses within the same individual or within the same outbreak have been detected, which may allow recombination between RNA genomes and may lead to genetic diversity.

Pathogenesis: Reversible histopathologic lesions appear in the small intestine within 24 h after virus challenge like broadening and blunting of villi, shortening of microvilli, vacuolization of the lining epithelium, crypt hyperplasia, and infiltration of lamina propria by lymphocytes and polymorphs. These changes are present during illness and persist for a variable period of time after the illness. The histopathologic changes generally clear within 2 weeks after the onset of illness, although some jejunal changes can be noticed as late as 6 weeks after challenge. Diarrhea induced by the Norwalk virus is associated with a transient malabsorption of D-xylose and fat and with decreased levels of brush border enzymes, including alkaline phosphatase and trehalase. Susceptibility to infection with certain norovirus strains has also been associated with the presence of H blood group carbohydrate antigens, which are also expressed in gastrointestinal epithelial cells (Hutson et al. 2002). Nausea and vomiting may be contributed to delayed gastric motor function.

Clinical features: The calicivirus gastroenteritis is generally mild and self-limiting that usually lasts 24–48 h, though prolonged infection may be seen in immune-compromised individuals. After an incubation period of around 24 h, the disease has a sudden onset with nausea, vomiting, diarrhea, abdominal cramps, fever, headache, chills, myalgia, or sore throat. Vomiting is more prevalent than diarrhea in children, whereas reverse is true in adults. The stools are typically loose and watery and do not contain mucus and leukocytes or red blood cells. Severe dehydration may occur in predisposed individuals like elderly patients with debilitating health conditions.

Immunity: Studies in human volunteers have shown that there are two forms of immunity to noroviruses: one is short term and the other is long term. Short-term immunity is protective, that is, serotype specific and protects against the infecting strain. While long-term immunity deviates from the traditional pattern; persons with higher levels of pre-existing antibody to NV are more susceptible to illness. Specific ABO, Lewis, and secretor blood group phenotypes may also influence susceptibility to NV.

Laboratory diagnosis: Currently available diagnostic tools cannot detect all the caliciviruses in single, because of genetic and antigenic diversity. The various tools available are as follows:

1. *Electron Microscopy*: It is of limited value in routine diagnosis of NV because particles are usually present in low concentration and may be difficult to distinguish from other round, small objects present in stool. Also, the timings of stool collection are important; samples that are collected within 72 h of onset of diarrhea have the greatest yield. The sapoviruses are easier to detect; these have a Star of David appearance, that is, a six-pointed star with a dark hollow in the center.

2. *Immunoassays*: Human calicivirus recombinant virus like particles rVLPs has been produced by expression of capsid proteins in baculovirus vectors. Hyperimmune antisera and now monoclonal antibodies are prepared against these VLPs which are employed in the development of ELISAs that can detect viral antigens in clinical specimens. Few diagnostic kits are commercially available.
3. *RT*-PCR: This is the most widely used technique for the detection of noroviruses in clinical samples like feces or vomitus, and contaminated food, water, or fomites. Quantitative real-time PCR has gained importance because it allows rapid detection as well as comparison of viral RNA levels.

Treatment: Not generally required because the disease is self-limited. Symptomatic treatment of headache, myalgias, and nausea with analgesics and antiemetics may provide relief. If severe dehydration develops, oral or intravenous therapy is indicated. Bismuth subsalicylate, loperamide, anticholinergic agents, adsorbents, or lactobacillus containing compounds are not recommended.

Prevention:

1. *Outbreak prevention*: Outbreak prevention should be situation specific, such as control of contamination of food and water, reduction in person–person spread by frequent hand washing, and effective environmental decontamination with disinfectants like hypochlorite at 5,000 ppm, hydrogen peroxide–based cleaners, and phenolic-based cleaners.
2. *Vaccines*: A role of rVLPs has been suggested as a subunit vaccine. NV VLPs produced in plants have undergone a phase I clinical trial in adult volunteers by the oral route, when they were found to be safe and immunogenic (Tacket et al. 2003). Alphavirus vectors, such as Venezuelan equine encephalitis (VEE) (Ball et al. 1999) virus, are also being tried for norovirus vaccine development. If a vaccine is feasible, it could reduce the incidence of epidemic viral gastroenteritis and would be of special importance to military personnel, nursing home residents, and malnourished infants.

Adenovirus

Adenoviruses were first cultured and reported as distinct viral agents in 1953 when Rowe et al. attempted to establish tissue culture lines from tonsils and adenoidal tissue, surgically removed from children. Similar viral agents were isolated from febrile military personnel with a variety of respiratory illnesses when these were called Acute Respiratory Disease (ARD) agents (Wold and Horwitz 2007). Till date, more than 30 different serotypes have been identified and have shown to cause several clinical syndromes, including upper and lower respiratory tract infections, keratoconjunctivitis, and infantile gastroenteritis.

Structure: Virions are nonenveloped, icosahedral particles, about 70–90 nm in diameter, and contain a linear 36-kb double-stranded DNA core complex. The outer capsid is composed of 252 capsomeres which are of two main types: hexon capsomeres, 240 in number and form most of the capsid surface, and penton capsomeres, 12 in number, one at each vertex of the icosahedron. Pentons contain a penton base, and a projecting fiber, composed of three domains, an N-terminal domain that binds to the penton base, a central shaft important for infection, and a globular C-terminal knob that binds the primary receptor on host cells. Most of the epitopes recognized by group- and serotype-specific antibodies are present on the hexon and fiber proteins.

Classification: Adenoviruses were first isolated from febrile military personnel with a variety of respiratory illnesses, when they were called ARD agents. Since then, four genera have been identified in adenoviruses that cause infection in a variety of hosts. Human adenoviruses (HAdV) (genus Mastadenovirus) have been classified into six species: A to F. Species B can be further divided into species B1 and B2. These six species are further divided into 51 serotypes based on their nucleic acid characteristics and homologies, hexon and fiber protein characteristics, and biologic properties (Jong et al. 1999). Recently, a novel human adenovirus (HAdV) species, the 52nd type, has been characterized which is proposed (Jones 2007).

Different adenoviruses are known to cause a variety of clinical manifestations. Respiratory tract infections are caused by adenoviruses of species C, B1, and E; conjunctivitis and pharyngoconjunctival fever by species C and Ad3; gastroenteritis by species F (Ad40 and Ad41) and the recently identified Ad52; acute hemorrhagic cystitis by species B2; epidemic keratoconjunctivitis by species D, especially Ad8, Ad19, and Ad37.

Epidemiology: Adenovirus infections occur worldwide and in humans and in a variety of animals. Adenovirus causes 5 % of all infectious diseases in infants and 3 % in the age of 2–4 years (Pacini et al. 1987). It is known that adenovirus is the second common agent in infantile gastroenteritis, after rotavirus (Blacklow and Greenherg 1991). Faeco-oral transmission is the most important route of transmission in young children. Initial spread can occur by the respiratory route, but feces are a more common source of transmission because prolonged secretion after acute infection occurs through the gastrointestinal tract and the virus is shed both during the acute illness and then intermittently thereafter (Fox et al. 1969). Also, these viruses have been reported to be nosocomially transmitted.

The incidence of adenovirus gastroenteritis varies worldwide, an Asian multicentric study has shown that PCR in stool for adenovirus was positive in 4.4 % of pediatric patients with diarrhea (Li et al. 2005), and adenovirus was responsible for 12.3 % of cases of diarrhea in Bangladesh outbreak (Jarecki-Khan et al. 1993); for 11.6 % pediatric gastroenteritis cases in Italy (Cevenini et al. 1987) and 2 % cases in Brazil (Filho et al. 2007).

About 80 % of fecal adenoviruses present in faeces of patients with acute gastroenteritis are caused by enteric adenovirus AD 40 and 41, but other serotypes have also been reported to cause gastroenteritis, like Ad9 and Ad10, and were found in Dhaka city, Bangladesh Dey et al. 2009; Ad2, Ad3, Ad8, and Ad31 were found in Japan, Korea, and Vietnam (Li et al. 2005).

HAdV is present with low prevalence throughout the year, with a slight but not significant increase in incidence in late summer and early autumn.

Clinical features: The predominant symptom of infections with enteric adenoviruses is watery diarrhea, with a mean duration of 8.6 days (Ad40) and 12.2 days (Ad41) accompanied by fever and vomiting (Van et al. 1992). Few children with Ad41 infections may experience prolonged symptoms (greater than or equal to 14 days). A low frequency of respiratory symptoms has also been reported (21 %) (Uhnoo 1984). The established adenoviruses may present a different clinical picture, characterized by diarrhea of shorter duration, higher fever, and significantly increased occurrence of respiratory symptoms.

Laboratory Diagnosis

1. *Agglutination Test:* For detection of adenovirus proteins in stool, latex agglutination tests are available. Though these tests may be useful for diagnosing enteric adenoviruses, they may give false-positive results with nonenteric adenoviruses and are less sensitive and nonspecific extracts of stool can cause a high number of invalid tests.
2. *Electron Microscopy:* Noncultivatable enteric adenoviruses can be detected in stool extracts by electron microscopy.
3. *ELISA:* Few commercial immunoassays are available for the detection of adenovirus antigen in stool samples. It is the test of choice to detect adenoviruses 40 and 41 in stool (Gleaves et al. 1993). Monoclonal antibodies can be used in both ELISA or IF format for differentiating serotypes 40 and 41 from other adenoviruses.
4. *Cell Culture:* Since enteric adenoviruses are difficult to grow in conventional cell lines used for respiratory adenoviruses, these have

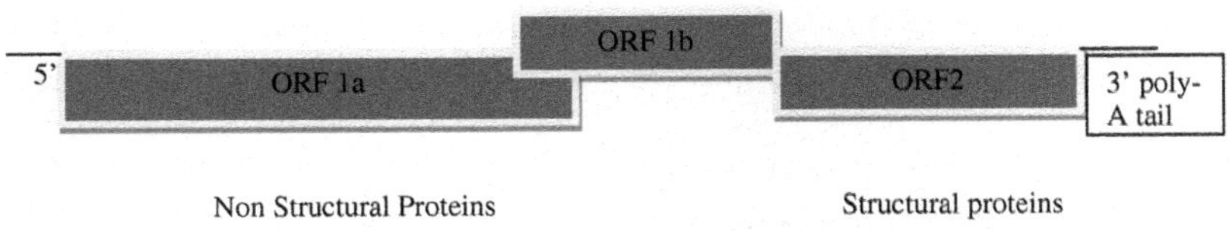

Fig. 4.6 Genomic organization of astroviruses. 5′ end and poly A tail are indicated. (Smits et al. 2010), 6.8-kb positive-sense RNA. Expression: subgenomic RNA

been referred to as noncultivatable adenoviruses, though these may be cultivated in alternate tissue culture host cell systems, for example, Graham 293 cells.

5. *PCR:* Currently, PCR is used for diagnosis, typing, and quantitation of adenovirus infections in stool samples.

Treatment: Generally, it is not required because the disease is often self-limited. No vaccine for enteric adenoviruses is available at present.

Astroviruses

Astroviruses were first observed by Madeley and Cosgrove in (1975) in stools of infants hospitalized with diarrhea and in outbreaks of gastroenteritis in newborn nurseries (Madeley and Cosgrove 1975). These RNA viruses, together with the Caliciviridae, account for much of the gastroenteritis that had previously been of unknown aetiology and was presumed to be caused by undetected viral agents.

Structure: Astroviruses are 28–30 nm particles with a smooth margin and a characteristic five- to six-pointed star on the surface (Greek: astron means star) as seen by electron microscope. These are nonenveloped and show icosahedral symmetry. Capsid consists of three proteins of approximately 30 kDa size, which are produced by proteolytic cleavage of a large polypeptide. The genome consists of positive-sense, single-stranded RNA of approximately 7.2 kb, is polyadenylated, and has both 5′ and 3′ nontranslated regions (NTR). It contains three ORFs, ORF1a, and ORF1b codes for nonstructural proteins, protease, and the RNA-dependent polymerase, respectively, while ORF2 encodes capsid proteins (Smits et al. 2010) (Fig 4.6). ORF two is translated from a subgenomic RNA and produces a polyprotein which is cleaved by cellular proteases (Macndez and Arias 2007).

Classification: Astroviruses have been isolated from humans and from a variety of animals and birds. Astroviruses are classified into two genera: Mamastrovirus, which infect mammals including humans, and Avastroviruses, which infect avian species. Human astroviruses (HAstVs) have been classified into eight serotypes, 1 to 8 on the basis of immunofluorescence and plaque neutralization assays, and by immunoelectron microscopy using hyperimmune sera against different culture-adapted strains. Genotypes defined by the amino acid sequence of the carboxyl-terminal region of the structural polyprotein correlate well with the eight HAstV serotypes that have been described (Mendez and Arias 2007).Worldwide, serotype 1 is the most common, although the predominant serotypes may vary by region and time (Guix et al. 2005).

Epidemiology: Astroviruses were found in less than 1 % of children with diarrhea when less sensitive assays such as electron microscopy were used for diagnosis. The prevalence increased to 2.5–9 % among patients hospitalized with diarrhea, when more sensitive assays like monoclonal antibody assay, enzyme immunoassays, and real-time PCRs were developed (Glass et al. 1996). Human astrovirus infections occur worldwide (Table 4.8), and infants and young children are most commonly affected, although adult population may also become infected. The incubation period for astrovirus disease is 1–5 days, and illness lasts 1–5 days in the immunocompetent host, although protracted diarrhea has been associated with serotype 3 strains (Caballero et al. 2003).

Table 4.8 Epidemiological features of astrovirus gastroenteritis outbreaks

Parameters	Li et al. (2010)	Oh et al. (2003)	Liu et al. (2007)
Place of study	Inner Mongolia, China	Germany	Wuhan City, China
Time period of study	9 October 2008–13 February 2009	February 2001–January 2002	June 2004–May 2005
Setting of outbreak	Nursery in maternity hospital	Children hospitalized with acute gastroenteritis	OPD of Wuhan children's hospital
Incidence of astrovirus	45.90 % (28/61)	4 % (5/129)	9.87 % (33/335)
Most common serotype detected	HAst V1b	HAst V1	HAst V1

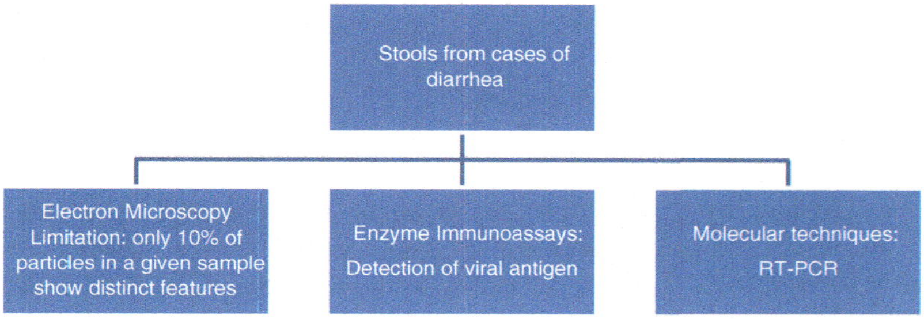

Fig. 4.7 Laboratory diagnosis of astrovirus infection

The main route of transmission of astroviruses is faeco-oral, though person-to-person spread, waterborne spread, and fomites have also been implicated in outbreaks of gastroenteritis.

Pathogenesis: The astroviruses are confined to mature epithelial cells near the villous tip in jejunum. No inflammation is associated with astrovirus infection. HAstVs have the ability to induce apoptosis in cultured cells. Astrovirus infection may result in decreased intestinal disaccharidase activity and subsequent osmotic diarrhea. Many other mechanisms may be used by these viruses to produce the disease.

Clinical features: Watery diarrhea may be associated with vomiting, abdominal pain, and fever. Dehydration may develop in patients with underlying gastrointestinal disease, poor nutritional status, and mixed infection. The incubation period of illness has been estimated to be 3–4 days. Immunocompromised patients such as those with HIV, chronic lymphocytic leukemia, or bone marrow transplant recipients are highly susceptible to infections with HAstV.

Laboratory diagnosis: Shown in Fig. 4.7. Illness associated with these agents is generally self-limited, and if required, treatment is supportive and directed at maintaining hydration and electrolyte balance.

Toroviruses

Toroviruses are 100–140 nm, enveloped viruses of family Coronaviridae that have a single-stranded, positive-sense linear RNA about 28 kb in size, capped, and polyadenylated. The tubular nucleocapsid is tightly coiled into an open torus or donut shape and hence the name "torovirus." The particles have a fringe of peplomers on their surface that are about 10 nm long. The virion RNA is infectious and serves as both the genome and viral messenger RNA. Genomic RNA encodes for ORF1a, and ORF1b is translated by a frame-shifting mechanism. Resulting proteins pp1a and pp1ab are processed into the viral polymerase (RdRp) and other nonstructural proteins involved in RNA synthesis. Structural proteins are expressed as subgenomic RNAs.

Replication: cytoplasmic

1. Virus attaches to host receptors through the S-protein and is endocytosed into vesicles in the host cell.
2. Fusion of virus membrane with the vesicle membrane (probably mediated by E2), ssRNA(+) genome, is released into the cytoplasm.
3. Synthesis and proteolysis of replicase polyproteins. Replicase produces the complementary ssRNA(−) first, which will serve as a template for the synthesis of genomic RNA and subgenomic mRNAs.
4. Synthesis of structural proteins encoded by subgenomic mRNAs.
5. Assembly and budding at membranes of the endoplasmic reticulum (ER), the intermediate compartments, and/or the Golgi complex.
6. Release of new virions.

The virus has been associated with gastroenteritis in horses (species Berne virus), cattle (species Breda virus), pig (species equine Torovirus), and probably humans (species Human Torovirus). A study from Canada demonstrated an association between torovirus excretion and nosocomial gastroenteritis in pediatric patients. Toroviruses were identified in 72 (35.0 %) of 206 gastroenteritis cases compared with 30 (14.5 %) of 206 controls ($P < 0.001$). Persons infected with torovirus were more frequently immunocompromised and nosocomially infected and experienced less vomiting and more bloody diarrhea than did those with rotavirus or astrovirus infection. An antibody response to torovirus was seen in more than half of the patients with torovirus in stool (Jamieson et al. 1998). Asymptomatic shedding is common, particularly in tropical climates and in populations living in poor hygienic conditions.

Diagnosis: Laboratory diagnosis of the human toroviruses depends entirely on electron microscopy of stool specimens and detection of characteristic particles in negatively stained specimens.

Picobirnavirus

These viruses are 30–40 nm in diameter and have icosahedral symmetry with triangulation number (T) equal to 3. These have a bisegmented, double-stranded RNA, and the length of the two parts of the genome is 1.7 and 2.5 kb. The capsid protein gene is encoded by the second open reading frame of the larger genomic segment (Chandra 1997). Although the virus has been detected mainly from the stools of gastroenteritis patients and several mammals and birds, the pathogenicity of the virus has not been established.

Coronavirus

Human coronaviruses have occasionally been implicated in enteric disease, particularly in newborns (Gerna et al. 1985). The evidence for human coronavirus in enteric disease is primarily based on visualization of coronavirus-like particles (VLPs) by electron microscopy, plus some reports of isolation of virus from children with enteric disease (Gerna et al. 1985; Payne et al. 1986; Resta et al. 1985). Because other particles in stool specimens (e.g., cellular membranes) can have similar morphology to coronaviruses, electron microscopic detection of coronavirus particles in stools is not diagnostic of infection.

Waterborne Outbreak of Viral Gastroenteritis

The epidemiologic prototype of viral gastroenteritis outbreaks has been the Norwalk virus infection. In the United States, Norwalk virus has been identified as the cause of 32–42 % of outbreaks of viral gastroenteritis (Hedberg and Osterholm 1993).

Laboratory confirmation of the cause of outbreaks of waterborne viral gastroenteritis requires either the detection of virus in stool or demonstration of a rise in specific antibody.

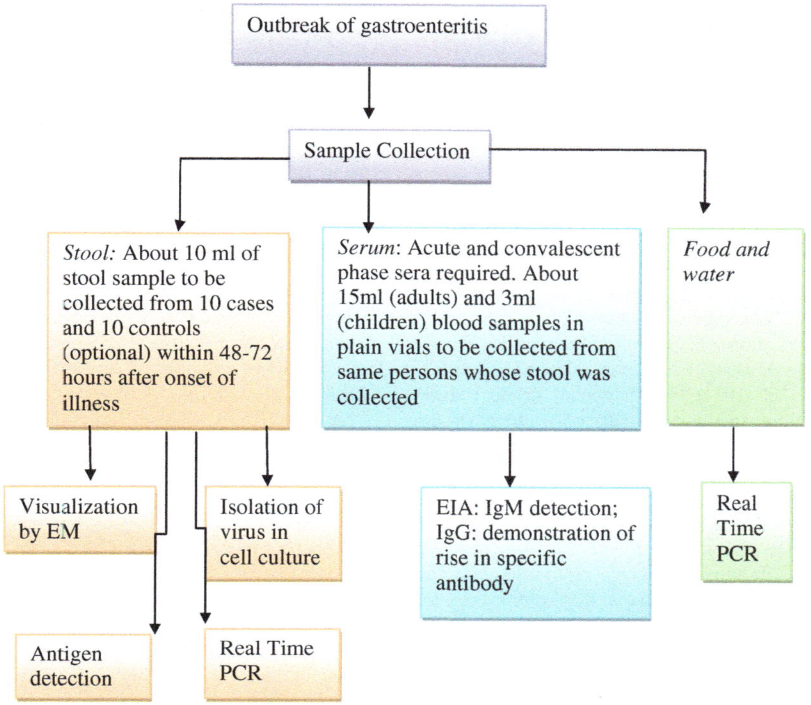

Fig. 4.8 Schematic representation of epidemiologic and laboratory methods for conducting an outbreak investigation. Hedberg and Osterholm (1993)

Algorithm for evaluating outbreaks of viral gastroenteritis is shown in Fig. 4.8. Novel approaches are being developed for syndrome surveillance of acute viral gastroenteritis like development of multiplex real-time reverse transcriptase PCRs (Svraka et al. 2009; Logan 2006).

References

Adler JL, Zickl R (1969) Winter vomiting disease. J Infect Dis 119:668–673

Atmar RL, Estes MK (2001) Diagnosis of noncultivatable gastroenteritis viruses, the human caliciviruses. Clin Microbiol Rev 14:15–37

Ball JM, Graham DY, Opekun AR et al (1999) Recombinant Norwalk virus-like particles given orally to volunteers: phase I study. Gastroenterology 117:40–48

Barnes GL, Lund JS, Mitchell SV, De Bruyn L, Piggford L, Smith AL, Furmedge J, Masendycz PJ, Bugg HC, Bogdanovic-Sakran N, Carlin JB, Bishop RF (2002) Early phase II trial of human rotavirus vaccine candidate RV3. Vaccine 20:2950–2956

Blacklow NR, Greenherg HB (1991) Viral gastroenteritis. N Engl J Med 325:252–264

Blacklow NR, Dolin R, Fedson DS, DuPont H, Northrup RS, Hornick RB, Chanock RM (1972) Acute infectious nonbacterial gastroenteritis: etiology and pathogenesis. Annu Intern Med 76:993–1008

Brandt CD, Kim HW, Rodriguez WJ, Arrobio JO, Jeffries BC, Stallings EP, Lewis C, Miles AJ, Chanock RM, Kapikian AZ, Parrott RH (1983) Pediatric viral Gastroenteritis during eight years of study. J Clin Microbiol 18:71–78

Bresee JS, Glass RI, Ivanoff B, Gentsch JR (1999) Current status and future priorities for rotavirus vaccine development, evaluation and implementation in developing countries. Vaccine 17:2207–2222

Caballero S, Guix S, El-Senousy WM, Calico I, Pinto RM, Bosch A (2003) Persistent gastroenteritis in children infected with astrovirus: association with serotype-3 strains. J Med Virol 71:45–50

Cannon RO, Poliner JR, Hirschhorn RB, Rodeheaver DC, Silverman PR, Brown EA, Talbot GH, Stine SE, Monroe SS, Dennis DT, Glass RI (1991) A multistate outbreak of Norwalk virus gastroenteritis associated with consumption of commercial ice. J Infect Dis 164:860–863

Carlson JA, Middleton PJ, Szymanski MT et al (1978) Fatal rotavirus gastroenteritis: an analysis of 21 cases. Am J Dis 132:477–479

Cevenini R, Mazzaracchio R, Rumpianesi F, Donati M, Moroni A, Sambri V, La Placa M (1987) Prevalence of enteric adenovirus from acute gastroenteritis: a five year study. Eur J Epidemiol 3(2):147-150

Chadwick PR, McCann R (1999) Transmission of a small round structured virus by vomiting during a hospital outbreak of gastroenteritis. J Hosp Infect 26:251–259

Chandra R (1997) Picobirnavirus, a novel group of undescribed viruses of mammals and birds: a mini-review. Acta Virologica 41(1):59–62

Chitambar SD, Lahon A, Tatt VS, Maniya NH, Tambe GU, Khatri KI, Desai HS, Ugare MR, Kulkarni SV, Waghmare AP (2011) Occurrence of group B rotavirus infections in the outbreaks of acute gastroenteritis from western India. Ind J Med Res 134:399–400

De Jong JC, Wermenbol AG, Verweij-Uijterwaal MW, Slaterus KW, Wertheim-Van Dillen P, Van Doornum GJ, Khoo SH, Hierholzer JC (1999) Adenoviruses from human immunodeficiency virus-infected individuals, including two strains that represent new candidate serotypes Ad50 and Ad51 of species B1 and D, respectively. J Clin Microbiol 37:3940–3945

Dennehy PH (2008) Rotavirus Vaccines: an Overview. Clin Microbiol Rev 21(1):198-208

Dey SK, Shimizu H, Phan TG, Hayakawa Y, Islam A, Salim AF, Khan AR, Mizuguchi M, Okitsu S, Ushijima H (2009) Molecular epidemiology of adenovirus infection among infants and children with acute gastroenteritis in Dhaka, Bangladesh. Infect Genet Evol 9(4):518–522

Dolin R, Reichman RC, Roessner KD, Tralka TS, Schooley RT, Gary W, Morens D (1982) Detection by immune electron microscopy of the Snow Mountain agent of acute viral gastroenteritis. J Infect Dis 146:184–189

Dowell SF et al (1995) A multistate outbreak of oyster associated gastroenteritis, implication for interstate tracing of contaminated shellfish. J Infect Dis 171:1497–1503

Estes MK, Kapikian AZ (2007) Rotaviruses. In: Knipe DM, Howley PM (eds) Fields virology, 5th edn. Lippincott Williams and Wilkins, Philadelphia, pp 981–1000

Filho EP, da Costa Faria NR, Fialho AM, de Assis RS, Almeida MM, Rocha M, Galvão M, dos Santos FB, Barreto ML, Leite JP (2007) Adenoviruses associated with acute gastroenteritis in hospitalized and community children up to 5 years old in Rio de Janeiro and Salvador, Brazil. J Med Microbiol 56:313–319

Fox JP, Brandt CD, Wassermann FE et al (1969) The virus watch program: a continuing surveillance of viral infections in metropolitan New York families.VI. Observations of adenovirus infections: virus excretion patterns, antibody response, efficiency of surveillance, patterns of infections, and relation to illness. Am J Epidemiol 89:25–50

Gerna G, Passarani N, Battaglia M, Rondanelli EG (1985) Human enteric coronaviruses: antigenic relatedness to human coronavirus OC43 and possible etiologic role in viral gastroenteritis. J Infect Dis 151:796–803

Gilger MA, Matson DO, Conner ME, Rosenblatt HM, Finegold MJ, Estes MK (1992) Extraintestinal rotavirus infections in children with immunodeficiency. J Pediatr 120:912–917

Glass RI, Noel J, Mitchell D, Herrmann JE, Blacklow NR, Pickering LK, Dennehy P, Ruiz Palacios G, de Guerrero ML, Monroe SS (1996) The changing epidemiology of astrovirus-associated gastroenteritis: a review. Arch Virol 12:287–300S

Glass RI, Bhan MK, Ray P et al (2005) Development of candidate rotavirus vaccines derived from neonatal strains in India. J Infect Dis 192(1):S30–S35

Gleaves CA, Militoni J, Ashley RL (1993) An enzyme immunoassay for the direct detection of adenovirus in clinical specimens. Diagn Microbiol Infect Dis 17:57–59

Graham DY, Jian X, Tanaka T et al (1994) Norwalk virus infection of volunteers: new insights based on improved assays. J Infect Dis 170:34

Guerrant RL, Hughes JM, Lima NL, Crane J (1990) Diarrhea in developed and developing countries: magnitude, special settings, and etiologies. Rev Infect Dis 12:S41–S45

Guix S, Bosch A, Pinto RM (2005) Under the microscope. Human astrovirus diagnosis and typing: current and future prospects. Lett Appl Microbiol 41:103–105

Hafliger D, Hubner PH, Luthy J (2000) Outbreak of viral gastroenteritis due to sewage contaminated drinking water. Int J Food Microbiol 54:123–126

Hedberg CW, Osterholm MT (1993) Outbreaks of food-borne and waterborne viral gastroenteritis. Clin Microbiol Rev 6(3):199–210

Hung T (1988) Rotavirus and adult diarrhea. Adv Virus Res 135:193–218

Hutson AM, Atmar RL, Graham DY et al (2002) Norwalk virus infection and disease is associated with ABO histo-blood group type. J Infect Dis 185:1335–1337

II Jones MS, Harrach B, Ganac RD, Gozum MM, Dela Cruz WP, Riedel B, Pan C, Delwart EL, Schnurr DP (2007) New adenovirus species found in a patient presenting with gastroenteritis. J Virol 81:5978–5984

Jamieson FB, Wang EEL, Bain C, Good J, Duckmanton L, Petric M (1998) Human Torovirus: a new nosocomial Gastrointestinal Pathogen. J Infect Dis 178(5):1263–1269

Jarecki- Khan K, Tzipori SR, Unicomb LE (1993) Enteric adenovirus infection among infants with diarrhea in rural Bangladesh. J Clin Microbiol 31:484–489

Jiang X, Wang M, Wang K et al (1993) Sequence and genomic organization of Norwalk virus. Virology 195:51–61

Jones MS, Kapoor A, Lukashov VV, Simmonds P, Hecht F, Delwart E (2005) New DNA viruses identified in patients with acute viral infection syndrome. J Virol 79:8230-8236

Kapikian AZ, Wyatt RG, Dolin R, Thornhill TS, Kalica AR, Chanock RM (1972) Visualization by immune electron microscopy of a 27 nm particle associated with acute infectious nonbacterial gastroenteritis. J Virol 10:1075–1081

Kapikian AZ, Hoshino Y, Chanock RM, Perez-Schael I (1996) Efficacy of a quadrivalent rhesus rotavirus-based human rotavirus vaccine aimed at preventing severe rotavirus diarrhea in infants and young children. J Infect Dis 174(Suppl. 1):S65–S72

Kelkar SD, Zade JK (2004) Group B rotaviruses similar to strain CAL-1, have been circulating in Western India since 1993. Epidemiol Infect 132:745–749

Kukkula M, Arstila P, Klossner ML, Maunula L, Bonsdorff CH, Jaatinen P (1997) Waterborne outbreak of viral gastroenteritis. Scand J Infect Dis 29(4):415–418

Lew JF, Kapikian AZ, Valdesuso J et al (1994) Molecular characterization of the Hawaii virus and other Norwalk-like viruses: evidence for genetic polymorphism among human caliciviruses. J Infect Dis 170:535–542

Li L, Phan TG, Nguyen TA et al (2005) Molecular epidemiology of adenovirus infection among pediatric population with diarrhea in Asia. Microbiol Immunol 49:121–128

Li CY, Liu N, Guo WD, Yu Q, Wang WR, Song ZZ, Yan H, Luo Y, Lu AT, Li HY, Zhu L, Duan ZJ (2010) Outbreak of Neonatal gastroenteritis associated with Astrovirus Serotype 1 at a Hospital in Inner Mongolia, China. J Clin Microbiol 48(11):4306–4309

Li CSY, Paul KS, Tang CJW (2009) Prevalence of diarrhea viruses in hospitalized children in Hong Kong in 2008. J Med Virol 81(11):1903-1911

Liu MQ, Yang BF, Peng JS, Zhou DJ, Tang L, Wang B, Liu Y, Sun SH, Ho WZ (2007) Molecular epidemiology of Astrovirus infection in infants in Wuhan, China. J Clin Microbiol 45:1308–1309

Logan C, O'Leary JJ, O'Sullivan N (2006) Real-time reverse transcription-PCR for detection of rotavirus and adenovirus as causative agents of acute viral gastroenteritis in children. J Clin Microbiol 44(9):3189–3195

Lopman BA, Reacher MH, van Duijnhoven Y, Hanon FX, Brown D, Koopmans M (2003) Viral gastroenteritis outbreaks in Europe, 1995–2000. Emerg Infect Dis 9(1):90–96

Macndez E, Arias CF (2007) Astroviruses. In: Knipe DM, Howley PM (eds) Fields virology, 15th edn. Lippincott Williams and Wilkins, Philadelphia, pp 981–1000

Madeley CR, Cosgrove BP (1975) Viruses in infantile gastroenteritis (letter). Lancet 2(7925):124

Mendez E, Arias CF (2007) Astroviruses. In: Knipe DM, Howley PM (eds) Fields Virology, 5th edn. Lippincott Williams and Wilkins, Philadelphia, pp 981–1000

Mendez E, Lopez S, Cuadras MA, Romero P, Arias CF (1999) Entry of rotaviruses in is a multistep process. Virology 263:450–459

Midthun K, Greenberg HB, Hoshino Y et al (1985) Reassortant rotaviruses as potential live rotavirus vaccine candidates. J Virol 53:949–954

Nelson EAS, Bresee JS, Parashar UD, Widdowson M-A, Glass RI (2008) Rotavirus epidemiology: the Asian Rotavirus surveillance network. Vaccine 26:3192–3196

Nguyen TA, Yagyu F, Okame M, Phan TG, Trinh QD, Yan H, Hoang KT, Cao AT, Le Hoang P, Okitsu S, Ushijima H (2007) Diversity of viruses associated with acute gastroenteritis in children hospitalized with diarrhea in Ho Chi Minh City. Vietnam. J Med Virol 79:582–590

Oh DY, Gaedicke G, Schreier E (2003) Viral agents of acute gastroenteritis in German children: prevalence and molecular diversity. J Med Virol 71:82–93

Okhuysen PC, Jiang X, Ye L et al (1995) Viral shedding and fecal IgA response after Norwalk virus infection. J Infect Dis 171:566–569

Oseto M, Yamashita Y, Hattori M, et al. (1993) Successful propagation of human group C rotavirus in a continuous cell line and characterization of the isolated virus. In: Abstracts of 27th joint viral diseases meeting of the US-Japan cooperative medical sciences program, pp 46–46

Pacini DL, Collier AM, Henderson FM (1987) Adenovirus infections and respiratory illnesses in children in group day care. J Infect Dis 156:920–926

Patel MM, Hall AJ, Vinje J et al (2009) Noroviruses: a comprehensive review. J Clin Virol 44:1–8

Payne CM, Ray CG, Borduin V, Minnich LL, Lebowitz MD (1986) An eight-year study of the viral agents of acute gastroenteritis in humans: ultrastructural observations and seasonal distribution with a major emphasis on coronavirus-like particles. Diagn Microbiol Infect Dis 5:39–54

Prasad BVV, Rothnagel R, Jiang X, Estes MK (1994) Three dimensional structure of baculovirus-expressed Norwalk virus capsids. J Virol 68:5117–5125

Resta S, Luby J, Rosenfeld C, Siegel J (1985) Isolation and propagation of a human enteric coronavirus. Science 229:978–981

Rippey SR (1994) Infectious diseases associated with molluscan shellfish consumption. Clin Microbiol Rev 7:419–425

Sanekata T, Ahmed MU, Kader A et al (2003) Human group B rotavirus infections cause severe diarrhea in children and adults in Bangladesh. J Clin Microbiol 41:2187–2190

Santos N, Hoshino Y (2005) Global distribution of rotavirus serotypes/genotypes and its implication for the development and implementation of an effective rotavirus vaccine. Rev Med Virol 15:29–56

Smits SL, van Leeuwen M, van der Eijk AA, Fraaij PLA, Escher JC, Simon JH, Albert D, Osterhaus ME (2010) Human astrovirus infection in a patient with new-onset celiac disease. J Clin Microbiol 48(9):3416–3418

Svraka S, van der Veer B, Duizer E, Dekkers J, Koopmans M, Vennema H (2009) Novel approach for detection of enteric viruses to enable syndrome surveillance of acute viral gastroenteritis. J Clin Microbiol 47:1674–1679

Tacket CO, Sztein MB, Losonsky GA et al (2003) Humoral, mucosal, and cellular immune responses to oral Norwalk virus-like particles in volunteers. Clin Immunol 108:241–247

Tang B, Gilbert JM, Matsui SM, Greenberg HB (1997) Comparison of the rotavirus gene 6 from different species by sequence analysis and localization of subgroup specific epitopes using site directed mutagenesis. Virology 237:89–96

Thornhill TS, Wyatt RG, Kalica AR, Dolin R, Chanock RM, Kapikian AZ (1977) Detection by immune electron microscopy of 26–27 nm virus-like particles associated with two family outbreaks of gastroenteritis. J Infect Dis 135:20–27

Uhnoo I, Wadell G, Svensson L, Johansson ME (1984) Importance of enteric adenoviruses 40 and 41 in acute gastroenteritis in infants and young children. J Clin Microbiol 20(3):365–372

Van R, Wun CC, O'Ryan ML et al (1992) Outbreaks of human enteric adenovirus types 40 and 41 in Houston day care centres. J Pediatr 120:516–521

Vega E, Barclay L, Gregoricus N, Williams K, Lee D, Vinjé J (2011) Novel surveillance network for norovirus gastroenteritis outbreaks, United States. Emerg Infect Dis 17(8):1389–1395

Werber SD, Laus Evic D, Mugos B, Vratnic Z, Ivanovic-Nikolic L, Alexandre-Bird A, Fiore L, Ruggeri FM, Di Bartolo I, Battistone A, Gassilloud B, Perelle S, Nitzan Kaluski D, Kivi M, Andraghetti R, Pollock KGJ (2009) Massive outbreak of viral gastroenteritis associated with consumption of municipal drinking water in a European capital city. Epidemiol Infect 137:1713–1720

Wilde J, Van R, Pickering L et al (1992) Detection of rotaviruses in the day care environment by reverse transcriptase polymerase chain reaction. J Infect Dis 166:507–511

Wold WSM, Horwitz MS (2007) Adenoviruses. In: Knipe DM, Howley PM (eds) Fields cirology, 5th edn. Lippincott Williams and Wilkins, Philadelphia, pp 981–1000

Cholera and Drinking Water

5

Brahm S. Srivastava

Abstract

Cholera is caused by intestinal infection of *Vibrio cholerae* O1 or O139 through oral route by ingestion of contaminated food and water. Symptoms of cholera include acute watery diarrhea, vomiting, cramps and dehydration. If not treated in time, patient can die of dehydration and hypovolemic shock. Effective therapy is rehydration replacement of salts and electrolytes orally and/or intravenously. Once diarrhea has commenced antibiotic administration only help reduce the bacterial burden and duration of illness. Pathogenesis of cholera is characterized by entry of vibrios in small intestine, adherence to epithelial surface, colonization and release of enterotoxin followed by secretion of isotonic fluid. Mucinase, motility, adherence, multiplication, colonization and secretion of one or more types of toxin are among important factors contributing to disease process. Lipopolysaccharide and flagellar antigens, cholera lectins, haemagglutinins and major outer membrane proteins have been identified as mediating adherence and colonization. Immune response after infection is exclusively humoral. Antibodies to cholera antigens have been detected in sera of experimental cholera and convalescent patients. Antibacterial and antitoxin immunities are synergistic and contribute in protection. A vaccine providing durable immunity has remained elusive despite tremendous efforts. However, judicious application of knowledge of mucosal immunity, pathogenesis, antigen formulations and vaccine delivery will yield the right cholera vaccine.

Keywords

Cholera · *Vibrio cholerae* · Pathogen · Diarrhea · Toxin · Cell-associated antigens · Vaccine

B. S. Srivastava (✉)
Nextec Lifesciences, 2/79 Vijay Khand,
Gomtinagar, Lucknow, 226010, India
e-mail: drbrahm@gmail.com

P. P. Singh and V. Sharma (eds.), *Water and Health*,
DOI: 10.1007/978-81-322-1029-0_5, © Springer India 2014

Cholera

Cholera is an acute intestinal infection caused by the toxigenic bacterium *Vibrio cholerae* of serogroups O1 or O139 (Ramamurthy et al. 1993). Both serogroups have caused widespread epidemics. *V. cholerae* O1 has two biotypes: classical and El Tor, and each biotype has two distinct serotypes: Ogawa and Inaba. The symptoms of infection by both serogroups are indistinguishable, although more people infected with the El Tor biotype remain asymptomatic or have only a mild illness. In recent years, infections with the classical biotype of *V. cholerae* O1 have become rare and are limited to parts of Bangladesh and India.

Since 1961, the seventh pandemic of cholera, caused by *V. cholerae* serogroup O1, biotype El Tor, has spread from Indonesia through most of Asia into Eastern Europe and Africa, and from North Africa to the Iberian Peninsula. In 1991, an extensive epidemic began in Peru and spread to neighbouring countries in the western hemisphere. Although few cases of cholera occur in South or Central America, *V. cholerae* O1 remains endemic in much of Africa and South and Southeast Asia. *V. cholerae* O139 spread rapidly through Asia in the early 1990s but has since remained localized to a few areas in Bangladesh and India.

Cholera infection is most often asymptomatic or results in a mild gastroenteritis. Severe cholera is characterized by acute, profuse watery diarrhoea, resembling slightly viscous rice water, and in some cases, diarrhoea may be associated with vomiting, cramps of arm and legs and dehydration. Signs and symptoms include tachycardia, loss of skin turgor, dry mucous membranes, hypotension and thirst. Additional symptoms, including muscle cramps, are secondary to the resulting electrolyte imbalances. If untreated, volume depletion can rapidly lead to hypovolemic shock and death. Cholera is a disease that affects humans only and infection takes place by ingestion of infected food or drinking water. In epidemics, the transmission of cholera takes place by the faecal-oral route through contaminated food, water and containers.

Cholera is confirmed through culture of a stool specimen or rectal swab. Cary-Blair medium is ideal for transport, and the selective thiosulfate-citrate-bile salts (TCBS) agar is ideal for isolation and identification. Further confirmation of *V. cholerae* and serogroups is achieved by agglutination test by specific antibodies easily available.

Rehydration is the cornerstone of therapy. Oral rehydration salts and, when necessary, intravenous fluids and electrolytes, if administered in a timely manner and in adequate volumes, will reduce case-fatality ratios to well under 1 %. Antibiotics reduce fluid requirements and duration of illness. Antimicrobial therapy is indicated for severe cases, which can be treated with tetracycline, doxycycline, furazolidone, erythromycin or ciprofloxacin. Readers are referred to two excellent books, a compilation of chapters on various aspects of cholera (Barua and Burrows 1974; Wachsmuth et al. 1994). The present review is limited to aspects of pathogenesis, virulence factors, nature of immune responses, cholera toxins and vaccines of cholera.

Pathogenesis

V. cholerae is strictly a non-invasive pathogen. Pathogenesis of cholera can be divided into four stages: (1) ingestion of viable vibrios with contaminated food and water, (2) entrance into small bowel and adhesion of vibrios to epithelial surface of small intestine, (3) bacterial multiplication and colonization of intestinal mucosa, (4) production and release of cholera toxin and (5) hypersecretion of toxin-induced isotonic fluid from the intestine.

Gastric acidity acts as a natural barrier to restrict infection of *V. cholerae* in their passage to intestine. The inoculum size which was required to produce disease in human volunteers was found to be 10^{11} live vibrios when given without bicarbonate (Cash et al. 1974). In contrast, volunteers who ingested 10^6 live vibrios with 2 g of $NaHCO_3$ developed cholera. If vibrios successfully escape the acidic environment of the stomach, they arrive in the small intestine where a series of pathogenic events occur that result in colonization of vibrios. Vibrios attempt to penetrate through the mucous layer and adhere to epithelium which may be a mechanism to escape clearance by peristalsis and ultimate colonization of epithelial surface. A number of factors have been identified making definite contribution in disease process.

Mucinase elaborated by vibrios helps the bacterial entrance into the mucus gel covering the mucosa (Burnet 1948; Freter 1955). Motility together with chemical substances present at the epithelial surface contributes towards movement of vibrios from the lumen of the gut to the epithelial surface. The role of motility in pathogenesis is well established (Freter et al. 1981; Freter and O'Brien 1981). *V. cholerae* possess a single polar flagellum which provides motile phenotype to this bacterium. Flagellum/motility is considered to be an important factor in the virulence of *V. cholerae* (Guentzel and Berry 1975). Non-motile mutants neither penetrate the mucous layer nor exhibit virulence in experimental cholera (Schrank and Verwey 1976; Bhattacharjee and Srivastava 1979; Srivastava et al. 1980). Non-motile aflagellate mutants poorly colonized the small intestinal epithelium. Inhibition of motility of vibrios by monoclonal and polyclonal antibodies in vitro and in vivo suggested that flagellar antigens might play a critical role in antibacterial immunity against cholera (Gustafsson and Holme 1985, Sinha et al. 1993). Little is known about the structure and composition of flagellum of *V. cholerae* and the antigens which are uniquely associated with the flagellum and likely to be exposed to the immune system of the host. In our laboratory, comparison of a non-flagellate non-motile mutant with its parent strain revealed two proteins of 40 and 38 kDa associated with flagellum (Sinha et al. 1993). Antibodies to these antigens bind to flagella and inhibit motility of *V. cholerae*. Few other studies on the antigenic composition of flagellar sheath and core of *V. cholerae* have been reported (Yang et al. 1977; Hranitzky et al. 1980; Richardson and Parker 1985).

Adhesion of vibrios to intestinal epithelium has been emphasized repeatedly. Motile strains of *V cholerae* appearing in large numbers in intervillous spaces and crypts adhere to the epithelial surface of the small intestine. The time course of adherence of vibrios to rabbit intestine and beginning of fluid outpouring in the ileal loop was studied (Nelson et al. 1976; Srivastava et al. 1980; Teppema et al. 1987). A good correlation was found between bacterial adherence and pathogenicity. Strains capable of adhering with high efficiency were pathogenic, whereas a non-adhesive mutant isolated in our laboratory was found to be attenuated in virulence and apathogenic (Srivastava and Srivastava 1980). Adherence of vibrios therefore to the mucosal cells of the intestinal epithelium is a very important step leading to colonization and release of toxin.

A number of antigens responsible for adherence and colonization have been described. The role of lipopolysaccharide antigens (Chitnis et al. 1982), flagellar antigens (Attridge and Rowley 1983; Hranitzky et al. 1980), cholera lectins (Finkelstein et al. 1983), various haemagglutinins (Bhattacharjee and Srivastava 1978; Finkelstein and Hanne 1982; Holmgren et al. 1983; Sharma et al. 2011), major outer membrane proteins (Kabir and Showkat 1983; Sperandio et al. 1995; Jonson et al. 1991; Ehara et al. 1987; Sharma et al. 2011, Sinha et al. 1993, 1994; Srivastava et al. 1980) has been reported to have possible relationship to *V. cholerae* adherence and colonization. In our laboratory, a non-adhesive mutant (CD11) of *V cholerae* was isolated from its parent strain (KB207). Although CDll was motile, chemotactic, and toxigenic, it adhered poorly, exhibited reduced virulence in experimental cholera and did not colonize the gut of infant mice (Srivastava and Srivastava 1980; Srivastava et al. 1980; Jacob et al. 1993). Analysis of the mutant revealed that a protein present in KB207 was absent in CDll. Antibodies to this antigen inhibited adherence of KB207 to rabbit intestinal mucosa and colonization in an infant mouse model (Jacob et al. 1993). More recently, the antigen was identified as DctP of 37 kDa which may be an important antigen involved in adherence and colonization of vibrios in the intestine and may be identified as mannose-sensitive haemagglutinin (Sharma et al. 2011). A pilus-colonizing factor, TCP, of 20.5 kDa from a classical strain of *V. cholerae* has been reported (Taylor et al. 1987).

As the *V. cholerae* adhere and colonize the intestinal mucosa, they elaborate cholera enterotoxin which causes diarrhoea (De 1959; Dutta et al. 1959). Cholera toxin is a protein whose structure, function and biological activity has been extensively studied (Kaper and Srivastava 1992). The role of cholera enterotoxin is undisputable in the pathogenesis of cholera. Other factors contributing to virulence were identified by testing a variety of mutants defective in motility, adherence and colonization (Bhattacharjee and Srivastava 1978, 1979; Levine et al. 1983). The studies suggested that immune mechanisms targeted towards inhibition of motility and adherence of vibrios could be important in the prophylactic control of cholera.

Immune Response After *V. cholerae* 01 Infection

Antibodies to various cholera antigens such as lipopolysaccharides, outer membrane proteins, cholera toxin and toxin-coregulated pilus have been detected in sera from individuals immunized with *V cholerae* O1 or from convalescent patients (Majumdar et al. 1981; Kabir 1983; Richardson et al. 1989). Body fluids, intestinal lavages and peripheral blood have been used to assess immune responses (Svennerholm et al. 1984). However, they do not represent the true level of immunoglobulins in the gut after an antigenic stimulus. A study of immune responses has been made in rabbits (Srivastava et al. 1979; Svennerholm et al. 1984; Kabir 1987a, b) A single-dose intraduodenal inoculation of live *V. cholerae* 01 produced antibodies to both LPS and cell surface proteins and cholera toxin and neuraminidase antigens in sera and intestinal extracts of rabbits, the latter containing predominantly IgA together with considerable amount of IgG. Although cholera is a toxin-mediated disease, the predominant protective immune mechanism appears to be antibacterial rather than antitoxic (Levine et al. 1979; Srivastava et al. 1979). While the role of LPS in protective immunity has been indicated from a field trial in Bangladesh, the contribution of other components of *V. cholerae* O1 in protection has not been definitely identified.

Immunity to cholera involves antibodies directed towards vibrios (antibacterial) and the toxin released by them (antitoxin). Rabbits immunized with bacterial vaccine or toxin, or both, showed resistance to challenge. Antibacterial or antitoxin antibodies, or both, appeared in the circulation depending on the vaccines given to rabbits. There is convincing evidence that antibacterial and antitoxin immunities work synergistically (Svennerholm and Holmgren 1976; Srivastava et al. 1979). Antitoxin antibodies neutralize cholera toxin which appears to be the mechanism of antitoxin immunity (Peterson et al. 1979). The data obtained in our laboratory strongly suggested that antibacterial immunity might act by inhibiting adherence of vibrios on the surface of the intestine (Srivastava et al. 1979, 1980; Jacob et al. 1993). A group of rabbits were immunized with killed whole-cell bacterial vaccine to induce antibacterial immunity and were then challenged with a pathogenic *V. cholerae* strain in the ileal loop model (De and Chatterji 1953). When the negative and positive loops were examined for the number of vibrios adherent to the intestine, fewer number of vibrios were adherent to epithelial surfaces in negative loops whereas large number of vibrios were adherent in the positive loops. Thus, there was a good correlation between low adherence and resistance to challenge. This finding suggested that antibacterial immunity provided protection mainly by interfering with the adherence of vibrios to intestine and that the adhesive antigen could be one of the key antigens in cholera immunity.

Cholera Toxins

V. cholerae secretes a toxic protein popularly named cholera toxin which is responsible for the diarrhoea and other associated clinical symptoms of cholera. Evidence that *V. cholerae* elaborate a toxin responsible for the onset of diarrhoea was simultaneously provided, independently, by two Indian investigators, viz., De (1959) and Dutta et al. (1959) who reported that outpouring of fluid in the gut of experimental animals could be produced by bacteria-free culture filtrates of *V. cholerae*, this very finding generated interest in purification, identification, elucidation of structure and genetics of cholera toxin leading to vaccine development.

Structure and Organization

Cholera toxin (CT) is a protein which is composed of two types of subunits called A (27.2 kDa) and B (11.6 kDa), and each CT molecule is composed of one subunit of A and five subunits of B. The A subunit contributes to the intracellular toxicity of CT molecule, whereas B subunits are required in the binding of CT molecules to the receptors on eukaryotic cell surface (GM_1 ganglioside) and thus facilitate the entry of A subunit into the target cell (Lospolluto and Finkelstein 1972; Cuatrecasas 1973; Gill 1976) The A subunit must be nicked to be enzymatically active and toxic (Mekalanos et al. 1979). The A subunit when cleaved yields two polypeptide chains: A1 peptide of 21.8 kDa and A2 peptide of 5.4 kDa. After cleavage, the A1 and A2 peptides are still linked by a disulphide bond before internalization (Gill and Rappaport 1979). The ratio of A to B subunits is 5:1 (Gill 1976). The crystal structure of CT has been determined which shows that B subunit forms a pentamer via interaction between β sheet I of one monomer and sheet II of an adjacent monomer. Each monomer thus forms a pentamer with a barrel in the centre creating a pore in which C-terminal end of the A2 binds to the B pentamer. The N-terminal of the A2 extends and interacts with the A1 peptide (Gill 1976). The proteolytic nicking catalyses the transfer of the ADP-ribose moiety from NAD to eukaryotic adenylate cyclase regulatory system resulting in elevation of adenylate cyclase activity and cAMP levels. The increase in cAMP alters the ion flux changes in epithelial cells resulting in the luminal discharge that triggers diarrhoea characteristic of cholera (Cassel and Pfeuffer 1978; Gill and Meren 1978).

Location, Molecular Cloning and Structure of Genetic Elements of Cholera Toxin

The structural genes encoding CT are located on the chromosome of *V. cholerae* and not on extrachromosomal genetic elements. The first evidence was provided by a mutant of *V. cholerae* strain 569B which produced 100-fold less toxin than 569B, and the mutation was mapped close to the *his* locus (Vasil et al. 1975). Although it turned out later to be a mutation in the regulatory gene rather than the structural gene of cholera toxin, it nevertheless suggested that the toxin encoding genes in *V. cholerae* might be located on the chromosome as opposed to the plasmid-borne genes of *Escherichia coli* enterotoxins (Smith and Halls 1968). Subsequent genetic and hybridization studies of *V. cholerae* indicated that the genes for CT are located on the chromosome (Kaper and Levine 1981; Pearson and Mekalanos 1982; Gennaro et al. 1982).

The genetic element encoding the cholera toxin is called ctxAB. The structural genes encoding A and B subunits of CT are designated as ctxA and ctxB (Mekalanos et al. 1983). It may be recalled that there are two biotypes of *V. cholerae*, viz., El Tor and classical. While structural *ctx* gene sequences are essentially identical between the El Tor and classical biotypes, the arrangement of *ctx* genetic element differs between the two biotypes. The ctxAB genes of an El Tor strain RV79 was found to map between *his* and *nal* loci on the chromosome whereas the classical strain 569B carries two copies of ctxAB, one of which is linked to the *nal* locus (Sporecke et al. 1984). El Tor and classical strains differ with respect to organization and copy number of ctxAB genes. Classical strains contain two complete copies of ctxAB which are unlinked on the chromosome. Most El Tor strains are found to have a single copy of the ctxAB genes but some El Tor may carry more than one copy. However, unlike classical strains, these *ctx* copies are arranged in tandem (Mekalanos 1983; Moseley and Falkow 1980).

No difference between two gene copies has been reported and all express active CT (Mekalanos et al. 1983). The genes encoding CT have been cloned from a number of *V. cholerae* strains by recombinant DNA techniques (Kaper and Levine 1981; Pearson and Mekalanos 1982; Gennaro et al. 1982; Mekalanos et al. 1983; Kaper et al. 1984; Srivastava et al. 1985) using E. coli eltAB gene (Moseley and Falkow 1980) or anti-CT antibody probes and by RP4: mini-Mu vector (Srivastava et al. 1989). A detailed nucleotide sequence analysis of the DNA sequence of cloned ctxAB genes revealed the structure of the whole operon (Mekalanos et al. 1983; Gennaro and Greenway 1983; Lockman and Kaper 1983; Lockman et al. 1984). The ctxAB genes form an operon with ctxA gene preceding the ctxB gene. The ctxA and ctxB genes predict translation products of 258 and 124 amino acids, respectively. Both A and B subunits are synthesized with N-terminal hydrophobic amino acid signal sequences of 18 (A subunit) and 21 (B subunits) residues. The ctxAB genes have been expressed in *E. coli* but the recombinant toxin is not as active as the native protein (Pearson and Mekalanos 1982; Srivastava et al. 1985). It was revealed that recombinant CT was not nicked in *E. coli* and if treated with trypsin, higher toxicity could be obtained. The promoter region of all the sequenced operons of ctxAB was found to be identical except for the number of tandem repeats (TTTTGAT) located 77 bp from the start of ctxA (Pearson and Mekalanos 1982).

The *ctx* genetic element reveals a very interesting substructure similar to transposition elements. The ctxAB sequences are located on a conserved 4.3-kb region called the "core region" (Mekalanos 1983). A 2.7-kb repetitive DNA sequence designated as RS1 is found upstream and in some cases downstream of the core region at the junction of *ctx* duplications. The RS1 element is involved in tandem duplication and amplification of the 4.3-kb core region. This recA-dependent process leads to increased copy number and gene dosage effect of the ctxAB gene (Mekalanos 1983; Goldberg and Mekalanos 1986). This tandem gene amplification has only been reported in El Tor, and not in classical strains since the latter biotype lacks an RS1 element downstream of the *ctx* locus.

The 4.3-kb core region and the 2.7-kb RS1 element upstream of the core region are conserved in all *V. cholerae* strains, both classical and El Tor (Mekalanos 1983). This conserved sequence including the ctxAB genes has not been found in naturally occurring non-toxigenic strains of *V. cholerae* isolated from the environment (Miller and Mekalanos 1984).

Regulation of ctx Genetic Element

A hypertoxigenic mutant of *V. cholerae* carrying a regulatory mutation *tox* was first isolated which produced 1000-fold less toxin (Vasil et al. 1975; Holmes et al. 1975). Other regulatory mutations, *htx* and *ltx*, were identified which mapped between *str* and *rif* loci on the chromosome, respectively, and conferred hyper- and hypo-toxin-producing phenotypes (Mekalanos et al. 1979; Mekalanos and Murphy 1980). Recently, a positive regulatory element, toxR, has been cloned from *V. cholerae* which activates transcription from the *ctx* promoter in *E. coli* and *V. cholerae* (Miller and Mekalanos 1984). The cloned toxR gene complements *tox* mutation described earlier (Holmes et al. 1975). The toxR is a transacting positive regulator of *ctx* which activates transcription and the level of activation varies for toxR alleles from different strains (Miller and Mekalanos 1985). The toxR gene product is a transmembrane protein which appears to bind to the repetitive sequence TTTT-GAT in the *ctx* promoter region and activates *ctx* promoter (Miller et al. 1987). The number of TTTTGAT repeat sequence may affect binding of toxR gene product and thus expression of *ctx*. In addition to *ctx*, toxR has been shown to regulate induction of a number of genes and this network of genes has been termed the toxR regulon (Taylor et al. 1987; Pearson and Mekalanos 1988).

Another regulatory gene called toxS has been identified which is found 45 bp downstream of toxR gene and can stimulate activity of toxR (Miller et al. 1989; DiRita and Mekalanos 1991). toxS encodes a 19-kDa transmembrane protein which interacts with the periplasmic portion of ToxR protein. It has been proposed that ToxS enhances dimerization of ToxR and thus activates *ctx* gene expression (DiRita and Mekalanos 1991). It is interesting and intriguing that the hypertoxigenic strain 569B is naturally deleted in 1.2-kb toxS sequence (Miller et al. 1989). A third regulatory gene called toxT has been cloned (DiRita et al. 1991) from 569B which might be activating transcription of *ctx* and possibly be responsible for hypertoxigenicity of 569B. There is also evidence that expression of toxR gene may be regulated by heat shock response (Parsot and Mekalanos 1990). A gene has been cloned from *V. cholerae* whose predicted amino acid sequence is strikingly similar to the *E. coli* heat shock protein HtpG. The gene is located immediately upstream of toxR. Significant drop in CT production by *V. cholerae* conjugative plasmid P was reported (Sinha and Srivastava 1978a; Khan et al. 1985). However, in a recent paper, the effect of P plasmid was attributed to poor colonization (Bartawsky et al. 1990).

Types of Toxin

The toxins potentially involved in diarrhoeagenicity of *V. cholerae* have been reported. The first toxin is encoded by ctxAB genes and has been the subject of extensive studies and reviews. The presence of a second toxin in *V. cholerae* has been reported (Sanyal et al. 1983). This toxin was produced by *ctx* gene negative *V. cholerae* O1 strains including 1074–78 (environmental isolate) and caused secretory diarrhoeal response in experimental animals. No antigenic relationship was found between the second cholera toxin and the CT produced by ctxAB genes (Saha and Sanyal 1990). A third toxin, distinct from CT, has been identified (Fasano et al. 1991) which alters the morphology of intracellular epithelial tight junctions. The gene encoding the third toxin, designated *zot* and located immediately upstream of ctxAB, has been cloned and sequenced (Baudry et al. 1992). The *zot* could potentially encode a 44.5-kDa protein. Another potential enterotoxin has been identified (Trucksis et al. 1993) called ace toxin (accessory cholera toxin). The gene encoding ace toxin is located immediately upstream of *zot* with a predicted molecular weight of 11.3 kDa. It is amazing that the conserved element of 7 kb together with the 4.3-kb core ctxAB element might be a repertoire for a number of toxin genes involved in diarrhoeal responses. Reports have been made of a heat-stable enterotoxin and Shiga-like toxin by *V. cholerae* O1 (Takeda et al. 1991; O'Brien et al. 1984).

Cholera Vaccines

An ideal vaccine against *V. cholerae* O1 should comprise a single dose offering a high degree of long-term protection against all biotypes and serotypes to the children and people of endemic areas. Besides, it should be economical so that the poor people, who are mostly the victims of cholera, can afford the vaccine.

Although both *V. cholerae* O1 and O139 elaborate similar cholera toxin (Hall et al. 1994), they differ in the composition of their surface components as *V. cholerae* O139 produces a polysaccharide capsule (Johnson et al. 1994). Thus, previous exposure to *V. cholerae* O1 does not confer immunity to attacks by *V. cholerae* O139 (Albert et al. 1993; Bhattacharya et al. 1994). Several cholera vaccines against *V. cholerae* O1 have been developed but only few have undergone field trials, and no vaccine against *V. cholerae* O139 has yet been subjected to a large-scale field trial to test its protective efficacy. Various vaccine trials involving *V. cholerae* O1 and effort to develop *V. cholerae* O139 vaccine will be discussed.

Ever since identification in 1883 of *V. cholerae* O1 as the aetiological agent of cholera, parenteral cholera vaccines were used in humans and interestingly, the concept of mass oral immunization with live *V. cholerae* O1 through drinking water supplies was introduced. It was found that *V. cholerae* O1, killed or live, produced similar serological response in the host. Therefore, killed cholera vaccines, which are easy to prepare and maintain, have been used parenterally since that time. Several controlled trials with parenteral whole-cell vaccines were carried out in the 1960s in a number of cholera-endemic countries such as India, Bangladesh (then East Pakistan) and the Philippines.

Non-Living Cholera Vaccines: Parenteral Whole-cell Vaccines with Adjuvants

A classical bivalent whole-cell (WC) vaccine containing Ogawa and Inaba serotypes adsorbed onto aluminium phosphate was subjected to a large-scale, double-blinded trial in Kolkata in 1975 (Pal et al. 1980). The vaccine offered 100 % protection to children under 5 years of age for 6 months, 89 % for 12 months and 92 % for 18 months. The overall protection rate for all age groups over a surveillance period of 1 and 2 years was 62 and 53 %, respectively. The vaccine produced mild reaction for 48–72 h. A conceptually similar vaccine was evaluated in a large-scale, double-blinded trial in Indonesia (Saroso et al. 1978), where the vaccine offered 88 % protection for 6 months and 68 % protection for around 1 year among children aged 1–4 years. No serious side effects were reported.

Non-Living Cholera Vaccines: Oral Vaccination of a Combination of *V. cholerae* O1 Components

The Killed Whole-cell-B Subunit Vaccine Trial in Bangladesh

Encouraged by results in rabbits that the parenteral administration of a combination of *V. cholerae* O1 antigens such as LPS and cholera toxin or the B subunit of cholera toxin (BS) induced 100-fold higher protection against challenge with live cholera vibrios than did vaccination with either of the two antigens alone (Svennerholm and Holmgren 1976), a vaccine comprising a combination of WC and BS (WC-BS) was prepared and tested in a large-scale field trial in Bangladesh in 1985 along with controls (Clemens et al. 1986, 1987, 1988, 1990). The WC-BS vaccine comprised killed cells from three strains Inaba (classical and El Tor) and Ogawa classical and 1 mg of BS. Each participant had received orally three doses of the vaccine with a total of 3×10^{11} cells. However, prior to the field trial, 11 healthy male American adults were immunized orally with three doses of the WC-BS vaccine and after 1 month, they were challenged with live *V. cholerae* O1 (Levine et al. 1983). Although they had received twice the amount of killed cells and five times more BS than those given to the Bangladeshi trial participants, 4 adults of 11 contracted cholera indicating a moderate protective efficiency of 64 % after 1 month.

The Bangladesh trial was massive and started with immense hope, but the results of the trial were disappointing. The initiation of the trial was followed by a pre-epidemic period (about 6 months) during which cholera incidence was low and the WC-BS offered to all age groups an efficacy of 79 and 85 % after 4 and 6 months, respectively (Clemens et al. 1986, Clemens et al. 1988). With the arrival of a cholera epidemic, however, protection of WC-BS fell to 62 and 58 % at 8 and 12 months, respectively. Then arrived the epidemic period during which the WC-BS vaccine had a negative contribution to cholera protection as 20-25 % more children (3–6 years) in the WC-BS group contracted cholera than those in the other groups including the placebo control group. The protective efficacy for WC-BS for all age groups during the third year was 17 % and that for WC was 43 % (Clemens et al. 1990). Hardly any protection (16 and 28 % for WC-BS and WC, respectively) was observed during the fourth year (van Loon et al. 1996). The vaccines enriched in

V. cholerae O1 of classical biotype offered less protection against El Tor infections as protection efficacy fell to 10 and 20 % for WC-BS and WC, respectively, at the end of 11 months. The WC-BS vaccine was practically ineffective in children. There are logistic problems as vaccine delivery is inconvenient, requiring stomach acid neutralization which can be problematic for those suffering from stomach ailments. Short-term efficacy, very high cost, reduced protection against El Tor cholera and the requirement for multiple doses make it an impractical vaccine for cholera control.

The Whole-cell-Recombinant BS (WC-rBS) Vaccine Trial in Peru

Since 1992, the BS prepared by recombinant DNA technology (rBS) has been incorporated into the oral combination vaccine. The WC-rBS was tested for reactogenicity and immunogenicity in Peru where it caused adverse reactions such as diarrhoea, stomach cramps, stomach gurgling, vomiting and fever (Sanchez et al. 1995). Two doses of WC-rBS given orally 2 weeks apart to 1426 adult male military recruits produced protective efficacy of 86 and 42 % against symptomatic and asymptomatic cholera infection, respectively, during a surveillance period of 4–6 months (Sanchez et al. 1994). Subsequently, the vaccine was tested in a larger field trial where participants received two spaced doses. Although the vaccine raised both vibriocidal and anticholera toxin titres, it offered hardly any protection against cholera (Taylor et al. 2000). These titres dropped to baseline levels within 1 year, increasing only after an additional booster dose. Recipients of the third booster dose 10 months after the second one got an overall protection of 61 %. The trial monitored for a period of 2 years indicated that a two-dose vaccination regime was inadequate in offering protection against cholera. Vaccine conferred reduced protection against the epidemic El Tor biotype.

The WC Vaccine Trial in Vietnam

A WC vaccine comprising four inactivated strains from both the serotypes and biotypes; one of them expressing the fimbrial colonization antigens such as mannose-sensitive haemagglutinin and TCP was subjected to an oral field trial in Vietnam during 1992–1993 (Trach et al. 1997). Determination of early vaccine efficacy was not possible due to the absence of cholera cases during the first 8 months. However, a cholera epidemic was observed 8 months after vaccination when the WC vaccine showed a reduction in the incidences of cholera in vaccinated versus unvaccinated households. Protection was about 66–68 %. The trial was neither double-blinded nor placebo-controlled and monitored for a period of 1 year after vaccination which did not provide enough information on long-term immunity.

Microencapsulated WC Vaccine

Vaccine development using controlled release technology is based on the slow and prolonged release of antigens encapsulated in biodegradable polymers such as polylactide-co-glycolides and polylactide/polyethylene glycol. The microencapsulation of antigens produces adjuvant effect where particle size plays an important role in determining immunogenicity. Killed WC encapsulated into microparticles and orally administered to mice produced higher levels of antibody in serum compared to those obtained by killed WC alone (Yeh and Chiang 2004). Mice immunized with encapsulated WC showed a high degree of protection when challenged intraperitoneally with live *V. cholerae* O1 about 8 weeks after immunization. Polylactide-co-glycolide-encapsulated *V cholerae* O1 protein antigens delivered to rabbits by the subcutaneous route offered better protection as determined by the ileal loop assay than that obtained by the oral route (Chandrasekhar et al. 1994).

Protein–Polysaccharide Conjugates of *V. cholerae* O1

Although *V. cholerae* 01 LPS is immunogenic in man, the lipid A part of LPS is pyrogenic and contributes to endotoxic properties that may cause adverse reactions among the recipients of the injectable whole-cell vaccines (Kabir 1982). A non-pyrogenic bivalent conjugate vaccine was prepared by chemically coupling a mixture of alkali-treated LPS from Ogawa and Inaba serotypes to the cell surface proteins of *V. cholerae* O1 (Kabir 1987a, b) which in rabbits induced high-titre antibodies that showed agglutinating and vibriocidal activities towards *V. cholerae* 01 strains of both biotypes and serotypes. Monovalent Inaba vaccines containing varying amount of cholera toxin chemically conjugated to base-treated Inaba LPS have also been prepared (Gupta et al. 1992, 1998).

Proteinases as Vaccine Candidate

Apart from cholera toxin, *V. cholerae* O1 strains secrete several biologically active products such as neuraminidase, mucinase, collagenase, lipase and proteinase (Shrivastava 1964). A recent study has shown that guinea pigs immunized with a combination of enzymes such as mucinases, proteinases and neuraminidase could be protected by the oral route when subjected to a live *V. cholerae* challenge (Stewat-Tull et al. 2004). This observation indicates the possibility of using such enzymes as acellular cholera vaccine candidates.

Live oral Vaccines

Several laboratories have been working to develop attenuated live *V cholerae* O1 strains as oral vaccine candidates. An ideal attenuated oral live vaccine should colonize the intestine without causing diarrhoea. It should multiply to simulate a natural cholera infection presenting all of the *V. cholerae* O1 antigens to the host. A single-dose live vaccine should evoke a strong immune response similar to that of natural cholera infection through a sustained presentation of antigens. This would be an economic way of mass immunization in contrast to the requirement for multiple doses of the expensive WC-BS vaccine used in the field trials of Peru and Bangladesh (Clemens et al. 1986; Taylor et al. 2000). However, there is a note of caution in the development of attenuated strains. As the cholera toxin genes are carried on a lysogenic phage (Waldor and Mekalanos 1996), there is a possibility that an attenuated vaccine strain could reacquire cholera toxin genes because of phage infection and become virulent.

Early studies on human volunteers with live naturally occurring non-toxigenic strains did not produce satisfactory results (Sanyal and Mukerjee 1969; Cash et al. 1974). By chemical mutagenesis, strains attenuated in virulence were isolated. Dwarf colony slow-growing strain (Bhaskaran and Sinha 1967) and M13 and Texas Star-SR were evaluated in animal models and in limited volunteers (Finkelstein et al. 1974; Honda and Finkelstein 1979). While Texas Star-SR was phenotypically A^{-} B^{-}, it induced diarrhoea in 25 % of the recipients, making it unsuitable as a vaccine candidate (Levine et al., 1984). Before the advent of recombinant DNA, in our laboratory during 1976–1978, a novel class of attenuated vaccine strain was obtained by transfer of P and V plasmids into pathogenic strains of *V. cholerae* (Sinha and Srivastava 1979). This was achieved by introducing P and V plasmids into virulent strains (Sinha and Srivastava 1978a). It was shown that cells harbouring these plasmids became attenuated and did not cause experimental cholera. The loss of virulence was found to be due to suppressed toxin production (Khan et al. 1985). Such attenuated strains appeared attractive candidate vaccine strains if it could be shown that the plasmids were stable and immunogenicity of plasmid harbouring strains was not altered. The stability of the plasmid was confirmed in vitro and in vivo and excellent protection from challenge was observed in experimental cholera (Sinha and Srivastava 1978b, 1979).

After cloning and sequencing of cholera toxin, a live attenuated strain (CVD103-HgR) was engineered by recombinant DNA technology. The strain was derived from the classical Inaba strain 569B by deleting 94 % of the enzymatically active A subunit of CT conserving the B subunit gene and inserting a mercury-resistance marker into the cryptic haemolysin (hly) locus of the gene for its differentiation from wild strains. Although the strain produced a significant increase in vibriocidal responses in most of the volunteers in the USA and displayed acceptable reactogenicity, it was a poor colonizer (Levine et al. 1988). In challenge studies on American volunteers, a single dose of the vaccine containing 5×10^6 c.f.u. demonstrated a 1-month protective efficacy of around 85 % against challenge with *V. cholerae* O1 of either serotype and 67 % against El Tor Inaba. However, this vaccine dose was poorly immunogenic among the socially less-privileged people of Peru (Gotuzzo et al. 1993), Indonesia (Suharyono et al. 1992) and Thailand (Suarehawarataha et al. 1992).

A single dose of CVD103-HgR was administered to participants in a double-blinded, randomized and placebo-controlled field trial in Indonesia in 1992 (Richie et al. 2000). Very few cholera cases were observed within 6 months of vaccination, making assessment of any short-term efficacy of the vaccine difficult. The serum vibriocidal response among the vaccinees was modest and observed in about two-thirds of the recipients. After 1 year, the vaccine's protective efficacy was 18 % for all age groups. The trial was monitored over a period of 4.5 years and efficacy was about 13 % demonstrating the vaccine's failure to confer protection in a cholera-endemic area.

Live Bivalent (Ogawa and Inaba) Attenuated Strains

CVD111 is an attenuated strain derived from a wild-type Ogawa El Tor strain by removing the virulence cassette containing the toxin genes (*ctx, zot, cep* and *ace*) and inserting the binding unit of cholera toxin (ctxB) and that for mercury resistance into the haemolysin locus (hlyA). A mixture of CVD111 and CVD 103-HgR representing both the biotypes and the serotypes of *V. cholerae* O1 was orally administered to a number of US military personnel who showed elevated levels of serum vibriocidal responses against both Ogawa and lnaba serotypes (Taylor et al. 1999). Although CVD111 is a good intestinal colonizer, it caused diarrhoea in a significant number of volunteers thus limiting the use of this combination vaccine without further attenuation of CVD111.

The Bivalent Cholera–Typhoid Strain

Different *V cholerae* O1 antigens have been cloned into an attenuated strain of *Salmonella typhimurium* Ty2la in order to construct bivalent, oral cholera–typhoid vaccines. One such hybrid strain expressing *V. cholerae* O1 Inaba LPS O antigen was subjected to human challenge studies in USA (Tacket et al. 1990). It elicited only a modest immune response and displayed a moderate level of protective efficacy suggesting the need for further work in the better expression of *V cholerae* O1 LPS in vivo.

The Strain Peru-l5

Peru-15 is a motility-defective attenuated strain derived from *V. cholerae* O1 El Tor Inaba by a series of genetic deletions and insertions in which the entire cholera toxin genetic element containing the genes for CT and other virulence determinants along with the attRS1 insertion like sequences were deleted (Kenner et al. 1995). The gene encoding the cholera toxin B subunit (ctxB) was fused with a heat shock promoter and inserted into the recA gene thereby making it unable to integrate exogenous DNA. The strain was subjected to a human challenge study in the USA in which it was orally administered to a few volunteers who got side reactions such as headache and in a few cases mild to severe abdominal cramps including diarrhoea (Cohen et al. 2002). While most showed vibriocidal response, anticholera toxin immune response was detected only in a few volunteers. When challenged with an El Tor Inaba strain within 3 months of immunization, a high degree of protection (96 %) was observed. However, this study had some limitations because 42 % of the placebo recipients did not get any diarrhoea and the trial period was of very short duration.

The Indian Vaccine VA1.3

An oral cholera vaccine strain was developed from a non-toxigenic clinical El Tor Inaba isolate which possessed toxR and tcpA genes but was devoid of the genes for CT and other virulence determinants (Thungapathre et al. 1999). Using recombinant DNA technology, the ctxB gene of *V. cholerae* was introduced into the cryptic haemolysin locus of the strain. The resulting strain, VA1.3, did not induce any diarrhoea when administered orally to rabbits and offered good protection against challenges with homologous and heterologous *V. cholerae* O1 strains performed 3 weeks after immunization. The strain was, however, not devoid of the attachment site (attRSl) of the CTX^+, thus raising a fear of its conversion to a toxigenic state. The strain was taken to clinical studies, however, it is wondered if the strain would be able to colonize the primed intestinal mucosa in Indian subcontinent.

The Cuban Strain *V. cholerae* 638

V. cholerae 638 (El Tor, Ogawa) has been developed in Cuba by deleting the CTXΦ prophage from the chromosome of a toxigenic clinical isolate. Subsequently, the hemagglutinin/protease (*hap*) gene was inactivated by the insertion of celA encoding *Clostridium thermocellum* endoglucanase A (Benitez et al. 1999). The strain was a good colonizer of the human bowel, and after 14 days, vaccinees showed a significant increase in vibriocidal activity towards Ogawa serotype. However, it induced adverse effects such as diarrhoea (10 %), abdominal cramps (30 %) and gurgling (33 %) among the vaccinees. The human sera inhibited significantly fluid accumulation in ligated rabbit intestine against the homologous strain and less significantly against the heterologous strains (Perez et al. 2000).

Vaccines Against *V. cholerae* O139

Attempts to develop vaccines against *V. cholerae* O139 are ongoing in three different directions. However, no vaccine to prevent *V. cholerae* O139 infection has yet been subjected to a field trial.

Live oral Whole-cell Vaccine

Two live attenuated *V cholerae* O139 strains, Bengal-15 and CVD112, have been subjected to human challenge studies in USA. Bengal-15 is a spontaneous non-motile derivative of an attenuated *V. cholerae* O139 strain constructed from an Indian epidemic strain MO10 (Coster et al. 1995). The strain carries deletions in four specific virulence determinants (*ctxA, ace, zot* and *cep*) and in other factors involved in site-specific and homologous recombination (RS1, attRSl and recA). A few volunteers in USA immunized orally with this strain did not develop diarrhoea. The protective efficacy of this vaccine one month after challenge with live homologous epidemic strain MO10 was 83 %. CVD112 was constructed from the wild-type strain *V. cholerae* O139 by recombinant DNA techniques similar to those used in the construction of the strain CVD111 (Tacket et al. 1995). A few volunteers were immunized orally with two different doses (10^6 and 10^8 c.f.u.) of this strain, the second dose inducing diarrhoea in some of the vaccinees. A short-term (5-week) evaluation of the vaccine against live oral challenge with the parent wild-type strain was 84 %.

Killed Oral Whole-cell Vaccine

Two doses of a bivalent vaccine comprising large amounts of killed *V. cholerae* O1 and O139 cells were administered orally to a small number of urban Vietnamese adults and children (Trach et al. 2002). Two weeks after the second immunization, around 31 % of the adults and 56 % of the children showed a significant rise in vibriocidal antibody titres against a partially encapsulated O139 strain, the immune response among the vaccinees to a capsule-deficient mutant strain MO10-T4 being much lower. Unlike Bengal-15 and CVD112, the vaccine's protective efficacy against cholera based in a field trial has not yet been reported.

Semi-Synthetic Protein–Polysaccharide Conjugate Vaccines

Three different laboratories have prepared conjugate vaccines by chemically coupling capsular polysaccharides (CPS) or the polysaccharide moiety of the LPS (pmLPS) of *V. cholerae* O139 with protein carriers such as tetanus toxoid (TT) and recombinant diphtheria toxin (rDT). Multiple immunization of the pmLPS-TT conjugate elicited IgG antibodies in mice which recognized both CPS and LPS, displayed vibriocidal activities, and offered significant protection in suckling mice against *V. cholerae* O139 infection (Boutonnie et al. 2001). A CPS-TT conjugate also induced protection in rabbits against experimental *V. cholerae* O139 infection as determined by the rabbit ileal loop assay. A CPS-rDT conjugate immunization in mice produced antibodies whose vibriocidal activities were in part mediated by anti-CPS IgG (Johnson et al. 1995).

Remarks

None of the vaccines met the requirements of an ideal cholera vaccine. Among all the vaccines evaluated in human volunteers, the bivalent whole-cell (WC) vaccine containing Ogawa and Inaba serotypes adsorbed onto aluminium phosphate has yielded by far the best results.

The importance of the route of vaccine administration in order to obtain maximum mucosal response in the gut has not been adequately addressed. The oral route of vaccine delivery was emphasized because colonization of the intestine by *V cholerae* was supposed to evoke mucosal immune response leading to the secretion of secretory IgA. However, in an endemic area where repeated, exposure to cholera bacteria is a common event the desired colonization of a live vaccine strain is doubtful. In recent years, a new concept has emerged that priming of one mucosal surface triggers immune response at most distal mucosa in the body. Delivery of the vaccine consisting of non-reactogenic subunit protective antigens from all serotypes and biotypes through nasal mucosa should be considered.

References

Albert MJ, Siddique AK, lslam MS, Faruque AS, Ansaruzzaman M, Faruque SM et al (1993) Large outbreak of clinical cholera due to *Vibrio cholerae* non-Ol in Bangladesh. Lancet 341:704

Attridge SR, Rowley D (1983) The role of flagellum in adherence of *Vibrio cholerae*. J Infect Dis 147:864–872

Bartowsky EJ, Attridge SR, Thomas CJ, Mayrhofer G, Manning PA (1990) Role of the P plasmid in attenuation of *Vibrio cholerae* O1. Infect Immun 58:3129

Barua D, Burrows W (eds) (1974) Cholera WB Saunders Co. Philadelphia, USA

Baudry B, Fasano A, Ketley J, Kaper JB (1992) Cloning of a gene encoding a new toxin produced by *Vibrio cholerae*: *zot*. Infect Immun 60:428–434

Benitez JA, Garcia L, Silva A, Garcia H, Fando R, Cedre B et al. (1999) Preliminary assessment of the safety and immunogenicity of a new CTXPhi-negative, hemagglutinin/protease-defective El Tor strain as a cholera vaccine candidate. lnfect lmmun 67:539–545

Bhaskaran K, Sinha VB (1967) Attenuation of virulence in *Vibrio cholerae*. J Hyg (camb) 65:135–148

Bhattacharjee JW, Srivastava BS (1978) Mannose-sensitive haemagglutinins in adherence of *Vibrio cholerae* eltor to intestine. J Gen Microbiol 107:407–410

Bhattacharjee JW, Srivastava BS (1979) Adherence of the wild-type and mutant strains of *Vibrio cholerae* to normal and immune intestinal tissue. Bull WHO 57:123–128

Bhattacharya SK, Goswami C, Bhattacharya MK, Dutta D, Deb A, Deb M et al. (1994) Epidemic of *Vibrio cholerae* O139 in Calcutta. Indian J Med Res 100:213–216

Boutonnier A, Villeneuve S, Nato F, Dassy B, Fournier JM (2001) Preparation, immunogenicity, and protective efficacy, in a murine model, of a conjugate vaccine composed of the polysaccharide moiety of the lipopolysaccharide of *Vibrio cholerae* O139 bound to tetanus toxoid. Infect Immun 69:3488–3493

Burnet FM (1948) The mucinase of *Vibrio cholerae*. Aust J Exp Biol Med 27:245–252

Cash RA, Music SI, Libonati JP, Schwartz AR, Hornick RB (1974) Live oral cholera vaccine: evaluation of the clinical effectiveness of two strains in human. Infect Immun 10:762–764

Cassel D, Pfeuffer T (1978) Mechanism of cholera toxin action: Covalent modification of the guanyl nucleotide- binding protein of the adenylate cyclase system. Proc Natl Acad Sci USA 75:2669

Chandrasekhar U, Sinha S, Bhagat HR, Sinha VB, Srivastava BS (1994) Comparative efficacy of biodegradable liposomes and microspheres as carriers for delivery of *Vibrio cholerae* antigens in the intestine. Vaccine 12:1348–1384

Chitnis DS, Sharma KD, Kamat RS (1982) Role of somatic antigen of *Vibrio cholerae* in adhesion to intestinal mucosa. J Med Microbiol 5:53–61

Clemens JD, Harris JR, Sack DA (1988) Field trial of oral cholera vaccines in Bangladesh: results of one year of follow up. J Infect Dis 15:860–869

Clemens JD, Sack DA, Harris JR, Van Loon F, Chakraborty J, Ahmed F et al (1990) Field trial of oral cholera vaccines in Bangladesh: results from three-year follow-up. Lancet 335(8684):270–273

Clemens JD, Sack DA, Harris JR, Chakraborty J, Khan MR, Stanton BF et al (1986) Field trial of oral cholera vaccines in Bangladesh. Lancet 2:124–127

Clemens JD, Stanton BF, Chakraborty J, Sack DA, Khan MR, Huda S et al. (1987) B subunit-whole cell and whole cell-only oral vaccines against cholera: studies on reactogenicity and Immunogenicity J Infect Dis 155:79–85

Cohen MB, Giannella RA, Bean J, Taylor DN, Parker S, Hoeper A et al (2002) Randomized, controlled human challenge study of the safety, immunogenicity, and protective efficacy of a single dose of Peru-15.a live attenuated oral cholera vaccine. Infect Immun 70:1965–1970

Coster TS, Killeen KP, Waldor MK, Beattie DT, Spriggs DR, Kenner JR et al (1995) Immunogenicity, and efficacy of live attenuated *Vibrio cholerae* O139 vaccine prototype. Lancet 345(8955):949–952

Cuatrecasas P (1973) Gangliosides and membrane receptors for cholera toxin. Biochemistry 12:3558–3566

De SN, Chatterjee DN (1953) An experimental study of the mechanism of action of *Vibrio cholerae* on the intestinal mucous membrane. J Path Bacteriol 66:559–562

De SN (1959) Enterotoxicity of bacteria-free culture filtrate of *Vibrio cholerae*. Nature 183:1533–1534

DiRita VJ, Mekalanos JJ (1991) Periplasmic interaction between two membrane regulatory proteins, ToxR and ToxS results in signal transduction and transcriptional activation. Cell 64:29

DiRita VJ, Parsot C, Jander G, Mekalanos JJ (1991) Regulatory cascade controls virulence in *Vibrio cholerae*. Proc Natl Acad Sci USA 88:5403

Dutta NK, Panse MV, Kulkarni DR (1959) Role of cholera toxin in experimental cholera; J Bacteriol 78:594–595

Ehara M, Ishibashi M, Ichinose Y, Iwanaga M, Shimotori S, Naito T (1987) Purification and partial characterization of fimbriae of *Vibrio cholerae* O1. Vaccine 5:283–288

Fasano A, Baudry B, Pumplin DW, Wasserman SS, Tall BD, Ketley JM, Kaper JB (1991) *Vibrio cholerae* produces a second enterotoxin, which affects intestinal tight junctions. Proc Natl Acad Sci USA 88: 5242

Finkelstein RA, Vasil ML, Holmes RK (1974) Studies on toxinogenesis in *Vibrio cholerae* 1 Isolation of mutants with altered toxinogenicity. J Infect Dis 129:117–123

Finkelstein RA, Hanne LF (1982) Purification and characterization of soluble haemagglutinin (cholera lectin) produced by *Vibrio cholerae*. Infect Immun 36:1199–1208

Finkelstein RA, Boesman-Finkelstein M, Holt P (1983) *Vibrio cholerae* haemagglutinin/lectin/protease hydrolyses fibronectin and ovomucin : F.M. Bernet revisited. Proc Natl Acad Sci USA 80:1092–1095

Freter R (1955) The serologic character of cholera vibrio mucinase. J Infect Dis 97:238–245

Freter R, Allweiss B, O'Brien PCM, Halstead SA, Macsal MS (1981) Role of chemotaxis in association of motile bacteria with intestinal mucosa: in vitro studies. Infect Immun 34:241–249

Freter R, O'Brien PCM (1981) Role of chemotaxis in association of motile bacteria with intestinal mucosa : chemotactic responses of *Vibrio cholerae* and description of motile non chemotactic mutants. Infect Immun 34:215–221

Gennaro ML, Greenway PJ (1983) Nucleotide sequences within the cholera toxin operon. Nucleic Acids Res 11:3855

Gennaro ML, Greenway PJ, Broadbent DA (1982) The expression of biologically active cholera toxin in *Escherichia coli*. Nucleic Acids Res 79:4883

Gill DM (1976) The arrangement of subunits in cholera toxin. Biochemistry 15:1242–1246

Gill DM, Rappaport RS (1979) Origin of the enzymatically active A1 fragment of cholera toxin. J Infect Dis 139:674–680

Gill DM, Meren R (1978) ADP-ribosylation of membrane proteins catalysed by cholera toxin: basis of the activation of adenylate cyclase. Proc Natl Acad Sci USA. 75: 3050

Goldberg I, Mekalanos JJ (1986) Effect of a recA mutation on cholera toxin gene amplification and deletion events. J Bacteriol 165:723

Gotuzzo E, Butron B, Seas C, Penny M, Ruiz R, Losonsky G et al (1993) Safety, immunogenicity and excretion pattern of single dose live oral cholera vaccine CVD 103-HgR in Peruvian adults of high and low socioeconomic levels. Infect Immun 61:3994–3997

Guentzel MN, Berry LJ (1975) Motility as a virulence factor for *Vibrio cholerae*. Infect Immun 11:90–97

Gupta RK, Taylor DN, Bryla DA, Robbins JB, Szu SC (1998) Phase 1 evaluation of *Vibrio cholerae* O1, serotype lnaba, polysaccharide-cholera toxin conjugates in adult volunteers. Infect lmmun 66:3095–3099

Gupta RK, Szu SC, Finkelstein RA, Robbins JB (1992) Synthesis, characterization and some immunological properties of conjugates composed of the detoxified lipopolysaccharide of *Vibrio cholerae* Ol serotype lnaba bound to cholera toxin. lnfect lmmun 60:3201–3208

Gustafsson B, Holm T (1985) Rapid detection of *Vibrio cholerae* O1 by motility inhibition and immunofluorescence with monoclonal antibodies. Eur J Clin Microbiol 4:291–294

Hall RH, Khambaty FM, Kothary MH, Keasler SP, Tall BD (1994) *Vibrio cholerae* non-Ol serogroup associated with cholera gravis genetically and physiologically resembles O1 El Tor cholera strains. lnfect lmmun. 62: 3859–3863

Holmes RK, Vasil ML, Finkelstein RA (1975) Studies on toxinogenesis in *Vibrio cholerae* III. Characterization of nontoxigenic mutants in vitro and in experimental animals. J Clin Invest 55:551

Holmgren J, Svennerhilm AM, Lindblat M (1983) Receptor like glycocompounds in human milk that inhibit classical and el tor *Vibrio cholerae* cell adherence (agglutination). Infect Immun 39:147–154

Honda T, Finkelstein RA (1979) Selection and characteristics of a *Vibrio cholerae* mutant lacking the A (A DP-ribosylating) portion of the cholera enterotoxin. Proc Natl A cad Sci USA 76:2052–2056

Hranitzky KW, Mulholland A, Larson AD, Eubanks ER, Hart LT (1980) Characterization of flagellar sheath protein of *Vibrio cholerae*. Infect Immun 27:597–603

Jacob A, Sinha VB, Sahib MK, Srivastava R, Kaper JB, Srivastava BS (1993) Identification of a 33 kDa antigen associated with an adhesive and colonizing strain of *Vibrio cholerae* EI Tor and its role in protection. Vaccine 11:376–382

Johnson JA, Joseph A, Morris JG (1995) Capsular polysaccharide protein conjugate vaccines against *Vibrio cholerae* O139 Bengal. Bull Inst Pasteur 93:285–290

Johnson JA, Salles CA, Panigrahi P, Albert MJ, Wright AC, Johnson RJ et al (1994) *Vibrio cholerae* O139 synonym Bengal is closely related to *Vibrio cholerae* El Tor but has important differences. Infect Immun 62:2108–2110

Jonson G, Holmgren J, Svennerholm AM (1991) Identification of a mannose-binding pilus on Vibrio cholerae El Tor Microb Pathog 11:433–441

Kabir S (1982) Characterization of the lipopolysaccharide from *Vibrio cholerae* 395 (Ogawa). Infect lmmun 38:1263–1272

Kabir S (1983) Immunochemical properties of the major outer membrane proteins of *Vibrio cholerae*. Infect Immun 39:452–455

Kabir S (1987a) Preparation and immunogenicity of a bivalent cell surface protein-polysaccharide conjugate of *Vibrio cholerae*. J Med Microbiol 23:9–18

Kabir S (1987b) Immunological responses in rabbits to various somatic and secreted antigens of *Vibrio cholerae* after intra-duodenal inoculation. J Med Microbiol 24:29–40

Kabir S, Showkat A (1983) Characterization of surface properties of *Vibrio cholerae*. Infect Immun 39:1048–1058

Kaper JB, Lockman H, Baldini MM, Levine MM (1984) A recombinant nontoxigenic *Vibrio cholerae* strains as attenuated cholera vaccine. Nature 308:655–658

Kaper JB Levine MM (1981) Cloned cholera enterotoxin genes in study and prevention of cholera. Lancet ii 1162–1163

Kaper JB, Srivastava BS (1992) Genetics of cholera toxin; Indian J Med. Res A 95:163–168

Kenner JR, Coster TS, Taylor DN, Trofa AF, Barrera-Oro M, Hyman T et al (1995) Peru-15, an improved live attenuated oral vaccine candidate for *Vibrio cholerae* Ol. J lnfect Dis 172:1126–1129

Khan AA, Srivastava R, Sinha VB, Srivastava BS (1985) Regulation of toxin biosynthesis by plasmids in *Vibrio cholerae*. J Gen Microbiol 131:2653–2657

Levine MM, Black RE, Clements ML, Lanata C, Sears S, Honda T Young CR Finkelstein RA (1984) Evaluation in humans of attenuated *Vibrio cholerae* El Tor Ogawa strain Texas Star-SR as a live oral vaccine. lnfect lmmun 43:515–522

Levine MM, Nalin DR, Craig JP, Hoover D, Bergquist ET, Waterman D et al (1979) Immunity to cholera in man: relative role of antibacterial versus antitoxin immunity. Trans R Soc Trop Med Hyg 73:3–9

Levine MM, Kaper JB, Black RE, Clements ML (1983) New knowledge on pathogenesis of bacterial enteric infections as applied to vaccine development. Microbiol Rev 47:510–530

Levine MM, Kaper JB., Herrington D, Ketley Losonsky C, Tacket CO, Tall B, Cryz S (1988) Safety, immunogenicity, and efficacy of recombinant live oral cholera vaccines, CVD 103 and CVD 103-HgR. Lancet; 2 (8609): 467–470

Lockman H, Kaper JB (1983) Nucleotide sequence analysis of the A2 and B subunits of *Vibrio cholerae* enterotoxin. J Biol Chem 158:13722

Lockman H, Gallen JE, Kaper JB (1984) *Vibrio cholerae* enterotoxin genes: nucleotide sequence analysis of DNA encoding ADP-ribosyltransferase. J Bacteriol 159:1086

Lospalluto JJ, Finkelstein RA (1972) Chemical and physical properties of cholera exo-enterotoxin (choleragen) and its spontaneously formed toxoid (choleragenoid). Biochim Biophys Acta 257:158

Majumdar AS, Dutta P, Dutta D, Ghose AC (1981) Antibacterial and antitoxin responses in the serum and milk of cholera patients. Infect Immun 32:1–8

Mekalanos JJ, Murphy JR (1980) Regulation of cholera toxin production in *Vibrio cholerae* genetic analysis of phenotypic instability in hypertoxigenic mutants. J Bacteriol 141:570

Mekalanos JJ (1983) Duplication and amplification of toxin genes in *Vibrio cholerae*. CeIl 35:253

Mekalanos JJ, Collier RJ, Romig WR (1979a) Enzymatic activity of cholera toxin. II. Relationships of proteolytic processing, disulfide bond reduction and subunit composition. J Biol Chem 254:5855

Mekalanos JJ, Sublett RD, Romig WR (1979b) Genetic mapping of toxin regulatory mutations in *Vibrio cholerae*. J Bacteriol 139:859

Mekalanos JJ, Swaartz DJ, Pearson GDN, Harford N, Groyne F, de Wilde M (1983) Cholera toxin genes: nucleotide sequence deletion analysis and vaccine development. Nature 306:551

Miller VL, Mekalanos JJ (1985) Genetic analysis of the cholera toxin-positive regulatory gene toxR. J Bacteriol 163:580

Miller VL, Mekalanos JJ (1984) Synthesis of cholera toxin is positively regulated at the transcriptional level by toxR. Proc Natl Acad Sci USA 81:3471

Miller VL, DiRita VJ, Mekalanos JJ (1989) Identification of toxS, a regulatory gene whose product enhances ToxR-mediated activation of the cholera toxin promoter. J Bacteriol 171:1288

Miller VL, Taylor RK, Mekalanos JJ (1987) Cholera toxin transcriptional activator ToxR is a transmembrane DNA binding protein. Cell 48:271

Moseley SL, Falkow S (1980) Nucleotide sequence homology between the heat labile enterotoxin gene of Escherichia coli and *Vibrio cholerae* deoxyribonucleic acid. J Bacteriol 144:444

Nelson ET, Clements JD, Finkelstein RA (1976) *Vibrio cholerae* adherence and colonization in experimental cholera: electron microscopic studies. Infect Immun 14:527–547

O' Brien AD, Chen ME, Holmes RK, Kaper JB, Levine MM (1984) Environmental and human isolates of *Vibrio cholerae* and Vibrio parahaemolyticus produce a *Shigella dysenteriae* I (Shiga)-like cytotoxin. Lancet 77:764

Pal SC, Deb BC, Sen Gupta PG, De SP, Sircar BK, Sen D, Sikdar SN (1980) A controlled field trial of an aluminum phosphate adsorbed cholera vaccine in Calcutta. Bull World Health Org 58:741–745

Parsot C, Mekalanos JJ (1990) Expression of ToxR, the transcriptional activator of the virulence factors in *Vibrio cholerae*, is modulated by heat shock response. Ptoc Natl Acad Sci USA 87:9898

Pearson GDN, Mekalanos JJ (1982) Molecular cloning of *Vibrio cholerae* enterotoxin genes in E. coli K-12. Proc Natl Acad Sci USA 79:2976–2980

Perez JL, Garcia L, Talavera A, Oliva R, Valmaseda T, Ano G, Perez O, Sierra G (2000) Passive protection of serum from volunteers inoculated with attenuated strain 638 of *Vibrio cholerae* Ol in animal models. Vaccine 19:376–384

Peterson JW, Hejtmancik EK, Markel DE et al (1979) Antigenic specificity of neutralizing antibody to cholera toxin. Infect Immun 24:774–779

Peterson KM, Mekalanos JJ (1988) Characterization of *Vibrio cholerae* ToxR regulation: identification of novel genes involved in intestinal colonization. Infect Immun 56:2822

Ramamurthy T, Garg S, Sharma R, Bhattacharya, SK,Nair GB, Shimada T et al. (1993) Emergence of novel strain of *Vibrio cholerae* with epidemic potential in southern and eastern India. Lancet. 341:7O3–704

Richardson K, Parker CD (1985) Identification and. occurrence of *Vibrio cholerae* flagellar core proteins in isolated outer membrane. Infect Immun 47:647–679

Richardson K, Kaper JB, Levine MM (1989) Human immune response to *Vibrio cholerae* O1 whole cell and outer membrane antigens. Infect Immun 57:495–501

Richie EE, Punjabi NH, Sidharta YY, Peetosutan KK, Sukandar MM, Wasserman SS et al (2000) Efficacy trial of single-dose live oral cholera vaccine CVD103-HgR in North Jakarta, Indonesia, a cholera-endemic area. Vaccine 18:2399–2410

Saha S, Sanyal SC (1990) Immunobiological relationships of the enterotoxins produced by cholera toxin gene-positive and negative strains of *Vibrio cholerae* O1. J Med Microbiol 32:33

Sanchez JL, Hayashi KE, Kruger HF, Meza R, English CK, Vidal W et al (1995) lmmunological response to *Vibrio cholerae* O1 infection and an oral cholera vaccine among Peruvians. Trans R Soc Trop Med Hyg 89:542–545

Sanchez JL, Vasquez B, Begue RE, Meza R, Castellares G, Cabezas C et al (1994) Protective efficacy of oral whole-cell/recombinant-B-subunit cholera vaccine in Peruvian military recruits. Lancet 344(8932):1273–1276

Sanyal SC, Mukerjee S (1969) Live oral cholera vaccine: report of a trial on human volunteer subjects. Bull World Health Org 40:503–511

Sanyal SC, Alam K, Neogi PKB, Huq M1 Al-Mahmud KA (1983) A new cholera toxin. Lancet 11:337

Saroso JS, Bahrawi W, Witjaksono H, Budiarso RL, Brotowasisto Bencic Z et al (1978) A controlled field trial of plain and aluminium hydroxide-adsorbed cholera vaccines in Surabaya, Indonesia, during 1973–75. Bull World Health Org 56:619–627

Schrank GD, Verwey WF (1976) Distribution of cholera organisms in experimental *Vibrio cholerae* infections:

proposed mechanisms of pathogenesis and antibacterial immunity. Infect Immun 13:195–203

Sharma SK, Moe TS, Srivastava R, Chandra D, Srivastava BS (2011) Functional characterization of VC1929 of *Vibrio cholerae* El Tor: role in mannose-sensitive haemagglutination, virulence and utilization of sialic acid. Microbiology, UK 157:3180–3186

Shrivastava DL (1964) Enzymes in *Vibrio cholerae*. Indian J Med Res 52:801–808

Singh SN, Srivastava R, Sinha VB, Srivastava BS (1994) A 53 kDa protein of *Vibrio cholerae* classical strain 0395 involved in intestinal colonization. Microb Pathog 17:69–78

Sinha VB, Srivastava BS (1979) Attenuation of virulence by P and V plasmids in *Vibrio cholerae*: strains suitable for oral immunization. Bull WHO 57:643–647

Sinha VB, Srivastava BS (1978a) Suppression of pathogenicity by P and V plasmids in *Vibrio cholerae*. J Gen Microbial 104:251–255

Sinha VB, Srivastava BS (1978b) Plasmid induced loss of virulence in *Vibrio cholerae*. Nature 276:708–709

Sinha VB, Jacob A, Srivastava R, Kaper JB, Srivastava BS (1993) Identification of the flagellar antigens of *Vibrio cholerae* EI Tor and their role in protection. Vaccine 11:372–375

Smith HW, Halls S (1968) The transmissible nature of the genetic factor in *Escherichia coli* that controls enterotoxin production. J Gen Microbiol 58:319

Sperandio V, Giron JA, Silveira WD, Kaper JB (1995) The OmpU outer membrane protein a potential adherence factor of *Vibrio cholerae*. Infect Immun 63:4433–4438

Sporecke I, Castro D, Mekalanos JJ (1984) Genetic mapping of *Vibrio. cholerae* enterotoxin structural genes. J Bacteriol 157:253

Srivastava R, Srivastava BS (1980) Isolation of a nonadhesive mutant of *Vibrio cholerae* and chromosomal localization of the gene controlling mannose-sensitive adherence. J Gen Microbiol 117:275–278

Srivastava R, Khan AA, Srivastava BS (1985) Immunological detection of cloned antigenic genes of *Vibrio cholerae* in *Escherichia coli*. Gene 40:267

Srivastava R, Sinha VB, Srivastava BS (1980) Events in the pathogenesis of experimental cholera: role of bacterial adherence and multiplication. J Med Microbial 13:1–9

Srivastava R, Sinha VB, Srivastava BS (1979) Reevaluation of antibacterial and antitoxin immunities in experimental cholera. Indian J Med Res 70:369–373

Srivastava R, Sinha VB, Srivastava BS (1989) Chromosomal transfer and in vivo cloning of genes in *Vibrio cholerae* using RP4 mini-Mu. Gene 75:3855

Stewart-Tull DE, Lucas C, Bleakley CR (2004) Experimental immunisation and protection of guinea pigs with *Vibrio cholerae* toxoid and mucinases, neuraminidase and proteinase. Vaccine 22:2137–2145

Su-Arehawaratana P, Singharaj P, Taylor DN, Hoge C, Trofa A, Kuvanont K et al (1992) Safety and immunogenicity of different immunization regimens of CVD103-HgR live oral cholera vaccine in soldiers and civilians in Thailand. J lnfect Dis 165:1042–1048

Suharyono SC, Witham N, Punjabi N, Heppner DG, Losonsky C, Totosudirjo H et al (1992) Safety and immunogenicity of single dose live oral cholera vaccine CVD 103-HgR in 5–9 year-old Indonesian children. Lancet 340(8821):689–694

Svennerholm AM Holmgren J (1976) Synergistic protective effect in rabbits of immunization with *Vibrio cholerae* lipopolysaccharide and toxin/toxoid. Infect Immun 13:735–740

Svennerholm AM, Jertborn M, Gothefors L, Karim A, Sack DA, Holmgren J (1984) Mucosal antitoxin and antibacterial immunity after cholera disease and after immunization with a combined B-subunit –whole cell vaccine. J Infect Dis 149:884–893

Tacket CO, Forrest B, Morona R, Attridge SR, LaBrooy J, Tall BD et al. (1990) Safety, immunogenicity, and efficacy against cholera challenge in humans of a typhoid-cholera hybrid vaccine derived from *Salmonella typhi* Ty21a. lnfect lmmun 58:1620–1627

Tacket CO, Losonsky G, Nataro JP, Comstock L, Michalski J, Edelman R et al (1995) Initial clinical studies of CVD112 *Vibrio cholerae* O139 live oral vaccine: safety and efficacy against experimental challenge. J lnfect Dis 172:883–886

Takeda T, Peina Y, Opwa A, Dohi S, Abc H, Nair GB, Pal SC (1991) Detection of heat-stable enterotoxin in a cholera toxin gene-positive strain of *Vibrio cholerae* O1. FEMS Microbiol Lett 64:23

Taylor DN, Cardenas V, Sanchez JL, Begue RE, Gilman R, Bautista C, Perez J, Puga R, Gaillour A, Meza R, Echeverria P, Sadoff JC (2000) Two-year study of the protective efficacy of the oral whole cell plus recombinant B subunit cholera vaccine in Peru. J Infect Dis 181:1667–1673

Taylor DN, Sanchez JL, Castro JM, Lebron C, Parrado CM, Johnson DE et al (1999) Expanded safety and immunogenicity of a bivalent, oral, attenuated cholera vaccine, CVD103-HgR plus CVD111, in United States military personnel stationed in Panama. Infect lmmun 67:2030–2034

Taylor RK, Miller VL, Furlong D, Band Mekalanos JJ (1987) Use of phoA gene fusions to identify a pilus colonizing factor coordinately regulated with cholera toxin. Proc Natl Acad Sci USA 84:2833–2837

Teppema JS, Guinee PAM, Ibrahim AA, Paques M, Ruitenberg EJ (1987) *In vivo* adherence and colonization of *Vibrio cholerae* strains that differ in hemagglutinating activity and motility. Infect Immun 55:2093–2097

Thunqapathra M, Sharma C, Gupta N, Ghosh RK, Mukhbpadhyay A, Koley H et al (1999) Construction of a recombinant live oral vaccine from a non-toxigenic strain ol *Vibrio cholerae* O1 serotype Inaba biotype El Tor and assessment of its reactogenicity and immunogenicity in the rabbit model. Immunol Lett 68:219–227

Trach DD, Cam PD, Ke NT, Rao MR, Dinh D, Hang PV et al (2002) Investigations into the safety and

immunogenicity of a killed oral cholera vaccine developed in Vietnam. Bull World Health Org 80:2–8
Trach DD, Clemens JD, Ke NT, Thuy HT, Son ND, Canh DG et al. (1997) Field trial of a locally produced, killed, oral cholera vaccine in Vietnam. Lancet. 349 (9O47):231–235
Trucksis M, Galen JE, Michalski J, Fasano A, Kaper JB (1993) Accessory cholera enterotoxin (ACE), the third toxin of a *Vibrio cholerae* virulence cassette. Proc Natl Acad Sci USA 90:5267–5271
van Loon FP, Clemens JD, Chakraborty J, Rao MR, Kay BA, Sack DA et al (1996) Field trial of inactivated oral cholera vaccines in Bangladesh: results from 5 years of follow-up. Vaccine 14:162–166
Vasil ML, Holmes RK, Finkelstein RA (1975) Conjugal transfer of chromosomal gene determining production of enterotoxin in *Vibrio cholerae*. Science 187:849
Wachsmuth I., Blake P.A., and Olsvik O. (eds) 1994. *Vibrio cholerae* and cholera: molecular to global perspective. American Society for Microbiology, Washington D.C. USA.
Waldor MK, Mekalanos JL (1996) Lysogenic conversion by a filamentous phage encoding cholera toxin. Science 272:1910–1914
Yang GCH, Schrank GD, Freeman BA (1977) Purification of flagellar cores of *Vibrio cholerae*. J Bacterial 129:1127–1128
Yeh MK, Chiang CH (2004) Inactive *Vibrio cholerae* whole-cell vaccine-loaded biodegradable microparticles: in vitro release and oral vaccination. J Microencapsulation 21:91–106

Author Biography

B. S. Srivastava Former Scientist and Head, Microbiology Division, Central Drug Research Institute, Lucknow and Professor and Chairman, Biotechnology Centre, Jawaharlal Nehru University, New Delhi.

Typhoid Fever and Water

6

Varsha Gupta and Ritu Garg

Abstract

Salmonella enterica serotype Typhi and *Salmonella enterica* serotype Paratyphi A are the main causes of enteric fever in India, with *S.* Typhi being predominant. The principal habitat of Salmonella is the intestinal tract of humans and animals. Humans infected with salmonellae can carry the bacteria in the gut without signs of disease. Typhoid fever has been virtually eliminated from the advanced countries during the last several decades, mainly as a result of improvement in water supply and sanitation, but it continues to be endemic in the poor nations of the world. The control of paratyphoid fever has not been so successful. Infants, children, and adolescents in south-central and south-eastern Asia experience the greatest burden of illness. Studying interactions between the infected host and Typhi would improve our understanding of typhoid fever. Data on human typhoid are still scarce since Typhi uniquely infects humans, and there are no suitable animal models available. The hallmark of typhoid fever is prolonged, persistent fever. The definitive diagnosis still depends on isolation of the bacilli from the patient by blood culture . The prevalence of resistance to multiple first-line oral drugs has been rising. The wide distribution and high prevalence of MDR among *Salmonella* has led to fluoroquinolones assuming a primary role in the therapy for invasive salmonellosis. However, their widespread use has also been associated with decreased susceptibility and documented resistance to this class of drugs. Despite the availability and WHO's recommendation for the use of vaccines among school children in endemic areas, the use is quite limited. Typhoid fever thus continues to remain one of the important water and food borne diseases.

Keywords

Salmonella Typhi · Fluoroquinolone · Diagnosis

V. Gupta (✉) · R. Garg
Department of Microbiology, Govt. Medical
College and Hospital, Chandigarh 160031, India
e-mail: varshagupta_99@yahoo.com

P. P. Singh and V. Sharma (eds.), *Water and Health*,
DOI: 10.1007/978-81-322-1029-0_6, © Springer India 2014

Introduction

The genus *Salmonella* was designated by Lignières in 1900 (Le Minor 2003; Popoff and Le Minor 2005). The genus *Salmonella*, a member of the family *Enterobacteriaceae*, includes Gram-negative motile straight rods. *S. enterica* subsp. *enterica* serovar Enteritidis is the most frequently isolated serovar from humans all over the world. Salmonellae pathogenic to humans can cause two types of disease: typhoid and paratyphoid fever and gastroenteritis (Le Minor 2003; Fenwick 2006).

Enteric fever is a disease of major public health problem in our country. *Salmonella enterica* serotype Typhi and *S. enterica* serotype Paratyphi A are the main causes of enteric fever in India, with *Salmonella* Typhi being predominant (Typhoid fever) (House et al. 2001). Salmonella usually causes an intestinal infection that often lasts 1 week or longer. The principal habitat of *Salmonella* is the intestinal tract of humans and animals (Le Minor 2003). Salmonellae are constantly found in environmental samples, because they are excreted by humans, pets, farm animals, and wild life. Municipal sewage, agriculture pollution, and storm water runoff are the main sources of these pathogens in natural waters (Popoff and Le Minor 2005; Arvanitidou et al. 2005). Salmonellae do not seem to multiply significantly in the natural environment, but they can survive several weeks in water and in soil if conditions of temperature, humidity, and pH are favorable (Le Minor 2003).

Salmonellae isolated from environmental sources are predominantly non-Typhi or Paratyphi serovars. The great majority of the strains are isolated from poultry, red meat, milk and dairy products, vegetables, and fruits.

Unlike cholera, humans infected with salmonellae can carry the bacteria in the gut without signs of disease. Infected humans can harbor the bacteria for considerable periods of time. About 5 % of patients clinically cured from typhoid fever remain carriers for months or even years. These people can be chronic holders of the bacterium in the gut and constitute the main reservoir of the bacteria in the environment (Popoff and Le Minor 2005).

The salmonellosis cycle in the environment can involve shellfish also. Salmonellae survive sewage treatments if suitable germicides are not used in sewage processing. If effluent from the sewage plant passes into a coastal area, edible shellfish (mussels, oysters) can become contaminated. Shellfish concentrate bacteria as they filter several liters of water per hour. Ingestion by humans of these seafoods (uncooked or superficially cooked) may cause typhoid fever or other salmonellosis (Popoff and Le Minor 2005).

Thus, human illness is linked with foods of animal origin. It is also transmitted by direct contact with animals, non-animal foods, by water and occasionally by human contact. So, typhoid fever is one of the important waterborne and foodborne diseases (Swaminathan et al. 2005). It has been observed that the incidence of typhoid fever decreases when the level of development of a country increases (i.e., controlled water sewage systems, pasteurization of milk and dairy products). Where these hygienic conditions are missing, the probability of fecal contamination of water and food remains high and so is the incidence of typhoid fever (Cabral 2010).

Typhoid fever is a distinctive acute systemic febrile infection of the mononuclear phagocytes. *S. enterica* subsp. *enterica* serotype Typhi is the etiological agent of typhoid fever (House et al. 2001). The name typhoid was given by Louis in 1829, as a derivative from typhus (Udeze et al. 2010). In India, though, *S. enterica* serotype Typhi remains the predominant *Salmonella* species causing enteric fever but it may be caused by several serovars like *Salmonella* Typhi, *Salmonella* Paratyphi A, *Salmonella* Paratyphi B, and occasionally *Salmonella typhimurium* (Bhattacharya et al. 2011). Isolation of *Salmonella* serotype Paratyphi A causing the same disease (paratyphoid fever) has also been reported increasingly (Chandel et al 2000; Sood et al. 1999). Salmonellae currently comprise above 2000 serotypes or species, all of them potentially pathogenic. Infections in humans include typhoid fever, focal systemic infections, septicemia, and most commonly diarrhea, varying from acute watery diarrhea to bloody diarrhea or dysentery. Salmonellae are ecologically entrepreneurial and exist in a multiplicity of habits; this characteristic adaptability, which accounts for the ubiquity of the organisms in nature and for the many ways in which they encounter potential new human hosts, is related to their genetic plasticity (Keusch GT 1998). The risk factors such as poor sanitation, lack of a safe drinking water supply, and low socioeconomic conditions in resource-poor countries are amplified by the evolution of multidrug-resistant *salmonellae* which is associated with increased mortality and morbidity associated with typhoid fever (Kanungo et al. 2008).

Morphology

The organism is a short bacillus about 2–4× 0.6 μm in size. It is motile due to its peritrichous flagella (Arora 2008).

Culture Characteristics

Salmonella Typhi is an aerobic and facultative anaerobic bacteria. It grows readily on simple media. It can grow at pH 6–8 and temperature requirement is 15–41 °C (optimum 37 °C). On MacConkey and deoxycholate citrate media, colonies are colorless due to the absence of lactose fermentation. On Wilson and Blair bismuth sulfite medium, jet black colonies with a metallic sheen are formed due to production of H_2S (Ananthanarayan and Jayaram Paniker 2005).

Biochemical Reactions

Salmonella Typhi ferments glucose, mannitol, and maltose, forming acid without gas. Lactose, sucrose, and salicin are not fermented. Indole is not reduced. They are MR positive, VP negative, and citrate negative. Urea is not hydrolyzed. H_2S is produced (Ananthanarayan and Jayaram Paniker 2005).

Sources of Infection

The sources of infection are food and drink that have been contaminated with salmonellae. The following sources are important:

1. Water—Contamination with feces often results in explosive epidemics.
2. Milk and other dairy products (ice cream, cheese, custard)—Contamination with feces and inadequate pasteurization or improper handling. Some outbreaks are traceable to the source of supply.
3. Shellfish—From contaminated water.
4. Dried or frozen eggs—From infected fowl or contaminated during processing.
5. Meats and meat products—From infected animals (poultry) or contamination with feces by rodents or humans.
6. Recreational drugs—Marijuana and other drugs.
7. Animal dyes—Dyes (carmine) used in drugs, foods, and cosmetics.
8. Household pets—Turtles, dogs, cats, etc. (Brooks et al. 2004).

Transmission

S. Typhi has no non-human vectors. The following are modes of transmission:

- Oral transmission via food or beverages handled by an individual who chronically sheds the bacteria through stool or, less commonly, urine.
- Hand-to-mouth transmission after using a contaminated toilet and neglecting hand hygiene.
- Oral transmission via sewage-contaminated water or shellfish (especially in the developing world (Earampamoorthy and Koff 1975).

An inoculum as small as 100,000 organisms causes infection in more than 50 % of healthy volunteers (Levine et al. 2001).

Epidemiology

Worldwide, 16 million cases are estimated to occur annually with 600,000 deaths. The proportion of typhoid to paratyphoid A is about 10:1. Paratyphoid B is rare and C very rare. In the year 2000, it was estimated that typhoid fever caused 21.7 million illnesses and 217,000 deaths and paratyphoid fever 5.4 million illnesses worldwide. Infants, children, and adolescents in south-central and south-eastern Asia experience the greatest burden of illness (Crump et al. 2004). The age incidence is related to the endemicity of the disease and the level of sanitation.

Typhoid fever has been virtually eliminated from the advanced countries during the last several decades mainly as a result of improvement in water supply and sanitation but it continues to be endemic in the poor nations of the world. The control of paratyphoid fever has not been so successful. The distribution of paratyphoid bacilli shows marked geographical differences. *S.* Paratyphi A is prevalent in India and other Asian countries, Eastern Europe, and South America; *Salmonella* Paratyphi B in Western Europe, Britain, and North America; and *Salmonella* Paratyphi C in Eastern Europe and Guyana.

Epidemiologic Trends

Despite the limitations of currently available epidemiologic data, a number of recent trends in enteric disease epidemiology have emerged in the African, Asian, and Latin American regions. In sub-Saharan Africa where the burden of enteric fever is the least well characterized, hospital-based studies indicate that non-Typhi serotypes of *Salmonella*, particularly *Salmonella enteritidis* and *Salmonella. typhimurium*, greatly outnumber *Salmonella* Typhi and *Salmonella* Paratyphi as causes of bloodstream infection (Shaw et al. 2008; Mweu and English 2008). In Asia, a large population-based prospective study using standardized surveillance methods has estimated typhoid fever incidence in China, India, Indonesia, Pakistan, and Vietnam. This study confirmed the high incidence of typhoid fever in the region, particularly among children and adolescents, but also demonstrated that substantial variation in incidence occurs between surveillance sites in the same region (Ochiai et al. 2008). Simultaneously, *Salmonella* Paratyphi A appears to be responsible for a growing proportion of enteric fever in a number of Asian countries, sometimes accounting for 50 % of *Salmonella* bloodstream isolates among enteric fever patients.

Enteric fever is endemic in all parts of India. There have been two large-scale studies in India on the incidence of blood culture-confirmed typhoid fever, one among individuals under 40 years old (Sinha et al. 1999) and another among children 6–17 years old (Chuttani et al. 1977), but as yet, none on paratyphoid fever. Thus, the actual burden of paratyphoid fever in India and its incidence and characteristics relative to typhoid fever are poorly understood. In a study conducted in Punjab that examined 340 enteric fever cases, 334 *Salmonella* Typhi and 6 *Salmonella* Paratyphi A isolates were identified (Pathania and Sachar 1965). This scenario, however, has changed as recent studies have highlighted the increasing occurrence of paratyphoid fever (Rodrigues et al. 2003).

A large-scale community study performed in an Indian urban slum showed incidence as high as 2 per 1,000 population per year for children under five, and 5.1 per 1,000 population per year for children under ten (Sur et al. 2006). Another study in Northern India showed that the majority of cases occurred in children aged 5–12 years and 24.8 % of cases were in children up to 5 years of age (Walia et al. 2006). *Salmonella* serovars showed an age-related bias, with paratyphoid fever more common in adults. One study from Kolkata showed the incidence of paratyphoid fever was lower (0.8/1000/year), and the mean age of paratyphoid patients was older

(17.1 years) compared to typhoid fever (incidence 1.4/1000/year, mean age 14.7 years) (Sur et al. 2007; Kanungo et al. 2008).

The source of infection is a patient or, far more frequently, a carrier. Patients who continue to shed typhoid bacilli in feces for 3 weeks–3 months after clinical cure are called convalescent carriers. Those who shed the bacilli for more than three months but less than a year are called temporary carriers and those who shed the bacilli for over a year are called chronic carriers. About 2–4 % of patients become chronic carriers. The development of the carrier state is more common in women and in the older age groups. Some persons may become carriers following inapparent infections (symptomless excretor). The shedding of bacilli is usually intermittent. The bacilli persist in the gall bladder or kidney and are eliminated in the feces or urine. Urinary carriage is less frequent.

The feces of persons who have unsuspected subclinical disease or are carriers are a more important source of contamination than frank clinical cases that are promptly isolated, for example, when carriers working as food handlers are shedding organisms. The best known of such typhoid carriers was Mary Mallon (Typhoid Mary), a New York cook who, over a period of 15 years, caused at least seven outbreaks affecting over 200 persons.

Typhoid fever occurs in two epidemiological types. The first is endemic or residual typhoid that occurs throughout the year though seasonal variations may sometimes be apparent. Some studies show a peak of the disease from July to September, as it coincides with the rainy season when the chance of water contamination is high, especially in crowded areas (Sur et al. 2006). The second is epidemic typhoid, which may occur in endemic or non-endemic areas. Typhoid epidemics are usually water, milk, or foodborne (Ananthanarayan and Jayaram Paniker 2005).

Pathogenicity

Infection is initiated by oral ingestion of organisms, which must pass the gastric acid barrier to establish infection. Salmonella exhibits a genetic adaptive acid tolerance response. Exposure to acid leads to synthesis of at least 40 proteins, some of which may play a role in pathogenesis. Bacteria successfully evading "acid death" in the stomach pass on to the distal ileum and colon, where they penetrate the mucosal barrier.

Initial bacterial invasion results in transient asymptomatic bacteraemia, as organisms are rapidly ingested by mononuclear phagocytes within which they survive and multiply. This process is favored by a lack of bactericidal antibodies in the susceptible host; in contrast, opsonized salmonellae are taken up by neutrophils and killed. Opsonophagocytosis is limited by the capsular Vi polysaccharides of *Salmonella* Typhi which enhances resistance to complement activation and bacterial lysis by the alternative pathway as well as to peroxide-mediated killing. The fate of organisms within macrophages depends both on microbial factors that promote resistance to killing and on specific host T lymphocyte-activated, cell-mediated immune mechanisms that are under the genetic control of the pathogenicity locus in mice. *S. enterica* serovar Typhi uses SPI-1 TSS3 to invade the intestinal epithelium and SPI-2 TSS3 to survive in macrophages and employs additional virulence factors to overcome the mucosal barrier and to colonize internal organs (Casadesús 2011). When intracellular multiplication has proceeded enough to permit the initiation of persistent bacteremia, the clinical phase of typhoid fever begins with invasion of the gallbladder and Peyer's patches of the intestine (Parry et al. 2002). The sustained bacteremia is responsible for the persistent fever of clinical typhoid, while inflammatory responses to tissue invasion determine the pattern of clinical expression (cholecystitis, intestinal hemorrhage, or perforation).With invasion of the gallbladder and Peyer's patches, bacteria regain entry to the bowel lumen and may be recovered in stool cultures beginning in the second week of clinical disease. Seeding of the kidney leads to positive urine cultures. The lipopolysaccharide of *Salmonella* Typhi may contribute to fever, leucopenia, and other systemic symptoms, but the occurrence of such symptoms in individuals rendered tolerant to endotoxin supports a role for other factors, such as cytokines released from infected mononuclear phagocytes, that can mediate inflammation (Keusch GT 1998).

Studying interactions between the infected host and Typhi would improve our understanding of typhoid fever. Typhi has developed remarkable persistence mechanisms within the host that help ensure its survival and transmission. However, data on human typhoid collected by using modern immunological and molecular techniques are scarce since Typhi uniquely infects humans, and there are no suitable animal models available. As survival within macrophages is an essential step for *Salmonella* pathogenesis, macrophages represent a useful model to study Typhi. Elucidating the bacterial genes expressed in the host and those underlying typhoid pathogenesis should lead to the development of new strategies including novel anti-bacterial treatments and identification of novel vaccine candidates to control the disease (Daigle 2008).

Clinical Manifestations

The incubation period of typhoid fever is variable, that is, 3–60 days and depends on both the inoculum size and the state of the host's defense. The disease classically presents with a step like daily increase in temperature associated with headache, malaise, and chills. The hallmark of typhoid fever is prolonged, persistent fever (4–8 weeks in untreated patients). The illness may also be mild and brief. In some cases, acute, severe infection with disseminated intravascular coagulation and central nervous system involvement rapidly results in death. In other instances, necrotizing cholecystitis or intestinal bleeding and perforation can occur in the third or fourth week of illness, when the patient is otherwise improving. In most cases, the onset of these complications is dramatic and clinically obvious. Intestinal perforation appears to be less common among children under 5 years of age.

Early intestinal manifestations include constipation (especially in adults) or mild diarrhea (in children) associated with abdominal tenderness. Mild hepatomegaly is detectable in the majority of patients. Bradycardia relative to the height of the fever may be clinical clue to typhoid but is present in only a minority of patients. Epistaxis may be noted in the early stages of illness. "Rose spots" appearing as small, pale red, blanching, slightly raised macules are occasionally seen on the chest and abdomen during the first week. They can evolve into non-blanching small hemorrhages that are difficult to see in dark skinned patients. The major characteristics of untreated typhoid are persistent high fever, severe anorexia, weight loss, and changes in sensorium, but a variety of other complications also may develop including hepatitis, meningitis, nephritis, myocarditis, bronchitis, pneumonia, arthritis, osteomyelitis, parotitis, and orchitis. Except for that or relapse, the frequency of all these complications, including hemorrhage and perforation, is reduced by prompt use of appropriate antibiotics. Typhoid, long considered to be uncommon and typically mild in young children, may in reality be frequent and severe in this group. Neonatal typhoid, which can be acquired vertically from infected mother or from exogenous sources, is often a severe, life-threatening septicemic illness with a high case fatality rate.

As multidrug-resistant *Salmonella* Typhi is becoming more prevalent in many endemic countries, the patients infected with resistant strains present with more severe illness, look toxic, and have a higher incidence of disseminated intravascular coagulation and hepatomegaly and a threefold higher mortality rate that is thought to be related to the longer duration of disease and to prior ineffective oral antibiotic therapy.

Around 3–5 % of patients become long-term asymptomatic carriers, some for life unless treated. Many carriers give no history of typhoid fever and probably have an undiagnosed mild infection (Keusch GT 1998).

Mortality/Morbidity

With prompt and appropriate antibiotic therapy, typhoid fever is typically a short-term febrile illness requiring a median of 6 days of hospitalization. When treated, it has few long-term sequelae and a 0.2 % risk of mortality (Lynch et al. 2009). Untreated typhoid fever is a life-threatening illness of several weeks' duration with long-term morbidity often involving the central nervous system. The case fatality rate in the United States in the pre-antibiotic era was 9–13 % (Crump et al. 2008).

Immunity

Infections with *Salmonella* Typhi or *Salmonella* Paratyphi usually confer a certain degree of immunity. Reinfection may occur but is often milder than the first infection. Circulating antibodies to O and Vi are related to resistance to infection and disease. However, relapse may occur in 2–3 weeks after recovery in spite of antibodies, secretory IgA antibodies may prevent attachment of salmonellae to intestinal epithelium.

Persons with S/S hemoglobin (sickle cell disease) are exceedingly susceptible to salmonella infections, particularly osteomyelitis. Persons with A/S hemoglobin (sickle cell trait) may be more susceptible than normal individuals (Brooks et al. 2004).

Laboratory Diagnosis

In around 25 % of patients, leucopenia and neutropenia are evident. In most patients, the white blood cell count is normal or low in relation to the degree of fever; this pattern may be a clue to the diagnosis. Whereas definitive diagnosis still depends on isolation of the bacilli from the patient, yield of culture is disappointingly low. The yield is affected by the overall several variables: when the cultures are performed, what is cultured, and whether the patient has taken antibiotics. Other tests for diagnosis are the demonstration of antibodies in the patient serum. A positive blood culture is diagnostic. Demonstration of antibodies is not conclusive evidence of current infection. A third method is the demonstration of typhoid bacillus antigen in blood or urine (Keusch GT 1998).

Blood culture: Bacteremia occurs early in the disease and blood cultures are positive in approximately 90 % of cases in the first week of fever. The popular belief that blood culture for diagnosis of typhoid fever is useful only in the first week is not true. Blood culture is positive in approximately 75 % of cases in the second week, 60 % in the third week, and 25 % thereafter till the subsidence of pyrexia. Blood culture rapidly becomes negative on treatment with antibiotics.

About 5–10 ml of blood is collected by venipuncture after proper preparation of site and inoculated into a culture bottle containing 50–100 ml of 0.5 % bile broth. Blood contains substances that inhibit the growth of the bacilli and hence it is essential that the broth be taken in sufficient quantity to provide at least fourfold dilution of blood.

After incubation overnight at 37 °C, the bile broth is subcultured on MacConkey agar. Pale non-lactose fermenting colonies that may appear on this medium should be identified. Confirm the isolate by further putting up biochemical reactions and performing slide agglutination. For identification of unusual serotypes, the help of the National Salmonella Reference Centre in India is located at the Central Research Institute, Kasauli. The reference centre for salmonellae of animal origin is at the Indian Veterinary Research Institute, Izatnagar.

If salmonellae are not obtained from the first subculture from bile broth, subcultures should be repeated every other day till growth is obtained. Cultures should be declared negative only after incubation for 10 days. An alternative to blood culture is the clot culture. Clot cultures yield a higher rate of isolation than blood cultures as the bactericidal action of the serum is obviated. Another advantage is that a sample of serum also becomes available. Even though agglutinins may be absent in the early stages of the disease, a Widal test provides a baseline titer against which the results of tests performed later may be evaluated.

Feces culture: Salmonellae are shed in feces throughout the course of the disease and even in convalescence, with varying frequency. Hence, fecal cultures are almost as valuable as blood cultures in diagnosis. A positive fecal culture, however, may occur in carriers as well as in patients. The use of enrichment and selective media and repeated sampling increase the rate of isolation. Fecal culture is particularly valuable in patients on antibiotics as the drug does not eliminate the bacilli from the gut as rapidly as it does from the blood.

Fecal samples are plated directly on MacConkey, DCA and Wilson-Blair media. For enrichment, specimens are inoculated into one tube each of selenite and tetrathionate broth and incubated for 12–18 h before subculture onto plates.

Urine culture: Salmonellae are shed in the urine irregularly and infrequently. Hence, urine culture is less useful than the culture of blood or feces. Cultures are generally positive only in the second and third weeks and then only in about 25 % of cases. Repeat sampling improves the rate of isolation.

Other materials for culture: Bone marrow culture is valuable as it is positive in most cases even when blood cultures are negative. Culture of bile obtained by duodenal aspiration is usually positive and may be employed for the detection of carriers. Other materials which may yield isolation at times are rose spots, pus from suppurative lesions, CSF, and sputum (Ananthanarayan and Jayaram Paniker 2005).

Serologic Methods

Widal test (Tube dilution agglutination test): Serum agglutinins rise sharply during the second and third weeks of salmonella infection. At least two serum specimens, obtained at intervals of 7–10 days, are needed to prove a rise in antibody titer. Serial (twofold) dilutions of unknown serum are tested against antigens from representative salmonellae. The results are interpreted as follows: (1) High or rising titer of O (>1:160) suggests that active infection is present. (2) High titer of H (>1:160) suggests past immunization or past infection. (3) High titer of antibody to the Vi antigen occurs in some carriers. Results of serologic tests for salmonella infection must be interpreted cautiously. The possible presence of cross-reactive antibodies limits the use of serology in the diagnosis of salmonella infections (Brooks et al. 2004). According to studies in developing countries with high incidence of typhoid fever, coagglutination is much more reliable than culture because so many patients have already taken antibiotics before being seen by a physician. New enzyme immunoassays that detect *S.* Typhi outer membrane proteins are being evaluated. A sensitive and specific multiplex polymerase chain reaction for Vi antigen promises to be highly sensitive and specific, but its value has not been proved in clinical studies, nor is it available yet (Keusch GT 1998).

Detection of carriers: The detection of carriers is important for epidemiological and public health purposes. Laboratory tests are also useful in screening food handlers and cooks to detect carrier state.

The identification of fecal carriers is by isolation of the bacillus from feces or from bile. The frequency and intensity of bacillary shedding vary widely, and it is essential to test repeated samples. Cholagogue purgatives increase the chance of isolation. For the detection of urinary carriers, repeated urine culture should be carried out.

The Widal reaction is of no value in the detection of carriers in endemic countries. The demonstration of Vi agglutinins has been claimed to indicate the carrier state. While this is useful as a screening test, confirmation should be made by culture.

The tracing of carriers in cities may be accomplished by the "sewer-swab" technique. Gauze pads left in sewers and drains are cultured, and by tracing positive swabs, one may be led to the house harboring a carrier. Another technique of isolating salmonellae from sewage is filtration through Millipore membranes and culturing the membranes on highly selective media such as Wilson and Blair media.

Bacteriophage Typing

Intraspecies classification of *S.* Typhi for epidemiological purposes was made possible by bacteriophage typing, first developed by Craigie and Yen in 1937. Apart from helping in tracing the source of epidemics, phage typing also provides information on the trends and pattern in the epidemiology of typhoid at the local, national, and international levels. Phage typing is carried out at the National Phage Typing Centre and is coordinated by the International Reference Centre. The National Salmonella Phage Typing Centre for India is located at the Lady Hardinge Medical College, New Delhi. Phage types A and F1 are the most common and are present throughout India (Ananthanarayan and Jayaram Paniker 2005).

Differential Diagnosis

When all the classic clinical manifestations are present, including rose spots, prolonged fever, relative bradycardia, and leucopenia, the diagnosis of typhoid will be strongly suggested. However, most cases do not fit this "typical" profile. Differential diagnosis includes infections associated with prolonged fevers, such as the rickettsioses, brucellosis, tularemia, leptospirosis, military tuberculosis, viral hepatitis, infectious mononucleosis, CMV, and malaria, as well as non-infectious causes of fever, such as lymphomas (Keusch GT 1998).

Treatment

Specific antibacterial therapy for enteric fever became available only in 1948 with the introduction of chloramphenicol, which continued as gold standard antimicrobial till the 1970s, after that resistance became common to this drug. No drug has been better in promoting a favorable clinical response, which usually becomes apparent within 24–48 h of the start of treatment in the appropriate dosages. Other effective oral regimens include amoxicillin, trimethoprim-sulfamethoxazole or for patients over 17 years of age, fluoroquinolones such as ciprofloxacin or ofloxacin.

A variety of intravenous drugs are also effective. Both chloramphenicol and trimethoprim-sulfamethoxazole can be given intravenously to patients who cannot take oral medications. Other effective parenteral antimicrobials include high-dose ampicillin, cefotaxime, cefoperazone, and 4-fluoroquinolones. However, none has been as rapidly acting or as effective as ceftriaxone. In addition, compared with that for other drugs, the relapse rate for ceftriaxone appears lower.

The prevalence of resistance to multiple first-line oral drugs has been rising among strains of *Salmonella* Typhi in developing countries, especially in the Indian subcontinent and Southeast Asia, due to the acquisition of plasmids encoding inactivating ß-lactamases and chloramphenicol acetyl transferases. Where multidrug resistance is a problem, ceftriaxone or a 4-fluoroquinolone should be administered initially to adults over 17 years of age, and ceftriaxone is the best choice for children because of concerns about quinolone induced arthropathy and cartilage damage in this age group. Short-course quinolone therapy has been shown to be effective against multidrug-resistant typhoid in children, and its use further reduces the likelihood of drug toxicity. Alternative oral agents that reportedly are effective for this indication include furazolidine and cefixime. Recent clinical trials suggest that azithromycin 500 mg once daily for 7 days in adults or azithromycin 20 mg/kg/day up to a maximum of 1,000 mg/day for 7 days in children is useful for the management of uncomplicated typhoid fever (Effa and Bukirwa 2008).

Eradication of the chronic carrier state, especially in the presence of gallstones, is notoriously difficult. Traditional regimens have used ampicillin or amoxicillin plus probenecid or trimethoprim-sulfamethoxazole plus rifampicin for at least 6 weeks. Recent studies suggest that a 4 week course of a 4-fluoroquinolone is at least as good and probably much better because the organism is exquisitely sensitive in vitro and the drugs reach the gut lumen, liver, gall bladder, and bile in active form. The new quinolones provide the best chance of eradicating *S.* Typhi in the presence of gallstones (Keusch GT 1998).

Antimicrobial Resistance

Antimicrobial resistance is a major public health problem in both *Salmonella* Typhi and *Salmonella* Paratyphi, and timely treatment with appropriate antimicrobial agents is important for reducing the mortality of enteric fever (Edelman and Levine 1986).

Resistance to chloramphenicol did not pose any problem in typhoid fever till 1972, when resistant strains emerged in Mexico and in Kerala (India). In Mexico; the resistant strain caused an explosive epidemic, with high mortality. Travelers who got infected in Mexico had, on occasion, conveyed the resistant strain to North America and Europe but it did not get established in these areas. Chloramphenicol-resistant typhoid fever has become a problem in many countries in Asia.

In India, chloramphenicol-resistant typhoid fever appeared in epidemic form first in Calicut (Kerala) in early 1972. It became endemic and was confined to Kerala till 1978. Subsequently, such strains carrying drug resistance plasmids appeared in many other parts of India. Though resistant to chloramphenicol, such strains were initially sensitive to ampicillin, amoxicillin, co-trimoxazole, and furazolidone, which were successfully used for treatment. By late 1980s, typhoid bacillus strains resistant to many or all of these drugs began to spread in most parts of India. At present, the drugs useful in treatment of such multiresistant typhoid cases are the later fluoroquinolones (such as ciprofloxacin, pefloxacin, ofloxacin) and the third-generation cephalosporins (such as ceftazidime, ceftriaxone, cefotaxime). Furazolidone is still active against most isolates but its action is too slow for it to be used alone in treatment. Recently, many strains have become resistant to fluoroquinolones, but several isolates of typhoid bacilli are now sensitive to chloramphenicol (Ananthanarayan and Jayaram Paniker 2005; Gupta et al. 2009).

Multiple Drug Resistance

Resistance to the traditional first-line antimicrobial agents ampicillin, chloramphenicol, and trimethoprim-sulfamethoxazole defines multiple drug resistance (MDR) in *Salmonella*. The MDR phenotype has been shown to be widespread among *Salmonella* Typhi for many years (Rowe et al. 1997) and is present, albeit at lower rates, among *Salmonella* Paratyphi (Gupta et al. 2008; Parry and Threlfall 2008). Surveillance studies demonstrate considerable geographic variation in the proportion of *Salmonella* Typhi isolates that are MDR within the same region, with sites in India, Pakistan, and Vietnam having higher rates of MDR than sites in China and Indonesia (Ochiai et al. 2008). Furthermore, longitudinal studies at the same site demonstrate marked changes in the proportion of *Salmonella* Typhi and *Salmonella* Paratyphi A with MDR over time, including reductions in the proportion of isolates with MDR (Maskey et al. 2008).

The wide distribution and high prevalence of MDR among *Salmonella* have led to fluoroquinolones assuming a primary role in the therapy for invasive salmonellosis. Some investigators have noted increase in the prevalence of more susceptible *Salmonella* Typhi and *Salmonella* Paratyphi strains coinciding with a switch from traditional first-line antimicrobials to fluoroquinolones for the management of enteric fever (Zaki and Karande 2011; Sood et al. 1999). However, the widespread use of fluoroquinolones has also been associated with decreased susceptibility (Lynch et al. 2009) and documented resistance to this class of drugs (Brown et al. 1996). A single chromosomal mutation in the quinolone resistance determining region (QRDR) of the *gyrA* gene may be sufficient to result in decreased ciprofloxacin susceptibility (DCS). Nalidixic acid resistance in the presence of ciprofloxacin susceptibility had been thought to be a reliable indicator of DCS, but this is now known not to be the case and many have suggested that DCS is most reliably determined by measurement of the ciprofloxacin minimum inhibitory concentration (MIC) (Crump et al. 2003; Threlfall et al. 1999). Patients with enteric fever due to isolates with DCS are more likely to have prolonged fever clearance times and higher rates of treatment failure (Crump et al. 2008). In the United States, MDR and DCS *Salmonella* Typhi are associated with travel to the Indian subcontinent (Lynch et al. 2009). In addition to DCS, ciprofloxacin resistance has been reported among both *Salmonella* Typhi (Chuang et al. 2009) and *Salmonella* Paratyphi A (Maskey et al. 2008).

As fluoroquinolone use continues to expand and as DCS and fluoroquinolone resistance drive the use of third-generation cephalosporins and other agents for the management of enteric fever, new patterns of antimicrobial resistance can be anticipated. Patterns of antimicrobial resistance seen in non-Typhi *Salmonella* and Enterobactericeae may emerge in *Salmonella* Typhi and *Salmonella* Paratyphi. Although quinolone resistance among Enterobactericeae usually arises due to mutations in the QRDR of *gyrA*, plasmid-mediated resistance is increasingly recognized. Plasmid-mediated quinolone resistance is associated with *qnr* genes that encode a protein that protects DNA gyrase from ciprofloxacin and by *aac(6′)-Ib-cr*, an aminoglycoside-modifying enzyme with activity against ciprofloxacin (Parry and Threlfall 2008). Plasmids bearing *qnr* or *aac(6′)-Ib-cr* may also contain an extended-spectrum cephalosporin resistance gene, which would pose a threat to the success of two major antimicrobial classes for the management of invasive salmonellosis. *Salmonella* Typhi isolate producing an SHV-12 extended-spectrum beta-lactamase (ESBL) (Al Naiemi et al. 2008) and ESBL-producing *Salmonella*Paratyphi A have recently been reported (Pokharel et al. 2006). Of further concern, rare non-Typhi *Salmonella* isolates have been described containing the carbapenemase, *bla*IMP-4 as well as *qnrB4* conferring both meropenem resistance and DCS (Nordmann et al. 2008).

Prevention and Control

Since the main route of typhoid transmission is feco-oral by contaminated food and water, the disease remains a serious problem in India where it is confounded by low socioeconomic conditions and overcrowding. Food and water safety are complex issues that depend on a number of interrelated environmental, cultural, and socioeconomic factors. Education of consumers and food handlers about basic principles of safe food handling is an important aspect of prevention but is insufficient by itself. The general strategy of prevention of foodborne and waterborne disease is to understand the mechanisms by which contamination and disease transmission occur and to institute appropriate prevention measures. As humans are the only reservoir of this feco-orally transmitted disease, preventive measures include improvement of water supply and sanitation facilities. Water treatment, waste disposal, and protecting the food supply from contamination are important public health measures. Carriers of typhoid must not be allowed to work as food handlers (Giannella 2010; Lima and Guerrant 2009). Where this approach is not yet possible and for travelers, immunization has been used.

Vaccines

The concept of vaccination against typhoid began in the 1960s when field trials showed the effectiveness of a killed vaccine, reporting a protection rate of approximately 70 % after two doses (Levine et al. 1999). Traditional heat killed, phenol preserved, whole-cell typhoid and paratyphoid vaccine constituting *Salmonella* Typhi, *Salmonella* Paratyphi A, and *Salmonella* Paratyphi B is no longer recommended because of its limited efficacy and duration of protection and the high frequency of local reactions and fever.

One option for children over 6 years of age and adults consists of three doses of a first generation the live, attenuated oral vaccine, Ty21a, which is invasive but metabolically defective and dies after a few cycles of replication. This vaccine is safe, provides as much protection as the killed vaccine, and continues to be protective for at least several years.

One dose of purified Vi polysaccharides vaccine has proved as effective and long lasting as multiple doses of Ty21a and may be used in children over 2 years of age and in at-risk HIV-infected patients.

Vi stimulates the IgG antibody, while Ty21a induces humoral and cell-mediated immune responses but not the Vi antibody (Levine 2001).

New genetically engineered live typhoid vaccine strains are being developed, not only for immunization against typhoid but also for use as live vectors into which extraneous genes can be cloned for oral delivery of protective antigens from unrelated species. In addition, Vi protein conjugates are being evaluated as immunogens suitable for infants, especially in endemic regions where infantile typhoid is prevalent and remains a dangerous disease (Keusch 1998). The immunologic properties of Vi polysaccharide (Vi) were improved by binding it to a recombinant *Pseudomonas aeruginosa* exoprotein A (rEPA). Vi-rEPA was 89 % effective at preventing blood culture-confirmed typhoid fever in 2–5-year-olds and induced high levels of serum IgG anti-Vi. A minimal protective level of 3.5 enzyme-linked immunosorbent assay units [ELISA units (EU)] was inferred from the level of anti-Vi 46 months after immunization. The authors conclude that Vi-rEPA was safe, induced protective anti-Vi levels, and was compatible with EPI vaccines, and it can be used in infants (Thiem et al. 2011).

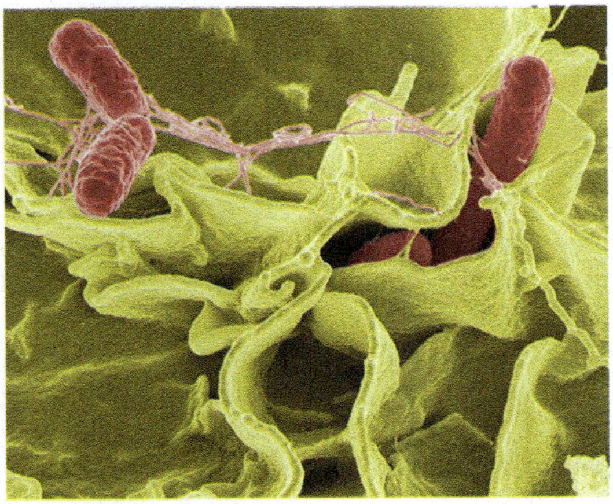

Fig. 6.1 Color-enhanced scanning electron micrograph showing *Salmonella typhimurium* (*red*) invading cultured human cells *Credit*: Rocky Mountain Laboratories, NIAID, NIH Source: [http://www2.niaid.nih.gov/biodefense/public/images.htm NIAID] This image is a work of the National Institutes of Health, part of the United States Department of Health and Human Services. As a work of the U.S. federal government, the image is in public domain.

Despite the availability of these vaccines and the WHO's recommendation for the use of vaccines among school children in endemic areas, the use is quite limited because of cost, lack of proper data, and the vaccine's ineffectiveness in children under 2 years of age. In view of the increasing number of infections with *Salmonella* Paratyphi A, development of a suitable vaccine against *Salmonella* Paratyphi A is urgently needed (Kanungo et al. 2008).

Conclusions

The existence of multidrug-resistant bacteria is a serious and growing problem in the treatment of typhoid, especially in the developing world. Isolation and characterization of MDRST from all regions of the world for effective epidemiologic surveillance and control should continue with intensive scrutiny of *Salmonella* Typhi strains from developing countries. Drug resistance in bacteria results in increase in morbidity and mortality rates associated with the disease. So, in view of the re-emergence of sensitivity to first-line drugs, large-scale systematic studies are required to determine whether these drugs can again be used for the treatment of typhoid fever in developing countries in future times to come. Finally, acceleration of the conjugate vaccine programmes should be done for global adoption (Fig. 6.1).

References

Al Naiemi N, Zwart B, Rijnsburger MC, Roosendaal R, Debets-Ossenkopp YJ, Mulder JA et al (2008) Extended-spectrum-beta-lactamase production in a *Salmonella enterica* serotype Typhi strain from the Philippines. J Clin Microbiol 46:2794–2795

Ananthanarayan R, Jayaram Paniker CK (2005) Enterobacteriaceae III: Salmonella. In: Ananthanarayan R, Jayaram Paniker CK (eds) Textbook of Microbiology, 7th edn. Orient Longman Ltd, Chennai, pp 290–304

Arora DR (2008) Salmonella. In: Arora DR, Arora B (eds) Text book of microbiology, 3rd edn. CBS Publishers and Distributors, New Delhi, pp 375–388

Arvanitidou M, Kanellou K, Vagiona DG (2005) Diversity of *Salmonella spp.* and Fungi in Northern Greek Rivers and their Correlation to Faecal Pollution Indicators. Environ Res 99:278–284

Bhattacharya SS, Das U, Choudhury BK (2011) Occurrence & antibiogram of *Salmonella* Typhi & *S.* Paratyphi A isolated from Rourkela. Orissa Indian J Med Res 133:431–433

Brooks GF, Butel SB, Morse SA (2004) Jawetz, Melnick, Adelberg's: medical Microbiology, 23rd edn. McGraw-Hill, Singapore, pp 256–260

Brown JC, Thomson CJ, Amyes SGB (1996) Mutations of the *gyrA* gene of clinical isolates of *Salmonella Typhimurium* and three other Salmonella species leading to decreased susceptibilities to 4-quinolone drugs. J Antimicrob Chemother 7:351–356

Cabral JPS (2010) Water microbiology. Bacterial pathogens and water. Int J Environ Res Public Health 7:3657–3703

Casadesús J (2011) Salmonella: from basic science to clinical issues. Future Microbiol 6:133–135

Chandel DS, Chaudhry R, Dhawan B, Pandey A, Dey AB (2000) Drug-resistant *Salmonella* enteric a serotype Paratyphi A in India. Emerg Infect Dis 6:420–421

Chuang CH, Su LH, Perera J, Karlos J, Tan BH, Kumarasinghe G et al (2009) Surveillance of antimicrobial resistance in *Salmonella enterica* serotype Typhi in seven Asian countries. Epidemiol Infect 137:266–269

Chuttani CS, Prakash K, Gupta P, Grover V, Kumar A (1977) Controlled field trial of a high-dose oral killed typhoid vaccine in India. Bull World Health Org 55:643–644

Crump JA, Barrett TJ, Nelson JT, Angulo FJ (2003) Reevaluating fluoroquinolone breakpoints for *Salmonella enterica* serotype Typhi and for non-Typhi salmonellae. Clin Infect Dis 37:75–81

Crump JA, Kretsinger K, Gay K, Hoekstra RM, Vugia DJ, Hurd S et al (2008) Clinical response and outcome of infection with *Salmonella enterica* serotype Typhi with decreased susceptibility to fluoroquinolones: a United States FoodNet multicenter retrospective cohort study. Antimicrob Agents Chemother 52:1278–1284

Crump JA, Luby SP, Mintz ED (2004) The global burden of typhoid fever. Bull World Health Org 82:346–353

Daigle F (2008) Typhi genes expressed during infection or involved in pathogenesis. J Infect Developing Countries 2:431–437

Earampamoorthy S, Koff RS (1975) Health hazards of bivalve-mollusk ingestion. Ann Intern Med 83:107–110

Edelman R, Levine MM (1986) Summary of an international workshop on typhoid fever. Rev Infect Dis 8:329–349

Effa EE, Bukirwa H (2008) Azithromycin for treating uncomplicated typhoid and paratyphoid fever (enteric fever). Cochrane Database of Systematic Reviews; 4 Art. No.: CD006083

Fenwick A (2006) Waterborne diseases: could they be consigned to history? Science 313:1077–1081

Giannella RA (2010) Infectious enteritis and proctocolitis and bacterial food poisoning. In: Feldman M, Friedman LS, Brandt LJ, (eds) Sleisenger & Fordtran's gastrointestinal and paratyphoid fever liver disease, 9th ed. Saunders Elsevier, Philadelphia: chapter 107

Gupta SK, Medalla F, Omondi MW, Whichard JM, Fields PI, Gerner Smidt P (2008) Laboratory-based surveillance of in the United States: travel and antimicrobial resistance. Clin Infect Dis 46:1656–1663

Gupta V, Kaur J, Kaistha N (2009) Re-emerging chloramphenicol sensitivity and emerging low level ciprofloxacin resistance among *Salmonella enterica* serotype Typhi isolates in North India. Trop Doct 39:28–30

House D, Wain J, Ho VA, Diep TS, Chinh NT, Bay PV (2001) Serology of typhoid fever in an area of endemicity and its relevance to diagnosis. J Clin Microbiol 39:1002–1007

Kanungo S, Dutta S, Sur D (2008) Epidemiology of typhoid and paratyphoid fever in India. J Infect Developing Countries 2:454–460

Keusch GT (1998) Salmonellosis. In: Harrison's principles of internal medicine, 14th edn, Vol 1. McGraw, USA, pp 950–956

Le Minor LE (2003) The genus *Salmonella*. In Dworkin M, Falkow S, Rosenberg E (eds) The Prokaryotes: an evolving electronic resource for the microbiological community, electronic release 3.14, 3rd edn. Springer, New York

Levine MM, Tacket CO, Sztein MB (2001) Host-Salmonella interaction: human trials. Microbes Infect 3:1271–1279

Levine MM, Ferreccio C, Abrego P, Martin OS, Ortiz E, Cryz S (1999) Duration of efficacy of Ty21a, attenuated *Salmonella* typhi live oral vaccine. Vaccine 17:S22–S27

Levine M .2001. Centre for Vaccine Development, UMB, 9th Conf on diarr and nutrition :17-18

Lima AA, Guerrant RL (2009) Inflammatory enteritides. In: Mandell GL, Bennett JE, Dolin R (eds) Principles and practice of infectious diseases, 7th edn. Elsevier Churchill Livingston, chapter 97

Lynch MF, Blanton EM, Bulens S, Polyak C, Vojdani J, Stevenson J et al (2009) Typhoid fever in the United States, 1999–2006. JAMA 302:8

Maskey AP, Basnyat B, Thwaites GE, Campbell JI, Farrar JJ, Zimmerman MD (2008) Emerging trends in enteric fever in Nepal: 9124 cases confirmed by blood culture 1993–2003. Trans Royal Soc Trop Med Hyg 102:91–95

Mweu E, English M (2008) Typhoid fever in children in Africa. Trop Med Int Health 13:1–9

Nordmann P, Poirel L, Mak JK, White PA, McIver CJ, Taylor P (2008) Multidrug-resistant *Salmonella* strains expressing emerging antibiotic resistance determinants. Clin Infect Dis 46:324–325

Ochiai RL, Acosta CJ, Danovaro-Holliday MC, Baiqing D, Bhattacharya SK, Agtini MD et al Domi Typhoid Study Group (2008) A study of typhoid fever in five Asian countries:disease burden and implications for control. Bull World Health Organ 86:260–268

Parry CM, Hien TT, Dougan G, Whitw NJ, Farrar JJ (2002) Typhoid fever. N Engl J Med 347:1770–1782

Parry CM, Threlfall EJ (2008) Antimicrobial resistance in typhoidal and nontyphoidal salmonellae. Curr Opin Infect Dis 21:531–538

Pathania NS, Sachar RS (1965) Typhoid and paratyphoid fever in Punjab (India): a study of 340 cases. Am J Trop Med Hyg 14:419–423

Pokharel BM, Koirala J, Dahal RK, Mishra SK, Khadga PK, Tuladhar NR (2006) Multidrug-resistant and extended-spectrum beta-lactamase (ESBL)-producing *Salmonella enterica* (serotypes Typhi and Paratyphi A) from blood isolates in Nepal: surveillance of resistance and a search for newer alternatives. Int J Infect Dis 10:434–438

Popoff MY, Le Minor LE (2005) Genus *Salmonella*. In Brenner DJ, Krieg NR, Staley JT (eds) Bergey's manual of systematic bacteriology, Vol 2, Part B, 2nd edn. Springer, New York, pp 764–799

Rodrigues C, Shenai S, Mehta A (2003) Enteric fever in Mumbai, India: the good news and the bad news. Clinical Infect Dis 36:535

Rowe B, Ward LR, Threlfall EJ (1997) Multidrug-resistant *Salmonella* Typhi: a worldwide epidemic. Clin Infect Dis 24:S106–S109

Shaw AV, Reddy EA, Crump JA (2008) Etiology of community-acquired bloodstream infections in Africa. 46th Annual Meeting of the Infectious Diseases Society of America. Infectious Diseases Society of America, Washington, DC

Sinha A, Sazawal S, Kumar R, Sood S, Reddaiah VP, Singh B, Rao M, Naficy A, Clemens JD, Bhan MK (1999) Typhoid fever in children aged less than 5 years. Lancet 354:734–737

Sood S, Kapil A, Dash N, Das BK, Goel V, Seth P (1999) Paratyphoid fever in India: an emerging problem. Emerg Infect Dis 5:483–484

Sur D, von Seidlein L, Manna B, Dutta S, Deb AK, Sarkar BL et al (2006) The malaria and typhoid fever burden in the slums of Kolkata, India: data from a prospective community based study. Trans R Soc Trop Med Hyg 100:725–733

Sur D, Ali M, Von SL, Manna B, Deen JL, Acosta CJ et al (2007) Comparisons of predictors for typhoid and paratyphoid fever in Kolkata. India. BMC Public Health 7:289

Swaminathan B, Beebe J, Besser J (2005) Investigation of foodborne and waterborne disease outbreaks. In Manual of Clinical Microbiology, 7th edn. American Society of India, Washington, pp 174–190

Thiem VD, Lin FYC, Canh DG, Son NH, Anh DD, Mao ND et al (2011) The Vi Conjugate Typhoid Vaccine Is Safe, Elicits Protective Levels of IgG Anti-Vi, and Is Compatible with Routine Infant Vaccines. Clin Vaccine Immunol 18:730–735

Threlfall EJ, Ward LR, Skinner JA, Smith HR, Lacey S (1999) Ciprofloxacin-resistant *Salmonella* Typhi and treatment failure. Lancet 353:1590–1591

Walia M, Gaind R, Paul P, Mehta R, Aggarwal P, Kalaivani M et al (2006) Age related clinical and microbiological characteristics of enteric fever in India. Trans R Soc Trop Med Hyg 100:942–948

Udeze AO, Abdulrehman F, Okonko IO, Anibijubon II (2010) Seroprevelence of salmonella typhi and salmonella paratyphi among the first year student of university of llorin, Nigeria. Middle East J Sci Res 6:257–262

Zaki SA, Karande S (2011) Multidrug-resistant typhoid fever: a review. J Infect Developing Countries 5:324–337

Shigellosis: An Emerging Water-Related Public Health Problem

7

Usha K. Baveja

Abstract

Shigellosis, an old disease in new clothes, caused by *Shigellae*, has emerged as one of the major public health problems of particularly the developing regions. *Shigellae* are a group of microorganisms, which are spread through various routes including contaminated water and food. *Shigellae* are the third of most common pathogens transmitted through food. These cause dysentery, a clinical condition characterized by tenesmus (intestinal colicky pain) and the frequent passage of blood-stained mucopurulent stools. The Genus *Shigella* is grouped into four species: *Shigella dysenteriae*, *Shigella boydi*, *Shigella flexneri*, and *Shigella sonnei*. These species differ in geographic distribution, virulence, and drug susceptibility. *Shigellosis* is common in tropics and developing world due to poor hygiene, sanitation, high-risk behavior, and scarcity of water. The global disease burden due to *Shigellosis* is some 120 million cases, majority of these occur in the developing countries and involve children less than 5 years of age. 60 % of the deaths out of total of 1.1 million deaths each year occurring in children under 5 years of age are due to *Shigella* infection. Currently, *Shigellosis* disease burden is estimated to be at 90 million episodes and 108,000 deaths per year. About 500,000 cases of shigellosis are reported in addition, each year among military personnel and travellers from developed countries. Estimates of *Shigellosis* by ICCDRB, Bangladesh, show a small reduction in number of infections but marked decrease (90 %) in mortality compared to previous estimates. However, shigellosis has become a major health problem due to emergence of drug-resistant organisms, low infectivity dose (10 *S. dysenteriae* organisms), poor hygiene, and virulence. Humans and possibly some other primates are the reservoir for *Shigellae*. Currently, there is no effective vaccine, and the organism is reportedly increasingly

U. K. Baveja (✉)
Department of Clinical Lab Medicine, Medanta,
The Medicity, Sector-38, Gurgaon, India
e-mail: ubaweja@gmail.com

P. P. Singh and V. Sharma (eds.), *Water and Health*,
DOI: 10.1007/978-81-322-1029-0_7, © Springer India 2014

developing resistant to the drugs like amoxicillin, cotrimoxazole, ampicillin, and chloramphenicol, etc. This is further adding to the problem of *Shigellosis*. Hence, it is prudent to include a chapter on *Shigellae* and *Shigellosis* in this tome on water and health.

Keywords
Disease burden • Drug resistance • Epidemiology • Prevention • Risk factors • Vaccine

Introduction

Shigellosis includes a spectrum of illness ranging from self-limiting mild diarrhea to bacillary dysentery to complications like hemolytic-uremic syndrome (HUS). *Shigellosis* is caused by the members of the Genus *Shigella*. The severity of disease depends upon the causative serotype of *Shigella*. *Shigella dysenteriae* Type 1 causes relatively severe disease compared to other *Shigellae* serotypes. *Shigellosis* out breaks are more common in developing world (poor hygiene, poor sanitation, overcrowding, and malnutrition) and in children below 4 years of age. Outbreaks caused by *S. dysenteriae* Type 1 are associated with high attack rates and high mortality. *S. dysenteriae* serotype 1 is responsible for the most virulent form of the disease. The multidrug resistance of this serotype to antibiotics, including fluoroquinolones, further adds to the problem particularly in the absence of an effective vaccine. Concerted efforts in the form of surveillance, safe drinking water, good sanitation, and good personal hygiene in addition to development of a successful vaccine will help to prevent the *Shigellosis* outbreaks and contain the disease.

History

The Japanese bacteriologist Kiyoshi Shiga, working as a research assistant in the Institute for Infectious Diseases, Japan, under the direction of Dr. Shibasaburo Kitasato, was asked to investigate cases of "*Sekiri.*" *Sekiri* is a Japanese word and means "red diarrhoea." Dr. Shiga studied these patients and isolated the bacteria from the intestinal tissue of a dysentery patient. This bacterium was cultivated in vitro and when fed to dogs produced the disease. The "red diarrhea" was discovered to be caused by a bacterium which was named *Bacillus dysenteriae* (now *S. dysenteriae)* in 1897. Many, many cases of "*sekiri*" occurred in 1897 with a mortality rate of more than 20 %. The bacteria produced toxins, later named as Shiga toxins. Dr. Shiga also tried to develop a vaccine and tried the first killed vaccine on himself (Trofa et al. 1999). However, we still have no successful vaccine. The *Bergey's Manual of Determinative Bacteriology* renamed the bacteria *Shigella* to honor Dr. Shiga in 1930, and the four species of *Shigella* (group A, B, C, and D) were assembled together taxonomically in 1950s (Ananthnarayan and Paniker 2009).

Shigella has history of being used as a potential biological weapon agent due to its high infectiousness/low-dose infectivity, mortality, antibiotics resistance, and the lack of a vaccine against the disease. It is reported that Japanese experimented with *Shigellosis* on prisoners of war during the World War II; Iraq developed *Shigellae* as a possible offensive biological agent, and this organism has also been used as a terrorist to willfully contaminate food and cause harm (Pike 2007).

Global Alert and Response (GAR) WHO have reported many outbreaks of *Shigellosis* from all over the developing world: *Shigellosis* in Sudan (Nov, 2004), *Shigellosis* in Central African Republic and bloody diarrhea in Liberia (Nov 2003), dysentery in Sierra Leone (Feb 2000) and many more.

Disease Burden

Shigellosis is endemic throughout the world, and outbreaks of *Shigellosis* have been reported from all over the world. *Shigellosis* is held responsible for some 120 million cases of severe dysentery with blood and mucus in the stools, and majority of cases occur in developing countries and involve children less than 5 years of age with 60 % of 1.1 million deaths due to *Shigellae* occurring in children under 5 years of age. The prevalence is more in developing countries compared to developed countries, and also, the species of *Shigellae* causing infections in developing and developed world are different (Niyogi 2005; WHO 2005).

WHO (2009) estimated global *Shigellosis* disease burden to be 90 million infection episodes and 108,000 deaths per year. Children between the ages of 1–4 and particularly those in child care facilities/homes/closed communities are the most affected. Outbreaks/epidemics in developing countries are primarily caused by *S. dysenteriae,* though the scenario of the serotype causing the outbreaks keeps changing every 10 years (WHO 2009).

It is estimated that approximately 450,000 cases of *Shigella* with 14,000 confirmed cases occur in US each year. Two-thirds of the reported cases are caused by *Shigella sonnei* and the rest by *Shigella flexneri.* Most of the *Shigellosis* outbreaks reported in USA are associated with contaminated food. Illinois reported an outbreak of *Shigellosis* caused by *S. sonnei* at a Subway restaurant in March 2010; an outbreak of *Shigellosis* was reported in 2000 affecting 10 states including California, Washington, and Oregon. Contaminated bean dip possibly due to an ill employee was implicated for the outbreak (US National Library of Medicine and the National Institute of Health 2010; National Center for Zoonotic, vector Borne, and Enteric Diseases 2009; Kimura et al. 2004).

International Vaccine Initiative (IVI) under the Diseases of the Most Impoverished (DOMI) Programme conducted *Shigellosis* disease-burden studies in Bangladesh, China, India, Indonesia, Pakistan, Thailand, and Vietnam by analysis of retrospective data, prospective surveillance, hygienic practices, and economic status. This was a two-year study on a population size of 568,000. Estimates of *Shigellosis* were 91 million cases and 414,000 deaths. The crude incidence rates of *Shigellosis* ranged from 1 to 12/1,000 per year, and the case-fatality rate was 1.5 per 1,000 *Shigella*-associated episodes. *S. flexneri* was found to be the predominant species, except in Thailand where it was *S. sonnei. S. dysenteriae* type 1 was not isolated at any surveillance site. Resistance to Ampicillin and co-trimoxazole was seen in 50 % of isolates. Bangladesh and China reported high resistance to Nalidixic acid (von Seidlein et al. 2006; Legros 2004).

We do not have such surveillance data on *Shigella* disease burden in India even though *Shigellosis* is endemic in India. Outbreaks of dysentery caused by multidrug resistant *S. dysenteriae* type 1 with high fatality were reported from different parts of India during 1984–1985. Again during 2002–2003, fluoroquinolone-resistant *S. dysenteriae* reemerged caused several outbreaks in Kolkata. Subsequently, there were outbreaks caused by *S. flexneri* in Kolkata. Currently, in 2009 and 2010, food-related outbreaks caused by *S. sonnei* were reported from Kerala and Maharashtra (Nandy et al. 2009).

Shigellosis outbreaks are common among travelers to foreign countries and refugee camp workers, in areas with poor sanitation, contaminated food and water, and crowded areas. About 500,000 cases of *Shigellosis* are reported each year among military personnel and travellers from industrialized countries (Berger 2010).

Transmission

All humans are susceptible to *Shigellae,* but in endemic areas, infants, young children, and the elderly are more susceptible. Immuno-compromised and, debilitated, malnourished adults are also at a higher risk.

Known significant reservoir of Shigella species is humans. Most common route of transmission is foecal-oral route. Most outbreaks of shigellosis are associated with consumption of contaminated food and water (food handlers with poor personal/hand hygiene are involved). *Shigellosis* outbreaks are also common in conditions of overcrowding and poor sanitation such as prisons, refugee camps, daycare homes, etc. Infection is also common among travellers and military troops deployed in camps with less than optimal hygiene conditions.

Transmission mostly from humans-to-humans by the following.

- Consumption of contaminated food and water
- Eating contaminated vegetables harvested from a field with contaminated sewage
- Inadequate basic hygiene and hand washing habits, poor toileting behavior
- Drinking/swimming in contaminated water
- Person to person among MSM (gay bowel syndrome)
- Direct foeco-oral among toddlers
- Transmission by houseflies: infected feces to food

Infectivity dose is 10–100 microorganisms and incubation period is 1–3 days, but it can range from 12 to 96 h; for *S. dysenteriae,* it can be even up to one week (Todar 2011).

Epidemiology

Humans are the only known significant reservoir of this microorganism. All humans are susceptible to get infection as infectivity dose is low (10–100 organisms) and virulence of some of the serotypes (*S. dysenteriae* type 1) is high. The risk factors include poor sanitation, poor personal/hand hygiene, poor toileting habits, consumption of contaminated water and food, low economic status, extremes of age (elderly and children 1–4 years), debilitated/malnourished/immunocompromised status, and other high-risk behaviors (men having sex with men). Poor personal/hand hygiene has been implicated in *Shigellosis* outbreaks particularly in the conditions of overcrowding and poor sanitation. Overall, *Shigellosis* outbreaks tend to occur more frequently before and after the rainy season. Until recently, *Shigellae* sensitive to fluoroquinolones, such as ciprofloxacin and norfloxacin, were being isolated and reported. However, now outbreaks due to *S. dysenteriae* type 1 strain, resistant to these and other commonly used antibiotics, are being reported. Report of a workshop held at ICDDR, B: Centre for Health and Population Research, Dhaka, Bangladesh, on 16–18 February 2004 indicated that *S. dysenteriae* type 1 infection rates appear to follow a 10-year cycle, and with each new cycle, the epidemic strain seems to acquire resistance to the antibiotics that were commonly used for treatment. Data presented reported that more than 50 % of isolates were resistant to Ampicillin, and a similar proportion was resistant to co-trimoxazole. High resistance to nalidixic acid was observed in Bangladesh and China. Moreover, clinical efficacy of Nalidixic acid has been questioned for many years, even against sensitive strains of *S. dysenteriae* type 1. Surveillance data collected from 11 countries revealed a median incidence rate of bloody diarrhea of 10.2/1,000 per year. *S. dysenteriae* type 1 of all *Shigella* species continues to be a threat because of the severity of disease and its epidemic potential. Although few outbreaks have been reported over recent years, the problem should not be considered to be solved. Outbreaks of dysentery due to *S. dysenteriae* type 1, resistant to ciprofloxacin and the other newer quinolones, have been observed in India, Bangladesh, and Nepal. Data from environmental studies demonstrated that *Shigellae* can be found in environmental water samples, including at times when no cases of *S. dysenteriae* were being isolated from patients (von Seidlein et al. 2006; Legros 2004; Nandy et al. 2009).

Shigellosis outbreaks with multi drug resistant (resistant to fluoroquinolones) and *S. dysenteriae* type 1 with high mortality have been reported from India from time to time (1984–1985, 2002–2003). *S. dysenteriae* type 1 has not been isolated since 2005, though outbreaks of *Shigellosis* did occur. These were caused by *S. flexneri*. *Shigellosis* outbreaks with *S. sonnei* were reported from Kerala and Maharashtra in 2009 and 2010. These studies emphasize the need for good epidemiological surveillance systems to detect the outbreaks, determine the antibiotic sensitivity patterns, and prepare for effective interventions, and the need to give special attention to outbreaks among refugees, internally displaced populations with overcrowding and poor sanitation. Access to diagnostic capacities, affordable/effective antibiotic regimens, and food supplements needs to be provided to control the outbreaks (Taneja 2007; Pajhani et al. 2008).

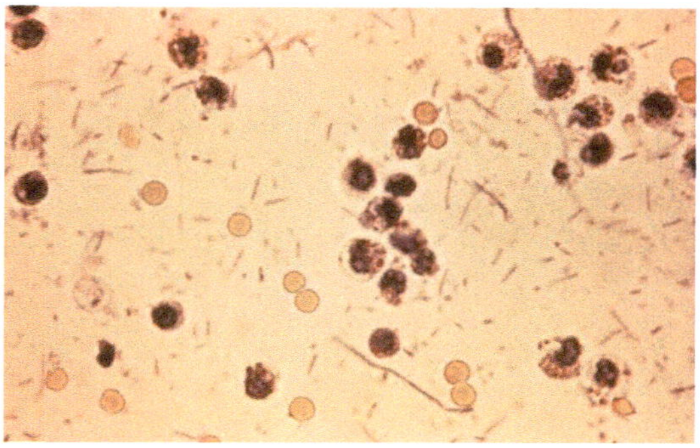

Fig. 7.1 Gram staining of stool smear from a patient showing Gram-negative bacilli—the Shigellae, RBCs and pus cells. *Source* http://microbewikikenyon.edu/images/2/29/shigella_stool.jpg

Bacteriology

Shigellae belong to the family Enterobacteriaceae, Tribe *Escherichiae,* and Genus *Shigella*. These are non-lactose-fermenting members of the family and are grouped into four species on the bases of biochemical and serological characteristics. The four species are *S. dysenteriae* (15 serotypes), *S. flexneri* (6 serotypes), *S. boydi* (23 serotypes), and *S. sonnei* (antigenically homogeneous one serotype, but occurs in two phases: phase I and II). *Shigellae* are Gram-negative slender rod-shaped, nonmotile, non-capsulated, nonsporing slender bacilli/coccobacilli, aerobes, and facultative anaerobes. *Shigella* species are differentiated on the basis of biochemical reactions and confirmed by serotyping (Trofa et al. 1999; Ananthnarayan and Paniker 2009; Shigella and Shigellosis 2011) (Fig. 7.1).

Antigens and Toxins

Shigellae possess both major and minor somatic antigens and also fimbrial antigens. Antigenic structure of *Shigellae* is simple compared to *Salmonellae*. Hence, the identification of *Shigellae* is done on the basis of both biochemical and serological characteristics and not solely on basis of slide agglutination test.

The **Shiga toxin,** also called the **verotoxin**, is produced by *S. dysenteriae* and enterohemorrhagic *Escherichia coli* (EHEC), of which the strain O157:H7 has become the best known. Shiga toxin causes the syndromes of dysentery, hemorrhagic colitis, and hemolytic-uremic syndrome. HUS patient may present with severe diarrhea, abdominal pain, vomiting, and bloody urine. The onset of symptoms is generally within a few hours, with higher doses leading to more rapid onset. Supportive care requires maintenance of fluid and electrolyte levels, and monitoring and support of kidney function. Immunoassays are available for rapid detection of the toxin. Inactivation of the toxin is achieved by steam treatment, oxidizing agents, such as bleach, and chemical sterilizing agents such as glutaraldehyde. Table 7.1 gives details of various genes that encode production of Shiga toxin.

Structure of the Toxin

The Shiga toxin, a holotoxin, is a protein made of two subunits A and subunit B. Subunit A is a single molecule, and Subunit B is made of five molecules. A subunit (32,000 MW) is responsible for the toxic action of the protein, and five molecules of the B subunit (7,700 MW) are responsible for binding to a specific cell type Gb 3 receptor on the cell surface.

Table 7.1 The toxin has been given several names depending on the bacterium that produces it and the gene that encodes it

Source organism	Gene designation	Toxin name	Older names
Shigella dysenteriae, type I	stx	Shiga toxin (Stx)	Shiga toxin
Escherichia coli	stx1	Shiga toxin 1 (Stx1)	Shiga-like toxin I, Verotoxin 1
	stx2	Shiga toxin 2 (Stx2)	Shiga-like toxin II, Verotoxin 2

Mechanism of Action

The toxin acts on the lining of the blood vessels, the vascular endothelium. The B subunits of the toxin bind to a component of the cell membrane known as Gb 3, and the complex enters the cell. When the protein is inside the cell, the A subunit interacts with the ribosomes to inactivate them. The A subunit of Shiga toxin is an N-glycosidase that modifies the RNA component of the ribosome to inactivate it and so bring a halt to protein synthesis leading to the death of the cell. This killing of cells leads to a breakdown of the lining and to hemorrhage. The first response is commonly a bloody diarrhea. This is because Shiga toxin is usually taken in with contaminated food or water. The toxin is effective against small blood vessels, such as found in the digestive tract, the kidney, and lungs. A specific target for the toxin appears to be the vascular endothelium of the glomerulus. Destroying glomeruli leads to kidney failure and the development of the often deadly and frequently debilitating hemolytic-uremic syndrome (Todar 2011).

Pathophysiology of Shigellosis

Pathology

The infection is acquired by foeco-oral route by consumption of contaminated food and water or through person-to-person contact. The minimum infective dose is 10–100 bacilli. Once ingested, the bacteria survive the gastric environment of the stomach and move on to the large intestine. Pathogenicity mechanism resembles enteroinvasive *E. coli*. The bacilli adhere to the epithelial cell of the villi of large intestine and are internalized via an endosome through a process which is similar to the mechanism of phagocytosis, penetrate the cells and multiply inside the cells. The multiplying bacteria spread laterally to the adjacent cells and deep into mucosa and penetrate the lamina propria. There is degeneration of the epithelium and inflammation of the lamina propria, capillary thrombosis resulting in necrosis and sloughing of patches of mucosa and formation of ulcers. Because of the ulcers, there is leakage of blood, inflammatory elements, and mucus into the intestinal lumen. Patients suffering from *Shigella* infection will therefore have fever; pass frequent, scanty, and dysenteric stool mixed with blood and mucus, since under these conditions, the absorption of water by the colon is inhibited. Bactereamia may occur at this stage particularly in malnourished, immunocompromised AIDS patients depending upon the virulence of the infecting type. Mostly, *S. dysenteriae* type I is the most virulent and *S sonnei* is the least virulent among the four species of *Shigellae*.

It is possible that prostaglandin interactions induced by the inflammatory response to bacterial invasion contribute to diarrhea in patients with *Shigella* colitis.

Mechanism of Adherence and Internalization

Entry of *S. flexneri* organism into the epithelial cell is aided by a number of proteins encoded by bacterial DNA and Plasmid DNA. These proteins are invasion plasmid antigens (Ipa), surface presentation antigens (Spa), membrane excretion proteins (Mxi), and virulence proteins (Vir). The VirF protein induces the expression of the VirB protein, which activates *ipa, mxi,* and *spa* promoters leading to expression of the *spa* and *mxi* operons. In the bacteria growing at 37 °C, the expression of Ipa, Spa and Mxi leads to assembly of a protein called the Mxi-Spa translocon. When *S. flexneri* adheres to the mucosal epithelial cell, this translocon becomes activated and secretes pre-synthesized Ipa proteins. IpaB, IpaC and IpaA associate to form a complex which interacts with the host epithelial cell membrane to induce a cascade of cellular signals which will lead to the internalization of the bacterium via an endosome.

Intracellular and Intercellular Spread

Extracellular *S. flexneri* cells are nonmotile, but intracellular bacteria move to occupy the entire cytoplasm of the infected cell, and they are able to spread between cells. The genes necessary for intracellular and intercellular spreading are *virG (icsA)* and *icsB*. For spread, the bacterium expresses both an Olm ("organelle-like movement") phenotype and an alternative Ics phenotype.

Olm phenotype allows the bacteria to "slide" along actin stress cables inside the host cell (intracellular), and Ics phenotype allows the bacteria to "spread" or infect adjacent cells (intercellular).

The *mxiG* gene is required for Ipa protein secretion and is also essential for entry. This gene and others in the Mxi-Spa translocon are also required for intercellular dissemination.

Pathological Effects

Once the *Shigellae* are internalized in the epithelial mucosal cells of the colon, depending upon the virulence of the infecting strain, there may be inflammation, degeneration of cells leading to desquamation and ulceration of the mucosa, and subsequent leakage of blood, inflammatory elements, and mucus into the intestinal lumen. Patients suffering from *Shigella* infection will therefore pass frequent, scanty, and dysenteric stool mixed with blood and mucus. This is not seen in patients suffering from diarrhea due to Shigellae (e.g., *S. sonnei*). It is possible that prostaglandin interactions induced by the inflammatory response to bacterial invasion contribute to diarrhea in patients with *Shigella* colitis (Todar 2011; Wolf and Gianella 1996; Moralez et al. 2011; ICDDR 2004).

Virulence mechanism:

- Adhesion to target cells (M cells and/or epithelial intestinal cells) via adhesins
- Penetration into target cell via endocytosis and subsequent bursting of endosome
- Migration to neighboring cells by remodeling the actin in the cytoskeleton of target cells
- Antigenic variation that produces the various serological strains of *Shigella*
- Evasion of phagocytosis by living within and consequently killing the phagocytes
- Enterotoxin production

Clinical Spectrum: Shigellosis

The term *Shigellosis* includes the whole spectrum of diseases from diarrhea to bacillary dysentery to bacteraemia to various complications including HUS, all caused by the members of the Genus *Shigella*. The onset and course of symptoms/disease vary depending upon the virulence of the infecting strain. The usual incubation period is 48 h; it can range from 1 to 7 days. The infecting dose is 10–100 bacilli. Human beings are the only known natural host of *Shigellae*.

Bacillary dysentery is characterized by passage of frequent, scanty, and dysenteric stool mixed with blood and mucus, along with abdominal cramps and tenesmus. Most cases of *Shigellosis* resolve with effective antibiotic therapy within days. Complications can occur in

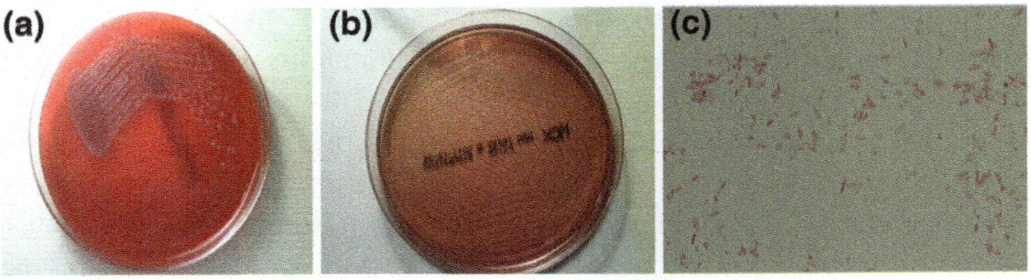

Fig. 7.2 **a** Blood Agar with *S. flexneri*. **b** McConkey Agar with *S. flexneri*. **c** *Shigella bacilli*, Gram-negative rods

cases infected with *S. dysenteriae* type 1. These complications can be intestinal and can include intestinal perforation, hemorrhage, and extra-intestinal and can include HUS, meningitis, vaginitis, arthritis, rash, hypoglycemia, and high white blood cell count, toxic neuritis and intussusception, etc. The case-fatality rate with *S. dysenteriae* type 1 can be between 10 and 25 %. The overall mortality rate in developed countries, where other *Shigella* subtypes are more common, is about 1 % (WHO 2005; National Center for Zoonotic, vector Borne, and Enteric Diseases 2009).

Diagnosis

Bacteriological examination of stool sample from cases suffering from dysentery suspected to be due to *Shigellae* should be done within 2–4 h after collection of specimen. In case, delay is anticipated and the samples should be placed in transport medium which can be either buffered glycerol in saline or Cary-Blair medium. Usually, enrichment is not required for stool samples. Enrichment is required for processing the suspect food/water samples to isolate and identify *Shigellae*. Selenite F broth and Tetra thionate broths can be used for enrichment. At least two differential media should be used for plating directly from stool and from transport medium. These are MacConkey and Xylose Lysine Desoxycholate (XLD) Agar. Colonies of *Shigellae* appear colorless on MacConkey Agar, while red or colorless on XLD Agar (Fig. 7.2).

Sensitivity of detection from stools is increased by use of immunomagnetic separation, which results in the enrichment of bacteria from stools.

According to the CDC, Kligler Iron Agar (KIA) or Triple Sugar Iron (TSI) Agar, screening media are the most useful biochemical test. *Shigellae* primarily produce an alkaline (red) slant, acidic (yellow) butt, little or no gas production, and no H_2S. If desired, other biochemical tests may be performed, including: Urea and Motility testing. The results of these tests are both negative.

Final confirmation is done by serotyping. Slide agglutination using both polyvalent (somatic antigen) and monovalent antisera is done for grouping and speciation of the isolated *Shigellae*.

Polymerase chain reaction (PCR) and immunoassays can be used to detect viable, but non-cultivable forms in environmental samples. Real-time PCR is an efficient, rapid detection method but is expensive (WHO 2005; von Seidlein et al. 2006; Legros 2004).

Treatment

Mild form of Shigellosis runs a natural course, and usually, there is spontaneous recovery. In case of colitis and diarrhea, oral rehydration salts (containing sodium chloride, potassium chloride, citrate and glucose) and fluid suffice to compensate for the water loss. Appropriate antibiotic therapy based on local sensitivity pattern is given in severe cases of *Shigellosis*. The commonly used antibiotics now are TMP-SMX, Ciprofloxacin, Fluoroquinolones, Azithromycin, Ceftriaxone, etc., as the organism has developed resistance to the earlier used drugs like Nalidixic acid, Ampicillin, and Chlorumphenicol, etc.

It is recommended that any antidiarrheal drugs should not be used. It should be kept in mind that severe cases of *Shigellosis* may require hospitalization (Ananthnarayan and Paniker 2009; Legros 2004) and (Status 2010).

Surveillance Techniques for Shigellosis

Surveillance for *shigellosis* includes collection and processing of stools, food, environmental, and relevant individual specimens for detection of *Shigellae*. The method will differ in each case because of different bacterial loads and the presence of different types of competitive organisms in different samples.

- Bacteriologic analysis of specimens from patients is done within 2–4 h of collection of samples. In case of delay, the specimens need to be stored at 4 °C in transport media (buffered glycerol in saline or Cary-Blair).
- Enrichment of food and environmental samples may be necessary due to the low numbers of viable organisms present.
- Use two differential media for direct plating from stools and for plating from enrichment broths.
- One of the two media is usually selective, while the other one is a non-inhibitory medium, such as MacConkey agar.
- Detection of viable but non-cultivable forms in environmental samples is best achieved by PCR or immunoassays. Real-time PCR is the most rapid detection method but is expensive.

Ongoing surveillance and antibiogram of the isolates in endemic areas help to take measures for prevention, control, and administration of appropriate antibiotic treatment of Shigellosis (Legros 2004; Berger 2010).

Prevention and Control

Most important practice for prevention and control of *Shigellosis* is observing and practicing "Hand Hygiene." Hands should be washed with soap and water following WHO recommendations. Interventions for prevention of *Shigellosis* include the following: [National Center for Zoonotic, vector Borne, and Enteric Diseases (2009); Kimura (2004); Nandy et al. (2009)].

- Ensuring quality of drinking water with point-of-use water disinfection and safe water storage
- Thoroughly washing hands after changing and disposing of an infant's diaper
- Improving sanitation conditions, altering toileting practices to minimize contact between hands and stool
- Preventing infected individuals from handling food
- Disinfecting surfaces handled by infected individuals
- If traveling, consuming boiled or filtered water, fruits peeled by self, and hot meals
- Fly control where appropriate
- Proper storage of food
- Not allowing infected children to play in day schools and community swimming areas

Some tips for preventing the spread of shigellosis:

- Wash hands with soap carefully and frequently, especially after going to the bathroom, after changing diapers, and before preparing foods or beverages
- Dispose of soiled diapers properly
- Disinfect diaper changing areas after using them
- Keep children with diarrhea out of child care settings
- Supervise hand washing of toddlers and small children after they use the toilet
- Do not prepare food for others while ill with diarrhea
- Avoid swallowing water from ponds, lakes, or untreated pools

Vaccination and Preventive Immunization

As of today, there is no successful vaccine available for *Shigellosis*. Antigenic complexity of *Shigella* species, the lack of cross-protective epitopes among the different species, and the lack of the understanding of the protective immune response are some of the factors impeding the development of a successful vaccine. Several studies in volunteers with candidates based on *S. flexneri* 2a and one based on *S sonnei* have shown protection and immunity to challenge with homologous strain. The protective immunity to *Shigellae* is most probably directed to the O-somatic antigen and is narrowly type-specific. In addition, cell-mediated immunity mechanisms, including IFN-?-secreting T cells, seem to play a role in recovery and immunity.

Several techniques that may provide protection and immunity from the disease are being studied. Parenteral O-specific polysaccharide conjugate vaccines; nasal proteosomes producing *Shigella* LPS and live, attenuated, multivalent mucosal vaccines administered orally. These methods are based on studies that have shown individuals infected with a certain strain are better protected against future infections with the same serological strain. However, results have shown that live, attenuated vaccines are effective and immunogenic when challenged with different serological strains in animal models.

Three conjugate vaccines developed and evaluated in phase I and II volunteer studies have shown immunogenicity and efficacy in adults. Whatever the method and whichever the candidate vaccine, the important thing is that it should prevent the adhesion and internalization of *Shigellae* of all types to and inside the target intestinal mucosal epithelial cell. However, in spite of the ongoing research, so many dilemmas still remain. These are as follows: which strains of *Shigellae* to include in a vaccine candidate for global use?; what are the attenuating mutations needed in the live strains?; what are the protective immunological mechanisms?; which vaccine will be effective globally and in all ages?; and the correlates of protection? Another problem is that this vaccine may belong to "low demand" category of vaccines so research is limited.

Vaccines Under Development

Polysaccharide conjugate vaccines developed by NIH were shown to be 74 % efficacious against disease when tested in field trials with Israeli military volunteers and demonstrated safety and immunogenicity in 4–7-year-old children. This and other conjugate candidate vaccines being developed and tried by NIH hold some promise. The synthetic oligosaccharides mimicking the O-antigen protective epitopes, conjugated to different protein carriers, offer promise for a better and cheaper generation of conjugate *Shigella* vaccines

Live Attenuated Vaccines

The main problem with development of candidate live oral attenuated *Shigellosis* vaccines is the degree of attenuation responsible for excessive reactogenicity of the vaccine, especially in children, and over-attenuation leading to insufficient immunogenicity in human subjects, especially in developing countries. Hybrid attenuated *Shigella* strain vaccines are also being developed using different serotypes. These have shown protection in field studies in China. However, the three-dose vaccination regimen with high doses of live vaccine strain remains a problem at this time.

The problems encountered to develop a live attenuated vaccine that has the right balance between robust immunogenicity, especially in young children, optimal colonization, and shedding patterns, and clinical tolerance of the attenuated strains as well as being multivalent vaccine to cover a spectrum of *Shigellae* species still remain.

Other Candidate Vaccines

Johns Hopkins University in Baltimore, MD, USA, developed a formalin-inactivated *S sonnei* killed, whole-cell oral vaccine. Antex (USA) is developing a *Shigella* inactivated whole-cell vaccine as well as an oral travellers' diarrhea vaccine (ActivaxTM) that will be effective against ETEC and *Shigella*. These candidate vaccines will shortly undergo clinical testing.

IVI is also involved in developing parenteral nuclear protein/ribosomal subunit *Shigella* vaccines. Walter Reed Army Institute of Research (WRAIR) is involved in developing a nasally administered proteosome vaccine consisting of *Shigella* LPS linked to micelles of the outer membrane protein of group B *Neisseria meningitidis*, still at a preclinical stage. A bacterial extract invasion complex named invaplex, which contains IpaB, IpaC, and LPS from *S. flexneri* and *S. sonnei*, was found to elicit protection against challenge in the guinea pig model (Ananthnarayan and Paniker 2009; WHO 2009; Legros 2004; Berger 2010).

References

Ananthnarayan R, Paniker CKJ (2009) Enterobacteriaceae II shigella. In: Ananthnarayan R, Paniker CKJ (eds) Ananthnarayan and Paniker's textbook of microbiology, 8th edn. University Press India Pvt. Ltd., India, pp 283–286

Berger SA (2010) Shigellosis: Global Status, 2010 Gideon e-book series, http://www.gideononline.com/ebooks/disease/shigellosis-global-status/outbreaksamongtravellers

ICDDR (2004) Meeting report-Shigellosis-pathophysiology and clinical management of Shigellosis. B Period J Health, Popul Nutr 22(4)

Kimura AC et al (2004) Multistate Shigellosis outbreak and commercially prepared food, United States. Emerg Infect Dis 10(6):1147–1149. http://www.cdc.gov/ncidod/EID/vol10no6/pdfs/03-0599.pdf. Accessed 9 May 2010

Legros D (2004) Shigellosis: Report of a Workshop-held at ICDDR, B: Center for Health and Population Research, Dhaka Bangladesh, on 16–18 Feb 2004, pp 445–449

Moralez EI, Lofland D (2011) Shigellosis with resultant septic shock and renal failure. Clin Lab Sci 24(3):147–152

Nandy S, Dutta S, Ghosh S et al (2011) Foodborne-associated *Shigella sonnei*, India, 2009 and 2010. Emerg Infect Dis 17(11):1–5. http://wwwnc.cdc.gov/eid/article/17/11/11-0403_article.htm

National Center for Zoonotic, vector Borne, and Enteric Diseases of the CDC (2009) Shigellosis. (16 Nov 2009): 6 May 2010. http://www.cdc.gov/nczved/divisions/dfbmd/diseases/shigellosis/

Niyogi SK (2005) Shigellosis. J Microbiol 43:133–143

Pajhani GP, Niyogi SK, Singh AK et al (2008) Molecular characterization of multidrug resistant Shigella species isolated from epidemic and endemic cases of Shigellosis. J Med Mic 57(7):856–863

Pike J (2007) Shigellosis-biological weapons. Global security. Global security, 23 Oct 2007. http://www.Globalsecurity.org/wmd/intro/bio-shigellosis.html. Accessed 7 Nov 2010

Baveja UK Shigellosis: National Institute of Allergy and Infectious Diseases, Department of Health and Human Services, NIH USA, http://www.niaid.nih.gov/topics/shigellosi/pages/

Taneja N (2007) Changing epidemiology of Shigellosis and emergence of Ciprofloxacin resistant Shigellae in India. J Clin Mic 45(2):678–679

Todar K (2011) Shigella and Shigellosis. 1–4, http:\\www.textbookofbacteriology.net

Trofa AF, Ueno-Olsen H, Oiwa R, Yoshikawa M (1999) Dr Kiyoshi Shiga: Discovery of the dysentery bacillus. Clinical Infect Dis 29(5):1303–1306. 9 May 2010 http://www.jstor.org/pss/4481998

US National Library of Medicine and the National Institute of Health (2010) Shigellosis. Medline Plus. (23 Mar 2010): 6 May 2010. http://www.nlm.nih.gov/medlineplus/ency/article/000295.htm

von Seidlein L, Kim DR, Ali M, Lee H, Wang X, et al (2006) A multicentre study of shigella diarrhoea in six asian countries: disease burden, clinical manifestations, and microbiology. PLoS Med 3(9):e353

WHO (2005) Shigellosis: disease burden, epidemiology and case management. Wkly Epidemiol Rec 80:94–99

WHO Initiative for Vaccine Research. Diarrhoeal Disease Burden. Shigellosis Feb 2009. Section 7

Wolf DC, Gianella RA (1996) Invasive pathogens. Consultations in Gastroenterology 381–384

Campylobacteriosis and Water: An Overview

8

Kishan K. Nyati and Kashi N. Prasad

Abstract

Campylobacteriosis is an acute infectious diarrheal disease caused by the genus *Campylobacter*. The infection is mainly due to consumption of raw or undercooked poultry meat, unpasteurized milk, and contaminated water. In recent decades, *Campylobacter* (especially *Campylobacter jejuni*, *Campylobacter coli*, *Campylobacter lari*, and more recently *Campylobacter upsaliensis*) have acquired great public health importance worldwide. The incidence of *Campylobacter* infection varies in both industrialized and developing countries. *Campylobacter* species are isolated more frequently than combined isolation rates by *Salmonella* and *Shigella* species from diarrheic stool samples. Serological and nucleic acid-based assays are used to study the epidemiology of *Campylobacter* infection. Antibiotic resistance in *C. jejuni* is considered as an emerging public health problem. Various virulence factors such as flagellin, lipopolysaccharides (LPSs), adhesins, and invasins have been implicated in the pathogenesis of *Campylobacter* infection. In this review, we present information available through literature search on epidemiology, clinical presentations, and bacteriology of campylobacteriosis.

Keywords

Antibiotic resistance · *Campylobacter* · Campylobacteriosis · Diarrhea · Pathogenesis

Introduction

Campylobacteriosis is an infectious disease caused by bacteria belonging to the genus *Campylobacter* that usually infect the intestinal tract and rarely the bloodstream and other extra-intestinal organs after consumption of contaminated food or water. Most people with campylobacteriosis clinically present diarrhea,

K. K. Nyati · K. N. Prasad (✉)
Department of Microbiology, Sanjay Gandhi Postgraduate Institute of Medical Sciences, Rae Bareli Road, Lucknow, 226014, India
e-mail: kashinprasad@gmail.com

P. P. Singh and V. Sharma (eds.), *Water and Health*,
DOI: 10.1007/978-81-322-1029-0_8, © Springer India 2014

abdominal cramp/pain, and fever within 2–5 days after exposure to the microorganism. Sometimes, severe bloody diarrhea with nausea and vomiting may occur. The illness usually lasts 1 week; however, some infected individuals may remain asymptomatic. In individuals with compromised immune system, *Campylobacter* occasionally spreads to the bloodstream and causes a serious life-threatening infection. Sanitation, personal and food hygiene as well as safe water supply are important in its prevention. In recent decades, thermotolerant species *Campylobacter* (*Campylobacter jejuni*, *Campylobacter coli*, *Campylobacter lari*, and more recently *Campylobacter upsaliensis*) have acquired great public health importance, particularly as agents of infectious diarrhea in human (WHO 2001). *C. jejuni* is the most frequently isolated both in developing and developed countries followed by *C. coli* in 5–10 % of all diarrhea cases (Goosens and Butzler 1992). *C. jejuni* alone accounts for more than 90 % of human campylobacteriosis cases in the USA (Zhao et al. 2001). The frequency of isolation of *Campylobacter* species is higher than the combined isolation rates of *Salmonella* and *Shigella* in the USA. The disease burden is underestimated in most countries due to difficulty in isolation of *Campylobacter* species and under-reporting. Antibiotic resistance in *Campylobacter* species, poor hygiene and sanitation, and unsafe drinking water, especially in developing countries, are considered to be the major reasons of persistence and transmission (Tauxe 2002). In this brief review, we present an overview on epidemiological, clinical, and bacteriological aspects of *Campylobacter* infection.

Historical Aspect of *Campylobacter*

Campylobacter species have long been associated with the cause of veterinary diseases, such as diarrhea and septic abortions in cattle and sheep. It was first described by Theodor Escherich in the late nineteenth century in the colon of infants, and he called it "*cholera infantum*" (Escherich 1886). Later, Smith and Taylor (1919) described the vibrionic abortion in cattle and named the agent *Vibrio fetus*. Jones et al. (1931) isolated a "vibrio"-like organism from cattle without selective media by serially diluting the material. The organism was similar to *V. fetus* but antigenically distinct, and it was named *V. jejuni*. Vinzent and coworkers isolated *V. fetus* from blood cultures of women with abortion (Vinzent et al. 1947). The association of *V. fetus* with enteric disease was first reported by Elizabeth King in 1957, and she preferred to name as "related vibrios". Originally placed in the genus *Vibrio* due to their spiral appearance, a new genus name *Campylobacter* (meaning "a curved rod") was proposed to reflect the fundamental differences from the vibrios (Sebald and Veron 1963). In 1970s, *Campylobacter* was successfully isolated from the stools of humans with acute enterocolitis (Dekeyser et al. 1972; Butzler et al. 1973; Skirrow 1977). Consequently, *Campylobacter* species emerged as a significant health problem throughout the world.

Classification

At present, there are 14 valid recognized *Campylobacter* species (Vandamme 2000). These are as follows:

Campylobacter genus
- *C. fetus* (Sheep)
- *C. hyointestinalis* (Pigs)
- *C. sputorum* (Human oral cavity & genital tract of bulls)
- *C. mucosalis* (Pigs)
- *C. concisus* (Human)
- *C. rectus* (Human gingival crevices)
- *C. gracilis* (Human gingival crevices)
- *C. showae* (Human dental plaque)
- *C. jejuni* (Human)
- *C. coli* (Human)
- *C. lari* (Gulls and other animals, humans)
- *C. upsaliensis* (Human, dogs and cats)
- *C. curvus* (Human)
- *C. helveticus* (Dogs and cats)

The genus *Campylobacter* has following general characteristics:

- Single polar unsheathed flagellum at one or both ends.
- Menaquinones-6 and a methyl-substituted menaquinone-6 are the only respiratory quinine detected.
- G + C content of the chromosomal DNA ranges from 29 to 47 mol %.

Epidemiology

Campylobacters are the frequently encountered agents of human bacterial gastroenteritis worldwide, and *C. jejuni* is identified as the most common species causing diarrhea in human globally (Zilbauer et al. 2008). The epidemiology of *Campylobacter* infection in developing countries differs markedly from developed countries. In developed countries, grossly bloody stool and abdominal pain, often with fever, are the characteristic presentations of campylobacteriosis, while in developing countries *Campylobacter* infections are usually milder and present with watery/mucoid diarrhea (Jain et al. 2005). The annual incidence of *Campylobacter* infections in the USA is reported to be 5–7 % of patients with diarrhea corresponding to 2.5 million cases per year (Linton et al. 1997). About 15 of every 100,000 people are diagnosed with campylobacteriosis every year; many cases are unreported, and up to 0.5 % of the general population may unknowingly harbor *Campylobacter* in their gut. As *Campylobacter* enteritis rarely requires hospitalization, under-reporting probably occurs and the true incidence is likely to be much higher (Tauxe 1997). *Campylobacter* is hyper-endemic in developing countries, owing to poor sanitation and close human contact with animals (Tauxe 1997). A study from India had shown that *Campylobacter* infections were more frequent in rural community than combined *Salmonella* and *Shigella* infections (47/348 vs. 15/348; $P < 0.001$) in subjects with diarrhea (Jain et al. 2005).

Campylobacter is considered to be a food-borne disease with infection being derived from a variety of foods and water-based environmental sources. The route of transmission is most commonly via the surface of poultry meat as a result of fecal contamination. Other sources include pets and other animals, untreated water and milk, and sewage contamination (Fig. 8.1) (Ketley 1997). It occurs in all age groups. Some studies have suggested a possible bimodal distribution with peaks in young adults and the

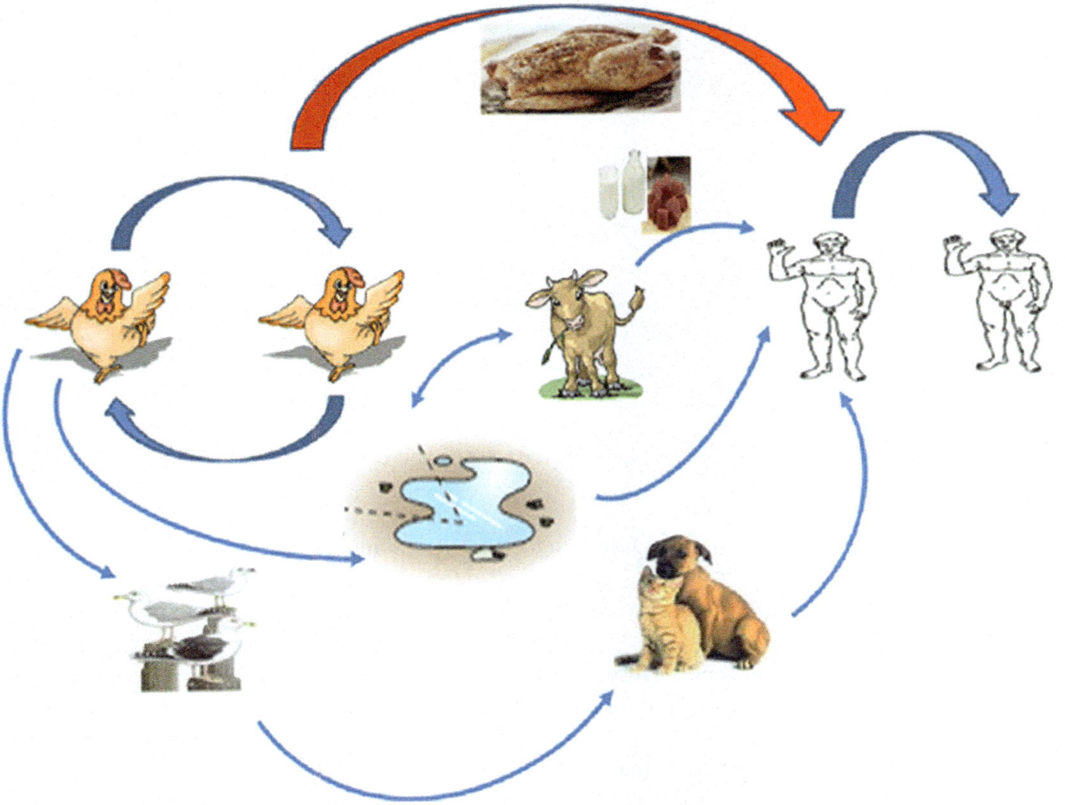

Fig. 8.1 Transmission route of *Campylobacter*

elderly (Stafford et al. 1996). The community-based studies estimated 0.4 episode per child (aged 1–4 year) per year, that is, 40,000/100,000 children in Thailand, Mexico, and Peru (Oberhelman and Taylor 2000). An epidemiological study in a rural community of North India reported *Campylobacter* infection rate 13.5 % in patients with diarrhea irrespective of age; however, the infection rate was significantly higher in children aged less than 5 years, families engaged in agriculture, and persons who did not wash their hands with soap after perianal washing following defecation (Jain et al. 2005). Some of the studies showed seasonal variation with increasing latitude (Nylen et al. 1998). Summer time peaks do occur in Europe (Nylen et al. 1998), New Zealand (Brieseman 1990), and Australia (Stafford et al. 1996). However, there have been some anecdotal reports from Israel and Hong Kong during winter and spring (Shmilovitz et al. 1982; Ho and Wong 1985).

Transmission

Campylobacteriosis usually occurs in single, sporadic cases, but it can also occur in outbreaks, when a number of people become ill at one time. Most cases of campylobacteriosis are associated with consumption of raw or undercooked poultry meat or from cross-contamination of other foods by these items (Zhao et al. 2010). Infants may get the infection by contact with poultry packages in shopping carts. Outbreaks of *Campylobacter* infection are usually associated with unpasteurized milk or contaminated water. Animals can also be infected, and some people have acquired their infection from contact with the stool of an ill dog or cat. The organism usually does not spread from one person to another, but this can happen if the infected person is producing a large volume of diarrhea. Many chicken flocks are infected with *Campylobacter* but show no signs of illness. In a recent study, *Campylobacter* was recovered mainly from chicken breast samples (50 %), and most isolates (70.8 %) were *C. jejuni*. In another report from the USA, a more limited survey of meats collected around the Washington, DC, area in 1999–2000 found *Campylobacter* in chicken (70 %), turkey (14.5 %), pork (1.7 %), and beef (0.5 %) samples (Zhao et al. 2001). Various recovery rates are reported in other studies from the USA and abroad, but all have demonstrated a higher contamination rate in chicken than in turkey, pork, and beef retail products (Atanassova and Ring 1999; Ono and Yamamoto 1999; Hariharan et al. 1990). Infected chicken feces may contain up to 10^9 bacteria per 25 gm, and due to the installations, the bacteria are rapidly spread to other chickens. *Campylobacter* can easily spread from bird to bird through a common water source or through contact with infected feces. When an infected bird is slaughtered, *Campylobacter* can be transferred from the intestine to the meat. Unpasteurized milk gets contaminated if the cow has an infection with *Campylobacter* in her udder or the milk is contaminated with manure. Surface water and mountain streams become contaminated by infected feces from cows or wild birds. This infection is common in the developing world, and travelers to foreign countries are also at risk of becoming infected with *Campylobacter* (de la Cabada Bauche and Dupont 2011).

Morphology

Spiral Form

Campylobacters are slender, spirally curved rods that are 0.2–0.8 μm wide and 0.5–5.0 μm long. They are Gram-negative and non-spore forming (Fig. 8.2a–c). Most species are motile with a characteristic cork screw-like motion by means of a single polar unsheathed flagellum at one or both ends. The flagellin gene is mainly characterized by *fla A* and *fla B*. These two flagellins have an approximate predicted molecular weight of 59 kDa and are >93 % homologous to each other (Patricia et al. 2000). The cell envelope has an inner bipolar lipid cell membrane, a thin peptidoglycan layer, an outer bipolar lipid layer with the lipid moiety of a lipopolysaccharide (LPS) layer embedded in it, and the carbohydrate portion extending to the surface of the cell. Membrane proteins are interspersed in the outer membrane, some of which are exposed to the surface and are antigenic for infected hosts.

Coccoid Form

Cells in old culture may form spherical or coccoid bodies, which are considered degenerative forms rather than a dormant stage of the organism (Hazelegerh et al. 1994). The degeneration appears to be an autolytic process leading to the dissolution of the cells.

Growth Requirement

Most pathogenic *Campylobacter* species require micro-aerophilic environment containing an oxygen concentration of 5 % and carbon dioxide concentration of 10 % and nitrogen of 85 % for optimal growth. Some species such as *Campylobacter sputorum*, *Campylobacter concisus*, *Campylobacter mucosalis*, *Campylobacter curvus*, *Campylobacter rectus*, and *Campylobacter hyointestinalis* require hydrogen for their growth. A gas mixture of O_2 6, CO_2 6, H_2 3, and N_2 85 % supports the growth of both micro-aerophilic as well as hydrogen requiring *Campylobacter* species.

Biochemical Characteristics

The useful tests for the identification of *Campylobacter* include growth temperature studies (e.g., growth at 25, 37 and 42 °C) and different

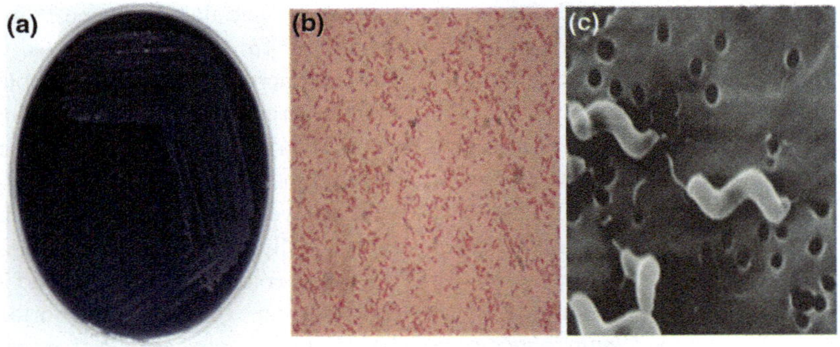

Fig. 8.2 Morphology of *Campylobacter jejuni*. **a** Typical colonies on CCDA medium **b**. Culture smear on CCDA medium (modified Gram stain × 1,000) **c**. Scanning electron microscope image of the *bacterium*

Table 8.1 Biochemical tests for the differentiation of *Campylobacter* species of human importance

Species	Growth at		Growth in			Hydrolysis		Sensitive to			Catalase
	25 °C	43 °C	0.04 % TTC	1 % Gly	1.5 % NaCl	Hipp.	Inox.	NA	Cpl.	Cpz.	
C. jejuni	–	+	–	+	–	+	+	+	–	–	+
C. coli	–	+	+	+	–	–	+	+	–	–	+
C. lari	–	+	d	+	+	–	–	–	–	–	+
C. fetus	+	–	–	+	–	–	–	–	+	+	+
C. upsaliensis	–	d	–	d	–	–	+	+	+	+	w/-
C. concisus	–	d	–	–	+	–	–	–	+	+	–

TTC 2,3,5 triphenyl tetrazolium chloride, *Gly* glycine, *Hipp* hippurate, *Inox* indoxyl acetate, *NA* nalidixic acid, *Cpl* cephalothin, *Cpz* cefoperazone, *d* differential

biochemical tests such as catalase, oxidase, hippurate hydrolysis, indoxyl acetate hydrolysis, TTC reduction, nitrate reduction, production of H_2S, and antibiotic sensitivity by the disk method (Table 8.1).

Virulence Factors

The virulence determinants required to cause disease are multi-factorial in nature: motility, chemotaxis, adhesion, invasion, iron acquisition, and the production of toxins (Ketley 1997; Wassenaar and Blaser 1999).

Flagella

Campylobacters are motile bacteria with twisted cell morphology. They exhibit both directional swimming and tumbling motility (Szymanski et al. 1995). The presence of flagellum and unique cell shape enables *Campylobacter* to achieve high level of motility in viscous media. This behavior has relevance to the penetration of the mucus that overlays the intestinal epithelium. The exact mechanism by which the flagellar apparatus contributes to *Campylobacter* adhesion and invasion is unknown. Flagella are likely to be involved in directed migration toward the host cell as found in *Helicobacter pylori* (Bjorkholm et al. 2000). The flagellin gene contains two adjacent genes, *flaA* and *flaB*. Both are of approximately equal size (1.7 Kbp). The *flaA* and *flaB* gene share a high level of base sequence identity (>93 %) and are independently transcribed and regulated by different promoters. The *flaA* is expressed at higher levels than *flaB*. In the absence of *flaA*, the *flaB* gene encodes a flagellin protein that forms a short, truncated non-functional flagellum (Wassenaar et al. 1991).

Chemotaxis

The role of chemotaxis has been demonstrated by testing chemically mutagenized, non-chemotactic *Campylobacter* mutants in an animal model (Takata et al. 1992). Such mutants failed to colonize in the suckling mouse intestine. *In vitro* studies demonstrated various chemoattractants such as mucin, L-serine, and L-fucose (Hugdahl et al. 1988) and chemorepellents such as bile acids. Very little is known about the molecular basis of *Campylobacter* chemotaxis. However, till date only one regulatory component, *che*Y, has been identified (Marchant et al. 2002). Mutation in this gene does not affect motility or invasion but results in loss of chemotaxis.

Adhesins

Several potential *C. jejuni* adhesins have been described, but the role of these determinants is still not clear.

- 43-kDa major outer membrane protein (OMP), which binds to INT407 membrane (Schroder and Moser 1997).
- Cell-binding factor-1 (CBF-1; 27 kDa) and cell-binding factor-2 (CBF-2; 29 kDa), which bind preferentially to HeLa cells (Fauchere et al. 1989).
- Four major *C. jejuni* OMPs, with apparent molecular masses of 28, 32, 36, and 42 kDa that bind to host cells. (De Melo and Pechere 1990).
- Four proteins, termed PEB1 through PEB4, with molecular masses of 28, 29, 30, and 31 kDa were identified. PEB1 binds to host cells and the gene coding for this potential adhesin has been cloned (Blaser and Pei 1993). The PEB1 protein has amino acid sequence similarity to transporter systems of other bacteria; however, PEB4 may participate in the folding of secreted proteins outside the cytolasmic membrane but are rarely involved in adhesion of *C. jejuni* to host cells (Burucoa et al. 1995).

Invasion

Penetration of the epithelial mucosa is an important virulence mechanism in *Campylobacter*-mediated disease. Although epithelial cell invasion in vivo is very rare, host cell invasion has been observed in both experimentally infected infant macaque monkeys (Russell et al. 1993) and in the colon of the patients (van Spreeuwel et al. 1985). Apart from this, there are also well-documented evidences that *C. jejuni* readily invades primary swine intestinal cells (Babakhani and Joens 1993) and tissue culture cells (DeMelo and Pechere 1990). The invasive phenotype is dependent on bacterial flagellin expression and flagellin-mediated motility. *Campylobacter* cross the intestinal epithelium by translocation or epithelial cell invasion followed by cell lysis.

Lipopolysaccharide

LPS possesses potent immune-modulating and immune-stimulating activities, principally due to lipid A (Moran 1994) that harbors binding sites for antibodies and non-immunoglobulin serum factors and contributes to bacterial virulence (Rietschel et al. 1990).

Ferritin

The iron storage protein, ferritin is produced by *C. jejuni* (Wai et al. 1995). A *C. jejuni* mutant in the gene encoding ferritin *cf* was found to grow poorly in iron-deficient media and sensitive to oxidative stress (Wai et al. 1996). Thus, production of ferritin may facilitate the colonization of the host by *C. jejuni* and may also help to protect the bacterium from high O_2 levels.

Campylobacter Infection and Water

Contaminated surface water is recognized as a source of *Campylobacter* outbreaks in human and may also play a role in contamination of farm animals. The first outbreak that occurred in Burlington, USA, through contaminated community water supply was the largest affecting an estimated 3,000 persons (Vogt et al. 1982). *C. jejuni* was implicated in 52 % of total outbreaks due to contaminated water consumption (Moore et al. 2005). The main factors implicated in these outbreaks are temporary or transient populations, lack or failure of water treatment, the presence of animals and heavy rains. The organism can also be isolated from various other water environments such as stream, sea water, and other recreational water. However, this type of water is not intended for human consumption, but accidental ingestion may pose a risk. *Campylobacter* contamination of surface water is likely to originate from fecal contamination by wild bird or domestic animals or from sewage effluent (Koenraad et al. 1997). Poor water quality, sanitation, and hygiene account 1.7 million deaths per year worldwide, mainly through infectious diarrhea (Moore et al. 2005).

Several investigations for the detection of *Campylobacter* in different types of seafood have also been carried out. Contaminated shellfish have been implicated as a vehicle in the dissemination of campylobacteriosis. Harvesting shellfish from *Campylobacter*-contaminated waters would be the most likely cause of infection (Wilson and Moore 1996). Shellfish beds located near potential contamination sources such as sewage effluents, farmlands runoffs, and waterfowl reservoirs were reported to present a health risk to persons consuming raw oysters (Abeyta et al. 1993). The public health problem associated with contaminated water supplies can be prevented by the identification and understanding of risk factors, proper protection of water sources, use of chlorinated drinking water, and adequate treatment and maintenance.

Campylobacter Infection in Food

Campylobacter infections in humans are considered to be mainly foodborne diseases. Live poultry, including broilers, laying hens, turkeys, and ducks, are often found to be colonized by large numbers of *Campylobacter* without any clinical symptoms. Colonized birds enter the slaughter house with heavy amount of these organisms on their feathers, skin, and their intestinal tracts, which leads to the contamination of equipments, working surfaces, water, and air. A large amounts of water used during poultry processing contribute to the spread and survival of campylobacters (Berndtson et al. 1992). Several studies showed that *Campylobacter* survived on egg shells (Doyle 1984; Jacobs-Reitsma 1994), and these eggs also presumed to be the cause of outbreaks. Laying hens often excrete *Campylobacter* in large numbers, which may lead to externally contaminated eggs.

Milk

Unpasteurized milk is a well-documented cause of a number of outbreaks of campylobacteriosis. A large outbreak took place in 1979 in the UK where over 2,500 children became infected by milk (Jones et al. 1991). The organism can be isolated from the feces of healthy dairy cows and raw milk. Similarly, raw cows' milk, unpasteurized goats' milk may transmit *Campylobacter* infection from the animals to humans. However, milkborne infection can be adequately controlled by proper pasteurization and prevention of recontamination after the heat treatment.

Vegetables and Meat

Meat-producing cattle also excrete *C. jejuni* in their feces with higher incidence in summer. Also swine are frequently found to harbor *Campylobacter*, and approximately 91.7 % of lambs slaughtered carry this pathogen (Stanley et al. 1998). Relatively few studies are carried out on the presence or absence of *Campylobacter* in vegetable food staff. Several studies showed that 1.6–3.1 % vegetables were contaminated with *Campylobacter*; however, it was completely removed after thorough washing with chlorinated water (Park and Sanders 1992). Reduction in the potential risk of contaminated poultry products must be achieved by application of good hygienic practices during harvest. Elimination of this pathogen in the poultry reservoir is an essential step in minimizing the public health problem. Meat products from cattle, swine, and lamb should be chilled overnight to remove the microorganism.

Clinical Spectrum

Campylobacter species have been the focus of attention since more than last 30 years because of the increasing frequency with which they have been isolated from human, animals, food, and water. The most common *Campylobacter* species causing human disease are *C. jejuni* and *C. coli* (Lastovica and Skirrow 2000). A total of 111 outbreaks of *Campylobacter* infections (99 gastroenteritis and two bacteremia outbreaks) were reported between 1978 and 1999 affecting 9,913 individuals. All but 3 of them were due to *C. jejuni,* one gastroenteritis outbreak by *Campylobacter fetus* subspecies *fetus,* and two bacteremia outbreaks in cancer patients by mixed species of *C. jejuni* and *C. fetus* subspecies *fetus.*

Clinical illness ranges from mild self-limiting, non-inflammatory diarrhea to severe inflammatory bloody diarrhea associated with bacteremia (Sack et al. 2001). The disease is usually characterized by diarrhea, fever, and abdominal cramps, the most common presentations of *C. jejuni*/*C. coli* infection. However, this infection occurs primarily in infants, elderly people, and patients with underlying diseases. Immunocompromised patients are at higher risk of infection that requires effective antimicrobial therapy. Despite being self-limiting in majority, post-infectious complications can be severe and occasionally life-threatening. An important sequel of *C. jejuni* infection is the development of the Guillain–Barré Syndrome (GBS), an acute neurological disease marked by ascending paralysis and has long been considered to be immunologically mediated (Nachamkin et al. 1998).

Intestinal Infection

The worldwide most frequent cause of enteric infection is *C. jejuni*, and *C. coli*. *C. jejuni* is responsible for 80–85 % of all enteric *Campylobacter* infections while *C. coli* ranks second (10–15 %). The other campylobacters such as *C. lari*, *C. upsaliensis*, and *C. fetus* are occasionally isolated depending on geographical locations. The main symptoms include diarrhea that varies from a mild, non-inflammatory, watery to severe and bloody form (Zilbauer et al. 2008). The other symptoms are cramping abdominal pain [sometimes difficult to distinguish from acute appendicitis (Blakelock and Beasley 2003)], fever, vomiting, and headache (Allos 2001). *Campylobacter* enteritis may occur in all age groups, but clinical presentation can vary according to age. In infants, the risk of dehydration or convulsion exists. In hyper-exposed individuals, immunity develops and infection may remain subclinical. The risk of infection may be severe in immunocompromised individuals (Moore et al. 2005).

Extra-Intestinal Infection

The overall incidence of extra-intestinal infections associated with *Campylobacter* is less compared to intestinal infections; however, they are more severe and can be life-threatening if they occur. The most important extra-intestinal or post-infectious disease is GBS, an acute demyelinating disease affecting myelin and axons of the peripheral nervous system and sometimes requiring artificial ventilation in severely affected individuals (Kuwabara 2007). *C. jejuni* is identified as a major triggering agent of GBS. It is estimated that approximately one in

Table 8.2 *Campylobacter* species and their associated disease in human

Species and subspecies	Disease in human
C. jejuni	Gastroenteritis, septicemia, meningitis, appendicitis, abortion, proctitis, peritonitis, toxic megacolon, segmental necrotizing syndrome, reactive arthritis, Reiter's syndrome, Guillain–Barré syndrome (GBS), irritable bowel syndrome, hemolytic uremic syndrome (HUS)
C. coli	Gastroenteritis, septicemia
C. lari	Gastroenteritis, septicemia
C. fetus subsp. fetus	Gastroenteritis, septicemia, abortion, meningitis
C. concisus	Periodontal disease, gastroenteritis
C. upsaliensis	Gastroenteritis, septicemia, abscess, GBS, HUS

every 1,058 reported *Campylobacter* illnesses leads to GBS (Nachamkin et al. 2000b). Development of GBS after *C. jejuni* infection is due to the triggering of humoral and cellular immune responses against antigenic epitopes on lipo-oligosaccharides (LOS) of *C. jejuni* that cross-react with shared epitopes in nerve myelin cells (Nyati et al. 2011; Komagamine and Yuki 2006). *C. jejuni* PEN19 serogroup in Japan was first identified and associated with GBS; subsequently, other serogroups were also described in Europe (Endtz et al. 2000). Another post-infectious manifestation is irritable bowel syndrome (IBS), which has been increasingly linked to enteric *C. jejuni* infection and characterized by abdominal pain, sometimes associated with altered bowel habit following bacterial enteritis (Spiller 2007). Indeed, campylobacteriosis has been recognized as one of the most common risk factors for post-infectious IBS. In May 2000, the municipal water supply of Walkerton, Ontario (Canada), was contaminated with livestock waste containing *Escherichia coli* O157:H7 and *Campylobacter* species. Subsequent waterborne infections caused acute enteritis in more than 2,300 people, of whom more that 36 % subsequently developed post-infectious IBS compared to 10 % of subjects without gastroenteritis (Marshall et al. 2006). The complete mechanisms by which this pathogen incites inflammatory disorders in the bowel remain unknown.

Bacteremia is probably a transient event in early stage of *Campyobacter* gastroenteritis characterized by high fever and rigor. The survey by the Communicable Disease Surveillance Centre, London, showed an average bacteremia rate of 1.5 per 1,000 intestinal infections with wide variations related to age [5.9 per 1,000 in patients aged $\geq$65 years and 3 per 1,000 in children aged 1–4 years (163)]. The rates were nearly twice as high in males as in females. The disease spectrum by different *Campylobacter* species in human is summarized in Table 8.2.

Immune Response

Immunoglobulin G (IgG) and IgM levels in serum rise in response to infection and vary both in their sensitivity and specificity for detecting the *Campylobacter* infection. Antibodies to *Campylobacter* antigens appear in the serum from about day 5 of illness, peak within 2–4 weeks, and then decline over several months (Blaser and Duncan 1984). Studies showed that *Campylobacter* infection conferred short-term immunity to the homologous strains (Black et al. 1992), but how long it lasts or how broad is the immunity from a single infection is yet not clear. However, in developing countries, where repeated infection is common in early childhood, infection rate declines with age. *Campylobacter* enteritis is virtually absent in older children and adults, but it is not yet clear whether sustaining this immunity depends on repeated re-exposure to the organism. Blaser and coworkers used a glycerol–HCl-extracted surface antigen of serotypes O:1, O:2, and O:3 in their ELISA. The sensitivity and specificity of IgG antibodies were 59 and 74 %, respectively.

However, for IgM, the sensitivity was 74 % and specificity was 68 % (Blaser and Duncan 1984). Another study from Denmark showed 71 and 60 % specificity for IgG and IgM, respectively, whereas a combined sensitivity of 90 % was found for both IgG and IgM (Strid et al. 2001). *Campylobacter*-specific IgA antibody is secreted in breast milk that protects infants against infection (Nachamkin et al. 1994). However, detailed investigation of *Campylobacter*-mediated intestinal immunopathology is hampered by lack of appropriate vertebrate models. Recently, Bereswill et al. developed a novel *C. jejuni* infection model using gnotobiotic mice in which the intestinal flora was eradicated by antibiotic treatment. They observed stable *C. jejuni* colonization with a pro-inflammatory immune response indicated by increased numbers of T- and B-lymphocytes, regulatory T-cells, neutrophils, and apoptotic cells, as well as increased concentrations of TNF-α, IL-6, and MCP-1 in the colon mucosa of mice. Detection of *C. jejuni*-LPS and *C. jejuni*-CpG-DNA by host TLR4 and TLR9, respectively, plays a key role in immunopathology. This study suggested that gnotobiotic and "humanized" mice represent excellent novel *C. jejuni* infection and inflammation models and provide deep insights into the immunological and molecular interplays between *C. jejuni* microbiota and innate immunity in human campylobacteriosis (Bereswill et al. 2011).

Detection of *Campylobacters*

Campylobacter species, particularly *C. jejuni* and *C. coli*, are the major causative agents of gastroenteritis and diarrheal disease in humans worldwide. Most laboratories do not routinely distinguish and diagnose these organisms. However, on the basis of culture-based methods, 80–90 % of *Campylobacter* infections in industrialized countries are possibly due to *C. jejuni* followed by *C. coli* in 5–10 % of cases (Nachamkin et al. 2000a). These methods were effective for the isolation of *Campylobacters* from human feces but less suitable for animal and environmental samples. This led to the development of the more selective medium for the isolation of *Campylobacters* from food and environmental samples.

Culture-Based Methods

Enrichment broth cultures have been used to enhance the recovery of *Campylobacter* from stool samples. Several media such as Preston enrichment broth (Bolton and Robertson 1982), Campy-thio (Blaser et al. 1979), and *Campylobacter* enrichment broths (Martin et al. 1983) have been regularly used. Enrichment cultures may be beneficial where small numbers of organisms may be expected due to delayed transport to the laboratory or after acute stage of disease, when the concentration may be low (Nachamkin 1997).

A number of approaches for isolating *C. jejuni*, *C. coli*, and other species on primary selective media and by filtration techniques have been described. Selective media include blood-free media such as charcoal cefoperazone deoxycholate agar (CCDA; Hutchinson and Bolton 1984), charcoal-based selective medium (CSM; Karmali et al. 1986), semi-solid blood-free motility medium (SSM; Goossens et al. 1989), blood-containing media such as Campy-CVA medium (Reller et al. 1983), Blaser–Wang medium, and Skirrow medium (Skirrow 1977). However, culture is insensitive for the detection of bacteria in cases treated with antibiotics, or in cases having mild/subclinical infection or in patients with late reactive complications such as arthritis and GBS or long-lasting intestinal distress (Linton et al. 1997). Delayed hospital admission and intake of antibiotics by the patient may also account for low culture positivity.

Serological Methods

Since there are many limitations associated with culture-based methods, recent research primarily focuses on the development of alternative methods for the detection of *campylobacters* in food and fecal samples. For this, the strength has been provided by the production of monoclonal and polyclonal antibodies specific for campylobacters that further helped in the development of new antibody-based assays. Serology is mainly used to detect the presence of antibodies against *C. jejuni* infection in patient's serum. Latex agglutination test has been developed for rapid identification of *Campylobacter* from culture. An enzyme-linked immunosorbent assay (ELISA) was developed for the detection of *C. jejuni/C. coli* in stool samples of gastroenteritis patients with sensitivity of 96 % and a specificity of 99 % (Tolchin et al. 2000). ELISA with a crude antigenic extract prepared from geographically prevalent *C. jejuni* strains is being used in GBS patients for diagnosis of antecedent *C. jejuni* infection. Several drawbacks of serology are as follows: there is no consensus on the choice of antigens; most often crude antigenic extracts and single serum samples are used yielding low specificity, especially in endemic and hyper-endemic countries due to high titers of antibodies in the resident population (Nyati et al. 2010; Tsang 2002). Moreover, the antibody detection assays can vary considerably between different laboratories in terms of their performance (Koga et al. 2001).

Nucleic Acid-Based Detection Methods

After the discovery of polymerase chain reaction (PCR), all areas of microbiology, especially detection of microbial pathogens, largely depend on this molecular technique. The first application of PCR for the specific detection of *C. jejuni* and *C. coli* was reported in 1992 (Oyofo et al. 1992) that targeted the flagellin A gene of these organisms and successfully detected 30–60 bacteria per reaction performed in fecal samples. The report also demonstrated the potential of the PCR-based methods to detect very low numbers of *Campylobacter* in the samples. PCR was earlier used in few studies to detect *Campylobacter* species in stool from patients with gastroenteritis (Linton et al. 1997). A multiplex PCR assay suitable for mass screening to detect *Campylobacter* directly from chicken feces has been developed (Wiemer et al. 2011). Recent studies demonstrated the association of *C. jejuni* in GBS patients by PCR (19.0–22.5 %), and its sensitivity was higher than culture (Nyati et al. 2010; Sinha et al. 2004). Adaptation of PCR assays into a microplate hybridization format or PCR-ELISA increased the sensitivity and specificity of the method. PCR-ELISA assays were used in a large-scale survey for the detection of *Campylobacter* species in gastroenteritis cases and could differentiate between *C. jejuni*, *C. coli*, *C. upsaliensis*, *Campylobacter helveticus*, *C. fetus*, *Campylobacter hyointesitinalis*, and *C. lari*. These assays may help in large epidemiological studies to detect the incidence of infection by various *Campylobacter* species other than *C. jejuni* and *C. coli* in humans. There have been many reports on PCR assays for the detection of *Campylobacter* in a range of sample types including food, water, and environmental samples. These assays may be useful as an adjunct to enrichment culture, by reducing the total time of detection by two or more days. The introduction of real-time PCR method facilitated the development of quantitative PCR assays for the detection of *Campylobacter* in food (Sails et al. 2003), milk, and water (Yang et al. 2003). The quantitative detection of *Campylobacter* directly in raw-meat rinse fluid samples was also demonstrated, but the limit of detection was compromised by the presence of PCR inhibitors and the low bacterial count (Sails et al. 2003). Although PCR is a highly specific and sensitive method, its sensitivity varies among the laboratories and PCR cannot exclude the diagnosis of infection.

Drug Resistance

Campylobacteriosis is considered to be a zoonotic disease and occurs through close contact with domestic animals such as poultry, cattle, and pigs (Helms et al. 2003; Mead et al. 1999). The majorities of infections are mild or self-limiting and do not require antimicrobial therapy (Zhao et al. 2010). In some severe and recurrent infections, antibiotic susceptibility testing of *Campylobacter* species is important to facilitate appropriate and timely treatment where indicated and also for surveillance of emergence of drug resistance (Avrain et al. 2003; Rautelin et al. 2002). *C. jejuni* and *C. coli* are almost universally resistant to penicillins, cephalosporins, trimethoprim, rifampicin, and vancomycin. Although they are highly susceptible to erythromycin, fluoroquinolones, tetracyclines, aminoglycosides, and clindamycin, a significant increase in the prevalence of resistance to macrolides and fluoroquinolones among *Campylobacter* spp. have been reported, and this is recognized as an emerging public health problem in many countries (Engberg et al. 2001). In developing countries like India, Jain et al. reported the antibiotic resistance of *Campylobacter* species as follows: ampicillin 81.6 %, ciprofloxacin 71.4 %, tetracycline 26.5 %, furazolidine 14.3 %, gentamicin 10.2 %, and erythromycin 6.1 %, while 30.6 % of strains were multidrug resistant. Increased quinolone resistance and multidrug resistance pose major risks of treatment failure (Jain et al. 2005). Entry of these isolates into the food chain could represent a significant threat to public health.

Prevention and Treatment

Some simple food handling practices can help in preventing *Campylobacter* infections:

- Cook all poultry products thoroughly. All poultry products should be cooked to reach a minimum internal temperature of 165°F.
- Wash hands with soap before preparing food.
- Wash hands with soap after handling raw foods of animal origin and before touching anything else.
- Prevent cross-contamination in the kitchen by using separate cutting boards for foods of animal origin and other foods and by carefully cleaning all cutting boards, countertops, and utensils with soap and hot water after preparing raw food of animal origin.
- Avoid consuming unpasteurized milk and untreated surface water.
- Make sure that persons with diarrhea, especially children, wash their hands carefully and frequently with soap to reduce the risk of spreading the infection.
- Wash hands with soap after contact with pet feces.

Physicians who diagnose campylobacteriosis and clinical laboratories that identify this organism should report their findings to the local health department. If many cases occur at the same time, it may mean that many people are exposed to a common contaminated food item or water source, which may still be available to infect more people. When outbreaks occur, community education can be directed toward proper food handling techniques and avoiding consumption of raw milk. Almost all persons infected with *Campylobacter* recover without any specific treatment. Patients should drink extra fluids as long as diarrhea lasts. While most *Campylobacter* infections are self-limiting, occasionally a more invasive illness can occur that requires effective antimicrobial therapy. In more severe cases, antibiotics such as azithromycin or erythromycin and fluoroquinolones can shorten the duration of symptoms if given early in the illness. In some regions, tetracycline or doxycycline and selective beta-lactams have been used for treating intestinal infections.

Summary and Conclusions

Campylobacter is the most common bacterial enteric pathogen both in the developed and developing countries worldwide. Although the organism is usually not invasive, infection is occasionally life-threatening. In the last 20 years, *Campylobacter* infection rates have increased in many developed countries. Part of this increase may be due to improvements in detection and reporting, but part may reflect a true increase in infections. The vast majority of *Campylobacter* infections occur as sporadic individual infections. Consumptions of raw or undercooked poultry, untreated water, raw milk or milk products, and contact with pets are important source of infection. Outbreaks of

Campylobacter in several regions are caused predominantly by contaminated water, milk, and poultry. Disease burden is higher in developing countries where poor sanitation and unhygienic conditions are common. Outbreaks due to *Campylobacter* are less in the developed countries; this may be due to the disinfection of ground and surface water before distribution and public health warnings about hazards of raw milk consumption. Geographical differences in *Campylobacter* infections between developing and developed world may also be related to social culture, population density, climate, and ethnic background.

Acknowledgments This study was supported by Indian Council of Medical Research (ICMR; Ref. no. 5/3/3/2/2000-ECD-I and 5/3/3/24/2006-ECD-I), Government of India (GOI), New Delhi. Senior research fellow K. K. Nyati acknowledges the financial assistance from ICMR (Ref. no. 80/569/2007-ECD-I), GOI, New Delhi.

References

Abeyta C, Deeter FG, Kaysner CA, Stott RF, Wekell MM (1993) *Campylobacter jejuni* in a Washington state shellfish growing bed associated with illness. J Food Prot 56:323–325

Allos BM (2001) *Campylobacter jejuni* infections: update on emerging issues and trends. Clin Infect Dis 32:1201–1206

Atanassova V, Ring C (1999) Prevalence of *Campylobacter* spp. In poultry and poultry meat in Germany. Int J Food Microbiol 51:187–190

Avrain L, Humbert F, L'Hospitalier R, Sanders P, Vernozy-Rozand C, Kempf I (2003) Antimicrobial resistance in *Campylobacter* from broilers: association with production type and antimicrobial use. Vet Microbiol 96:267–276

Babakhani FK, Joens LA (1993) Primary swine intestinal cells as a model for studying *Campylobacter jejuni* invasiveness. Infect Immun 61:2723–2726

Bereswill S, Fischer A, Plickert R, Haag LM, Otto B, Kühl AA et al (2011) Novel murine infection models provide deep insights into the "Ménage à Trois" of *Campylobacter jejuni*, microbiota and host innate immunity. PLoS ONE 6:e20953

Berndtson E, Tivemo M, Engvall A (1992) Distribution and numbers of Campylobacter in newly slaughtered broiler chickens and hens. Int J Food Microbiol 15:45–50

Bjorkholm B, Zhukhovitsky V, Lofman C, Hulten K, Enroth H, Block M et al (2000) *Helicobacter pylori* entry into human gastric epithelial cells: potential determinants of virulence, persistence, and treatment failure. Helicobacter 5:148–154

Black RE, Perlman D, Clements ML, Levine MM, Blaser MJ (1992) Human volunteer studies with *Campylobacter jejuni*. In: Nachamkin I, Blaser MJ, Tompkins LS (eds) *Campylobacter jejuni*: current status and future trends. American Society for Microbiology press, Washington, DC, pp 472–479

Blakelock RT, Beasley SW (2003) Infection and the gut. Semin Pediatr Surg 12:265–274

Blaser MJ, Duncan D (1984) Human serum antibody response to *Campylobacter jejuni* infection as measured in an enzyme-linked immunosorbent assay. Infect Immun 44:292–298

Blaser MJ, Pei Z (1993) Pathogenesis of *Campylobacter fetus* infections: critical role of high-molecular-weight S-layer proteins in virulence. J Infect Dis 167:372–377

Blaser MJ, Berkowitz ID, LaForce FM, Craven J, Reller LB, Wang L (1979) *Campylobacter* enteritis: clinical and epidemiologic features. Ann Intern Med 91:179–185

Bolton FJ, Robertson L (1982) A selective medium for isolating *Campylobacter jejuni/coli*. J Clin Pathol 35:462–467

Brieseman MA (1990) A further study of the epidemiology of *Campylobacter jejuni* infections. N Z Med J 100:207–209

Burucoa C, Fremaux C, Pei Z, Tummuru M, Blaser MJ, Cenatiempo Y et al (1995) Nucleotide sequence and characterization of peb4A encoding an antigenic protein in *Campylobacter jejuni*. Res Microbiol 146:467–476

Butzler JP, Dekeyser P, Detrain M, Dehaen F (1973) Related vibrio in stools. J Pediatr 82:493–495

de la Cabada Bauche J, Dupont HL (2011) New developments in traveler's Diarrhea. Gastroenterol Hepatol (New York) 7:88–95

De Melo MA, Pechere JC (1990) Identification of *Campylobacter jejuni* surface proteins that bind to eukaryotic cells in vitro. Infect Immun 58:1749–1756

Dekeyser P, Gossuin-Detrain M, Butzler JP, Sternon J (1972) Acute enteritis due to related vibrio: first positive stool cultures. J Infect Dis 125:390–392

Doyle MP (1984) Association of Campylobacter with laying hens and eggs. Appl Environ Microbiol 47:533–536

Endtz HP, Ang CW, van Den Braak N, Duim B, Rigter A, Price LJ et al (2000) Molecular characterization of *Campylobacter jejuni* from patients with Guillain-Barré and Miller Fisher syndromes. J Clin Microbiol 38:2297–2301

Engberg J, Aarestrup FM, Taylor DE, Gerner-Smidt P, Nachamkin I (2001) Quinolone and macrolide resistance in *Campylobacter jejuni* and *C. coli*: resistance mechanisms and trends in human isolates. Emerg Infect Dis 7:24–34

Escherich T (1886) Articles adding to the knowledge of intestinal bacteria, III. On the existence of vibrios in the intestines and feces of babies. Münchener medizinische Wochenschrift 33:815–817

Fauchere JL, Kervella M, Rosenau A, Mohanna K, Veron M (1989) Adhesion to HeLa cells of *Campylobacter jejuni* and *C. coli* from human feces. Res Microbiol 140:379–392

Goosens H, Butzler JP (1992) Isolation and identification of *Campylobacter* spp. In: Nachamkin I, Blaser MJ, Tompkins LS (eds) *Campylobacter jejuni*: current status and future trends. American Society for Microbiology, Washington, DC, pp 93–109

Goossens H, Vlaes L, Galand I, Van den Borre C, Butzler JP (1989) Semisolid blood-free selective-motility medium for the isolation of *Campylobacters* from stool specimens. J Clin Microbiol 27:1077–1080

Hariharan H, Wright T, Long JR (1990) Isolation and antimicrobial susceptibility of *Campylobacter coli* and *Campylobacter jejuni* from slaughter hogs. Microbiologica 13:1–6

Hazelegerh W, Arkesteijn C, Toorop-Bouma A, Beumer R (1994) Detection of the coccoid form of *Campylobacter jejuni* in chicken products with the use of the polymerase chain reaction. Int J Food Microbiol 24:273–281

Helms M, Vastrup P, Gerner-Smidt P, Molbak K (2003) Short and long term mortality associated with foodborne bacterial gastrointestinal infections: registry based study. Br Med J 326:357–359

Ho BS, Wong WT (1985) A one- year survey of *Campylobacter* enteritis and other forms of bacterial diarrhea in Hong Kong. J Hyg 94:55–60

Hugdahl MB, Beery JT, Doyle MP (1988) Chemotactic behaviour of *Campylobacter jejuni*. Infect Immun 56:1560–1566

Hutchinson DN, Bolton FJ (1984) Improved blood free selective medium for the isolation of *Campylobacter jejuni* from faecal specimens. J Clin Pathol 37:956–957

Jain D, Sinha S, Prasad KN, Pandey CM (2005) *Campylobacter* species and drug resistance in a north Indian rural community. Trans R Soc Trop Med Hyg 99:207–214

Jaobs-Reitsma WF (1994) Epidemiology of Campylobacter in poultry. PhD thesis Wageningen Agricultural University, Washington, The Netherlands

Jones FS, Orcutt M, Little RB (1931) Vibrios (*Vibrio jejuni*, n. sp.) associated with intestinal disorders of cows and calves. J Exp Med 53:853–864

Jones PH, Willis AT, Robinson DA, Skirrow MB, Josephs DS (1991) Campylobacter enteritis associated with the consumption of free school milk. J Hyg 87:155–162

Karmali MA, Simor AE, Roscoe M, Fleming PC, Smith SS, Lane J (1986) Evaluation of a blood-free, charcoal-based, selective medium for the isolation of *Campylobacter* organisms from feces. J Clin Microbiol 23:456–459

Ketley JM (1997) Pathogenesis of enteric infection by *Campylobacter*. Microbiology 143:5–21

Koenraad PMFJ, Rombouts FM, Notermans SHW (1997) Epidemiological aspects of thermophilic Campylobacter in water related environments: a review. Water Environ Res 69:52–63

Koga M, Yuki N, Hirata K (2001) Antecedent symptoms in Guillain-Barré syndrome: an important indicator for clinical and serological subgroups. Acta Neurol Scand 103:278–287

Komagamine T, Yuki N (2006) Ganglioside mimicry as a cause of Guillain-Barré syndrome. CNS Neurol Disord: Drug Targets 5:391–400

Kuwabara S (2007) Guillain-Barré syndrome. Curr Neurol Neurosci Rep 7:57–62

Lastovica AJ, Skirrow MB (2000) Clinical significance of *Campylobacter* and related species other than *Campylobacter jejuni* and *Campylobacter coli*. In: Nachamkin I, Blaser MJ (eds) Campylobacter. American Society for Microbiology press, Washington, DC, pp 89–120

Linton D, Lawson AJ, Owen RJ, Stanley J (1997) PCR detection, identification to species level, and fingerprinting to *Campylobacter jejuni* and *Campylobacter coli* direct from diarrheic samples. J Clin Microbiol 35:2568–2572

Marchant J, Wren B, Ketley J (2002) Exploiting genome sequence: predictions for mechanisms of *Campylobacter* chemotaxis. Trends Microbiol 10:155–159

Marshall JK, Thabane M, Garg AX, Clark WF, Salvadori M, Collins SM (2006) Incidence and epidemiology of irritable bowel syndrome after a large waterborne outbreak of bacterial dysentery. Gastroenterology 131:445–450

Martin WT, Patton CM, Morris GK, Potter ME, Puhr ND (1983) Selective enrichment broth for isolation of *Campylobacter jejuni*. J Clin Microbiol 17:853–855

Mead P, Slutsker L, Dietz V, Mc Caig L, Bresee J, Shapiro C et al (1999) Food related illness and death in the United States. Emerg Infect Dis 5:607–625

Moore JE, Corcoran D, Dooley JS, Fanning S, Lucey B, Matsuda M et al (2005) Campylobacter. Vet Res 36:351–382

Moran AP (1994) Structure-bioactivity relationships of bacterial endotoxins. J Toxicol Toxin Rev 14:47–83

Nachamkin I, Allos BM, Ho TW (1998) *Campylobacter* species and Guillain-Barré syndrome. Clin Microbiol Rev 11:555–567

Nachamkin I, Allos BM, Ho TW (2000a) *Campylobacter jejuni* infection and the association with Guillain-Barré syndrome. In: Nachamkin I, Blaser MJ (eds) Campylobacter. American Society for Microbiology press, Washington, DC, pp 155–175

Nachamkin I, Engberg J, Aarestrup FM (2000b) Diagnosis and antimicrobial susceptibility of *Campylobacter* species. In: Nachamkin I, Blaser MJ (eds) Campylobacter. American Society for Microbiology press, Washington, DC., pp 45–66

Nachamkin I, Fischer SH, Yang XH, Benitez O, Cravioto A (1994) Immunoglobulin A antibodies directed against *Campylobacter jejuni* flagellin present in breast-milk. Epidemiol Infect 112:359–365

Nachamkin I (1997) Microbiologic approaches for studying *Campylobacter* in patients with Guillain-Barré syndrome. J Infect Dis 176:S106–S114

Nyati KK, Prasad KN, Rizwan A, Verma A, Paliwal VK, Pradhan S (2010) Lymphocyte transformation test detects a response to *Campylobacter jejuni* antigens in patients with Guillain-Barré syndrome. Med Microbiol Immunol 199:109–116

Nyati KK, Prasad KN, Rizwan A, Verma A, Paliwal VK (2011) Th1 and Th2 response to *Campylobacter jejuni* antigen in Guillain-Barre syndrome. Arch Neurol 68:445–452

Nylen G, Palmer SR, Bager F, Feierl G, Molbak K, Ruutu P et al (1998) *Campylobacter* seasonality in Europe. Paper presented at the 4th world congress on food borne infections and intoxications. Berlin, Germany, pp 293–297

Oberhelman R, Taylor DE (2000) *Campylobacter* infections in developing countries. In: Nachamkin I, Blaser MJ (eds) Campylobacter. American Society for Microbiology press, Washington, DC, pp 139–153

Ono K, Yamamoto K (1999) Contamination of meat with *Campylobacter jejuni* in Saitama, Japan. Int J Food Microbiol 47:211–219

Oyofo BA, Thornton SA, Burr DH, Trust TJ, Pavlovskis OR, Guerry P (1992) Specific detection of *Campylobacter jejuni* and *Campylobacter coli* by using polymerase chain reaction. J Clin Microbiol 30:2613–2619

Park CE, Sanders GW (1992) Occurrence of thermotolerant *Campylobacters* in fresh vegetables sold at farmers' outdoor markets and supermarkets. Cancer J Microbiol 38:313–316

Patricia G, Alm R, Szymanski C, Trust JT (2000) Structure, function and antigenicity of *Campylobacter* flagella. In: Nachamkin I, Blaser MJ (eds) Campylobacter. American Society for Microbiology press, Washington, DC, pp 405–421

Rautelin H, Vierikko A, Hanninen M, Vaara M (2002) Antimicrobial susceptibilities of *Campylobacter* strains isolated from Finnish subjects infected domestically or from those infected abroad. Antimicrob Agents Chemother 47:102–105

Reller LB, Mirrett S and Reimer LG (1983) Controlled evaluation of an improved selective medium for isolation of *Campylobacter jejuni*, abstr. C-274. Paper presented at the 83rd annual meeting of the American Society for Microbiology. Washington, DC, p 357

Rietschel ET, Brade L, Schade U, Seydel U, Zahringer U, Lindner B et al (1990) Chemical structure and biological activity of lipopolysaccharide. In: Baumgartner JD, Calandra T, Carlet J (eds) Endotoxin-from pathophysiology to therapeutic approaches. Flammarion Medicine Sciences, Paris, France, pp 5–18

Russell RG, O'Donnoghue M, Blake DC Jr, Zulty J, DeTolla LJ (1993) Early colonic damage and invasion of *Campylobacter jejuni* in experimentally challenged infant Macaca mulatta. J Infect Dis 168:210–215

Sack DA, Lyke C, McLauglin C and Suwanvanichkij V (2001) Antibacterial resistance in Shigellosis, Cholera and campylobacteriosis, vol 8. WHO/CDS/CSR/DRS, pp 1–56

Sails AD, Fox AJ, Bolton FJ, Wareing DR, Greenway DL (2003) A real-time PCR assay for the detection of *Campylobacter jejuni* in foods after enrichment culture. Appl Environ Microbiol 69:1383–1390

Schroder W, Moser I (1997) Primary structure analysis and adhesion studies on the major outer membrane protein of *Campylobacter jejuni*. FEMS Microbiol Lett 150:141–147

Sebald M, Veron M (1963) DNA base content in the classification of vibrios. Annales de l'Institut Pasteur 105:897–910

Shmilovitz M, Kretzer B, Rotman N (1982) *Campylobacter jejuni* as etiological agent of diarrheal diseases in Israel. Isr J Med Sci 18:935–940

Sinha S, Prasad KN, Pradhan S, Jain D, Jha S (2004) Detection of preceding Campylobacter jejuni infection by polymerase chain reaction in patients with Guillain-Barré syndrome. Trans R Soc Trop Med Hyg 98:342–346

Skirrow MB (1977) *Campylobacter* enteritis: a "new" disease. Br Med J 2:9–11

Smith T, Taylor M (1919) Some morphological biological characters of the spirilla (*Vibrio fetus*, n. sp.) associated with disease of the fetal membranes in cattle. J Exp Med 30:299–311

Spiller RC (2007) Role of infection in irritable bowel syndrome. J Gastroenterol 42:41–47

Stafford R, Tenkate T, McCall B (1996) A five year review of *Campylobacter* infection in Queensland. Commun Dis Intell 20:478–482

Stanley KN, Wallace JS, Currie JE, Diggle PJ, Jones K (1998) Seasonal variation of thermophilic Campylobacters in lambs at slaughter. J Appl Microbiol 84:1111–1116

Strid MA, Engberg J, Larsen LB, Begtrup K, Molbak K, Krogfelt KA (2001) Antibody responses to *Campylobacter* infections determined by an enzyme-linked immunosorbent assay: 2-year follow-up study of 210 patients. Clin Diagn Lab Immunol 8:314–319

Szymanski CM, King M, Haardt M, Armstrong GD (1995) *Campylobacter jejuni* motility and invasion of Caco-2 cells. Infect Immun 63:4295–4300

Takata T, Fujimoto S, Amako K (1992) Isolation of nonchemotactic mutants of *Campylobacter jejuni* and their colonization of the mouse intestinal tract. Infect Immun 60:3596–3600

Tauxe RV (1997) Emerging foodborne diseases: an evolving public health challenge. Emerg Infect Dis 3:425–434

Tauxe RV (2002) Emerging foodborne pathogens. Int J Food Microbiol 78:31–41

Tolcin R, LaSalvia MM, Kirkley BA, Vetter EA, Cockerill FR 3rd, Procop GW (2000) Evaluation of the Alexon-trend ProSpecT *Campylobacter* microplate assay. J Clin Microbiol 38:3853–3855

Tsang RS (2002) The relationship of *Campylobacter jejuni* infection and the development of Guillain-Barré syndrome. Curr Opin Infect Dis 15:221–228

van Spreeuwel P, Duursma GC, Meijer CJLM, Bax R, Rosekrans PCM, Lindeman J (1985) *Campylobacter coli*: histological, immunohistochemical and ultra structural findings. Gut 26:945–951

Vandamme P (2000) Taxonomy of the family *Campylobacteraceae*. In: Nachamkin I, Blaser MJ (eds) Campylobacter. American Society for Microbiology press, Washington, DC, pp 155–175

Vinzent R, Dumas J, Picard N (1947) Serious septicemia during pregnancy due to a vibrio, followed by abortion. Bull de l'Academie Nationale de Med 131:90–92

Vogt RL, Sours HE, Barret T, Feldman RA, Dickinson RJ, Whitherell L (1982) *Campylobacter enteritis* associated with contaminated water. Ann Intern Med 96:292–296

Wai SN, Nakayama K, Umene K, Moriya T, Amakok (1996) Construction of a ferritin- deficient mutant of *Campylobacter jejuni*: contribution of ferritin to iron storage and protection against oxidative stress. Mol Microbiol 20:1127–1134

Wai SN, Takata T, Takade A, Hamasaki N, Amako K (1995) Purification and characterization of ferritin from *Campylobacter jejuni*. Arch Microbiol 164:1–6

Wassenaar TM, Blaser MJ (1999) Pathophysiology of *Campylobacter jejuni* infections of humans. Microbes Infect 1:1023–1033

Wassenaar TM, Bleumink-pluym NM, van der Zeijst BA (1991) Inactivation of *Campylobacter jejuni* flagellin genes by homologous recombination demonstrates that flaA but not flaB is required for invasion. EMBO J 10:2055–2061

Wiemer D, Loderstaedt U, von Wulffen H, Priesnitz S, Fischer M, Tannich E et al (2011) Real-time multiplex PCR for simultaneous detection of *Campylobacter jejuni*, Salmonella, Shigella and Yersinia species in fecal samples. Int J Med Microbiol 301:577–584

Wilson IG, Moore JE (1996) Presence of Salmonella spp. and Campylobacter spp. in shellfish. Epidemiol Infect 116:147–153

World health organization (2001) The increasing incidence of human campylobacteriosis. Report and Proceedings of a WHO Consultation of Experts. Copenhagen. WHO/CDS/CSR/APH/2000.4

Yang C, Jiang Y, Huang K, Zhu C, Yin Y (2003) Application of real-time PCR for quantitative detection of *Campylobacter jejuni* in poultry, milk and environmental water. FEMS Immunol Med Microbiol 38:265–271

Zhao C, Ge B, De Villena J, Sudler R, Yeh E, Zhao S et al (2001) Prevalence of *Campylobacter* spp., *Escherichia coli*, and *Salmonella* serovars in retail chicken, turkey, pork, and beef from the greater Washington, D.C., area. Appl Environ Microbiol 67:5431–5436

Zhao S, Young SR, Tong E, Abbott JW, Womack N, Friedman SL et al (2010) Antimicrobial resistance of *Campylobacter* isolates from retail meat in the United States between 2002 and 2007. Appl Environ Microbiol 76:7949–7956

Zilbauer M, Dorrell N, Wren BW, Elliott MA (2008) *Campylobacter jejuni*-mediated disease pathogenesis: an update. Trans R Soc Trop Med Hyg 102:123–129

Pathogenic Mycobacteria and Water

9

Sukhraj Kaur

Abstract

The spectrum of waterborne infectious disease is changing due to dramatic changes in our society and environment. In 1990s, the major concern of the water utilities was to deal with coliform bacteria in the drinking water. But more recently, waterborne mycobacterial spp have assumed significance owing to their role in causing disseminated infections in immunocompromised people, such as in AIDS patients. *Mycobacterium avium* is a free-living environmental mycobacterium that can cause localized disease in the lungs, lymph glands, skin, wounds or bones. Drinking water is the major source of *M. avium* infection because the practice of disinfecting water using chlorine allows *M. avium*, which is chlorine resistant, to flourish in the absence of any competition. *M. avium* also exists in a symbiotic relationship with certain species of amoebae that may be fundamental to its transmission. *M. bovis*, on the other hand, is not a free-living environmental mycobacterium, but has a broad host range including wildlife, domestic livestock and humans. There are evidences of it being transmitted between the wildlife, livestock and humans, specifically under water-scarce environment. More recently, a link between Buruli ulcer (BU), a disease caused by *M. ulcerans*, and a biting aquatic insect has been reported. This leads to the speculation that BU is a vector-borne disease and that water may be an important environment for the proliferation of the vector. This chapter examines water as a route of infection for different types of mycobacterial diseases in humans and their health impacts on humans. Thus, appropriate public health measures must be developed to control the spread of mycobacterial diseases through water.

S. Kaur (✉)
Department of Microbiology,
Guru Nanak Dev University,
Amritsar, 143005, India
e-mail: kaursukhraj@rediffmail.com

P. P. Singh and V. Sharma (eds.), *Water and Health*,
DOI: 10.1007/978-81-322-1029-0_9,

Keywords
Bovine tuberculosis · Buruli ulcer · Mycobacteria · Nontuberculous mycobacteria

Introduction

Water has a major impact on the quality of life. However, in both developed and developing countries, water quality continues to be a major health concern. As a result, waterborne microbial diseases remain the leading causes of deaths worldwide with expanding spectrum and increasing incidences. One of the factors that is causing an increase in the spectrum of waterborne pathogens is the changing dynamics of relationship between humans, pathogens and their environment. Exploiting the changing interface between humans and the environment, the environmental mycobacteria such as *Mycobacterium avium* and *M. ulcerans* have emerged as pathogenic bacteria associated with water.

Tuberculosis (TB), primarily caused by pathogenic *M. tuberculosis*, remains an important cause of death from an infectious agent and second only to the human immunodeficiency virus (HIV; WHO 2004). It is a major health problem with over eight million new cases reported annually in the world and three million deaths (WHO 1994). With the spread of HIV epidemic, there has been a striking increase both in the incidence of TB and in the problem of drug resistance to TB. Thus, TB control is high on the international public health agenda. Environmental mycobacterium *M. avium* is the most frequent mycobacterial subspecies isolated from patients with AIDS that often causes serious disseminated disease. On the other hand, *M. intracellulare* is more common in immunocompetent people (Guthertz et al. 1989).

The diseases caused by environmental mycobacteria also called as nontuberculous mycobacteria (NTM) are rising and are believed to rise in future. One important factor that has an impact on NTM disease involves overlap between human ecology and mycobacterial ecology. Both NTM and humans share the ecology of water supply. The need to reduce the incidence of water-associated gastrointestinal disease has led to a widespread implementation of disinfection of drinking water. One consequence of this effort is the selection for disinfectant-resistant mycobacteria that can grow in the limited organic matter in water made available by the death of microbial competitors. The second important factor of NTM ecology and disease epidemiology is the presence of a lipid-rich outer cell wall. The unique cell wall components of NTM contribute to the hydrophobicity, impermeability and slow growth of mycobacteria (Brennan and Nikaido 1995). These features, in turn, lead to the preferential attachment to surfaces as biofilms and resistance to disinfectants and antibiotics. NTM are oligotrophs and able to grow on a variety of organic compounds including some found in water and soil, for example humic and fulvic acids.

The current increasing incidences of TB in humans, particularly in immunocompromised persons, have given rise to a renewed interest in the zoonotic TB. *M. bovis*, the aetiological agent of bovine TB, can cause human TB that resembles clinically to the TB caused by *M. tuberculosis*. In industrialized countries, bovine TB control and elimination programs, together with milk pasteurization, have drastically reduced the incidences of bovine TB in both cattles and humans. However, in developing countries, animal TB is widely prevalent as control measures are rarely applied, and as humans and animals share the same microenvironment and dwelling premises, especially in rural areas (Anaelom et al. 2010). Further, it is very difficult to control and eradicate bovine TB as *M. bovis* has a wide host range and is known to infect wildlife in some parts of the world.

Buruli ulcer (BU) is a neglected emerging disease that has recently been reported in some countries as the third most frequent mycobacterial disease in humans after TB and leprosy. Though the disease was first reported by Cook in Uganda in 1897, the aetiological agent of BU, that is, *M. ulcerans* was discovered in 1948 by MacCallum and associates (MacCallum et al. 1948). The disease often occurs in close proximity to water bodies.

In this chapter, the incidences, health impacts and control measures for TB due to NTM, *M. bovis* and *M. ulcerans* are discussed. Further, the physiological characteristic of environmental mycobacteria that makes it a successful aquatic emerging pathogen has also been discussed.

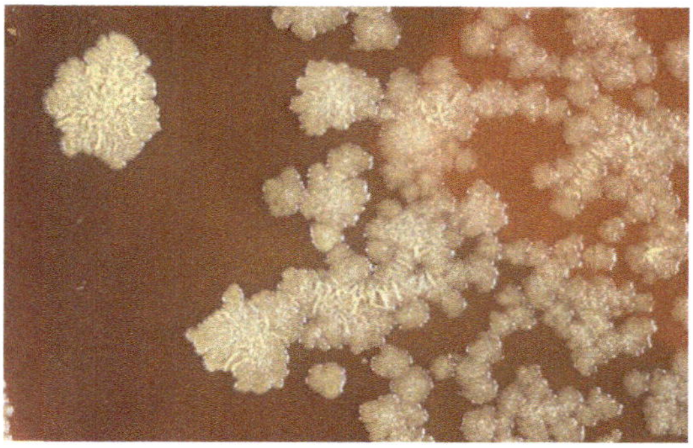

Fig. 9.1 A photograph depicting the characteristic colonies of *M. avium*

Mycobacteria

Mycobacteria are a large group of microorganisms that inhabit a diverse range of natural environments. It can be classified into three major groups for the purpose of diagnosis and treatment: *M. tuberculosis* complex, which can cause TB, includes species such as *M. tuberculosis*, *M. bovis*, *M. africanum*, and *M. microti*; *M. leprae*, which causes Hansen's disease or leprosy; third group includes NTM that are all the other mycobacteria, which can cause pulmonary disease resembling TB, lymphadenitis, skin disease or disseminated disease. NTM have been identified in numerous environmental sources, including water. There has been recent interest in the NTM species, due to their ability to cause disease in humans and animals after environmental exposures. CDC estimates that 72 % of the worldwide NTM diseases (non-AIDS related) are attributable to *M. avium* complex, whereas 25–50 % of individuals with AIDS will develop NTM diseases, primarily attributable to *M. avium* (Fig. 9.1).

M. bovis, the aetiological agent of bovine TB, is also responsible for human disease, which makes this bacterium an important zoonotic species. It has a broad range of mammalian hosts, including humans, cattle, deer, llamas, pigs, domestic cats, wild carnivores (fox, coyotes) and omnivores (possums, mustelids and rodents). Dual HIV and *M. bovis* infection have been reported in various industrialized countries (Cornuz et al. 1991). The epidemic of HIV infection in developing countries, particularly countries in which *M. bovis* infection is present in animals, could make zoonotic TB a serious public health threat to persons at risk. The re-emergence of TB and continued failure to eradicate the disease in livestock in many countries have been attributed to reservoirs of infection in wildlife populations.

M. ulcerans causes BU, which is the third most common human mycobacteriosis worldwide, after TB and leprosy. *M. ulcerans* is closely related to the fish pathogen *M. marinum*. Analysis of their full 16S rRNA sequences shows >99.8 % identity, and comparisons of other gene sequences show a similar high level of sequence conservation (Stinear et al. 2000). It has been classified as neglected emerging disease by WHO (URL: http://www.who.int/buruli/en/). Nearly, 30 tropical countries have reported the occurrence of the disease, and 5,000 cases are reported annually. The disease is characterized by large skin lesions that 80 % of the times are located on the limbs, most commonly on the lower extremities yet some variation exists. In Africa, all ages and sexes are affected, but most cases of the disease occur in children between the ages of 4–15 years (World Health Organisation 2008).

Mycobacteria in Water

Out of 100 taxonomically known mycobacterial species, nearly 33 can be found in drinking water distribution systems. The nonexhaustive list of various species of NTM in water has been very well documented in the review by (Vaerewijck et al. 2005).The species *M. avium*, *M. chelonae*, *M. fortuitum*, *M. gordonae*, *M. kansasii* and *M. xenopi* are the most frequently reported mycobacteria occurring in drinking water. Mycobacteria were isolated from 38 % (16 samples out of 42) of drinking water samples in the USA (Covert et al. 1999). Similarly in Greece and Paris, 21.3 % (42/197) and 72 % (104/144) of the samples from a drinking water distribution systems were positive for mycobacteria (Tsintzou et al. 2000; Dantec et al. 2002).

Physiological Adaptation of Mycobacteria to Persist in Aquatic Environment

Unique Hydrophobic Cell Wall

All *Mycobacterium* species share a characteristic cell wall, thicker than in many other bacteria, hydrophobic (van Oss et al. 1975), waxy and rich in mycolic acids. The cell wall consists of the hydrophobic mycolate layer and a peptidoglycan layer held together by a polysaccharide, arabinogalactan.

The high lipid content of the cell wall makes the cell wall of mycobacteria the most hydrophobic of all the other bacteria. The hydrophobicity of the cell wall contributes to their enrichment at the water–air interface, especially in the bubbles and aerosols. In the case of *M. avium*, the concentration of colony-forming units (CFUs) in the ejected droplets divided by the concentration in the bulk suspension can be as high as 1,000 (Parker et al. 1983).

Slow Growth

M. avium is a member of the slow-growing mycobacteria. Generation times in the laboratory medium are usually one day. Slow growth is a consequence of the presence of a single rRNA gene cluster (Bercovier et al. 1986) that slows down the process of protein synthesis, increased energy requirements for the synthesis of long chain fatty acids, lipids and waxes (Brennan and Nikaido 1995), and the impermeability of the lipid-rich cell wall that limits the transport of hydrophilic nutrients. As a consequence of slow growth, *M. avium* dies relatively slowly and therefore can survive starvation, and antimicrobial and disinfectant exposure.

Metabolism

Many *Mycobacterium* species adapt readily to growth on very simple substrates, using ammonia or amino acids as nitrogen sources and glycerol as a carbon source in the presence of mineral salts. For example, *M. intracellulare* persisted with only one log loss of viability after 1.4 years of its growth in deionized water. Thus, their ability to utilize many substances as nutrients enables them to grow successfully in many biotopes. Growth is stimulated by glycerol and fatty acids. In contrast to some pathogenic mycobacteria, *M. avium* grows well on minimal media such as Sauton's, which consists of nothing more than glycerol, citric acid, L-asparagine as a carbon nitrogen source and trace salts of potassium, magnesium, iron and zinc. Thus, environments mycobacteria have comprehensive biosynthetic capabilities, consistent with their ability to colonize dilute environments.

Environmental Factors Required for Growth

Mycobacteria are, in general, hardy organisms that survive in the dry as well as in the aquatic environments for long periods of time. NTM such as *M. avium* have the capacity to survive and multiply under a wide range of environmental conditions, including low pH (Chapman and Bernard 1962), extreme temperature (Schulze-Röbbecke and Buchholtz 1992), starvation (Archuleta et al. 2005), chlorine or ozone treatment (Norton and LeChevallier 2000) and low oxygen level (Brooks et al. 1984). Their adaptation for life in the diverse environments is linked to physiological characteristics of the mycobacteria such as the impermeable cell wall and slow growth (Gonzalez-y-Merchand et al. 1997).

M. avium can grow over a wide range of temperatures from 25°C to over 50°C. (George et al. 1980). Its ability to grow at 45°C is undoubtedly responsible for its presence in hot water systems (du Moulin et al. 1988). Not only can *M. avium* grow at 45°C, but also *M. avium* and a number of other environmental mycobacteria are relatively resistant to high temperature (Schulze-Röbbecke and Buchholtz 1992). *M. avium* was isolated from hot water systems (51.9–57.2°C) in hospitals. The number of organisms recovered by du Moulin et al. (1988) ranged from 1 to 500 CFU/100 ml. *M. avium* was found in hot tap and shower samples collected from two hospitals, in numbers ranging from 0.4 to 5.2 CFU/ml (von Reyn et al. 1994).

The optimal pH for the growth of most of the environmental NTM is towards acidic. In the case of *M. avium*, *M intracellulare*, *M. Kansasii* and *M xenopi*, the optimal pH for the growth is between 5.0 and 6.5. Furthermore, *M. avium* is resistant to acid and the acidic conditions of the human stomach (Bodmer et al. 2000).

Resistance to Disinfectants

The susceptibility of *M. kansasii*, *M. avium* and *M. fortuitum* to chlorine has been measured, and it was found that all are extremely resistant (Taylor et al. 2000). In fact, *M avium* is approximately 1,000 times more resistant to chlorine than is *Escherichia coli*, the standard for drinking water disinfection (Taylor et al. 2000). Disinfection of water can lead to the selection of *M. avium*, *M. intracellulare* and other mycobacteria because disinfection kills competitors permitting growth of *M. avium* on the available nutrients. The use of disinfectants in medicine, industrial settings, home spas and hot tubs (Embil et al. 1997 also leads to the predominance of mycobacteria in these habitats.

Genome

The genome sequence of *M. avium* has a high G+C content (68.5 %), and it is about 5.48 mb long. It has approximately 4,480 coding sequences that most likely appear to be genes. In contrast, the genome of *M. tuberculosis* is about 4.4 mb long and has 3,959 likely genes (Cole et al. 1998; http://genolist.pasteur.fr/TubercuList/). Approximately 385 genes of *M. avium* subspecies *avium* have no counterpart in *M. tuberculosis*. Presumably, some of these *M. avium*-specific genes confer the ability to live and grow in the environment, which *M. tuberculosis* lacks. *M avium* complex plasmids are likely to contribute significantly to the genetic make-up of that group because the plasmids are large (e.g. 15–300 kb), and single strains can harbour as many as six individual plasmids that comprise 30 % of the DNA of the strain. The presence of identical plasmids in *M avium*, *M intracellulare* and *M scrofulaceum* (Meissner and Falkinham 1986) indicates that the plasmids are transmissible between strains. The plasmid of *M. avium* has been shown to encode for mercury (Meissner and Falkinham 1984) and copper resistance (Erardi et al. 1987).

The genome sequence of *M. bovis* is >99.95 % identical to that of *M. tuberculosis*, but deletion of genetic information has led to a reduced genome size. The absence of any unique genes in the genome of *M. bovis* implies that differential gene expression may be the key to host tropisms in human and bovine bacilli (Garnier et al. 2003).

The *M. ulcerans* genome is estimated to be about 5.8 mb long, but a more accurate size assessment will come with further sequence assembly. The *M. ulcerans* genome is extraordinarily rich in insertion sequence (IS) with two different elements (IS2404 and IS2606) accounting for 10 % of the total genome. The biological consequences of the presence of these high copy number IS are as yet unknown, but it seems possible that they may contribute to the relatively fastidious nature of *M. ulcerans*.

Transmission and Route of *M. avium* Infection

NTM such as *M. avium* are not transmitted by the human-to-human route, but are instead transmitted from environmental sources. Healthy persons can become infected with *M. avium* after repeated exposure to *M. avium* either by inhalation of aerosols or by ingestion. This was shown in few studies where *M. avium* caused mycobacterial disease after frequent bathing or use of hot tubs.

Drinking Water

Potable water is considered as the primary source of *M. avium* infection in humans (Ristola 1999). *M. avium* strains from infected AIDS patients in Los Angeles have been shown to be genetically related to isolates recovered from water to which the patients were exposed through drinking or bathing (Von Reyn et al. 1994). Similarly, the presence of high numbers of *M. avium* in Finnish drinking water was associated with a high frequency of *M. avium* infection in Finnish AIDS patients (Ristola et al. 1999).

Mycobacteria have been referred to as the 'ducks of the microbial world' due to their thick, waxy, outer coating which enables them to thrive in aquatic environments. *M. avium* have been isolated from numerous water sources, including wastewater, surface water, recreational water, groundwater and tap water (Falkinham III 1996, 2002; Torkko et. al. 2000, 2001). Piped water supplies are readily colonized by mycobacterial biofilms, which may serve as a reservoir for these opportunistic pathogens. Some reports indicate that NTM have been recovered in 11–38 % of raw water samples at concentrations of <0.1–48 CFUs/ml. *M. avium* are relatively resistant to chlorine, monochloramine, chlorine dioxide and ozone (Taylor et al. 2000). The slower-growing organisms (e.g. *M. avium* and *M. paratuberculosis*) appear to be more resistant to chlorine disinfectants than the more rapidly growing organisms. *M. avium* numbers in raw water sources for drinking water systems correlate with the concentration of particulate matter, suggesting that *M. avium* is associated with particulates (Falkinham III et al. 2001). Thus, one approach for reducing the number of *M. avium* in a drinking water system is to remove particulates (e.g. turbidity) from the raw water during treatment.

Biofilms

Mycobacteria have the ability to form biofilms that may serve as a reservoir of opportunistic pathogens. A biofilm is a microbial community formed by cells that are irreversibly attached to a substrate or interface or to each other and are embedded in a matrix of extracellular polymeric substances produced by the cells. Biofilm formation is a universal strategy used by many species of bacteria to obtain the available nutrients. Bacteria, including mycobacteria, embedded in a biofilm are more protected against antimicrobial agents than planktonic cells (Lewis 2001). Restricted penetration of the antimicrobials into the biofilms, decreased growth rate and expression of possible biofilm-specific resistance genes are mechanisms which, alone or in combination, explain biofilm survival in a number of cases (Lewis 2001).

Mycobacteria lack surface structures like flagella, pili, fimbriae, surfactants, slime or capsules as surface translocation modes. Hydrophobic interactions between the exposed fatty acid tails of the mycobacterial cell wall glycopeptidolipids and the hydrophobic surface mediate the attachment and biofilm formation (Recht et al. 2000). Biofilm formation depends on the mycobacterial strain. In one study (Falkinham III et al. 2001), among the 267 biofilm mycobacterial species isolates, 131 were identified as *M. intracellulare*, four were identified as *M. avium*, and 30 isolates could not be assigned to the species level. The number of *M. intracellulare* in the mycobacterial biofilm was 1.3–2,900 CFU cm^{-2}, with an average number of 600 CFU cm^{-2}. Whereas in another study, mycobacterial counts up to 4.6×10^5 CFU cm^{-2} were found in drinking water distribution system biofilms in South Africa (September et al., 2004).

The type of surface apparently had an effect on mycobacterial biofilm numbers. Numbers of *M. intracellulare* were 4,400 CFU/cm^2 on brass or bronze surfaces compared to 70 CFU/cm^2 on plastic (Falkinham III 2001).

Amoebae

Amoebae are cell wall–free, unicellular aquatic eukaryotes that switch from a motile trophozoite phase towards an immobile cyst phase under harsh environmental conditions such as lack of nutrients, drying, hypoxia and fluctuation in temperatures. Trophozoites are professional phagocytes that engulf any particle with a diameter of 0.5 mm into phagocytic vacuole with further lysosome fusion and destruction. Encystment consists in the formation of a cell wall composed of polysaccharides with β-configuration, such as cellulose. Cysts are resistant to alterations in osmolarity and pH, desiccation, freezing, high concentration of hydrochloric acid, moist heat, chemical antimicrobial agents and biocides (Aksozek et al. 2002). Moreover, cysts have been shown to survive in the laboratory for at least 20 years (Sriram et al. 2008).

Amoebae normally feed on number of environmental bacteria. But some bacteria are able to survive within amoeba, for example mycobacteria ingested by amoebae grow intracellularly, acquiring an invasive phenotype (Cirillo et al. 1997). The phagocytic protozoa, *Tetrahymena pyriformis* and amoebae such as *Acanthamoeba polyphaga* (Adekambi et al. 2006) and *A. castellanii* (Cirillo et al. 1997) have been shown to harbour live mycobacteria and have been isolated from habitats overlapping with that of mycobacteria (Thomas and McDonnell 2007) such as cold drinking water systems (Thomas et al. 2008), hot water systems in hospitals (von Reyn et al. 2002) and cooling towers (Pagnier et al. 2008). Direct observation of mycobacteria within amoebae collected in such environments has not been reported; however, different species of mycobacteria have been isolated from these free-living amoebae collected in such environments. Once inside the amoeba, it has been shown that mycobacterium is protected from antibiotics and water disinfectants, such as chlorine. The various mycobacterial species that have been shown to be phagocytosed and to penetrate amoebal trophozoites into vacuoles include *M. tuberculosis* (Hagedorn et al. 2009), *M. bovis* and BCG strains (Taylor et al. 2003), *M. leprae*, *M. marinum*, *M. avium*, *M. avium subsp. paratuberculosis*, *M. kansasii*, *M. xenopi*, *M. fortuitum*, *M. smegmatis* and 26 additional nontuberculous species (Drancourt et al. 2007; Salah et al. 2009). All these *Mycobacterium* species have been shown to persist in trophozoites, with the exceptions of *M. smegmatis* and the two BCG strains (Pasteur and Japan) of *M. bovis*, which were killed within *A. castellanii* trophozoites, contrary to the *M. bovis* parent strain (Taylor et al. 2003).

It has been suggested that adaptation to amoebae served as a pre-adaptation stage to the macrophage internal environment, an important step in the process of becoming a human pathogen. Various molecular evidences have shown the similarities both in the infections of macrophages and protozoa, as well as life cycles of the mycobacteria within amoeba and macrophages (Cirillo et al. 1997). Thus, the amoeba operates as a 'Trojan horse', introducing the pre-adapted bacterium into the human host. The co-culture of amoebae with *M. avium* enhanced the latter's virulence, particularly its motility and ability to cross the murine intestinal epithelium (Cirillo et al. 1997).

Health impacts of *M. avium*

NTM have been reported to cause localized or disseminated disease, depending on local predisposition and/or degree of immune deficit. The *M. avium* complex with its two major species *M. avium* and *M. intracellulare* is almost frequently responsible for disease, primarily in elderly men with restrictive or obstructive pulmonary conditions, in cystic fibrosis patients. In non-HIV patients, different NTM may cause localized pulmonary disease, adenitis, soft tissue infections, infections of joints/bones, bursae, skin ulcers and generalized disease in individuals like leukaemia, transplant patients, etc.

In AIDS patients, the manifestations may range from localized to disseminated disease. Disseminated *M. avium* was one of the first opportunistic infections detected in AIDS patients (Greene et al. 1982). It has been estimated that *M. avium* infects nearly 50 % of AIDS patients with CD4 lymphocyte counts less than 100/μL (Inderlied et al. 1993). In such patients, infections caused by *M. avium* are fourfold more frequent than those due to *M. intracellulare*. Clinical symptoms include local organ-specific signs and symptoms to nonspecific symptoms such as persistent high-grade fever, night sweats, anaemia, anorexia, diarrhoea, myalgia, weight loss and occasional painful adenopathy.

Lymphadenitis in children

There has been a dramatic change in the causative agent of mycobacterial-related cervical lymphadenitis in children in England (Colville 1993), the United States (Wolinsky 1995) and Australia. Historically, the major mycobacterial species recovered from children with cervical lymphadenitis was *M. scrofulaceum* (Wolinsky 1979). Currently, however, *M. scrofulaceum* is almost never isolated and *M. avium* is isolated (Colville 1993; Wolinsky 1995). Wolinsky (1995) estimated that the shift from *M. scrofulaceum* to *M. avium* occurred over the period 1975 to 1985. Most interestingly, the shift occurred over the same period of time in England, Australia and the United States. Thus, the possible hypotheses include the chlorination of drinking waters and changes in water treatment.

Hypersensitivity pneumonitis

Exposure to aerosols that contain environmental mycobacteria has been reported to cause hypersensitivity pneumonitis. Hypersensitivity pneumonitis has been reported in automobile workers exposed to aerosols generated from metalworking fluid used in metal grinding and finishing operations (Kreiss and Cox-Ganser 1997) in life guards exposed to aerosols generated in indoor swimming pools and individuals at home exposed to aerosols from aerated hot tubs (Embil et al. 1997), spas, humidifiers, and water-damaged building materials. In a number of instances, mycobacteria, including *M. avium* and *M. chelonae* have been recovered from the fluid or water. In all instances, the fluid or water had been subjected to disinfection before symptoms appeared in exposed workers or individuals. As with municipal water chlorination, the disinfection procedure selected for the intrinsically resistant mycobacteria.

Lowers the efficacy of BCG vaccine and interferes with TB diagnosis

It has been suggested that a major reason for the poor efficacy of BCG vaccination against TB in the Indian Chingleput trial was due to prior exposure of humans in that area to environmental mycobacteria. In a laboratory study conducted with mice, it was observed that prior sensitization (2×10^6 CFU, subcutaneous, three infections at 2-week intervals) with environmental mycobacteria *M. avium*, *M. scrofulaceum* and *M. vaccae* exposure in mice inhibited BCG multiplication and thereby prevented the induction of an efficient BCG-mediated immune response and protection against TB challenge (Brandt et al. 2002).

On the other hand, exposure to rapidly growing environmental mycobacteria in Malawi (various unknown doses, likely low, via multiple routes) protected humans against TB and leprosy (Fine et al. 2001). Thus, two mechanisms have been proposed: 'blocking', which implies that the previous immunity induced by exposure to environmental mycobacteria restricts the growth of the BCG and 'masking', which implies that the BCG is unable to confer any additional immunity to that is already induced by the natural mycobacterial exposure.

Further, it is well established now that prior exposure to environmental mycobacteria makes an individual positive to TB purified protein derivative skin testing and thereby interferes with the diagnosis of TB using Mantoux test (von Reyn et al. 2001).

Control of NTM in drinking water

NTM are the most hydrophobic of the microorganisms (van Oss et al. 1975). NTM numbers in drinking water distribution systems are higher in systems with higher turbidity (Falkinham et al. 2001), likely because of the hydrophobicity-driven attachment of NTM to soil particulates. Thus, reduction in water turbidity by using filtration would be expected to reduce the NTM numbers in the drinking water. However, the filter is required to be changed regularly because filters provide an ideal habitat for NTM to attach and grow on the organic compounds concentrated on the filters, even if the filter is impregnated with an antimicrobial agent (e.g. silver; Rodgers et al. 1999).

Filters coated with hydrophobic materials (e.g. paraffin) could be used to selectively remove NTM from waters, aerosols or dusts because NTM are the most hydrophobic than any other microbes. NTM cells can be almost entirely removed (>99.9 %) from aqueous suspensions by partitioning into an organic solvent such as hexadecane (Stormer and Falkinham 1989).

Copper–silver ions has shown efficacy in reducing *Legionella* numbers in a substantial number of hospitals and public buildings. They can also kill *M. avium* (Lin et al. 1998) and other NTM, albeit requiring higher dosages compared with Legionella.

Mycobacterium species are very sensitive to ultraviolet light. The UV sensitivity is generally in the range for other vegetative bacteria (http://www.iuva.org). Application of ultraviolet light for microbial inactivation is gaining increased attention, especially for control of *Cryptosporidium*. However, UV irradiation is mutagenic, and it will be important to determine whether UV disinfection leads to an increase in mutants among the survivors.

Bovine TB

A number of zoonotic TB studies, published between 1954 and 1970 and carried out in various countries around the world, have estimated that the proportion of human cases due to *M. bovis* accounted for 3.1 % of all forms of TB, 2.1 % of pulmonary forms and 9.4 % of extrapulmonary forms (Cosivi et al. 1998). Preliminary studies conducted in Africa indicate that approximately 5–7 % of human TB cases are caused by *M. bovis*. In Tanzania, 10.5 % of people with stomach or lymph gland TB were infected with *M. bovis* (Bolognesi 2007).

Wildlife as Reservoir of *M. bovis*

M. bovis, unlike any other member of *M. tuberculosis* complex, has a broad host range. Therefore, it causes disease in a wide range of domestic, free-ranging and farmed wildlife animals, as well as in humans. The nonexhaustive list of various species of animals that are known to be infected by *M. bovis* has been reviewed previously (Biet et al. 2005). However, only a small proportion of these infected wild animal species can serve as reservoirs of infection for this organism. The factors that determine their role as reservoirs are the ability of excretion, ethology (for example, gregarious or not gregarious behaviour) and ecology (alimentary behaviour, population density and interactions with other species).

The control programme of *M. bovis* infection in domestic animals in various countries has received a setback because of these wildlife reservoirs that cannot be controlled and can reintroduce infection in livestock which in turn could transmit the infection to humans. For example in Great Britain and Ireland, badger (*Meles meles*) has been suggested to act as a significant source of infection to bovines (Clifton et al. 1995). In New Zealand, the eradication of bovine TB is threatened especially due to brushtail possum (*Trichosurus vulpecula*; Weyer et al. 1999). The presence of *M. bovis* infection in white-tailed deer (*Odocoileus virginianus*) in Michigan poses a serious menace to the control and eradication programmes for bovine TB in the United States (Payuer et al. 2002). Infection with *M. bovis* has also been described across a range of animals such as buffalo, kudu, lion, baboon and antelope in the Kruger National Park in South Africa, having severe consequences on the biodiversity of this region (Weyer et al. 1999). In France, a high proportion of *M. bovis*-infected wild deer (*Cervus elaphus*) were found in regions where cattle outbreaks were reported, opening up the suspicion of transmission from wildlife (Biet et al. 2005).

Transmission and Route of Infection of *M. bovis*

Infection of humans may occur by the inhalation of aerosols or through the consumption of contaminated milk. In countries where bovine TB is uncontrolled, or in developed countries before strict control campaigns and milk pasteurization, most human cases occur in young persons and result from drinking contaminated milk. This alimentary route of infection leads to extrapulmonary forms of TB, where infection can become established in the cervix and less frequently in the axillary lymph nodes leading to chronic skin TB. The aerosols produced by handling infected carcasses can also lead to infection in humans (Neill et al. 1989).

In bovines, the primary route of infection is through inhalation of *M. bovis*-infected respiratory excretion or through contaminated dust particles. The secondary route is through ingestion of contaminated water, through milk (Palmer et al. 2002) or through infected feed or pastures (Schmitt et al. 1997). The route of transmission of *M. bovis* can be deduced by the pattern of lesions observed in slaughtered animals. Animals with lesions restricted to the thoracic cavity are presumed to have been infected by the inhalation of aerosols, while those with lesions in mesenteric lymph nodes are thought to have acquired the infection by ingestion (Pollock and Neill 2002).

Water Scarcity is a Risk Factor for Contracting *M. bovis* Infection

M. bovis can survive for long periods outside an animal host in an environment directly or indirectly contaminated by discharges of infected animals, thereby suggesting that the direct contact between the reservoir of infection and cattle is not at all essential for its transmission (Fine et al. 2011). The role of *M. bovis*-contaminated environmental substrates in the interspecies transmission of bovine TB between cattle and deer has been investigated (Witmer et al. 2010). Although *M. bovis* was not identified from any of the environmental substrates tested, particular cattle management practices and environmental factors have been shown to cause the indirect transmission of bovine TB from deer to cattle on cattle farms in north-east Michigan (Kaneene et al. 2002). The factors and practices identified included the presence of ponds or open water in cattle areas, maintaining cattle outside more than 50 % of the time, feeding, and watering cattle outside and not protecting feed intended for cattle from deer.

Severe droughts in Mediterranean habitats and seasonal scarcity of water resources lead to crowding of watering sites by wild *M. bovis*-infected wild animals in the advanced stages of infection, which leads to the spread of infection to farm animals by ingestion or inhalation of nasal and oral excretions from contaminated vegetation, water, mud or fomites. Also, it was shown that households living closer to wildlife protected areas tended to have fewer small stock losses to disease, while water-restricted households reported more cattle and sheep losses. Survey results indicated that people in water-restricted households were more likely to share water with their livestock and that many households lacked awareness about the risks of contracting disease from livestock or from sharing contaminated water with livestock.

Health Impacts of Bovine TB

The human form of *M. bovis* infection is clinically, radiologically and pathologically indistinguishable from that caused by *M. tuberculosis*. Most contemporary studies (WHO 1994; Cosivi et al. 1998) agree that the most common clinical manifestation of *M.bovis* infection in man is associated with the extrapulmonary form of the disease, but about half of the cases of post-primary (reactivation) disease involve the lung and this raises the possibility of human-to-human transmission of TB due to *M. bovis* (WHO, 1994). Following ingestion of the organism, the primary infection in the intestine may heal, it may progress in the intestine, or it may disseminate to other organs (Grange and Collins 1987). Cervical lymphadenopathy, intestinal lesions, chronic skin TB (lupus vulgaris) and other nonpulmonary forms are particularly common (Cosivi et al. 1998). Bolognesi (2007) reported that young children infected with *M. bovis* typically have abdominal infections, whereas older patients suffer from swollen and sometimes ulcerated lymph glands in the neck.

Control of Bovine TB

The various general practices to prevent the transmission of bovine TB include pasteurization or boiling of milk prior to human consumption or further processing, inspection of meat before human consumption and regular PPD skin testing of farm animals. In fact, all the animals entering the food chain should be subjected to ante-mortem and post-mortem inspection. In case the cattle are found to be infected, it should not be treated at all rather it must be slaughtered. This is because the risk of shedding the organisms, hazards to humans and potential for drug resistance make treatment controversial.

However, the involvement in wildlife and environmental reservoirs makes the control and eradication programmes of bovine TB extremely complicated. The detection and diagnosis of mycobacteria infections in wildlife are extremely difficult, due to low sensitivity and specificity of the currently available diagnostic tests. Bacterial culture remains the gold standard for diagnosis; however, isolation of mycobacteria strains from the environment or wildlife is often particularly difficult.

Vaccine approaches to control the disease in wildlife reservoirs have been reviewed (Wedlock et al. 2002; Buddle et al. 2000). Vaccination of both farm and wild animals against *M. bovis* could be done to control bovine TB in wild. One of the successful use of a vaccine in wildlife has been in the control of fox rabies in Switzerland (Wandeler et al. 1988). In the vaccination programme, food baits impregnated with a modified live rabies virus were distributed in the countryside.

Buruli Ulcer

BU is a serious necrotizing cutaneous infection caused by *Mycobacterium ulcerans*, a slow-growing environmental mycobacterium (Walsh et al. 2008). It is a debilitating disease characterized by large necrotic skin ulcers (Fig. 9.2). Early lesions are closed, but as the tissue necrosis spreads, the overlying dermis and epidermis eventually ulcerate, with undermined edges. The large lesions often result in scarring, contractual deformities, amputations and disabilities. BU has been reported from 30

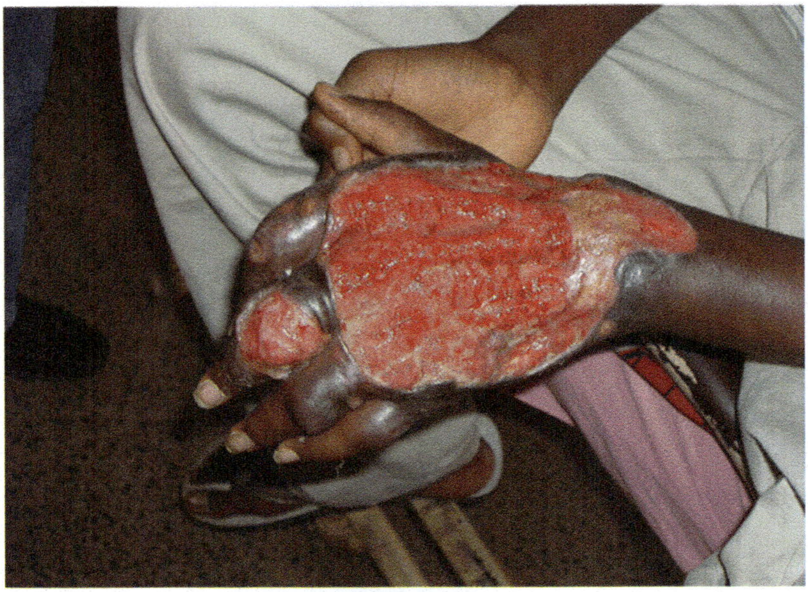

Fig. 9.2 A typical ulcerated hand lesion due to *M. ulcerans* infection on the left hand of Nigerian boy (From Wikipedia, the free encyclopaedia)

countries in Africa, the Americas, Asia and the Western Pacific, mainly in tropical and subtropical regions (Fig. 9.3). In Côte d'Ivoire, West Africa, approximately 24,000 cases have been recorded between 1978 and 2006. In Benin, nearly 7,000 cases have been recorded between 1989 and 2006; in Ghana, more than 11,000 cases have been recorded since 1993. In Australia, more cases of BU are being reported recently—25 in 2004, 47 in 2005 and 72 in 2006 (Merrit et al. 2010).

Bacteria produce mycolactone, a toxic lipid, which is cytotoxic and immune suppressive (George et al. 1999). It causes cellular necrosis, which is responsible for skin ulcers. The toxin damages peripheral nerve Schwann cells; therefore, the developed ulcers are almost painless.

The most important phenotypic characteristic of *M. ulcerans* is the low optimal growth temperature and the extremely restricted growth temperature range. Under the laboratory conditions, growth takes place between narrow temperature ranges of 28–34°C and optimal growth of most strains is found between 30 and 33°C. The restricted growth temperature of *M. ulcerans* is thought to play a substantial role in the pathogenesis of BU by limiting infection to the skin. The organism has never been isolated from internal organs of human patients or from bone in cases of osteomyelitis or from the internal organs or blood of experimentally infected animals (Demangel et al. 2009). It has been recently reported that many isolates of *M. ulcerans* survive at 37°C for 13 days, although numbers decline after the first few days. No one has isolated or derived a strain capable of growth at 37°C (Merritt et al. 2010).

Transmission of BU

Epidemiological evidence has not clearly supported person-to-person transmission. Secondly, the exact environmental source has not been identified as it is extremely difficult to isolate *M. ulcerans* from the environment. Portaels et al. (1999) were the first to suggest that aquatic bugs (Hemiptera) might be reservoirs of *M. ulcerans* in nature, and they described the isolation of pure culture of *M. ulcerans* from a water strider (Hemiptera: Gerridae, Gerris sp.) from Benin (Portaels et al. 2008). Under the laboratory conditions, the pathogen was found to localize in the salivary glands of aquatic creeping bug belonging to the family *Naucoridae* (Fig. 9.4. Masollier

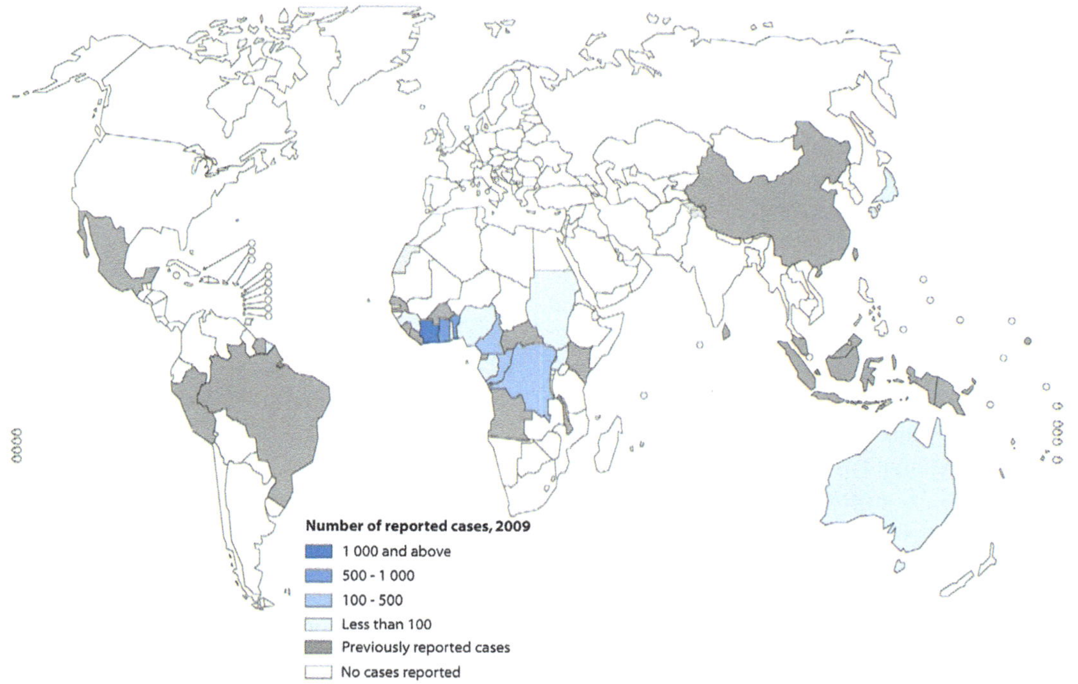

Fig. 9.3 A global map representing countries that have reported cases of Buruli ulcer disease as of 2009 (WHO). doi: 10.1371/journal.pntd.0000911.g003

Fig. 9.4 *Naucoris cimicoides*—a creeping water bug is believed to transmit *M. ulcerans* to humans through their bites

et al. 2002). They also described the experimental transmission of *M. ulcerans* from water bugs to mice, which further demonstrated the possibility of transmission of *M. ulcerans* to humans through the bites of aquatic insects. However, the fact that these species are not haematophagous and bite humans only accidentally casts doubt on the relevance of these associations for transmission of *M. ulcerans*. The role of mosquitoes as biological vector for *M. ulcerans* has also been studied, and it was found that larvae of various species of mosquitoes ingest the bacteria *M. ulcerans*; however, the bacteria are not carried over to pupae or adult mosquitoes (Wallace et al. 2010).

Ecological Associations with Water Bodies

BU has been widely associated with proximity to aquatic habitats. The disease is rare in the savanna regions of West Africa and drier areas of Australia. Its presence in Australia is notably coastal, however, where water is often saline. Nearly, all epidemiological studies have associated disease outbreaks with villages in close proximity to human-disturbed aquatic habitats, including both standing and moving water bodies (Merritt et al. 2010). Increased BU incidence has been reported in association with (1) unprecedented flooding of lakes and rivers during heavy rainfall; (2) the damming of streams and rivers to create impoundments and wetlands; (3) resorts that modify wetlands; (4) deforestation practices and increased agriculture leading to increased flooding; (5) construction of agricultural irrigation systems; (6) rice cultivation; (7) alluvial, pit and sand mining operations; and (8) population expansion, resettlement and migration closer to water bodies.

Further, deforestation is known to cause loss of riparian cover, resulting in increased water temperatures that may facilitate *M. ulcerans* growth at optimal temperatures of 30–33°C. UV light lowers *M. ulcerans* cell viability, and thus, deforestation and high-impact agriculture may promote increased nutrients, higher temperatures, UV attenuation and lower dissolved oxygen—environmental conditions that facilitate *M. ulcerans* growth.

Conclusions

The incidences of TB due to NTM is more likely to increase in the coming years owing to disinfection regimen of drinking water by using chlorine and by the use of disinfectants in medical and industrial settings, thereby selecting mycobacteria by reducing competition. Secondly, due to the increasing percentage of human population with predisposing conditions, most notably AIDS, age and immunosuppressive regimens, for example after organ transplantation. Humans are having a major impact on mycobacterial ecology as evident from the apparent loss of *M. scrofulaceum* from the environment and its replacement by *M. avium*. Further, focal outbreaks of BU disease have followed flooding, human migrations and man-made topographical modifications such as dams and resorts. Deforestation and increased basic agricultural activities may significantly contribute to the current marked increase in incidences of *M. ulcerans* infections, especially in West Africa, where the disease is rapidly emerging. Thus, research in understanding the physiological ecology of mycobacteria is needed to fully discover the effects that mycobacteria have on humans and to allow us to intervene when necessary. Efforts must be focused on actions that will specifically reduce or eradicate mycobacteria from habitats where humans or animals can be exposed.

References

Adekambi T, Ben SS, Khlif M, Raoult D, Drancourt M (2006) Survival of environmental mycobacteria in *Acanthamoeba polyphaga*. Appl Environ Microbiol 72:5974–5981

Aksozek A, McClellan K, Howard K, Niederkorn JY, Alizadeh H (2002) Resistance of *Acanthamoeba castellanii* cysts to physical, chemical, and radiological conditions. J Parasitol 88:621–623

Anaelom NJ, Ikechukwu OJ, Sunday EW, Nnaemeka UC (2010) Zoonotic tuberculosis: a review of epidemiology, clinical presentation, prevention and control. J Publ Health Epidemiol 2:118–124

Archuleta RJ, Hoppes PY, Primm TP (2005) *Mycobacterium avium* enters a state of metabolic dormancy in response to starvation. Tuberculosis 85:147–158

Bercovier H, Kafri O, Sela S (1986) Mycobacteria possess a surprisingly small number of ribosomal RNA genes in relation to the size of their genome. Biochem Biophys Res Commun 136:1136-1141

Biet F, Boschiroli ML, Thorel MF, Guilloteau LA (2005) Zoonotic aspects of *Mycobacterium bovis* and *Mycobacterium avium-intracellulare* complex (MAC). Vet Res 36:411–436

Bodmer T, Miltner E, Bermudez LE (2000) *Mycobacterium avium* resists exposure to the acidic conditions of the stomach. FEMS Microbiol Lett 182:45–49

Bolognesi N (2007) TB or not TB: the threat of bovine tuberculosis. http://www.scidev.net/en/features/tb-or-not-tb-thethreat-of-bovinetuberculosis.html

Brandt L, Cunha JF, Olsen AW, Chilima B, Hirsch P, Appelberg R et al (2002) Failure of the *Mycobacterium bovis* BCG vaccine: some species of environmental mycobacteria block multiplication of BCG and induction of protective immunity to tuberculosis. Infect Immun 70:672–678

Brennan PJ, Nikaido H (1995) The envelope of mycobacteria. Ann Rev Biochem 64:29–63

Brooks RW, George KL, Parker BC, Falkinham JO, Gruff H (1984) Recovery and survival of nontuberculous mycobacteria under various growth and decontamination conditions. Can J Microbiol 30:1112–1117

Buddle BM, Skinner MA, Chambers MA (2000) Immunological approaches to the control of tuberculosis in wildlife reservoirs. Vet Immunol Immunopathol 74:1–16

Chapman JS, Bernard JS (1962) Tolerances of unclassified mycobacteria. I. Limits of pH tolerance. Am Rev Respir Dis 86:582–583

Cirillo JD, Falkow S, Tompkins LS, Bermudez LE (1997) Interaction of *Mycobacterium avium* with environmental amoebae enhances virulence. Infect Immun 65:3759–3767

Clifton-Hadley RS, Wilesmith JW, Richards MS, Upton P, Johnston S (1995) The occurrence of *Mycobacterium bovis* infection in cattle in and around an area subject to extensive badger (*Meles meles*) control. Epidemiol Infect 114:179–193

Cole ST, Brosch R, Parkhill J, Garnier T, Churcher C, Harris D et al (1998) Deciphering the biology of Mycobacterium tuberculosis from the complete genome sequence. Nature 393:537–544

Colville A (1993) Retrospective review of culture-positive mycobacterial lymphadenitis cases in children in Nottingham, 1979–1990. Eur J Clin Microbiol Infect Dis 12:192–195

Cornuz J, Fitting JW, Beer V, Chave JP (1991) Mycobacterium bovis and AIDS. AIDS 5:1038–1039

Cosivi O, Grange JM, Daborn CJ, Raviglione MC, Fujikura T, Cousins D et al (1998) Zoonotic tuberculosis due to *Mycobacterium bovis* in developing Countries. Emerg Infect Dis 4:59–70

Covert TC, Rodgers MR, Reyes AL, Stelma GN Jr (1999) Occurrence of nontuberculous mycobacteria in environmental samples. Appl Environ Microbiol 65:2492–2496

Dantec LC, Duguet JP, Montiel A, Dumoutier N, Dubrou S, Vincent V (2002) Occurrence of mycobacteria in water treatment lines and in water distribution systems. Appl Environ Microbiol 68:5318–5325

Demangel C, Stinear TP, Cole ST (2009) Buruli ulcer: reductive evolution enhances pathogenicity of *Mycobacterium ulcerans*. Nat Rev Microbiol 7:50–60

Drancourt M, Adekambi T, Raoult D (2007) Interactions between *Mycobacterium xenopi*, amoeba and human cells. J Hosp Infect 65:138–142

du Moulin GC, Stottmeier KD, Pelletier PA, Tsang AY, Hedley-Whyte J. (1988) Concentration of *Mycobacterium avium* by hospital hot water systems. JAMA 260:1599–1601

Embil J, Warren P, Yakrus M, Stark R, Corne S, Forrest D (1997) Pulmonary illness associated with exposure to *Mycobacterium avium* complex in hot tub water: hypersensitivity pneumonitis or infection? Chest 111:813–816

Erardi FX, Failla ML, Falkinham JO III (1987) Plasmid encoded copper resistance and precipitation by *Mycobacterium scrofulaceum*. Appl Environ Microbiol 53:1951–1954

Falkinham JO III (2002) Nontuberculous mycobacteria in the environment. Clin Chest Med 23:529–551

Falkinham JO III (1996) Epidemiology of infection by nontuberculous mycobacteria. Clin Microbiol Rev 9:177–215

Falkinham JO III, Norton CD, LeChevallier MW (2001) Factors influencing numbers of *Mycobacterium avium*, *Mycobacterium intracellulare*, and other mycobacteria in drinking water distribution systems. Appl Environ Microbiol 67:1225–1231

Fine AE, Bolin CA, Gardiner JC, Kaneene JB (2011) A Study of the persistence of *Mycobacterium bovis* in the environment under natural weather conditions in Michigan, USA. Vet Med Int. doi: 10.4061/2011/765430

Fine PE, Floyd S, Stanford JL, Nkhosa P, Kasunga A, Chaguluka S et al (2001) Environmental mycobacteria in northern Malawi: implications for the epidemiology of tuberculosis and leprosy. Epidemiol Infect 126:379–387

Garnier T, Eiglmeier K, Camus J, Medina N, Mansoor H, Pryor M et al (2003) The complete genome sequence of *Mycobacterium bovis*. Proc Natl Acad Sci 100:7877–7882

George KL, Parker BC, Gruft H, Falkinham JO III (1980) Epidemiology of infection by nontuberculous mycobacteria. II. Growth and survival in natural waters. Am Rev Respir Dis 122:89–94

George KM, Chatterjee D, Gunawardana G, Welty D, John H, Lee T et al (1999) Mycolactone: a polyketide toxin from *Mycobacterium ulcerans* required for virulence. Science 283:854–857

Gonzalez-y-Merchand JA, Garcia MJ, Gonzalez-Rico S, Colston MJ, Cox RA (1997) Strategies used by pathogenic and nonpathogenic mycobacteria to synthesize rRNA. J Bacteriol 179:6949–6958

Grange JM and Collins CH (1987) Bovine tubercle bacilli and disease in animals and man. Epidemiol Infect 99:221–234

Greene JB, Sidhu GS, Lewin S, Levin JF, Masur H, Simberkoff MS et al (1982) *Mycobacterium avium–intracellulare*: a cause of disseminated life threatening infection in homosexuals and drug abusers. Ann Intern Med 97:539–546

Guthertz LS, Damsker B, Bottone EJ, Ford EG, Midura TF, Jamda JM (1989) *Mycobacterium avium* and *Mycobacterium intracellulare* infections in patients with and without AIDS. J Infect Dis 160:1037–1041

Hagedorn M, Rohde KH, Russell DG, Soldati T (2009) Infection by tubercular mycobacteria is spread by nonlytic ejection from their amoeba hosts. Science 323:1729–1733

Inderlied CB, Kemper CA, Bermudez LEM (1993) The *Mycobacterium avium* complex. Clin Microbiol Rev 6:266–310

Kaneene JB, Bruning-Fann CS, Granger LM, Miller R, Porter-Spalding BA (2002) Environmental and farm management factors associated with tuberculosis on cattle farms in northeastern Michigan. J Am Vet Med Assoc 221:837–842

Kreiss K, Cox-Ganser J (1997) Metalworking fluid-associated hypersensitivity pneumonitis: a workshop summary. Am J Ind Microbiol 32:423–432

Lewis K (2001) Riddle of biofilm resistance. Antimicrob Agents Chemother 45:999–1007

Lin Y, Vidic RD, Stout JE, McCartney CA, Yu VL (1998) Inactivation of *Mycobacterium avium* by copper and silver ions. Water Res 32:1997–2000

MacCallum P, Tolhurst JC, Buckle G, Sissons HA (1948) A new mycobacterial infection in man. I. Clinical aspects. II. Experimental investigations in laboratory animals. III. Pathology of the experimental lesions in the rat. IV. Cultivation of the new mycobacterium. J Pathol Bacteriol 60:93–122

Marsollier L, Robert R, Aubry J, Saint-André JP, Kouakou H, Legras P et al (2002) Aquatic insects as a vector for *Mycobacterium ulcerans*. Appl Environ Microbiol 68:4623–4628

Meissner PS, Falkinham JO III (1984) Plasmid-encoded mercuric reductase in *Mycobacterium scrofulaceum*. J Bacteriol 157:669–672

Meissner PS, Falkinham JO III (1986) Plasmid DNA profiles as epidemiological markers for clinical and environmental isolates of *Mycobacterium avium*, *Mycobacterium intracellulare*, and *Mycobacterium scrofulaceum*. J Infect Dis 153(325–3):31

Merritt RW, Walker ED, Small PLC, Wallace JR, Johnson PDR, Benbow ME et al (2010) Ecology and transmission of Buruli ulcer disease: a systematic review. PLoS Neglected Trop Dis 4:e911

Neill SD, Hanna J, O'Brien JJ, McCracken RM (1989) Transmission of tuberculosis from experimentally infected cattle to in-contact calves. Vet Rec 124:269–271

Norton CD, LeChevallier MW (2000) A pilot study of bacteriological population changes through potable water treatment and distribution. Appl Environ Microbiol 66:268–276

Parker BC, Ford MA, Gruft H, Falkinham JO III (1983) Epidemiology of infection by nontuberculous mycobacteria. IV. Preferential aerosolization of *Mycobacterium intracellulare* from natural water. Am Rev Respir Dis 128:652–656

Pagnier I, Raoult D, La SB (2008) Isolation and identification of amoeba resisting bacteria from water in human environment by using an *Acanthamoeba polyphaga* co-culture procedure. Environ Microbiol 10:1135–1144

Palmer MV, Waters WR, Whipple DL (2002) Milk-containing *Mycobacterium bovis* as a source of infection for white-tailed deer fawn. Tuberculosis 82:161–165

Payeur JB, Church S, Mosher L, Robinson-Dunn B, Schmitt S, Whipple D (2002) Bovine tuberculosis in Michigan wildlife. Ann New York Acad Sci 969:259–261

Pollock JM, Neill SD (2002) *Mycobacterium bovis* infection and tuberculosis in cattle. Vet J 163:115–127

Portaels F, Elsen P, Guimaraes-Peres A, Fonteyne P, Meyers WM (1999) Insects in the transmission of *Mycobacterium ulcerans* infection. The Lancet 353:986

Portaels F, Meyers WM, Ablordey A, Castro AG, Chemlal K, Rijk P et al (2008) First cultivation and characterization of *Mycobacterium ulcerans* from the Environment. PLoS Neglected Trop Dis 2:e178

Recht J, Martinez A, Torello S, Kolter R. (2000) Genetic analysis of sliding motility in *Mycobacterium smegmatis*. J Bacteriol 182:4348–4351

Ristola MA, von Reyn CF, Arbeit RDS, Soini H, Lumio J, Ranki A et al (1999) High rates of disseminated infection due to non-tuberculous mycobacteria among AIDS patients in Finland. J Infect 39:61–67

Rodgers MR, Blackstone BJ, Reyes AL, Covert TC (1999) Colonisation of point of use water filters by silver resistant non-tuberculous mycobacteria. J Clin Pathol 52:629–635

Salah B, Ghigo E, Drancourt M (2009) Free-living amoebae, a training field for macrophage resistance of mycobacteria. Clin Microbiol Infect 15:894–905

Schmitt SM, Fitzgerald SD, Cooley TM, Bruning-Fann CS, Sullivan L, Berry D et al (1997) Bovine tuberculosis in free-ranging white-tailed deer from Michigan. J Wildl Dis 33:749–758

Schulze-Röbbecke R, Buchholtz K (1992) Heat susceptibility of aquatic mycobacteria. Appl Environ Microbiol 58:1869–1873

September SM, Brozel VS, Venter SN (2004) Diversity of nontuberculoid Mycobacterium species in biofilms of urban and semiurban drinking water distribution systems. Appl Environ Microbiol 70:7571–7573

Sriram R, Shoff M, Booton G, Fuerst P, Visvesvara GS (2008) Survival of *Acanthamoeba* cysts after desiccation for more than 20 years. J Clin Microbiol 46:4045–4048

Stormer R, Falkinham JO III (1989) Differences in antimicrobial susceptibility of pigmented and unpigmented colonial variants of *Mycobacterium avium*. J Clin Microbiol 27:2459–2465

Stinear TP, Jenkin GA, Johnson PDR, Davies JK (2000) Comparative genetic analysis of *Mycobacterium ulcerans* and *Mycobacterium marinum* reveals evidence of recent divergence. J Bacteriol 182:6322–6330

Taylor RH, Falkinham JO III, Norton CD, LeChevallier MW (2000) Chlorine, chloramine, chlorine dioxide, and ozone susceptibility of *Mycobacterium avium*. Appl Environ Microbiol 66:1702–1705

Taylor SJ, Ahonen LJ, de Leij FA, Dale JW (2003) Infection of *Acanthamoeba castellanii* with *Mycobacterium bovis* and *M. bovis* BCG, and survival of *M. bovis* within the amoebae. Appl Environ Microbiol 69:4316–4319

Thomas V, McDonnell G (2007) Relationship between mycobacteria and amoebae: ecological and epidemiological concerns. Lett Appl Microbiol 45:349–357

Thomas V, Loret JF, Jousset M, Greub G (2008) Biodiversity of amoebae and amoebae-resisting bacteria in a drinking water treatment plant. Environ Microbiol 10:2728–2745

Torkko P, Suomalainen S, Iivanainen E, Suutari M, Tortoli E, Paulin L et al (2000) *Mycobacterium xenopi* and related organisms isolated from stream waters in Finland and description of *Mycobacterium botniense* sp. nov. Int J Syst Evol Microbiol 50:283–289

Torkko P, Suomalainen S, Iivanainen E, Suutari M, Paulin L, Rudback E et al (2001) Characterization of *Mycobacterium bohemicum* isolated from human, veterinary, and environmental sources. J Clin Microbiol 39:207–211

Tsintzou A, Vantarakis A, Pagonopoulou O, Athanassiadou A, Papapetropoulou M (2000) Environmental mycobacteria in drinking water before and after replacement of the water distribution network. Water Air Soil Pollut 120:273–282

Vaerewijck JM, Huys G, Palomino JC, Swings J, Portaels F (2005) Mycobacteria in drinking water distribution systems: ecology and significance for human health. FEMS Microbiol Rev 29:911–934

Van Oss CJ, Gillman CF, Neumann AW (1975) Phagocytic engulfment and cell adhesiveness as cellular surface phenomena. Marcel Dekker, Inc., New York, p 160

Von Reyn CF, Maslow JN, Barber TW, Falkinham JO III, Arbeit RD (1994) Persistent colonization of potable water as a source of *Mycobacterium avium* infection in AIDS. Lancet 343:1137–1141

von Reyn CF, Horsburgh CR, Olivier KN, Barnes PF, Waddell R, Warren C et al (2001) Skin test reactions to *Mycobacterium tuberculosis* purified protein derivative and *Mycobacterium avium* sensitin among health care workers and medical students in the United States. Int J Tuberc Lung Dis 5:1122–1128

Von Reyn CF, Arbeit RD, Horsburgh CR, Ristola MA, Waddell RD, Tvaroha SM et al (2002) Sources of disseminated *Mycobacterium avium* infection in AIDS. J Infect 44:166–170

Wallace JR, Gordon MC, Hartsell L, Mosi L, Benbow ME, Merritt RW et al (2010) Interaction of *Mycobacterium ulcerans* with mosquito species: implications for transmission and trophic relationships. Appl Environ Microbiol 76:6215–6222

Walsh D, Portaels F, Meyers W (2008) Buruli ulcer (*Mycobacterium ulcerans* infection). Trans R Soc Trop Med Hyg 102:969–978

Wedlock DN, Skinner MA, de Lisle GW, Buddle BM (2002) Control of *Mycobacterium bovis* infections and the risk to human populations. Microbes Infect 4:471–480

Wandeler AI, Capt S, Kappeler A, Hauser R (1988) Oral immunization of wildlife against rabies: concept and first field experiments. Rev Infect Dis 10(Suppl 4):S649–S653

Weyer K, Fourie PB, Durrheim D, Lancaster J, Haslov K, Bryden H (1999) *Mycobacterium bovis* as a zoonosis in the Kruger National Park, South Africa. Int J Tuber Lung Dis 3:1113–1119

World Health Organisation (1994) Zoonotic Tuberculosis (*M. bovis*): a Memoranda from WHO Meeting (with participation of FAO). Bulletin of the WHO 72:851–857

World Health Organisation (2004) The World Health Report 2004—changing History

World Health Organisation (2008) Buruli ulcer: progress report, 2004–2008. In: WHO (ed) Weekly epidemiological record, vol 83. World Health Organization, Geneva, Switzerland, pp 145–156

Witmer AE, Fine J Glonfriddo, Pipas M, Shiveli K, Piccolo K et al (2010) Epizootiologic survey of *Mycobacterium bovis* in wildlife and farm environments in northern Michigan. J Wildl Dis 46:368–378

Wolinsky E (1995) Mycobacterial lymphadenitis in children: a prospective study of 105 nontuberculous cases with long-term follow-up. Clin Infect Dis 20:954–963

Wolinsky E (1979) Nontuberculous mycobacteria and associated diseases. Am Rev Respir Dis 119:107–159

Water, Mosquitoes and Malaria

10

V. P. Sharma

Abstract

There are >3,000 species of mosquitoes in the world. Mosquitoes need water to complete their aquatic life from egg, larva, pupa to adult emergence of both sexes. Adults are terrestrial and feed on nectar and fruit juices, and females feed on blood of warm-blooded animals. Females bite and transmit the diseases. Male mosquitoes are short lived (about a week or 10 days), mate, and die, while females feed on blood, lay eggs on alternate days, may live for 4–6 weeks, and die. During the process of feeding on man, they transmit dreadful diseases such as malaria, filariasis, dengue, chikungunya, yellow fever (in Africa), and many other viral diseases. There are innumerable mosquitoes' breeding sites created by the rainfall, irrigation, stagnant water bodies, and man-made water collection in the rural and urban areas. Water bodies produce a variety of mosquitoes including disease vectors and nuisance mosquitoes. Anophelines breed in clean water and some species breed in brackish water, whereas Culex mosquitoes breed in polluted water with organic matter. Rainy seasons are the main transmission season as vector breeding and longevity are favorable for disease transmission. Since the discovery of malaria transmission by mosquitoes, malaria wastelands reclaimed mainly through drainage have produced enormous wealth and prosperity to the malaria-endemic countries in the world. Classical examples include malaria control in Pontine Marshes near Rome, Tennessee Valley Authority in USA, drainage in Assam during the World War II, installation of siphons to wash away mosquito breeding, Banbasa Head works in India, drainage in Malaysia, Indonesia, Philippines, and many other countries. The paper provides strong link between malaria mosquitoes and water and brings out the importance of drainage in human welfare.

V. P. Sharma (✉)
Centre for Rural Development and Technology,
Indian Institute of Technology, Hauz Khas,
New Delhi, 110 016, India
e-mail: vinodsharma1938@gmail.com

P. P. Singh and V. Sharma (eds.), *Water and Health*,
DOI: 10.1007/978-81-322-1029-0_10,

Keywords
Drainage • Environmental management • Malaria • Mosquitoes • Vectors • Water

Introduction

Malaria is endemic in India with nearly a billion populations at risk of contracting the disease. An estimated 1.31 million parasite-positive cases and 753 malaria deaths (NVBDCP Web site; see Fig. 10.1) are reported by the National Vector-borne Disease Control Programme (NVBDCP) in 2011. WHO estimates 30 million cases and 19,500 deaths due to malaria in India. Other estimates based on in-depth reviews, anti-malaria drug consumption, and periodical surveys also bring out gross under-reporting of malaria cases in the country. Dhingra et al. (2010) estimated 205,000 (125,000–277,000) deaths due to malaria in India during 2001–2003. Malaria is unevenly distributed in time and space. Bulk of malaria cases are found in flood plains of northern India, coastal plains, the northeastern region, and the forests occupied by tribal communities. Malaria epidemics are often widely spread and cover the entire ecotype like the epidemic of malaria in western Rajasthan in 1994 (Mathur et al. 1990; Bouma and van der Kaay 1994). Malaria situation is alarming in Thar Desert, stated to be sitting on the tip of the malaria iceberg (Sharma 2004). The country is witnessing major ecological changes albeit protected mosquito breeding under the 5-year plans (Sharma 1996a, b). Rainfall and irrigation encourage mosquito proliferation and provide huge opportunities for breeding all over in the rural, urban, and industrial settings. The situation is further compounded by excessive rainfall, poor drainage, and climate change (Bouma et al. 1994). Malaria till the 1950s was a rural disease, but over the last 5–6 decades, new malaria ecotypes have emerged. These have been identified as the forest malaria, irrigation malaria, rural malaria, urban and peri-urban malaria, industrial malaria, migration malaria, and border malaria (Pattanayak et al. 1994). Along with new ecotypes, many new problems emerged in the control of malaria, viz. widespread vector resistance, so that spraying now produces diminishing returns: emergence of drug resistance in *P. falciparum* and more recently in *P. vivax* (Singh et al. 1989a, b; Sinha et al. 1989; Dua et al. 2003). Deaths due to malaria that had been completely eliminated during the eradication era surfaced again throughout the country. Resurgent malaria was widespread and did not respond to spraying (Sharma 1996a, b; Dev et al. 2004; Yadav et al. 2003) as was the case in early 1960s. The return of malaria entered all receptive areas at one time freed from the disease. Country now faces formidable challenges in malaria control, viz. high cost of malaria control year after year, environmental pollution, resurgence of drug resistant malaria, enormous losses in agriculture, industry, tourism, difficulties in the development of the hinterland, rising trend of man-made malaria, erratic and heavy rainfall and malaria outbreaks/epidemics, etc. (Tyagi and Yadav 1997). Malaria-endemic area remains poverty stricken, and agriculture and industrial production is compromised; malaria causes anemia, adversely affecting health of children and pregnant women, produces low-birth-weight babies, affects cognitive development in children, destroys household economy, and arrests all developmental activities. Malaria is the single most important cause of the vicious cycle of poverty, hunger, and ill-health. The paper briefly discusses the malaria epidemiology in the background of the continuing challenges in malaria control in India (Sharma 2012).

Water is an absolutely essential requirement for sustaining plant and animal life. Water is also the source of many unwanted and lethal organisms. Mosquitoes require water for development of aquatic life. Survival and longevity of adult mosquitoes depend on humidity as in the dry climate they tend to desiccate and die. Figure 10.1

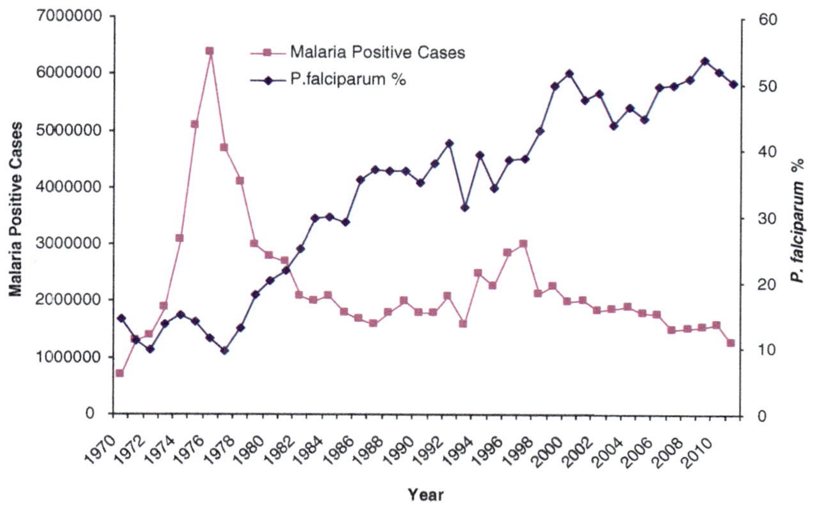

Fig. 10.1 Malaria cases in India (1970–2011) as recorded by the NVBDCP. Cases started rising in 1970, peaking in 1976 to 6.45 m cases, and thereafter following the implementation of the modified plan of operation in 1977, malaria cases declined, but mainly P. vivax malaria due to its sensitivity to chloroquine. *P. falciparum* % was about 10 in 1977 but due to fall in vivax malaria, this % has risen to about 50 and the parasite has become mono- to multi-drug resistant (*data source* NVBDCP)

gives the life cycle of the mosquitoes. For the development of malaria parasite in the body of the Anopheles mosquitoes, minimum 55–60 % or more humidity is required to complete the sporogony. Water is also required for the gravid females to lay eggs and thus ensure continuity of subsequent generations. The paper provides a link between malaria and mosquitoes with water, challenging the survival of humankind, and how this challenge was converted to opportunity for human welfare.

Mosquitoes

There are >3,000 species of mosquitoes in the world, and about 430 species belong to genus Anopheles. All mosquitoes breed in water and complete their aquatic life from eggs to larval instars (I to IV) and pupae. Aquatic stages last for 1–2 weeks depending on the air temperature. From pupal stage, adult mosquitoes emerge and disperse in the environment to mate, feed, and thus complete their life cycle (Fig. 10.2). After adult emergence, both sexes feed on nectar or fruit juices, but females also feed on the blood of warm-blooded animals. Blood feeding is required to provide proteins for the ovarian development to lay eggs. The first batch of eggs require 2 or 3 blood meals but subsequent batches of eggs are laid on second or third day, and the process of feeding and egg laying continues throughout life of the mosquitoes from 4 to 6 weeks. Each time 70–100 or more eggs are laid on water. Males survive for about a week to 10 days and die after mating. Mosquitoes usually mate during flight. Human malaria transmission is from man to man through the bite of female Anopheles mosquitoes. The capacity to transmit malaria varies from species to species, some are highly efficient malaria vectors, for example *An. baimaii* (=*dirus*), or poor vectors, for example *An. subpictus*. Even within the species, there is great variation in the transmission capacity of malaria, for example *An. culicifacies* species B (non-vector), while all other sibling species (A, C, D, and E) are vectors.

In the chain of disease transmission, mosquitoes play a key role in maintaining malaria endemicity. Absence of mosquito vectors would result in the absence of the diseases they transmit. Therefore, for the disease to occur and

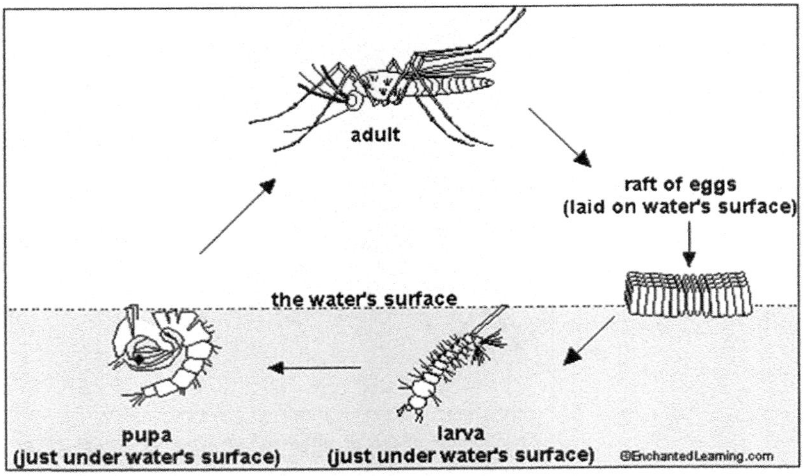

Fig. 10.2 Life cycle of a Culex mosquito. *Source* http://www.enchantedlearning.com/subjects/insects/mosquito/lifecycle.shtml

multiply, it is necessary that vectors are either present or introduced in a given area. The pathogen (e.g., Plasmodium species for malaria) develops in the body of the mosquito. The bite of a mosquito after successful incubation period results in a patent infection that causes the disease. Environment plays a vital role in the transmission of malaria. It regulates the life cycle and biting behavior of the mosquitoes, and development of the pathogen is profoundly affected by the prevailing conditions of the environment, such as the temperature and humidity. Water bodies regulate the humidity, rainfall, etc. Malaria is transmitted by the bite of the female Anopheles mosquitoes. Globally, 60 Anopheles species are the known vectors of malaria. In India, Anopheles fauna comprises of 55 species, but malaria is transmitted by 6 major vectors, viz. *Anopheles culicifacies, An. stephensi, An. fluviatilis, An. minimus, An. baimaii* (=*dirus*), and *An. sundaicus*. Natural distribution of these vectors is given in Fig. 10.3. *Anopheles culicifacies* is widely present in the plains throughout rural India. *Anopheles culicifacies* generates annually an estimated 65 % malaria cases and 55 % *P. falciparum* cases. However, in some hilly terrain and areas with flowing streams, *An. fluviatilis* participates in transmission along with *An. culicifacies* contributing an estimated 15 % malaria cases and 30 % *P. falciparum* cases. In addition, *An. stephensi*, the vector of urban malaria, also generates malaria cases in semiarid and desert climate and in the peri-urban settlements. In the main township, *An. stephensi* is the sole vector of malaria. *An. stephensi* accounts for 12 % cases. In the plains of India, *An. culicifacies* and *An. stephensi* together transmit malaria in the industrial and peri-urban areas. *An. fluviatilis* is widely distributed predominantly in the foothill with streams and contributes about 15 % cases, and *P. falciparum* 30 %. *An. minimus* breeds in streams in the northeastern states and *An. dirus* (=*baimaii*) in the puddles in jungles in the northeastern states, and together, they contribute about 7–8 % cases. *An. sundaicus* breeds in the brackish waters, and it has a strain that colonized sweet water. It has retracted from the mainland and now occurs in the Andaman and Nicobar islands, contributing about a few thousand cases. Rural malaria control is based on spraying residual insecticides (e.g., DDT, HCH, malathion, and synthetic pyrethroid). The main problem faced in the control of malaria is either that the vectors have become resistant to one or more insecticide and/or exhibit exophilic/exophagic behavior, thus avoiding contact with the sprayed walls. The choice of insecticides is very limited. DDT is under restricted use due to possible harmful effects on the human health. HCH has been banned in the country. Malathion has pungent smell, and therefore refusal rates are

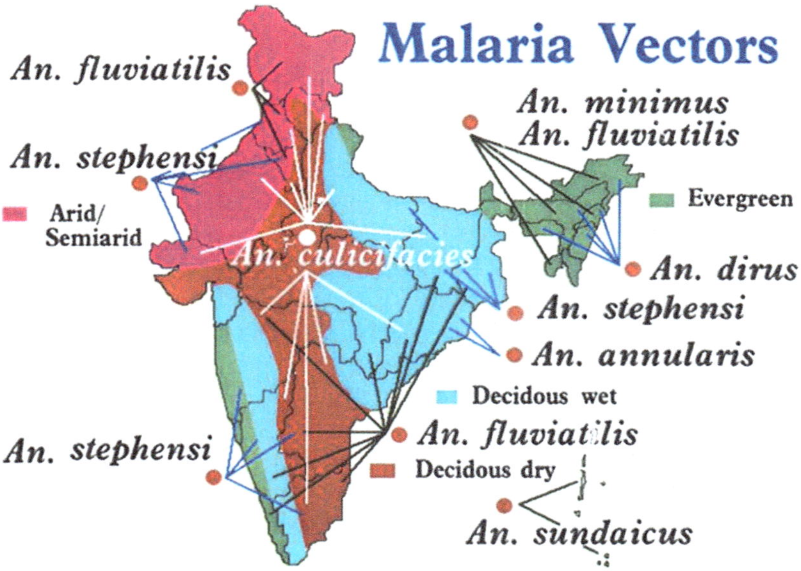

Fig. 10.3 Distribution of major malaria vectors of India

very high. Synthetic pyrethroids (deltamethrin, cyfluthrin, deltacyhalothrin, etc.) are being sprayed selectively to control in areas with epidemic situations. There are no other insecticides currently available for large-scale spraying to control malaria. Malaria control by insecticide spraying is therefore at the crossroads and requires innovative methods for large-scale application through the primary healthcare system.

Link to Water

Mosquitoes have aquatic and terrestrial life. They complete their aquatic life, and adult mosquitoes emerge from pupae and disperse in the environment to mate, feed, and reproduce and thus complete their life cycle (see Fig. 10.2). Water quality influences the attraction of mosquitoes to lay eggs. In general, clean water such as from the rains or rivers will promote the breeding of Anopheles and Aedes mosquitoes, whereas polluted water will attract Culex mosquitoes. For example, major mosquito breeding habitats for malaria vectors are: *An. culicifacies*—the vector of rural malaria (rainwater, tube well water, stored domestic water, variety of ponds and ditches, agricultural field, wells); *An. stephensi*—malaria vector of urban and peri-urban malaria (rain water collection, leaking water supply in urban and peri-urban areas, wells, over head and underground tanks, water holding receptacles, solid waste, domestic water, etc.) *An. fluviatilis*—malaria vector in plains and foot hills (slow-flowing streams in plains and hills, rice fields, rain water, etc.); *An. minimus*—malaria vector in the degraded forest (breeds in slow-flowing streams predominantly in the northeastern region and Orissa); *An. baimaii (=dirus)*—malaria vector in the northeastern region, particularly breeds in deep jungles in small rain water collection, elephant hoof prints; *An. sundaicus*—malaria vector in Andaman and Nicobar islands (breeds in brackish water, etc., particularly in creeks where sea and rain water mix, wells with brackish water), and a sibling species of *An. sundaicus* breeds in sweet water. A variety of other mosquitoes are encountered in the urban and rural areas that transmit deadly diseases and breed in receptacles and discarded containers. These are the *Aedes aegypti* (breeds in houses in small water collections in the urban areas, for example, ant trap, solid waste, coconut shells, discarded tins and containers, mud pots, etc.). *Aedes albopictus* principally breeds in forests in stagnant water, tree holes, leaf axils, and often migrates to human habitations and has been incriminated in the transmission of dengue and chikungunya fever along with or independent of *Aedes aegytpi*. *Culex quinquefasciatus* mosquito—the vector of lymphatic filariasis—breeds in polluted water in open drains, dis-used wells, gutters, stagnant polluted water all over the rural and urban areas. This mosquito is the major cause of mosquito nuisance all over the country. Culicine mosquitoes, mainly *Culex vishnui* group (*Culex tritaeniorhynchus*, Culex vishnui and *Culex pseudovishnui*), are the chief vectors of JE. A very large number of water bodies are found all over in the rural and urban areas. Most of these water bodies can support mosquito breeding. However, some of the most common sources of mosquito production and disease transmission are: (a) irrigation, (b) rainfall, and (c) water supply. A brief description of these water sources and their link to mosquitoes and malaria is given below.

Irrigation

Canal and tank irrigation remains a major contributor of mosquitoes. Irrigation is essential for food production. For example, globally irrigated rice fields account for 150 million hectares, the largest wetland environment. Of these, 77 million hectares are irrigated, and 95 % wetlands are in developing countries. Wetlands produce the largest number of mosquito population, and some of these are important vectors of malaria, for example, *An. culicifacies* in India and Sri Lanka, *An. aconitus* in Indonesia. The Indian green revolution relied on high-yielding variety of seeds, fertilizers, and water. Irrigation in India has taken a quantum jump from 22.6 m ha in 1950 to >100 m ha. Irrigation produces most favorable and extensive mosquito breeding grounds (see Fig. 10.4). Irrigation provides prolific breeding opportunities and also enhances humidity producing profound effect on the basic reproduction rate of malaria and thus causes increase in malaria season and transmission (Sharma and Mehrotra 1982a, b). Sharma and Uprety (1982) reported high incidence of malaria in canal irrigated villages compared to rain-fed agriculture. A similar observation has been reported recently by Kibret and Alemu (2010) from central Ethiopia. Unfortunately, adequate health safeguards were not taken at the time of construction to prevent vector proliferation. Therefore, there is intense vector breeding for extended periods due to rise in water table leading to water logging, extensive seepage, poor drainage, damage to canals due to absence of sufficient number of bridges, creation of borrow pits, poorly maintained banks of canals and sub-canals and their beds, silted canals, defective and poorly maintained distribution chambers, general water untidiness, absence of a controlled system of field channels, flood irrigation, lack of drainage channels in agriculture fields and minor engineering aberrations, etc. Most of these problems are the result of continued neglect and lack of inter- and intra-sectoral coordination between various agencies working at the district level.

Intensive irrigation has changed the ecology and made the irrigated areas more receptive to malaria. Terai in Uttar Pradesh (now Uttarakhand state) in 1930s was holoendemic for malaria. *P. falciparum*, *P. vivax* and *P. malariae* were the prevalent malaria parasites, and the prevalence of *P. malariae* was about 30 %. The vector of malaria was *An. minimus*. This region was death trap for labor entering the deep forests. Several attempts to colonize Terai failed because of high mortality in the labor force (Issaris et al. 1953). DDT spraying made it possible to control malaria. Terai was deforested in 1950s and opened up for agriculture. Gobind Ballab Pant University of Agriculture was established in Pantnagar in Terai. Eight water reservoirs were constructed and a network of canals and sub-canals was laid in the region. These changes brought about major changes in the mosquito fauna of Terai. *An. minimus* slowly retracted and disappeared from the region. Instead, its place was occupied by another efficient vector of malaria, *An. fluviatilis*. Also *P. malariae* disappeared. Surveys in the 1960s revealed that *An. culicifacies* invaded Terai, and its distribution was widespread. *An. fluviatilis* was restricted to the irrigation channels.

Fig. 10.4 Canal under construction in Hardwar (Uttarakhand state) for Tehri dam. It will remain a source of mosquito production and malariogenic conditions till the completion of the project. Thereafter, breeding will occupy seepage and agricultural fields

The region became hyper-endemic for malaria, and clear cycles of *P. vivax* and *P. falciparum* were established. A further survey in the 1990s revealed that *An. fluviatilis* has further retracked and was not found to be involved in malaria transmission. Investigations revealed that vector sibling species S has disappeared leaving small populations of non-vector sibling species T and U. *An. culicifacies* assumed the role of a sole vector of malaria in Terai. Green revolution brought prosperity to the region and Terai witnessed the growth of towns in the region. These ecological changes also brought about the introduction of *Culex quinquefasciatus* that breeds profusely in the drains and other polluted water (Fig. 10.5 see Culex larvae in stagnant water). This is a well-known vector of *Wuchereria bancrofti*, and Terai has become endemic for lymphatic filariasis. In the rice fields, breeding of *Culex vishnui* group of mosquitoes has made the region prone to Japanese encephalitis (JE). *An. stephensi*, the vector of urban malaria that was completely absent during the 1930s, entered the towns and now breeds in potable and rainwater collections. Along with it, *Aedes aegypti*, the vector of dengue, dengue hemorrhagic fever, and Chikungunya, has entered the urban areas and breeds in receptacles and containers. At present, Terai region is endemic for malaria and filariasis and has high receptivity for JE and dengue fever (Sharma et al. 1987). Indira Gandhi Nahar Pariyojana (IGNP) in Thar Desert converted the low-malaria-endemic desert to perennial transmission of malaria (Oomen et al. 1994). *An. stephensi* was present in low numbers and malaria was at the low ebb. Canal network changed the terrain ecology providing opportunity for *An. culicifacies* to multiply. Malaria is now transmitted by two vector systems. Since the 1980s, irrigation has changed the cropping pattern and water is mismanaged and wasted leading to mosquito breeding. *A. culicifacies* has penetrated deep into the desert along with the rise in

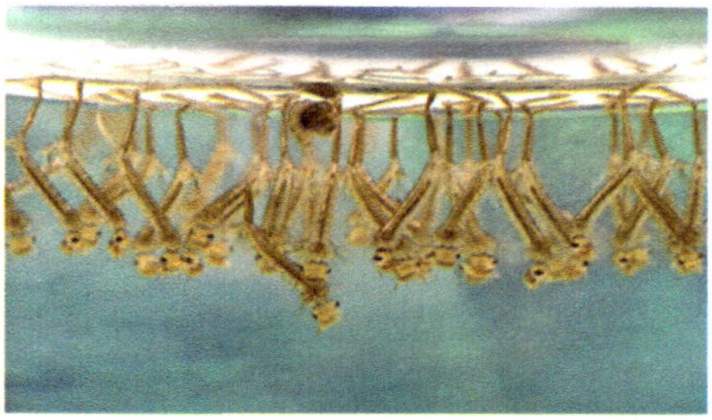

Fig. 10.5 Culex mosquito larvae hang down in water and a pupa in the center (*Source* http://www.edupics.com/photo-mosquito-larvae-i14781.html#image)

P. falciparum-dominated malaria in the Thar Desert (Tyagi 2004). The region is now visited by epidemics that occupy the entire ecotype, for example 1990 malaria epidemic in western Rajasthan (Mathur et al. 1992)

A study in Narmada valley during the construction of Bargi dam in Jabalpur, Madhya Pradesh, revealed that malaria cases were rare in Narayanganj PHC, Mandla district, before the start of the work. After the construction of dam, malaria epidemic claimed hundreds of lives in 1996, and malaria spread to new villages in 1997. The SPR increased >7.45-fold and SfR > 32-fold between 1979 and 1997. NVBDCP data reported doubling of epidemiological indices. This rise in malaria was attributed to the increase in the densities of *An. fluviatilis* (Singh et al. 1999). There was an outbreak of falciparum malaria in submerged villages of the same PHC (Singh et al. 1997). Similar rise in malaria cases has been reported from Sardar Sarovar Project in Gujarat (Kalra 1992) and many other irrigation schemes. The examples where irrigation development has resulted in the spread of malaria are as follows: Narmada river development in Madhya Pradesh, India, led to heavy breeding of *An. culicifacies* and epidemic malaria (Singh et al. 1999). Chulka hydroelectric project in Bhutan promoted the breeding of *An. minimus* and *An. fluviatilis* causing fourfold increase in malaria (Kalra and Prakash). A similar situation has been witnessed in Nepal where east–west highway construction led to obstruction in natural flow of water, and this led to breeding of *An. fluviatilis* and malaria transmission in villages along the highway. In Bangladesh, construction of embankments and laying of canals and sub-canals led to 14-fold increase in malaria. In Sri Lanka, Mahaweli irrigation project caused heavy mosquito breeding, and malaria transmission was enhanced to epidemic proportions (Amersinghe 2003). In addition, irrigation water became the source of Culex vishnui group of mosquitoes in rice fields. Faulty drainage, rainwater harvesting, and multistory housing have increased the potential of *Aedes aegypti* in Maldives. Tidy irrigation and proper maintenance of the canals and sub-canals, seepage control, improving drainage, intermittent irrigation in rice fields are all important mosquito prevention and control measures. Canals should be so designed so that these can be drained dry when not in use. Rao (1948) observed that malaria has followed the development of each irrigation project in the state, the earliest dating back some centuries in the Cauvery and Hemvati basins. Large tracts of land with only seasonal irrigation are hyper-endemic for malaria in Mysore and Mandaya districts. More recently, each new project has brought malaria with it into areas where it was practically unknown before. Tidy irrigation and proper maintenance of the canals and sub-canals, seepage control, improving drainage, intermittent irrigation in rice fields are all important mosquito prevention and control measures. Farmers can be trained in many simple techniques of prevention of mosquito breeding and fever diagnosis and early reporting of fever cases through the Farmers Field Schools and Krishi Vigyan Kendras of the Indian Council of Agricultural Research (ICAR). There are many other areas of opportunity such as the epidemic forecasting using the facilities available with the agriculture department on monitoring the meteorological conditions of the area, community participation in malaria control, and the health system reforms to improve access to treatment.

Rainfall

Rainfall is the primary cause of malaria as rains provide huge potential for the breeding of mosquitoes. Rainfall also leads to water logging. Almost all malaria vectors breed during the rainy season. Breeding potential, however, differs on the water preferences of each vector species. With particular reference to malaria in India, the most important vector in terms of total cases generated each year is *An. culicifacies*. This is a vector found throughout the plains of the country and breeds profusely in the rain water pools and puddles on the ground. In addition, *An. culicifacies* also breed in disused wells, grassy margins in the channels, canal seepage, river beds, etc. In India, malaria is a rainy season phenomenon, and *An. culicifacies* alone maintains high transmission year after year. *An. culicifacies* often is the primary cause of epidemic malaria, both in years of rainfall or the failure of rains. During drought years, *An. culicifacies* breeding is scanty, but drying streams leave puddles in the river bed which support moderate to heavy breeding of *An. culicifacies*. The villages adjacent to the river suffer heavily from malaria fevers. This situation was particularly pronounced in Sri Lanka. *An. stephensi* type form is a major vector of urban malaria, and it breeds profusely in the rain water collections in ponds, ditches, water tanks, wells, containers, and receptacles, etc., often found inside and around the houses and other structures producing the populations of *An. stephensi* mainly in the urban and peri-urban areas. In northeastern states, rains create innumerable streams for breeding of *An. fluviatilis* and *An. minimus*. *An. baimaii* breeds in rain water collection in puddles, hoof prints, and similar situation. It is the most efficient vector of malaria. It breeds in deep jungles where access to its control is very limited. *An. sundaicus*, an important vector of malaria in Andaman and Nicobar islands, breeds in the creeks in brackish water. Rain water mixes with the sea water and creates ideal conditions for the breeding of *An. sundaicus*. All malaria in this island is caused by *An. sundaicus*. Installation of one way slice gates prevents this mixing of water, and as a consequence, *An. sundaicus* breeding stops without any additional interventions.

Rainfall leads to water logging, and in poorly drained lands, water logging leads to mosquito production on sustained basis. Link to mosquito production and malaria resulted in organizing drainage as a means of malaria control. Since bulk of the mosquito breeding of vectors and non-vectors that cause mosquito nuisance occurs in stagnant water, a systematic planning of drainage of the marsh lands can eliminate most mosquito breeding sites. Drainage in the agricultural fields will eliminate major mosquito production sites. Wells in the villages and agricultural fields are another source of mosquito production. Wells should either be capped or filled if not in use or treated with chemicals or fishes to control breeding. Building bye laws will help in protecting the structures from mosquito nuisance and disease transmission. Drains should be constructed with cunnette to prevent water stagnation and carry small amount of water easily. Drains should be periodically cleaned and de-silted before the rainy season. All water hyacinths that cover the water bodies should be manually removed and destroyed or converted to organic manure or paper boards, etc. Care should be taken that water hyacinths do not grow again to become nuisance. Often negligence and lack of maintenance cause the growth of water hyacinth. They constitute an important site for the resting of mosquitoes. *Cx gelidus* mosquitos are often the dominant species resting under the leaf of water hyacinth plants and implicated in the transmission of JE.

Piped Water Supply

Water is supplied in towns and cities by the local government/municipalities/municipal corporation. Water supply is erratic, and in many cities, water has been rationed. People have no recourse but to store water in various containers, and in the overhead tanks, underground tanks, etc. Water pipes are corroded, leaking pipes lead to water stagnation, water chambers contain leaking water, flooding for gardening, open taps, tree holes, water for birds et al., create a large number of mosquito breeding habitats. Water is stored in the underground tanks and overhead tanks. Stored water is the major source of mosquito breeding. Most tanks are open to sky and collect filth in course of time. *An. stephensi* (malaria vector) and *Aedes aegypti* (dengue vector) breed in stored water. These two mosquitoes are the major source of vector-borne diseases in the towns. Malaria in urban areas has become an important malaria ecotype in the last 2 decades. At the time of launching of the National Malaria Control Programme (NMCP) in 1953, urban malaria was not recognized as a problem that required to be tackled by the then NMEP. Rather, its control was given to the local government. Gradually, malaria in the rural areas was seen declining sharply under the National Malaria Eradication Programme (NMEP), but it was rising in the urban areas. The malaria problems were going out of control in all urban areas, and therefore, the Government of India launched an Urban Malaria Scheme (UMS) in 1971–1972 (Sharma 2012). The scheme identified 131 towns with malaria, and it took nearly 3 decades to cover the towns identified by the NMEP. Since then, more towns have been added, and malaria is a serious emerging problem in most cities, but the interventions are largely wanting in most urban situations. Construction activities and water storage practices have multiplied enormously. Water is in short supply and rationed in most towns. Households have to store water for domestic needs, drainage is poor, and most drains are silted. The result is enormous mosquito population build–up, and people depend on repellents to ward off mosquitoes. In addition, *Culex quinquefasciatus* breeding is intense in water contaminated with organic matter, open drains, underground sewage, ditches, ponds, septic tanks, and a variety of water collection that should be drained to avoid the breeding. Vast areas in all cities are the perennial source of *Cx. quinquefasciatus* breeding. In areas with high humidity such as the new sources of water for irrigation, humidity around the dams increases to favor mosquito survival. High humidity also promoted lymphatic filariasis transmission.

Environmental Management

The discovery of malaria transmission by mosquitoes led to *inter alia* drainage of marshes and water-logged lands, often referred to as malaria wastelands. These environmental management methods (modification, manipulation, and habitat management) received a setback with the use of indoor residual insecticides, for example spraying of DDT and other residual insecticides, to control malaria. The multifaceted problems in the use of insecticides in malaria control led to a realization that instead of depending on the chemicals alone, integrated vector management (IVM) would be a more scientific and sustainable strategy for malaria control (Berg et al. 2012). Some classical examples of drainage in malaria control are described below.

Realizing the importance of drainage in malaria control, "In 1897 a sanitation officer in the Indian Medical Service (IMS) recorded his perceptions of the links between irrigation and the catastrophic fevers of Bengal: drainage only will probably ever produce a large decrease in the prevalence". Russel (1938) commented in his paper that malaria persisted due to defective and untidy irrigation lacking drainage. Although Sir Ronald Ross established the role of mosquitoes in malaria transmission in India, yet British engineers did not create drainage channels in the flooded fields to rid of surplus water. The Army Sanitary Commission report (1901) similarly states: "The methods hitherto found to be the most effective in preventing and curing malarial fevers do not appear likely to lose any of their importance as a consequence of recent discoveries [i.e., Ross]. On the contrary it seems likely that drainage and the use of quinine may in future have more value than ever assigned to them in the conflict with this disease" (British Library 1809). "Today (in 1938) the situation as regards irrigation and malaria is probably worse than ever before.... For natural laws still obtain': Each malarial area had its own special problems: there are more than a dozen anopheles types capable of bearing a lethal plasmodium. Some breed and bite after the flooding of a region through irrigation; others bite after rainfall and natural flooding; others bite during dry seasons" (Watts 1999).

Historically, drainage has played a major role in malaria control that eventually led to the expansion of agriculture, industries, and sustenance of human and animal life. For example, malaria in Pontine Marshes near Rome was the major cause of disease, poverty, and deprivation. Benito Mussolini in 1932 deployed troops to drain Pontine Marshes and successfully controlled malaria. The agricultural economy received a boost and Mussolini became a popular national leader. In 1943, Adolf Hitler used malaria as biological war weapon. His army destroyed pumps and filled the Pontine Marshes with water, and malaria returned with vengeance. Long afterward in 1962, Pontine Marshes were again drained that finally eliminated malaria from Italy. Maintenance of good drainage network has seen the flourishing of the region with good agriculture and urbanization (Frost 1934; Caprotti 2006). Panama Canal construction (1905–1910) was possible with the simultaneous attack on malaria and yellow fever. For example in 1906, 21,000 out of 26,000 worker force were hospitalized due to malaria and yellow fever. By 2012, out of 50,000 workers, only 5,600 were hospitalized. Integrated methods of malaria and yellow fever control led to the construction of Panama Canal. Drainage of marshes remained the principal method of mosquito control. Similarly Tennessee River's potential for irrigation, hydroelectric power generation, and waterways was going waste while communities suffered from food shortages, poverty, and malaria; for example, about 30 % people of the areas suffered from malaria and poverty (TVA 1974). Mosquito breeding was eliminated by regulating water levels, drainage, and insecticide application (Derryberry and Gartrell 1952). Malaria control in Tennessee Valley Authority transformed the American economy (Mills 1984)

An. gambiae, one of the most notorious mosquito responsible for the highest number of malaria cases in the world, entered Brazil. Rockfeller foundation eliminated *An. gambiae* from northeast coast of Brazil in 1939–1940 (Soper and Wilson 1943). *An gambiae* from Nile Valley of Egypt was eradicated in 1942–1945 (Shousha 1948). Panama Canal construction was not even attempted because of the high incidence of malaria and yellow fever. *Inter alia*, drainage for the Panama Canal construction led to the successful completion of the Panama Canal. In India, seepage control and drainage in Banbasa headworks (Uttarakhand state) was a successful malaria control activity that helped in colonization of Terai and generating wealth through agriculture (Barrett 1913). Eline et al. (2002) provide several examples of drainage in malaria control from India, Sri Lanka (Worth and Subrahmanyam 1940), Malaysia (Williamson and Scharff 1936), and African countries. Siphons have been used to flush the mosquito breeding in many countries in the world (WHO 1973; Ramsay and Anderson 1940.)

Keiser et al. (2005) reported that about 40,000 large and 800,000 small dams have been built in over 50-year period worldwide. This has brought 272 million hectares of land under irrigation. These irrigated tracts have become receptive to malaria, often faced with epidemics. Construction of dams and irrigation systems ought to have included health impact assessment and remedial measures put in place concurrently along with the development in irrigation. Unfortunately, as we see it today bulk of malaria is localized in the irrigation tracts. Crops requiring irrigation are given flood irrigation and water along the flooded fields stagnates to produce harmful insects, some of which are the vectors of human diseases. The more insecticide we use to control mosquitoes, the problem of resistance and environmental pollution magnifies to dangerous levels. Malaria control now requires sound water management practices and integrated malaria control measures to mitigate malaria burden in locations near the dam sites and in proximity to irrigation tracts (Lindsay et al. 2004). Since bulk of the mosquito breeding of vectors and non-vectors that cause mosquito nuisance occurs in stagnant water, a systematic planning of drainage of the cities can eliminate most mosquito breeding sites. Drainage in the agricultural fields will eliminate a major mosquito productions site. Wells in the villages and agricultural fields is another source of mosquito production. Wells should either be capped or filled if not in use or treated with chemicals or fishes to control breeding. Building bye laws will help in protecting the structures from mosquito nuisance and disease transmission. Drains should be V shaped to allow water flow without stagnation, periodically cleaned, and de-silted before the rainy season. All water hyacinths that cover the water bodies should be removed and destroyed or converted to organic manure, etc. Care should be taken that water hyacinths do not grow again to become nuisance. Often negligence and lack of maintenance cause the growth of water hyacinth. They constitute an important site for the resting of mosquitoes. *Cx gelidus* mosquitos are often the dominant species resting under the leaf of water hyacinth plants and implicated in the transmission of JE. Health impact assessment in all development projects must become an important legal requirement before undertaking any construction or dealing with water. Remedial measures must be incorporated at the design stage and applied along with the construction to ensure that there would be no adverse impact of the project. This was done in the construction of Konkan railway, and the vulnerable railway track area of about 800 Km is free from any mosquito or disease problem for almost three decades now (Ashwani Kumar NIMR personal communication).

In the background of major ecological changes related to agricultural practices, it is essential that agriculture sector is closely associated in the planning of malaria control. Land use pattern can mitigate many situations that otherwise would be receptive and vulnerable to malaria. This is an area of great opportunity and should not be missed. For example, Malnad region in southern India (Karnataka–Kerala) measuring about 50,000 sq km. was hot bed of malaria. *An. fluviatilis* was the only malaria vector in Malnad. Aquatic stages of *An. fluviatilis* used to breed profusely in slow-running perennial streams in the forests. The source of perennial streams was rainwater absorbed by the thick vegetation cover on the ground. Arrival of DDT paved the way for colonization of Malnad and converted the malaria wasteland to coffee plantations. Jungle clearance removed thick forest. As a result, litter on the ground gradually disappeared and so also the seepage. This natural change in the ecology of the region removed the breeding habitats of *An. fluviatilis*, the only vector of malaria prevalent in that area. Since then, the region has become healthy, and there are no cases of indigenous malaria transmission in Malnad (N.L. Kalra, personal communication).

Malaria was rampant in the rural areas in Karnataka. Streams with big boulders form a natural ecology of the area. Natural populations of *An. culicifacies* sibling species A and B are encountered in Karnataka. Control of breeding in streams was very problematical. Studies revealed that sibling species B breeds in streams exclusively, and in the wells and other stagnant water, sibling species A was present. Based on this information, mosquito control in streams was not attempted as sibling species B is a non-vector. Introduction of fishes to control mosquito control in wells led to successful malaria control at negligible cost (Ghosh et al. 2004). Studies on bioenvironmental malaria control successfully eliminated mosquito nuisance and led to malaria control in many ecological settings in India (Sharma 1987a, b, 1988; Sharma et al. 1988; Sharma and Sharma 1989; Singh et al. 1989a, b). Studies on bioenvironmental malaria control have revealed that convergence of various agencies, for example agriculture, irrigation, and drainage, and public health can give heavy dividends in malaria control, and the process can be accelerated and made sustainable with community participation.

References

Amersinghe FP (2003) Irrigation and mosquito borne diseases. J Parasitol 89(suppl): 14–22

Barrett J (1913) Panama canal : what it is, what it means. Pan American Union Publisher, Washington, D.C

Berg H, van den Mutero CM, Kazuyo Ichimori K (2012) Guidance on policy-making for integrated vector management. World Health Organization, Geneva

Bouma MJ, van der Kaay HJ (1994) Epidemic malaria in India and the El Niño Southern oscillation. Lancet 344(8937):1638–1639

Bouma M, Sondorp HE, vander Kaay HJ (1994) Climate change and periodic epidemic malaria. Lancet 343:1440

Caprotti F (2006) Malaria and technological networks: medical geography in the Pontine marshes, Italy, in the 1930s. Geogr J 172(2):145–155

Derryberry OM, Gartrell EE (1952) Trends in malaria control program of the Tennessee valley authority. Am J Trop Med Hyg 1:500–507

Dev V, Phookan S, Sharma VP, Anand SP (2004) Physiographic and entomologic risk factors of malaria in Assam, India. Am J Trop Med Hyg 71(4):451–456

Dhingra N, Jha P, Sharma VP et al (2010) Adult and child malaria mortality in India: a nationally representative mortality survey. Lancet 376(9754): 1768–1774

Dua VK, Dev V, Phookan S, Gupta NC et al (2003) Multi-drug resistant plasmodium falciparum malaria in Assam, India. Timing of recurrence and antimalarial drug concentration in whole blood. Am J Trop Med Hyg 69:555–557

Frost RS (1934) The reclamation of Pontine marshes. Geographical Review. American Geographical Society 24(4):584–595

Boelee E, Konradsen F, van der Hoek W (eds) (2002) Malaria in irrigated agriculture. Working paper 47.M water management international. Institute SIMA document 2. Papers and abstracts for the SIMA special seminar at the ICID 18th international congress on irrigation and drainage, Montreal, 23 July 2002

Ghosh SK, Tiwari SN, Sathyanarayan TS et al (2004) Larvivorous fish in wells target the malaria vector sibling species of the *Anopheles culicifacies* complex in villages in Karnataka, India. Trans Roy Soc Hyg Trop Med 99:101–105

Issaris PC, Rastogi SN, Ramakrishna V (1953) Malaria transmission in the Tarai Nainital district, Uttar Pradesh, India. Bull World Health Organ 9:311–333

Kalra NL (1992) Status report on malaria and other health related aspects of the (SSP) projects and recommendations regarding short-term and long-term remedial measures. World Bank, New Delhi

Keiser J, De Castro MC et al (2005) Effect of irrigation and large dams on the burden of malaria on a global and regional scale. Am J Trop Med Hyg 72(4): 392–406

Kibret S, Alemu Y (2010) The impact of a small-scale irrigation scheme on malaria transmission in Zimwabway area, central Ethiopia. Trop Med Int Health 15:41–50

Lindsay S, Kirby M, Baris E et al (2004) Environmental management for malaria control in the east Asia and Pacific regions. The International Bank for Reconstruction and Development/The World Bank

Mathur KK, Harpalani G, Kalra NL et al (1990) Epidemic of malaria in Barmer district (Thar desert) of Rajasthan during. Indian J Malariol 29:1–10

Mathur KK, Harpalani G, Kalra NL et al (1992) Epidemic of malaria in Barmer District (Thar desert) ofRajasthan during 1990. Indian J Malariol 29:1–10

Mills DB (1984) TVA handbook revised. Tennessee Valley Authority, Knoxville

Oomen JMV, de Wolf Jobin WR (1994) Health and irrigation. Incorporation of disease-control measures in irrigation, a multi-faceted task in design construction, operation. Publication 45. International Institute for Land Reclamation and Improvement/ILRI.P.O. Box 45,6700 AA Wageningen, The Netherlands

Pattanayak S, Sharma VP, Kalra NL et al (1994) Malaria paradigms in India and control strategies. Indian J Malariol 31:141–195

Ramsay GC, Anderson IR (1940) Investigation on the use of automatic siphon sluices on a group of tea estates in Northern Bengal. J Malaria Inst India 3(1):93–97

Rao BA (1948) A short account of the malaria problem in Mysore state. Indian J Malariol 2:285–305

Russell PF (1938) Malaria due to defective and untidy irrigation: a preliminary discussion. J Malaria Inst India 1(341):343

Sharma D (2004) Thar desert: sitting on the tip of a malarial iceberg. Lancet 4:322

Sharma VP, Sharma RC (1989) Community based bioenvironmental control of malaria in Kheda district, Gujarat India. J Am Mosq Control Assoc 5(4): 514–521

Sharma VP, Dua VK, Sharma SK (1987) Bioenvironmental control of industrial malaria. Indian Counc Med Res Bull 17(7):59–62

Sharma VP (1988) Community based bioenvironmental control of malaria in India. Ann Nat Acad Med Sci (India) 24(3):157–169

Sharma VP (1987a) Community-based malaria control in India. Parasitol Today 3(7):222–226

Sharma VP (1987) The green revolution in India and ecological succession of malaria vectors. In: Effects of agricultural development on vector borne diseases. FAO Publication. *AGL/Misc*/12/87, pp 116–125

Sharma VP, Mehrotra KN (1982a) Return of malaria. Nature (London) 298:210

Sharma VP, Mehrotra KN (1982b) Malaria resurgence. Nature (London) 300:212

Sharma VP, Uprety HC (1982) Preliminary studies on irrigation malaria. Indian J Malariol 19:139–142

Sharma VP (1996a) Ecological changes and vector-borne diseases. Trop Ecol 37(1):57–65

Sharma VP (1996b) Re-emergence of malaria in India. Indian J Med Res 103:26–45

Sharma VP (2012) Continuing challenge of malaria in India. Curr Sci 102(5):678–682

Shousha AT (1948) The eradication of *A. gambiae* from Upper Egypt, 1942–1945. Bull World Health Organ 1:309–348

Singh N, Sharma VP, Mishra AK et al (1989) Bioenvironmental control of malaria in a tribal area of Mandla district, M.P., India. Indian J Malariol 26(2): 103–122

Singh N, Shukla MM, Sharma VP et al (1989) A focus of high degree chloroquine resistant *P. falciparum* in Mandla district (M.P.). Indian J Malariol 26(1):45–51

Singh N, Shukla MM, Chand SK, Sharma VP (1997) Outbreak of falciparum malaria in submerged villages of Narayanganj PHC, district Mandla due to Narmada irrigation project, central India (M.P.). Curr Sci 73(8):686–691

Singh N, Mehra RK, Sharma VP (1999) Malaria and the Narmada river development in India. A case study of the Bargi dam. Ann Trop Med Parasitol 93:477–488

Sinha S, Dua VK, Sharma VP (1989) Malaria relapses and chloroquine resistance at the BHEL industrial complex, Hardwar, India. Trans R Soc Trop Med Hyg 83:606

Soper FL, Wilson DB (1943) *Anopheles gambiae* in Brazil, 1930–1940. The Rockefeller Foundation, New York

TVA (Tennessee Valley Authority) (1974) Vector control program environmental statement. Technical report, TVA, Knoxville, Tennessee
Tyagi BK (2004) A review of the emergence of *Plasmodium falciparum*-dominated malaria in irrigated areas of the Thar desert. India Acta Tropica 89:227–239
Tyagi BK, Yadav SP (1997) Correlation of irrigation and flood water management with malaria in the Thar desert. Reg Health Forum 2(2):6–13
Watts Sheldon (1999) British development policies and malaria in India 1897-c. 1929
WHO (1973) Manual on larval control operations in malaria programmes, Geneva, p 195
Williamson KB, Scharff JW (1936) Anti-larval sluicing. Malayan Med J 11:123–150
Worth HN, Subrahmanyam K (1940) Anti-larval flushing of rivers and streams in Ceylon. J Malar Inst India 3:81–92
Yadav RS, Bhatt RM, Kohli VK, Sharma VP (2003) The burden of malaria in Ahmedabad city, India—a retrospective analysis of reported cases and deaths. Ann Trop Med Parasitol 97:793–802

Water, Amoebiasis and Public Health 11

Prati Pal Singh and Abhiruchi Galhotra

Abstract

Human amoebiasis causes highest number of deaths due to any parasitic disease after malaria and schistosomiasis. *Entamoeba histolytica/Entamoeba dispar*, the causative parasites, exist in two forms: the highly active and invasive trophozoites and the infective cysts. The disease is transmitted by faecal-oral rout, oral-rectal contamination and by the ingestion of infective cysts in drinking water and food etc. Metronidazole, the most commonly used drug for the treatment of symptomatic amoebiasis, is fraught with several problems including emergence of resistance by *E. histolytica*. The prevention of the disease through improved sanitation and drinking water quality are some of the most effective means to control the disease. Additionally, the role(s) of mothers, proper hand washing and maintenance of good personal hygiene are also important in the containment of the disease.

Keywords

Cysts · Drinking water · *Entamoeba histolytica/E. dispar* · Hand washing · Metronidazole · Sanitation

Introduction

Amoebiasis, a worldwide extended disease of poverty, can be referred to as a condition in which a person despite harbouring protozoan parasites *Entamoeba histolytica/Entamoeba dispar* may or may not show clinical symptoms of the disease. There are only four protozoan parasites of the genus *Entamoeba* (*E. coli, E. dispar, E. hartmani* and *E. histolytica*) which are found in the human gastrointestinal tract, and only *E. histolytica* is pathogenic, causes invasive amoebiasis and thus has distinct and established medical importance

P. P. Singh (✉)
Centre of Infectious Diseases, Department of Pharmacology and Toxicology, National Institute of Pharmaceutical Education and Research, S. A. S. Nagar, Punjab, 160 062, India
e-mail: drppsingh2005@gmail.com

A. Galhotra
Department of Community and Family Medicine, All India Institute of Medical Sciences, Raipur, India

P. P. Singh and V. Sharma (eds.), *Water and Health*,
DOI: 10.1007/978-81-322-1029-0_11, © Springer India 2014

(Diamond and Clark 1993). Invasive amoebiasis is a major fatal parasitic disease with nearly 75,000–100,000 deaths per year worldwide, and *E. dispar*, morphologically similar to *E. histolytica,* is responsible for nearly 90 % of 500 million cases previously thought to be caused by *E. histolytica* (Walsh 1984; Li and Stanley 1996). Amoebiasis is a major public health problem in developing countries (World Health Organization 1969, 1997). The incidence of human intestinal parasitic infections shows a wide variation in the range of 25.2 % (India) to 98.9 % (Ecuador; Gatti et al. 2002). The symptomatic amoebiasis (which can be further subdivided into intestinal and extra-intestinal amoebiasis) occurs in <10 % of infected individuals.

History

One of the earliest records is possibly that from the Sanskrit document Bhrigu-Samhita, written about 1000 BC, which refers to bloody dysentery. Assyrian and Babylonian texts from the library of King Ashurbanipal refer to blood in faeces, which suggests the prevalence of amoebiasis in Tigris–Euphrates basin before the sixth century BC, and it is possible that hepatic and peri-anal abscesses described in both Epidemics and Aphorisms in the Corpus Hippocratorum refer to amoebiasis. Because the epidemics of dysentery by themselves are likely to result from bacterial infections, and dysentery associated with disease of the liver is likely to be amoebic, the later records are easier to interpret. In second century AD, both Galen and Celsus described liver abscesses which were probably amoebic, and the works of Aretaeus, Archigenes, Aurelanus and Avicenna towards the end of first millennium give good accounts of both dysentery and hepatic involvement. As amoebiasis became widespread in the developed world, there emerged numerous records of bloody flux in Europe, Asia, Persia and Greece, in middle ages. Amoebiasis is thought to have been introduced into the New World by Europeans sometime in the sixteenth century, and with the later development of European colonies and increased world trade, numerous clear descriptions of both intestinal and hepatic forms of the disease became available (Cox 2002). In nineteenth century, several books mainly concerned with the diseases prevalent in India, including *Researches into the Causes, Nature and Treatment of the Prevalent Diseases of India and of Warm Climates Generally* by James Annersley, clearly refer to both the forms of the disease. Budd (1857) very clearly demonstrated the link between amoebic dysentery and amoebic liver abscess. Though Lewis (1870) and Cunningham (1871) for the first time described human intestinal amoebae, it is now well established that Fedor Aleksandrovich Lösch (1875), a Russian physician, in St. Petersburg, Russia, discovered actively motile pathogenic amoebae (trophozoites), which invariably contained red blood cells, in the stools of a patient suffering from dysentery, and also established the relationship between the parasite and the disease in dogs experimentally infected with amoebae from patients. Schaudinn (1903) designated these amoebae as *E. histolytica*. Stephanos Kratulis, a Greek physician, in Egypt in 1885 and 1896, also found amoebae in the intestinal ulcers in patients suffering from dysentery and noted that he never found amoebae from non-dysenteric cases (1886). Kartulis (1887, 1891) also showed that cats could be infected with amoebae *per rectum* and developed dysentery. The report by William Thomas Councilman and Henri Lafleur, of John Hopkins Hospital in 1891, represents a definitive statement of what was known about the pathology of amoebiasis at the end of nineteenth century. Recently, Clark (1998) divided *E. histolytica* into non-pathogenic *E. dispar* and pathogenic *E. histolytica* and gave a new life to the view originally propounded by Brumpt (1925).

Distribution

Amoebiasis is a global disease and a major public health problem in developing countries (Walsh 1986). *E. histolytica* infects nearly 10 % of the human population, with the percentage being higher in poor and developing areas. *E. histolytica* is estimated to be the third leading parasitic causes of death worldwide (Stauffer 2008); malaria and schistosomiasis being the first and second, respectively (Walsh 1984; Li and Stanley 1996). It is a major health problem in China, South East and West Asia and Latin America (mainly Mexico). An estimate of worldwide burden of human infection due to *E. histolytica* was reported by Walsh (1986). This classical reference was based on the analysis of published data regarding the frequency of infection and disease, the results of which indicated that 10 % of the world's population was infected by the parasite and only 1 % of the infected individuals developed the invasive form of the disease. The mortality rate estimated at that time for the invasive form of amoebiasis was around 100,000 deaths/year. Globally, it is estimated that in 1997, 45 million people harboured *E. histolytica/E. dispar* in their intestinal tract, and approximately 1/10th of the infected people suffered from invasive amoebiasis (Park 2011). It is estimated that invasive amoebiasis accounted for 40,000–100,000 deaths worldwide (Walsh 1988; World Health Organization 1998; Wertheim et al. 2012). Prevalence rates vary from as low as 2 % to 60 % or more in areas devoid of sanitation (WHO Scientific Working Group 1980). In India, amoebiasis affects an estimated 15 % of the population throughout the country. The prevalence rate is about 15 %, ranging from 3.6 to 47.4 % in different areas. The reported variations in prevalence are attributed to the variations in clinical diagnostic criteria and to the technical difficulties in the establishment of a correct diagnosis and lack of sampling criteria (Park 2011).

Epidemiology

In areas of high prevalence, amoebiasis occurs in endemic forms due to high levels of transmission and continued re-infection. Epidemic water-borne infections are known to occur whether or not there is heavy contamination of drinking water supply (Markell 1986; Park 2011). In developed countries, amoebic infection and disease are concentrated in high-risk groups, such as those with prior exposure to an endemic environment or those more likely to have direct faecal-oral contamination because of unhygienic living conditions or sexual practices. *E. histolytica/E. dispar* is specifically known to spread within families, institutions (day-care centres) and may cause epidemics, occasionally.

Agent Factors

Symptomatic amoebiasis is caused by pathogenic strains of *E. histolytica/E. dispar*, which exists in two forms: the vegetative trophozoite form and the infective cystic form. Trophozoites dwell in the colon where they multiply and encyst. The cysts are excreted in stool. The ingested cysts undergo excystation and release trophozoites, which colonize the large intestine. Some of the trophozoites invade the bowel and cause ulceration, mainly in the caecum and ascending colon, then in the rectum and sigmoid colon. Some trophozoites may enter a vein and reach the liver and other organs. The trophozoites are short-lived outside the human body and do not play any role in the transmission of the disease. The cysts are infective to man and remain viable and infective for several days in faeces, water, sewage and soil, in moist and low-temperature conditions. The cysts are not affected by chlorine in amounts normally used in water purification, but they are readily killed if dried, heated (to about 55 °C) or frozen.

Life Cycle

The life cycle of *E. histolytica/E. dispar* is completed only in one host, the man (Fig. 11.1). *E. histolytica/E. dispar* exist in two morphologically distinct forms: the vegetative and motile trophozoites, and the infective and immotile cysts. The size of a trophozoite varies (10–60 μm), and their variability is affected by the changes in temperature, pH, osmolarity and feeding conditions, among others (Martinez-Palomo and Espinosa-Cantellano 1998). The cysts (10–12 μm in diameter) were first observed by

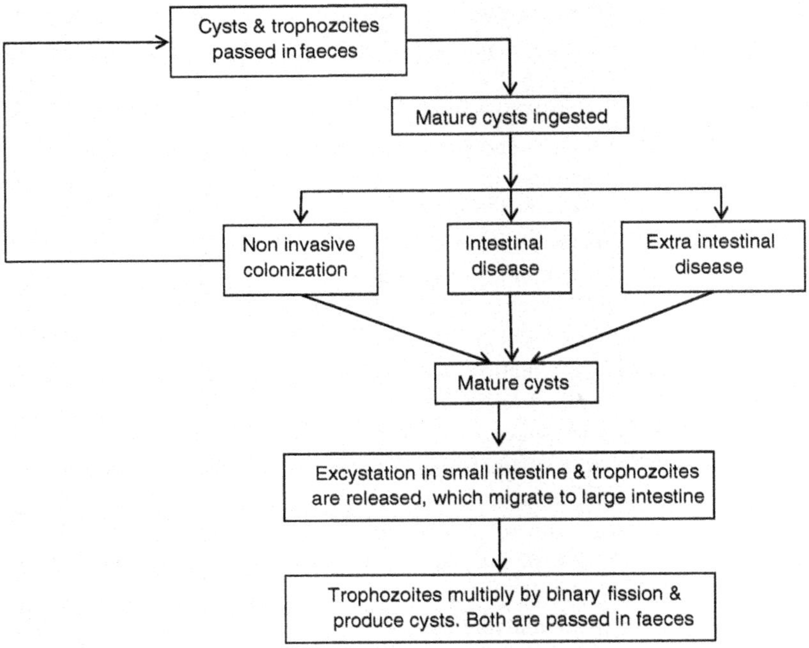

Fig. 11.1 Flow diagram showing the life cycle of *Entamoeba histolytica/Entamoeba dispar*

Quincke and Roos (1893) in the stool of a dysentery patient in Germany. The cysts are very rugged structures and can remain viable outside the body under very harsh and extreme conditions for several days. The infection usually occurs by the ingestion of cysts in water or food contaminated by faecal matter. The cyst wall (containing up to 70 % cellulose) is dissolved in the upper gastrointestinal tract, the excystation (release of daughter amoebae) occurs in the lumen of terminal ileum, and eight uninucleated trophozoites are then released in the lumen of the intestine. The trophozoites of *E. histolytica* are one of the most powerful tissue invaders. The trophozoites penetrate the intestinal mucosa, and then, dissemination to other organs (usually liver, lung, brain, spleen, skin and urogenital tract) can occur. The colon-dwelling trophozoites multiply, undergo encystation and are then passed out in the stool from where further spread of the disease is possible (Chatterjee 2011; Wertheim et al. 2012).

Reservoir of Infection

Though man is thought to be the only major reservoir of infection, *E. histolytica/E. dispar* infection has been reported in cats, dogs and non-human primates. According to Eichinger (1997), there is no zoonotic reservoir of *E. histolytica*. The immediate source of infection is the faeces containing infective cysts. Most of the individuals infected with *E. histolytica/E. dispar* remain asymptomatic and are the healthy carriers of the parasite(s). The carriers can continue to discharge up to 1.5×10^7 cysts/day. Amoebiasis appears to show some gender preferences: amoebic liver abscess is usually more frequently observed in males as compared to females (Seeto 1999). Further, amoebic liver abscess is 10-times more common in adults than in children (Seulveda and Trevino-Garcia 1986). Rivera et al. (1998) have reported that in Philippines, the 5–14-year-age group was more affected with *Entamoeba* (*E. dispar* 7 % and *E. histolytica* 1 %).

Period of Communicability

The cysts are excreted for long time; the period may be several years, if cases are unrecognized and untreated. Cysts remain viable in water for at least 2 or 3 weeks; however, they are easily killed by desiccation. Therefore, contamination of food must occur under conditions of sufficient moisture both on the carrying agent and in the food; the moisture prevents drying of the cysts. This contention emphasizes the importance of first proper washing and then drying of hands and of taking particular care to avoid handling of moist or liquid foods without proper procedures.

Modes of Transmission

Faecal-oral route: Transmission of amoebiasis occurs through the intake of water and/or food contaminated with faeces containing infective cysts. Epidemic water-borne infections can occur if there is heavy contamination of drinking water supply. Vegetables, especially those eaten raw, from fields irrigated with sewage-polluted water can readily spread the infection. Viable cysts have often been found on hands and under finger nails, and this may lead to direct hand-to-mouth transmission of the disease.

Sexual transmission: It is one of the most important routes for the transmission of amoebiasis, especially among those having preferences for both oral and anal sex.

Vectors: Insects such as flies and cockroaches, and rodents are capable of carrying cysts and thus can contaminate food items and drinks.

Clinical Features

Clinical features of amoebiasis range from asymptomatic colonization to amoebic dysentery and invasive extra-intestinal amoebiasis, which is manifested most commonly in the form of abscesses in liver and lungs.

Laboratory diagnosis: Demonstration of actively motile and erythrocyte-containing trophozoites in freshly passed stool is a diagnostic gold standard for amoebiasis. The trophozoites are most readily seen in fresh mucus passed through rectum. The methods for in vitro culture of amoebae have been developed and are quite efficient for the diagnosis but can be used only in better-equipped laboratories. Several new serological techniques have now been developed, including counter immuno-electrophoresis, fluorescent immunoassay and enzyme-linked immunosorbent assay, which are very useful for the diagnosis of the cases of invasive amoebiasis and for the evaluation of the point-prevalence of invasive infections. A technique has also been developed for the detection of *E. histolytica/E. dispar* antigens, based on the enzyme-linked reaction between specific *E. histolytica/E. dispar* antibodies immobilized on an immunosorbent disc and *E. histolytica/E. dispar* antigens from faeces. The occurrence of invasive amoebiasis is indicated by the finding of trophozoites (often containing ingested erythrocytes) in stool, a positive serological test for anti-amoebic antibodies, and the presence of ulcerative mucosal lesions observed by lower gastro-intestinal endoscopy. Serology for *E. histolytica* becomes positive within 1 week of the onset of the disease.

For patients with amoebic liver abscess, ultrasonography is highly sensitive, non-invasive and relatively inexpensive; however, computerized tomography (CT) and magnetic resonance imaging are not highly specific and are only slightly more sensitive. If infrastructural facilities for amoebic serologic testing are not available, ultrasonography or CT-guided-fine-needle-aspiration-cytology (FNAC) can be helpful. The serum anti-amoebic antibodies often develop after 7 days of the onset of symptoms. *E. histolytica/E. dispar* trophozoites or cysts can be found in the stool of only a small number of patients with amoebic liver abscess (Stauffer 2008).

Chemotherapy

Very much like many other parasitic diseases, the chemotherapy of amoebiasis also is deeply associated with its pathogenesis. As in asymptomatic carrier cases, the amoebae (*E. dispar*) do not invade the tissue of the intestinal wall, and live and multiply in the lumen of the intestine, the anti-amoebic drugs which kill these lumen-dwelling amoebae are known as luminal amoebicides. Drugs like diloxanide furoate, paromomycin and diidohydroxyquin are the commonly used luminal amoebicides; diloxanide furoate being the safest continues to be the mainstay for the treatment of asymptomatic amoebiasis (Di Perri et al. 1989). Whereas the treatment of luminal amoebiasis appears to be quite simple and largely effective, the treatment of invasive (symptomatic) amoebiasis is far from satisfactory and is often fraught with several difficulties. The patients of invasive amoebiasis pass bloody mucus containing highly motile amoebae (trophozoites; *E. histolytica*) in their stool, and these amoebae almost invariably contain the ingested erythrocytes. In such patients, the amoebae successfully invade the intestinal wall and live and multiply in the intestinal wall tissue and also migrate to other organs like liver, lungs and brain. Tissue-amoebicides are the drugs of choice for the treatment of symptomatic amoebiasis, and metronidazole (Powell et al. 1966), a 5-nitroimidazole, continues to be the most commonly used tissue amoebicide for the treatment of intestinal and extra-intestinal amoebiasis. For the standard treatment, an oral adult dose of 750 mg metronidazole, three-times-a-day, is considered best (Abramowicz 2002; Gilbert et al. 2002). However, the beginning of the emergence of *E. histolytica* strains resistant to metronidazole, its metallic taste, mutagenicity, and association with transient myopia, neuropathy and immunosuppression are some of the problems which are becoming its limitations (Goodman et al. 1992). It should be noted here that as metronidazole is the only tissue amoebicide available around, every possible care should be taken to ensure a long working life for this drug. It is, therefore, strongly recommended that metronidazole should not be used for prophylactic treatment. Because *E. histolytica* and *E. dispar* are known to co-exist in a host, it is always advisable to treat a patient of amoebiasis with a combination of luiminal- and tissue-amoebicides (Qureshi et al. 1997).

Prevention and Control

As the presently available treatment(s) of amoebiasis is far from satisfactory, chemoprophylaxis is not recommended and no suitable human anti-amoebic vaccine(s) is available, the prevention and control of amoebiasis assume greater significance. Because man is the only host of *E. histolytica/E. dispar*, apparently, there is no possibility of zoonotic transmission of amoebiasis. The avoidance of the consumption of food or water contaminated with faecal material constitutes an effective control strategy. The education and awareness of people regarding improving their personal hygiene, sanitary disposal of faeces and proper hand washing are some of the particularly effective means to control the transmission of amoebiasis. It is now well proven that preventing water supplies from getting mixed with sewage lines drastically reduces the endemicity and epidemics of amoebiasis. It should be noted that *E. histolytica/E. dispar* cysts are highly resistant to standard drinking water chlorine treatment, but are readily killed by iodine or boiling. Sedimentation and filtration processes are also quite effective in removing the cysts of *E. histolytica/E. dispar*.

Role(s) of Mother

Based on a careful survey of 5 families, comprising 27 individuals of whom 23 were known to be infected with *E. histolytica/E.dispar*, Meleny et al. (1932) concluded that the mother was probably the source of spread of the disease because she was the "chief food and child handler". Women should take particular care while changing baby's diapers. As the mother is the first doctor around, they take care of the sick children, and usually, while doing so, they often become more vulnerable to contracting infection themselves. Further, in many communities, often women of the house are more closely associated with drinking water as they bring the drinking water from long distances and use it for cooking, dish washing and other household purposes. And if the drinking water is contaminated with the cysts of *E. histolytica/E.dispar*, which more often than not is the case, they easily get infected. He observed that poor sanitation and hygiene were important factors. A fault in personal hygiene is invariably involved whenever there was an outbreak of amoebiasis. And where such faults are combined with low intelligence or sanitary ignorance, it has been observed that the problem of the infected food handler and housewife assumes larger proportions. If a mother or her child is suffering from amoebiasis, it is advisable that she should follow the following practices:

1. Wash her hand thoroughly with soap and hot water after going to toilet and then dry hands.
2. Wash her hands after changing the diapers/nappies.
3. Should not cook, serve and eat the food without properly washing and drying hands.
4. Should avoid going to work, school and any other places where she may come in contact with other persons, till complete recovery.
5. Should ensure that her towels, sheets and other linens are not shared by others.
6. Should avoid participating in community kitchens.
7. Those involved in the handing of food and other food related materials must take off till recommended by doctors to get back to job.

Role(s) of Hand Washing

It is now well recognized that proper hand washing is essential to protect against several infectious diseases. As contacts between hands and mouth allow easy ingestion of the infective stages of various diseases, keeping hands clean assumes greater significance. Because amoebiasis is a disease that is transmitted though faecal-oral route and by oral-rectal contamination, the practice of proper hand washing can play a very effective role(s), in the containment of the spread of the disease. In children and mothers who are taking care of infants and thus often have to change the diapers of the babies some of whom may be suffering from amoebiasis, hand washing is of even greater importance. Hand washing with soap removes transient potentially pathogenic organisms from hand including infective cysts of *E. histolytica/E.dispar*. Hand washing is one of the simplest, most cost-effective and efficacious preventive measure that can make a significant difference against challenges of infectious diseases, especially in those places where people do not have access to proper lavatories and sanitation facilities. Interventional studies done by Ejemot-Nwadiaro et al. (2012) suggest that interventions that promote hand washing can reduce diarrhoea episodes by about one-third. This significant reduction is comparable to the effect of providing clean water in low-income areas. Not only proper washing of hands, use of clean towels after hand washing is necessary.

Role of Drinking Water

The cysts, the infective stages of *E. histolytica/E.dispar*, are known to be abundantly present in drinking water contaminated with sewage. This kind of contamination usually occurs when drinking water supply gets mixed-up with sewage due to bursting of pipes (Markell 1986). The cysts in human excreta thus find a very easy entry into drinking water supply and thus are ingested by humans and usually cause infections. The drinking water thus constitutes a very effective and important source of human amoebic infections. Besides drinking, the drinking water is also used in cooking. There are several food items which do not require boiling, heating or frying. Under these conditions, the cysts present in the water get easy access to food materials and are ingested during consumption of such food items. The drinking water is also utilized for washing vegetables and salads, which are generally not cooked and are consumed as such. Such food items when eaten allow the ingestion of cysts resulting in infection. Further, drinking water is also largely used for doing the dishes and washing cooking utensils. The cysts present in drinking water may be left behind on the utensils and thus may find their way in food. For hand washing also, more often than not, drinking water is used. Hand washing with such water often results in the infestation of hands; the cysts left behind on the hands may end up being a source of infection. The drinking water is also used in bakeries, ice cream factories, dairies and milk processing plants, confectioneries, soft drinks and meat processing units etc., and if the water used is contaminated with cysts, it may be a potential source of amoebic infections.

References

Abramowicz M (2002) Drugs for parasitic infections. Med Lett Drug Ther 44:1–12

Brumpt E (1925) Etude sominaire de l "*Entamoeba dispar*" n sp Amibe a kystes quadrinuclees, parasite de l'homme. Bull Acad Med (Paris) 94:943–952

Budd G (1857) On diseases of the liver. Churchill, London

Chatterjee KD (2011) Parasitology, 13th edn. CBS Publishers, New Delhi, pp 18–41

Clark CG (1998) Amoebic disease: *Entamoeba dispar*, an organism reborn. Trans R Soc Trop Med Hyg 92:361–364

Cox FEG (2002) History of human parasitology. Clin Microbiol Rev 15:595–612

Cunningham DD (1871) A report on cholera. In: 7th annual report sanitation communications, India 1869, Appendix B, pp 141–234

Di Perri G, Strosselli M, Rondanelli EG (1989) Therapy of entamoebiasis. J Chemother 1:113–122

Diamond LS, Clark CG (1993) A redescription of *Entamoeba histolytica* Scahudin, 1903 (Emended Walker, 1911) separating it from *Entamoeba dispar* Brumpt, 1925. J Eukaryote Microbiol 40:340–344

Eichinger D (1997) Encystation of entamoeba parasites. Biol Essays 19:633–639

Ejemot-Nwadiaro RI, Ehiri JE, Meremikwu MM, Critchley JA (2012) Hand washing for preventing diarrhoea (Review), vol 2. The Cochrane Library, Wiley, New York

Gatti S, Swierczynski G, Robinson F, Anselmi M, Corrales J, Moreira J et al (2002) Amebic infections due to the *Entamoeba histolytica-Entamoeba dispar* complex: a study of the incidence in a remote rural area of Ecuador. Am J Trop Med Hyg 67:123–127

Gilbert DN, Moellering RC, Jr. Sande MA (eds) (2002) Treatment of parasitic infections. In: The sanford guide to antimicrobial therapy, 32 edn. Antimicrobial Therapy Inc., Hyde Park, pp 90–98

Goodman A, Rall TW, Niess AS, Taylor P (1992) The pharmacologic basis of therapeutics, 8th edn. Pregamon Press Inc, New York, p 1002

Kartulis S (1886) Zur Aetiologie der Dysenterie in Aegypten. Virchows Archieves fur Pathologie und Anatomie 105:521

Kartulis S (1887) Zur Aetiologie der Leberabscesse: lebende Dysenterie–Amoeben in Eiter der dysenterischen Leberabscesse. Zentralbl fur Bakt 1:745

Kartulis S (1891) Einiges uber die Pathogenese der Desenterieamoben. Zentralbl. fur Bakt. (Abt 1), 9:365

Lewis TR (1870) A report on the microscopic objects found in cholera evaluations. In: 6th annual report sanitation communications, India 1869, Appendix A, pp 125–178

Li E, Stanley SL Jr (1996) Protozoa: Amebiasis—in parasitic diseases of the liver and intestines. Gastroenterol Clin North Am 25:471–492

Markell EK (1986) The 1933 Chicago outbreak of amoebiasis. West J Med 144:750

Martinez-Palomo A, Espinosa-Cantellano M (1998) Intestinal amoebae. In: Cox FEG, Kreier JP, Wakeline D (eds) Topley and Wilson's microbiology and microbial infections, vol 5, 9th edn, pp 157–177

Meleny HE, Bishop EL, Leathers WS (1932) Investigations of *E. histolytica* and other intestinal protozoa in Tennessee. Am J Hyg 16:253

Park K (2011) Park's textbook of social and preventive medicine, 21st edn. Bhanot Publishers, Jabalpur

Powell SJ, Macleod L, Wilmot AJ, Elsdon-Dew R (1966) Metronidazole in amoebic dysentery and amoebic liver abscess. Lancet 2:1329–1331

Quincke HI, Roos E (1893) Ueber Amoben-Enteritis. Wien Med Wochschr 30:1089–1094

Qureshi H, Ali A, Baquai R, Ahmed W (1997) Efficacy of a combined diloxanide-furoate-metronidazole preparation in the treatment of amoebiasis and giardiasis. J Intern Med Res 25:167–170

Rivera WL, Tachibana H, Kanbara H (1998) Field study on the distribution of *Entamoeba histolytica* and *Entamoeba dispar* in the northern Philippines as detected by the polymerase chain reaction. Am J Trop Med Hyg 59:916–921

Schaudinn F (1903) Untersuchungen uber die Fortpflanzung einiger Rizhopoden. Arb Keiserl Gesundh Amte 19:547

Seeto RK (1999) Amoebic liver abscess: epidemiology, clinical features and outcome. West J Med 170:104–109

Seulveda B, Trevino-Garcia MN (1986) Clinical manifestations and diagnosis of amoebiasis. In: Martinez-

Palomo A (ed) Amoebiasis. Elsevier, Amsterdam, pp 169–188

Stauffer W (2008) Amoebiasis and amebic meningoencephalitis. In: Wallace Robert B (ed) Public health and preventive medicine, 15th edn. McGraw Hill, USA

Walsh JA (1984) Estimating the burden of illness in the Tropics. In: Warren KS, Mahmoud AAF (eds) Tropical and geographical medicine. McGraw-Hill, New York, pp 1073–1085

Walsh JA (1986) Problems in recognition and diagnosis of amoebiasis: estimation of the global magnitude of morbidity and mortality. Rev Infect Dis 8:228–238

Walsh JA (1988) Prevalence of *Entamoeba histolytica* infection. In: Ravdin JI (ed) Amebiasis: human infection by *Entamoeba histolytica*. Wiley, New York, pp 93–105

Wertheim HFL, Horby P, Woodall JP (eds) (2012) Atlas of human infectious diseases. Blackwell Publishing Ltd, First Edition

WHO Scientific Working Group (1980) Parasite related diarrhoeas. Bull World Health Organ 58:819–830 WHO, Geneva, Switzerland

World Health Organization (1969) Amoebiasis: report of a WHO expert committee. WHO technical report series 421, pp 1–52. WHO, Geneva, Switzerland

World Health Organization (1997) WHO/PAHO/UNESCO report: a consultation with experts on amoebiasis. Mexico City, Mexico. 28–29 Jan 1997. WHO, Geneva, Switzerland

World Health Organization (1998) Life in the 21st Century: a vision for all. The world health report 1998. World Health Organization, Geneva, Switzerland

Cyclosporiasis: An Emerging Potential Threat for Water Contamination

12

H. Ahmad and Sanjeev Sinha

Abstract

According to the World Health Organization (WHO), 3.4 million people die every year from water-related diseases. Almost half the world's population have no acceptable means of sanitation and availability of potable drinking water. Lack of improved domestic water supply leads to various waterborne diseases, such as cholera, diarrhea, viral hepatitis A, dysentery and typhoid, which are transmitted by contaminated drinking water. Hence, improved water quality, sanitation and personal hygiene can significantly reduce the spread of water-related diseases. Diarrhea is the major illness caused due to the consumption of unhygienic water. It can be of bacterial, viral or parasitic origin. Most of the bacterial and viral agents like coliforms, *Salmonella, Shigella*, caliciviruses that contaminating water are killed by routine chlorination, but the parasitic forms (dormant/resistant cysts and oocysts) are difficult to remove and need special treatment. According to the World Health Organization (WHO), *Cryptosporidium parvum* oocysts and *Giardia lamblia* cysts are introduced to waters all over the world by fecal pollution. Although chlorine is the primary disinfectant of choice in water treatment practice, parasites like *Cryptosporidium* and *Cyclospora* are resistant to chlorine treatment and hence pose a formidable threat to water health. *Cyclospora cayetanensis* is an emerging protozoan parasite and causes small intestine gastroenteritis. The disease has been implicated in many foodborne outbreaks worldwide, especially contaminated products (raspberries, basil and lettuce) imported from developing nations. After washing these products, *Cyclospora* oocysts were found in water. This intrigued researchers to define the epidemiological link of *Cyclospora* to water. In the USA, *Cyclospora* has been detected in several sporadic cases

H. Ahmad · S. Sinha (✉)
Department of Medicine, All India Institute of Medical Sciences, Ansari Nagar, New Delhi, 110029, India
e-mail: drsanjeevsinha2002@yahoo.com

P. P. Singh and V. Sharma (eds.), *Water and Health*,
DOI: 10.1007/978-81-322-1029-0_12, © Springer India 2014

associated with exposure to drinking or recreational sewage and water sources. There is apparently a worldwide distribution, including regions of endemicity, for example, Nepal, Haiti and Peru. Thus, there is an increased risk to travelers visiting these endemic areas. Moreover, because of the recent developments in the detection of protozoal parasites using acid-fast staining, the detection of *Cyclospora* cases has been raised worldwide. Due to the lack of a quantification technique, there is limited information on the prevalence of *Cyclospora* in water environments, necessitating the need for further research on pathways and transmission dynamics of cyclosporiasis and encouraging innovative research in water treatment for improving sanitation and public health.

Keywords
Cyclospora cayetanensis · Diarrhea · Water contaminant

Introduction

Cyclospora sp. infects enterocytes of the small intestine and can produce disease which is associated with a diarrheal illness (Bendall et al. 1993). The organism has been isolated from feces of children and immunocompetent and immunocompromised adults. Cyclosporiasis appears to be most common in tropical and subtropical regions of the world. Outbreaks in the USA and Canada have been linked to various types of imported fresh products and contaminated water. People of all ages are at risk of infection, and the travelers to developing countries can be at increased risk (Soave 1996). In Nepal, the organism was actually detected in water in a British camp where 12 of 14 developed diarrhea and oocysts were detected in 6 of 8 stool samples. The source of infection was identified as the drinking water which was a mixture of river water and chlorinated municipal water (Rabold et al. 1994). In the USA, several reports with stagnant water in storage tanks or broken water pump and those related to recreational water in swimming pools, lakes and ponds have implicated the possible role of *Cyclospora* in causing waterborne outbreaks (Huang et al. 1995).

The infective dose is low about (100 oocysts), and even in clinically ill patients, the organism is often present in low numbers, necessitating the need for concentration procedures like formalin ethyl acetate techniques. The oocysts are resistant to disinfection and are not inactivated by chlorination practices generally applied in the production of drinking water. According to Water and Sanitation Program (WSP), control measures that can be applied to manage potential risk from *Cyclospora* include the prevention of source water contamination by human waste, followed by adequate treatment and protection of water during distribution. Owing to the resistance of the oocysts to disinfectants, *E. coli* (or, alternatively, thermotolerant coliforms) cannot be relied upon as an index of the presence or absence of *Cyclospora* in drinking water supplies.

One of the major problems in evaluating waterborne risks presented by *Cyclospora* is that many fundamental data are missing, and the relative insensitivity and lack of standardization of waterborne detection methods have made the determination of exact modes of transmission difficult. However, there is both epidemiological and microbiological evidence for waterborne route. While outbreaks in developed countries have been primarily caused by contaminated

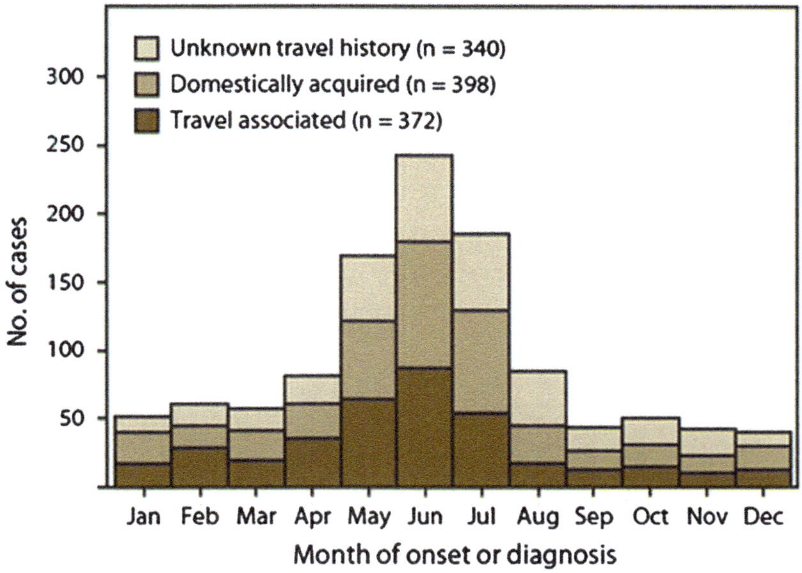

Fig. 12.1 The figure shows the number of reported laboratory-confirmed sporadic cases of cyclosporiasis reported in the USA for 1997–2008 by month of symptom onset or diagnosis (and for 14 cases, the month of report) and by international travel history. The greatest numbers of cases were reported for May, June and August. (Extracted from http://www.cdc.gov/mmwr/preview/mmwrhtml/ss6002a1.htm)

products, in developing countries, the main vehicle of infection is drinking water and the use of contaminated water in crop production is a biologically plausible route of contamination (Percival et al. 2004).

Epidemiology

Cyclospora was first reported in Papua New Guinea in 1979 as an oocyst-like body found in 3 patients with intestinal infections (Ashford, 1979). Previously thought to be blue-green algae, cyanobacterium-like or coccidia-like bodies (CLBs), or large species of *Cryptosporidium,* the observation that these organisms sporulated and contained typical sporocysts and sporozoites confirmed their coccidian nature. In 1986, coccidian-like oocysts of 8–10 μm in size were identified in the stools of patients with diarrhea, until the organism was identified and named *Cyclospora cayetanensis* by Ortega, Oilman and Sterling (Ortega et al. 1993). Infections are seasonal, correlating with rainy season (spring/summer) in temperate zones; in the USA, most cases have occurred April through August (Fig. 12.1). The first outbreak was reported from Nepal, where drinking water consisting of a mixture of river water and municipal water that chlorinated at a concentration of 0.3–0.8 ppm was associated with infections in 12 of 14 soldiers (Rabold et al. 1994). In the USA, several isolated cases of cyclosporiasis possibly associated with exposure to drinking or recreational water or to sewage have been reported (Ooi et al. 1995; Wurtz et al. 1993; Hale et al. 1994). The first documented US outbreak occurred in a physicians' dormitory in Chicago, USA, in 1990. The infections were associated with drinking tap water that had possibly been contaminated with stagnant water from a rooftop storage reservoir (Huang et al. 1995).

In January 1999, the Centre for Disease Control and Prevention (CDC) declared cyclosporiasis as a nationally notifiable disease, and

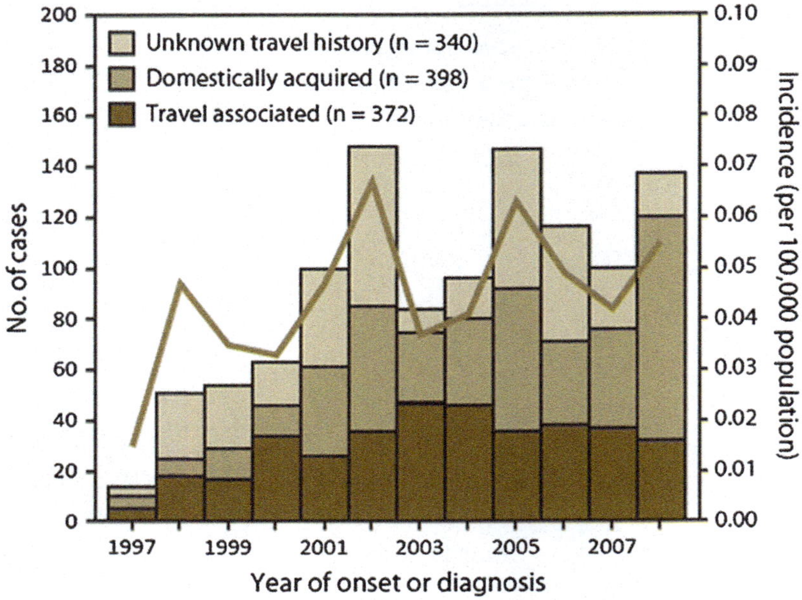

Fig. 12.2 The figure shows the number and incidence per 100,000 population of laboratory-confirmed sporadic cases of cyclosporiasis reported in the USA for 1997–2008 by year of symptom onset or diagnosis and by international travel history. Of the 1,110 patients whose cases were reported, 372 were associated with travel, 398 were acquired domestically, and 340 had an unknown travel history. (Extracted from http://www.cdc.gov/mmwr/preview/mmwrhtml/ss6002a1.htm)

CDC was notified of 1,110 laboratory-confirmed sporadic cases of cyclosporiasis that occurred during 1997–2008. The overall population-adjusted incidence rates ranged from a low of 0.01 cases per 100,000 persons in 1997 to a high of 0.07 in 2002. Approximately one-third of cases occurred in persons with a known history of international travel who might have become infected while traveling outside the continental United States. Domestically acquired cases were concentrated in time (spring and summer) and place (eastern and southeastern states; Fig. 12.2).

Pathogenesis

C. cayetanensis (Fig. 12.3) is a single-cell, obligate, intracellular, coccidian protozoan parasite, which belongs to the family Eimeriidae. Although there are many species within the genus, the only one believed to cause human infections is C. *catyetanensis* (Ortega et al. 1994). *Cyclospora* produces thick-walled oocysts of 8–10 μm in size that are excreted in the feces of infected individuals. Life cycle is initiated with the sporulation of oocysts in the external environment and oocytes mature containing two sporocysts, each containing two sporozoites (Fig. 12.4c). Sporozoites are released from the oocysts which are ingested and penetrate epithelial cells in the small intestine of susceptible individuals. *Cyclospora* oocysts are immature and noninfectious when excreted from feces. The oocyst is the environmentally resistant form of the organism which sporulates after 7–12 days at ambient temperature and more importantly moisture to sporulate and become infectious. Clinical symptoms of cyclosporiasis include watery diarrhea, abdominal cramping, weight loss, anorexia, myalgia and occasionally vomiting and/or fever. Relapsing illness often occurs.

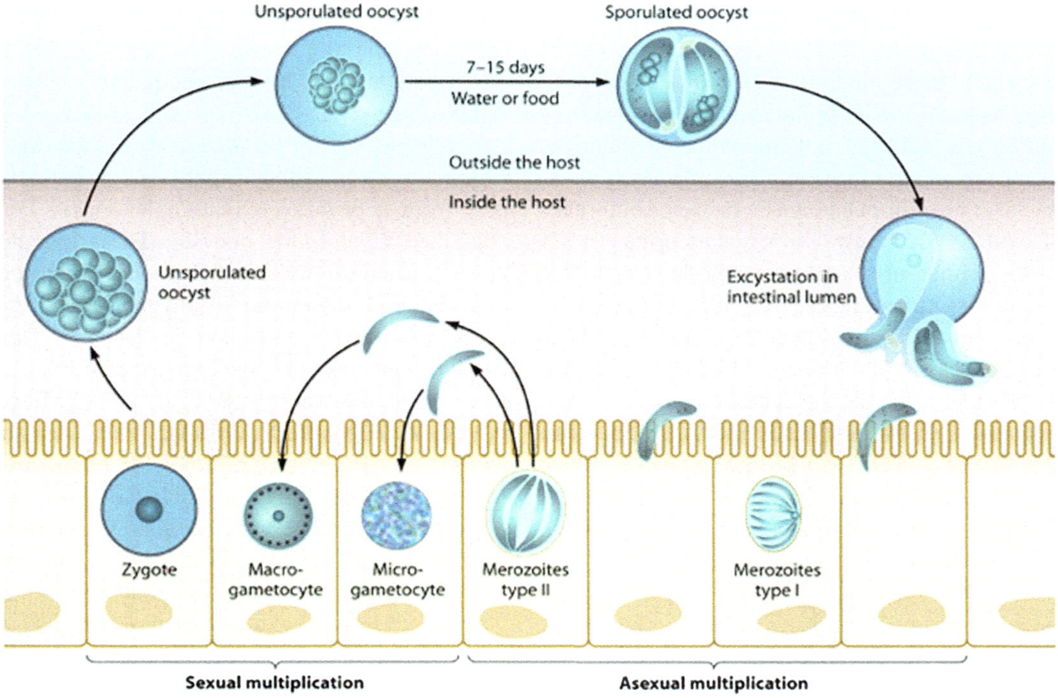

Fig. 12.3 Life cycle of *Cyclospora cayetanensis*. Unsporulated oocysts differentiate into sporulated oocysts, which undergo the excystation process. Sporozoites infect cells to form type I merozoites, and these form type II merozoites. The sexual-stage microgametocyte fertilizes the macrogametocyte to become a zygote and thus to differentiate as an unsporulated oocyst. (Extracted from Ortega and Sanchez 2010)

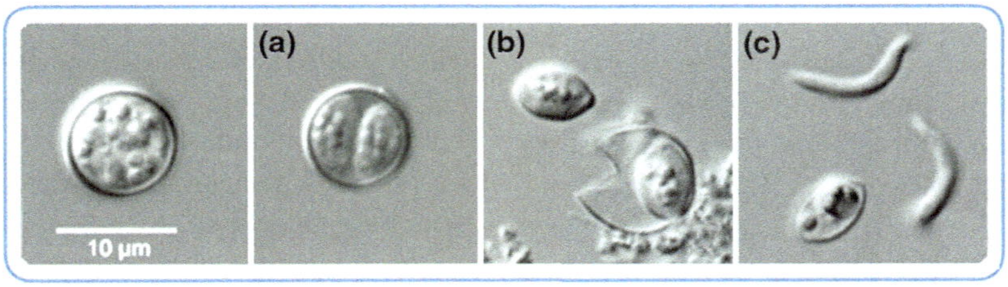

Fig. 12.4 Scanning electron microscope images of *Cyclospora cayetanensis* oocysts. They are excreted unsporulated (noninfective); they usually require at least 1 week under laboratory conditions to sporulate (become infective). An undifferentiated oocyst is shown next to a sporulating oocyst that contains two maturing sporocysts (**a**) An oocyst is mechanically ruptured and releases one of its two immature sporocysts (**b**) One free sporocyst is shown with two free sporozoites, the infective stage of the parasite (**c**) (Extracted from CDC/DPDM http://www.cdc.gov/parasites/cyclosporiasis/)

Humans are the only host identified for this parasite. The unsporulated oocysts pass into the external environment with feces and undergo sporulation, which is completed in 7–12 days, depending on environmental conditions. Only the sporulated oocysts are infectious (Fig. 12.4b).

Transmission

C. cayetanensis is mostly transmitted by the fecal–oral route. Person-to-person transmission is virtually impossible, because the oocysts must sporulate outside the host to become infectious. The primary routes of exposure are contaminated water, food and vegetables especially raspberry, basil and lettuce. The initial source of organisms in foodborne outbreaks has generally not been established, but contaminated water has been implicated in several cases. Drinking water has also been implicated as a cause of outbreaks (Dowd 2003). Transmission occurs primarily via water (either drinking or swimming) contaminated with human feces. Agricultural water may contaminate berries and other fresh products such as basil and lettuce. Shellfish have also been proposed to concentrate oocysts from contaminated waters. Laboratory studies with freshwater clams (*Corbicula fluminea*) have showed that 48–100 % of the clams retained *Cyclospora* oocysts for up to 13 days (Graczyk et al. 1998).

Waterborne Outbreaks and Transmission

From 1986 to 1991, several reports described that diarrhea was associated with a large "*Cryptosporidium*" or Cyanobacterium-like bodies in both immunocompetent and immunosuppressed patients from North, Central and South America; the Caribbean; Nepal; India; and Southeast Asia. Shlim et al. reported the largest series of cases from the CIWEC Clinic (1991), Travel Medicine Center in Katmandu, Nepal, where it is endemic. In Nepal, the *Cyclospora* season starts each year in May and peaks in June and July (Shlim et al. 1991). The risk of infection decreases from August until the organism disappears altogether in October. The risk of getting is fairly high in the *Cyclospora* season, 7–11 % of all expatriates living in Katmandu during this season are affected each year. The risk is highest among new foreign residents. In a study of 77 foreign residents living in Nepal for less than 1 year, 32 % developed *Cyclospora* during the first-season exposure. Recently due to the emergence of the AIDS pandemic, *Cyclospora* cases has also been increasingly observed in immunocompromised cases especially in individuals infected with HIV/AIDS. In endemic countries, the prevalence has been higher in AIDS patients than in diarrheic patients without AIDS (Pape et al. 1994).

Cyclospora has been detected in wastewater (Sturbaum et al. 1998). Eight of 11 water samples from a primary oxidation lagoon in Peru contained *Cyclospora* oocysts. These samples were concentrated using Envirocheck capsules and Hannifin filters. Environmental water collected from rivers and lakes in Vietnam, Guatemala and Egypt was also positive for *Cyclospora* (El-Karamany et al. 2005; Miegeville et al. 2003). In California, water from the Santa Clara River was positive for *Cyclospora* by the Relman PCR protocol; however, the examination of the same samples using other PCR–RFLP methods yielded negative results. These contradictions suggest that the nested PCR method gives false-positive results, as amplification of other coccidia present in river water or environmental samples occurs.

Lately, a number of sporadic cases have emerged in Turkey, Egypt and Vietnam. A study tested 27 sachets containing drinking water that were sold for human consumption in Accra, Ghana (Kwakye-Nuako et al. 2007). *Cryptosporidium* oocysts were identified in 63 % of the bags, while 59.3 % contained *Cyclospora* oocysts, suggesting fecal contamination. *Cyclospora* was also identified in 5 of 12 water sources used for human consumption in rural areas near Guatemala City (Dowd et al. 2003). In Dakahlia Governorate, Egypt, 0.24 % of 840 surveyed drinking water samples contained *Cyclospora* oocysts (Elshazly et al. 2007). In 2005, in a village close to Izmir, Turkey, 30 cases of abdominal pain, diarrhea and nausea were reported for school-age children (Aksoy 2007). Among 191 locals, 5 % had *Cyclospora* infection. Water used either for irrigation or for processing vegetables has also been found to contain *Cyclospora* oocysts. In a recent study at Vietnam, *Cyclospora* sp. were found in 34 of 288 (11.8 %) market water and herb samples and in 24 of 287 (8.4 %) farm samples.

Clinical Symptoms

Cyclospora infects the small intestine, causing an illness characterized by prolonged watery diarrhea, fatigue (sometimes profound) and anorexia in humans. The incubation period from infection to onset of symptoms averages 1 week. The stools of infected patients are typically watery, without blood or white blood cells, and may be explosive. Immunity is not protective, so that re-infection can occur. *Cyclospora* infects patients of all ages, both immunocompromised and immunocompetent. Travelers to endemic areas are at an increased risk of infection (Goodgame 2003).

Laboratory Diagnosis

Diagnosis is made by the identification of oocysts in the stool, duodenal/jejunal aspirate or small bowel biopsy specimen. The gold standard for diagnosis of cyclosporiasis is microscopic identification of oocysts, which is performed by a skilled and experienced microscopist. *Cyclospora* oocysts can easily be confused with fungal spores and *C. parvum,* or may be passed as artifact. In retrospect, some cases of gastroenteritis in 1985 (Babcock et al. 1985) might have mistakenly been attributed to *Blastocystis hominis* infection. In saline wet mount (400×), the protozoa appears as nonrefractile sphere with granular appearance (Fig. 12.5a). Since feces is the specimen of choice, multiple (2–3) stool samples are requested as the oocysts are shed intermittently in feces and hence can cause false-negative reporting.

Recovery and Detection of Cyclospora in Water Samples

Methods for the recovery of *Cyclospora* from rain water and other sample materials have often been based on those developed to recover *Cryptosporidium* oocytes. For example, membrane filtration and microscopy were used to detect *Cyclospora* in chlorinated water supplies to homes in cases during an outbreak in Nepal in 1994 (Rabold et al. 1994). *Cyclospora* have been isolated from irrigation and drinking water by filtration, using Hannifin polypropylene cartridge filters and Envirocheck capsules (Sturbaum et al. 1998). Particles trapped in the filters are released using an elution buffer (1 % sodium dodecyl sulfate, 1 % Tween 80, NaCl, KH_2PO_4, $Na_2HPO_4 \cdot 12H_2O$, KCl and antifoam A). After centrifugation, the pellets are stored in 2.5 % potassium dichromate and examined for the presence of *Cyclospora.* Autofluorescence, phase-contrast microscopy and PCR–RFLP have been used successfully to test water samples.

Isolation of *Cyclospora* from water sources requires concentration of a large volume of water by passage through a yarn-wound depth filter rated at 1 micron nominal porosity. Sample volume can range from 40 liters to 100 gallons depending upon the source water. The filter element is transported to the laboratory where it is washed, and the sediment is concentrated. After a series of clean-up steps, the sample concentrate is examined by the fluorescence method described above. The acid-fast method of detection in environmental samples is impractical due to limited numbers and interfering acid-fast substances. BioVir currently employs the fluorescence assay method of detection. For the detection in environmental samples, a portion of the oocysts which is in feces, water or 2.5 % potassium dichromate at temperatures between 22 and 37 °C kept for 14 days in the laboratory will sporulate (Ortega et al. 1993).

Light Microscopy

Cyclospora can be identified by phase contrast microscopy or bright-field microscopy during the examination of specimens for the ova and parasites. This is the method of choice for fecal samples and employs modified acid-fast staining (Kinyoun stain) using 1 % acid alcohol decolorizer and strong carbolfuchsin. Morphologically

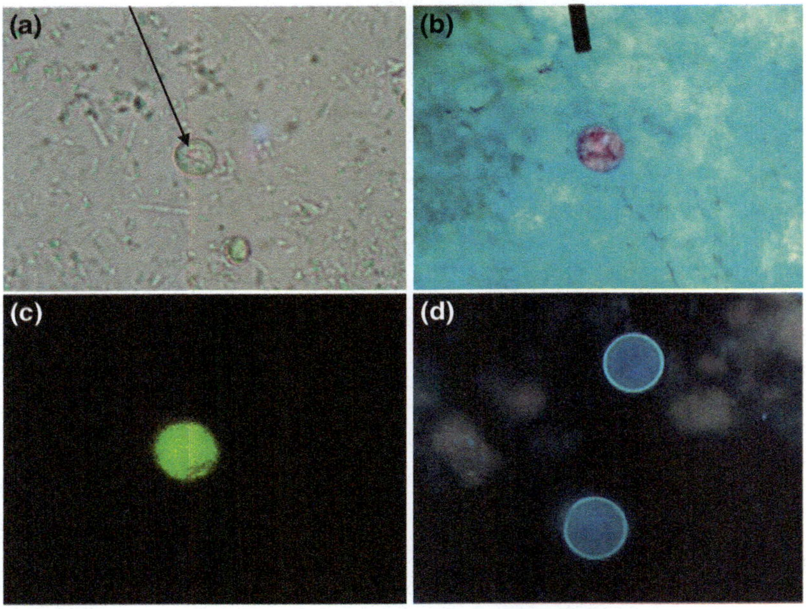

Fig. 12.5 *Cyclospora* oocysts demonstrated in HIV/AIDS patients at parasitology laboratory, AIIMS. The images show saline wet mount **a** modified acid-fast (Kinyoun)-stained oocysts **b** oocysts detected by Auramine–Rhodamine UV fluorescence **c** and oocyst demonstrated by autofluorescence **d** (Unpublished original pictures taken from fecal samples of HIV/AIDS patients). **a** Spherical double-walled, nonrefractile *Cyclospora* oocyst in direct wet mount measuring (8–9 μm in diameter). **b** Typical crumpled cellophane appearance of *Cyclospora* oocysts by modified acid-fast (Kinyoun staining). **c** *Bright green Cyclospora* oocysts stained by Auramine–Rhodamine and visualized with fluorescent microscope. **d** *Blue* and *spherical* oocysts seen as autofluorescence using UV fluorescence microscopy (excitation filter of 330–365 nm) (image extracted from CDC)

oocysts are spherical, 8–10 μm in size and with acid variable staining (faint pink, dark pink, to red). Colorless forms which do not take up the dye are also seen in clinical samples and often confound the identification (Ortega et al. 1993). A modified safranin technique with microwave heating was reported to uniformly stain the oocyst a brilliant orange. Stained oocysts can be easily visualized on 1000× magnification and has a characteristic "soap bubble" appearance. Many a times, dark black granules are also seen within the oocysts (Fig. 12.5b). Lately, Auramine–Rhodamine (Auramine O) fluorochrome staining has also been used for rapid screening and is less labor-intensive as well as cost-effective. The smears can also be stained by Kinyoun stain and diagnosis confirmed (Fig. 12.5c). *Cyclospora* oocysts autofluorescence white-blue under an epifluorescence microscope, using a 330–380 DM excitation filter, or fluorescent green when using an excitation filter of 450–490 DM (Fig. 12.5d). This characteristic has been used to confirm the diagnosis of *Cyclospora* and also to purify the parasite from human fecal samples by flow cytometry. Detection of fluorescence has also been suggested as a useful alternative for screening large numbers of samples in the event of an outbreak (Dixon et al. 2005). Lately, a single-step polymerase chain reaction (PCR) along with *Cyclospora*-specific primers from the ITS-1 region of the genome is also utilized for research and epidemiological studies (Ozdamar et al. 2010).

Micrometry

The conventional and simple technique plays an important role in distinguishing morphologically *C. cayetanensis* from *C. muris* which is 6–8 μm in size and overlaps between the size of *Cryptosporidium* and *Cyclospora*.

Treatment of Diarrhea Caused by Cyclosporiasis

Cyclosporiasis can be treated with a 7–10 day course of oral trimethoprim sulfamethoxazole, shown to cure about 90 % of cases (for adults, 160 mg trimethoprim plus 800 mg sulfamethoxazole twice daily; for children, 5 mg/kg trimethoprim plus 25 mg/kg sulfamethoxazole twice daily). Patients with HIV infection may require higher dosage and longer treatment. Ciprofloxacin is less effective than trimethoprim sulfamethoxazole, but is the treatment of choice for patients who cannot tolerate sulfa drugs. In all patients, fluid and electrolyte balance should be monitored and maintained. In patients who are not treated, illness can be protracted, with remitting and relapsing symptoms (Soave and Johnson 1995).

Inactivation and Removal of *Cyclospora* from Water

Several practices have been tested for the ability to inactivate or reduce the number of viable parasites in foods and water. Because of the lack of animal or in vitro infectivity models, oocyst sporulation has been used as an indicator of viability. Inactivation of *Cyclospora* oocysts in water can be accomplished by freezing at −20 °C for at least 2 days or at −70 °C for 60 min. Inactivation also occurs by heating oocysts in water at 70 °C for 15 min. Hence, boiled water is an easy method for purifying drinking water from *Cyclospora*. Sporulation at 50 °C is minimal (0.01–0.03 %). Microwave heating is also known to inactivate *Cyclospora* oocysts in water when the temperature reaches 80 °C or higher (Ortega and Sanchez 2010). Gaseous chlorine dioxide at 4.1 mg/l does not affect the sporulation of *Cyclospora*; however, this treatment does inactivate *Cryptosporidium* and microsporidia. Gamma irradiation (137Cs) of sporulated and unsporulated *Toxoplasma* oocysts was evaluated as a model system for the inactivation of *Cyclospora* oocysts. It was recommended that 0.5 kGy be used to kill coccidian oocysts on fruits and vegetables.

Preventive Measures

- To provide public education about personal hygiene, especially the sanitary disposal of feces and night soil, careful hand washing after defecation and sexual contact, and before preparing or eating food.
- To educate food handlers about proper food handling, preparation and hygiene.
- Although thorough washing of fresh products especially salad leaves and lettuce is encouraged prior to consumption, it is best to avoid uncooked or unpeeled vegetables in salads during *Cyclospora* season in endemic areas.
- Since iodine and chlorine do not kill *Cyclospora*, drinking water must be boiled or filtered with special filters because ordinary candle filters are unable to eliminate *Cyclospora*.
- All travelers should seek out and be provided with information regarding water, sanitation and food preparation while traveling as well as proper hand washing techniques.
- To ensure safe disposal of sewage, night soil, effluents from leather and manure industries away from water bodies like ponds, lakes and rivers.
- To ensure better quality of water used for agriculture and irrigation purpose.
- Avoid drinking untreated water, particularly in areas where hygiene and sanitation are poor.
- To develop appropriate water testing and purification guidelines for drinking water.
- To clean hands after: using the bathroom; before preparing or eating food; after changing a diaper; after blowing your nose, sneezing or coughing; after caring for a sick person; and after touching an animal.
- Adequate processing of water and foods and abstaining from consumption of raw products

when traveling to areas of endemicity will aid in reducing the risk of acquiring *Cyclospora* infection.

- Good agricultural practices would indeed contribute to reducing the burden of parasite contamination at the farm level. These practices would involve the use of properly treated irrigation water and the use of pathogen-free water for washing products.

Future Research Areas

- To explore the possibility of animal reservoirs and their plausible role in outbreaks.
- To develop molecular tools on the epidemiology of this emerging pathogen and study the various pathways and transmission dynamics of cyclosporiasis throughout the globe.
- To study the various conditions of exposure time, temperature, disinfectant concentration and doses of irradiation to kill oocysts in food or water during routine commercial handling or processing.
- To experimentally analyze the minimum time of sporulation and ranges of temperature and humidity favorable for sporulation.
- Lastly, the development of methods to assess the viability/infectivity of *C. cayetensis* has been hampered by the lack of an animal model, cell culture methods and other surrogate methods.

References

Aksoy U, Akisu C, Sahin S, Usluca S, Yalcin G, Kuralay F, Oral AM (2007) First reported waterborne outbreak of cryptosporidiosis with Cyclospora co-infection in Turkey. Europe Surveillance 12:E070215

Ashford RW (1979) Occurrence of an undescribed coccidian in man in Papua New Guinea. Ann Trop Med Parasitol 73:497–500

Babcock D, Houston R, Kumaki D, Shlim D (1985) Blastocystis hominis in Kathmandu, Nepal [letter]. N Engl J Med 313:1419

CIWEC Clinic. (1991) Travel Medicine Center in Katmandu, Nepal (http://www.ciwec-clinic.com/articles/cyclospora.php)

Bendall RP, Moody A, Chiodini PL, Lucas S, Tovey G (1993) Diarrhoea associated with cyanobacterium-like bodies: a new coccidian enteritis of man. Lancet 341(8845):590–592

Dixon BR, Bussey JM, Parrington LJ, Parenteau M (2005) Detection of *Cyclospora cayetanensis* oocysts in human fecal specimens by flowcytometry. J Clin Microbiol 43:2375–2379

Dowd SE (2003) Confirmed detection of *Cyclospora cayetanensis*, *Encephalitozoon intestinalis* and *Cryptosporidium parvum* in water used for drinking. J Water Health 1:117–123

El-Karamany EM, Zaher TI, el-Bahnasawy MM (2005) Role of water in the transmission of cyclosporiasis in Sharkia Governorate, Egypt. J Egypt Soc Parasitol 35:953–962

Elshazly AM, Elsheikha HM, Soltan DM, Mohammad KA, Morsy TA (2007) Protozoal pollution of surface water sources in Dakahlia Governorate, Egypt. J Egypt Soc Parasitol 37:51–64

Goodgame R (2003) Emerging causes of traveller's diarrhea: *Cryptosporidium*, *Cyclospora*, *Isospora* and microsporidia. Curr Infect Dis Rep 5:66–73

Graczyk TK, Ortega YR, Conn DB (1998) Recovery of waterborne oocysts of *Cyclospora cayetanensis* by Asian freshwater clams (*Corbicula fluminea*). Am J Trop Med Hyg 59:928–932

Hale D, Aldeen W, Carroll K (1994) Diarrhea associated with cyanobacteria like bodies in an immunocompetent host. JAMA 271:144–145

Huang P, Weber JT, Sosin DM et al (1995) The first reported outbreak of diarrheal illness associated with *Cyclospora* in the United States. Ann Intern Med 123:409–414

Kwakye-Nuako G, Borketey P, Mensah-Attipoe I, Asmah R, yeh-Kumi P (2007) Sachet drinking water in Accra: the potential threats of transmission of enteric pathogenic protozoan organisms. Ghana Med J 41:62–67

Miegeville M, Koubi V, Dan LC, Barbier JP, Cam PD (2003) *Cyclospora cayetanensis* presence in aquatic surroundings in Hanoi (Vietnam). Environmental study (well water, lakes and rivers). Bull Soc Pathol Exot 96:149–152

Pape JW, Verdier RI, Boncy M, Boncy J, Johnson WD Jr (1994) Cyclospora infection in adults infected with HIV. Clinical manifestations, treatment, and prophylaxis. Ann Intern Med 121:654–657

Soave R (1996) Cyclospora an overview. Clin Infect Dis 23(3):429–435

Soave R, Johnson WD Jr (1995) Cyclospora: conquest of an emerging pathogen. Lancet 345:667–668

Ooi WW, Zimmerman SK, Needham CA (1995) *Cyclospora* species as a gastrointestinal pathogen in immunocompetent hosts. J Clin Microbiol 33:1267–1269

Ortega Y, Sterling C, Oilman R, Cama V, Díaz F (1993) *Cyclospora* species-a new protozoan pathogen of humans. N Engl J Med 328:1308–1312

Ortega YR, Gilman RH, Sterling CR (1994) A new coccidian parasite (Apicomplexa: Eimeriidae) from humans. J Parasitol 80:625–629

Ortega YR, Sanchez R (2010) Update on *Cyclospora cayetanensis*, a Food-Borne and Waterborne Parasite. Clin Microbiol Rev 2010(23):218–234

Ozdamar M, Hakko E, Turkoglu S (2010) High occurrence of cyclosporiasis in Istanbul, Turkey, during a dry and warm summer. Parasites Vectors 3:39. http://www.parasitesandvectors.com/content/3/1/39

Percival SL, Chalmers RM, Embrey M, Hunters PR, Sellwood J, Wyn-Jones P (2004) Microbiology of waterborne diseases. Elsevier academic press, California, pp 267–290

Rabold JG (1994) *Cyclospora* outbreak associated with chlorinated drinking water [letter]. Lancet 344:1360–1361

Shlim DR, Cohen MT, Eaton M, Rajah R, Long EG, Ungar BL (1991) An alga-like organism associated with an outbreak of prolonged diarrhea among foreigners in Nepal. Am J Trop Med Hyg 45:383–389

Sturbaum GD, Ortega YR, Gilman RH, Sterling CR, Cabrera L, Klein DA (1998) Detection of *Cyclospora cayetanensis* in wastewater. Appl Environ Microbiol 64:2284–2286

Wurtz RM, Kocka FE, Peters CS, Weldon-Linne CM, Kuritza A, Yungbluth P (1993) Clinical characteristics of seven cases of diarrhea associated with a novel acid-fast organism in the stool. Clin Infect Dis 16:136–138

Neurocysticercosis: An Emerging Waterborne Parasitic Disease of Public Health Importance

13

Subhash Chandra Parija and Ira Praharaj

Abstract

Neurocysticercosis is a major health problem for a large proportion of the developing world where it has been found to be the most common cause of acquired epilepsy. This central nervous system parasite is also quite common in many parts of India because of lack of hygiene, clean drinking water and the lack of knowledge regarding modes of acquiring this infection. Diagnosis of this condition can be accomplished with a fair deal of accuracy only when various laboratory-based immunodiagnostic test results are used in conjunction with clinical features. Imaging methods like CT and MRI also play pivotal roles in the diagnosis of this condition. Both antibody detection as well as antigen detection tests have been found to be vital in the diagnosis of this condition, and each of them has their own advantages as well as short comings. Recombinant DNA technology has made it possible to develop many different recombinant antigens which have potential for designing various immunodiagnostic tests in the diagnosis of neurocysticercosis.

Keywords

Imaging methods · Laboratory diagnosis · Neurocysticercosis · Recombinant antigens · Serodiagnosis

Introduction

Neurocysticercosis (NCC) caused by *Cysticercus cellulosae* is one of the commonest parasitic infections affecting the central nervous system. It has emerged as a major public health problem for a large population of the developing world, especially those with low socio-economic status and poor hygiene. The clinical presentation, pathogenesis and management of NCC depend on

S. C. Parija (✉) · I. Praharaj
Department of Microbiology, Jawaharlal Institute of Postgraduate Medical Education and Research, Pondicherry, 605 006, India
e-mail: subhashparija@yahoo.co.in

P. P. Singh and V. Sharma (eds.), *Water and Health*,
DOI: 10.1007/978-81-322-1029-0_13,

a number of variables like number and location of the cysticerci in the brain and nature of the inflammatory reaction in the host. This condition has been found to be endemic in populations where pig rearing is common. Areas like Mexico, Central and South America, and many parts of the Indian subcontinent are found to be endemic for this condition (Del brutto and Sotelo 1988).

NCC may remain asymptomatic in a major proportion of cases and may be detected as a chance finding in many instances. However, it is a major cause of epilepsy in much of the developing world. WHO estimated that approximately 50,000 deaths occur due to NCC annually throughout the world and not less than 20,000 million people infected with cysticerci (Schantz et al. 1993; Pawlowski et al. 2006). The International League against Epilepsy has found NCC to be the single most common cause of acquired epilepsy in developing countries. Also, the prevalence rate of active cases of epilepsy has been found to be twice that of developed countries (Garcia and Del Brutto 2005).

Problem of NCC in India

In India, NCC is regarded as the second most important cause of intracranial space occupying lesions (ICSOL) following tuberculosis (Mahajan 1982) and is the most common cause of epilepsy (Sawhney et al. 1996). Also, NCC is the most frequent and widely disseminated human neuroparasitosis (Pittella 1997). Clinicians, neurosurgeons and neurologists have reported a number of case studies of NCC mostly from urban centres (Chandramukhi 2000; Malla and Mahajan 2001). Its prevalence has been found to be more in the northern states of Bihar, Uttar Pradesh and Punjab. NCC in humans is endemic in many parts of India, predominantly in areas where poor hygiene prevails and pig rearing is done (Prasad et al. 2008). In a recent study, based on 30 cluster sampling approach in a rural pig farming community in Uttar Pradesh, the prevalence of taeniasis was found to be 18.6 % (Prasad et al. 2007). In the same community, active epilepsy was identified and clinically confirmed in 5.8 % of the populations during door-to-door survey and 48.3 % of them fulfilled either definitive or probable diagnostic criteria of NCC. Although cases have been reported from all over the country, the prevalence has been found to show wide regional variation (Rajashekhar and Chandy 2000).

Several studies in different parts of the country have tried to measure the prevalence of this condition. The prevalence of intestinal taeniasis has been reported to be 0.36 % in Puducherry (Parija and Rao 1987), 0.9 % in Lucknow (Srivastava and Pandey 1986), 0.8 % in Chandigarh (Ramesh et al. 1991) and 0.5 % in Northern India (Malla et al. 1992). A more recent study from North India reported as much as 9.7 % of human subjects with intestinal taeniasis presenting with seizures (Prasad et al. 2002). However, the comparison of the prevalence data from different places is very difficult because no single study protocol is followed in the above studies and because of lack of continuous monitoring of the prevalence in any of these places to assess the increasing or decreasing trends in the prevalence over the years.

NCC as a Waterborne Disease

Cysticercosis in general and NCC in particular can be acquired by ingestion of *Taenia* eggs contaminating food and water. Infection with eggs of *Taenia* can also be acquired by faecal contamination of soil and water used for growing vegetables. It is thus that vegetarians who have never consumed pork can also acquire this condition. It is a common misconception that NCC can only be acquired by ingestion of pork and therefore should not be found in vegetarians.

Taeniasis and cysticercosis have been considered a poverty-related disease (WHO 2006). It has been neglected in much of the developing world because of lack of awareness regarding this problem and absence of appropriate sensitive tests for its diagnosis (Praet et al. 2010).

Developing countries like India suffer from lacunae like lack of potable drinking water and lack of effective public sanitation system and improper disposal of human faeces (Praet et al. 2010). Apart from this, many farmlands in India have been found to be irrigated with untreated sewage water. Fruits and vegetables cultivated in these fields have been found to be contaminated with many helminthic eggs including *Taenia* and help in the transmission of cysticercosis.

Risk factors for NCC in urban population of the developing world include poor household conditions such as raising pigs, earthen floors, lack of potable water, lack of sanitary toilets, low academic education and lack of knowledge about the parasite (Fayer 2000).

Diagnosis

The clinical signs and symptoms of NCC are not specific and can be found in many ICSOL like tuberculoma or even malignancies of the brain. Further, the protean manifestations of this disease make it difficult to diagnose it clinically. It has been found that using clinical manifestations, radiological or immunological criteria in isolation for this condition can lead to a confusing picture (Takayanagui and Odashima 2006). An accurate diagnosis is possible only by suspicion on epidemiological grounds, proper interpretation of clinical data, analysis of neuroimaging studies and specific immunological tests using CSF and serum.

It was for this reason that in 1996, Del Brutto et al. (1996) proposed diagnostic criteria for NCC based on clinical, epidemiological, immunological and imaging data. According to this, four categories of criteria have been defined based on their diagnostic strength—absolute, major, minor and epidemiologic. Using these criteria, NCC can be diagnosed with three degrees of certainty: definitive diagnosis, probable diagnosis and possible diagnosis. Patients with one absolute criterion or two major criteria or two major and one minor criterion comprise the "definitive diagnosis" group. Patients with one major plus two minor criteria or those having one major, one minor and one epidemiologic criteria belong to the "probable diagnosis" group. "Possible diagnosis" group patients should have one major criterion, two minor criteria or one minor plus one epidemiologic criterion (Del Brutto et al. 1996)

Histopathological Findings

Histopathological examination and a positive brain biopsy are considered as one of the absolute criteria for the diagnosis of NCC. Suspected cyst tissue can be identified as cysticerci with their characteristic oval or round appearance containing an invaginated scolex (Schmidt and Roberts 2001). A dying larva elicits pronounced cellular reaction in the brain which can also be detected by histopathology (Garcia et al. 2001). Direct immunofluorescence assay has been found to be quite sensitive in detecting cysticercus antigens in frozen or fixed brain sections (Garcia et al. 2001).

Imaging

Imaging methods have formed the cornerstone of NCC diagnosis in the recent years and provide one of the most important tools available to clinicians in diagnosing this condition. Imaging methods like the CT scan and magnetic resonance imaging (MRI) have also proved helpful in determining the modality of treatment for different forms of this disease (Garcia and Del Brutto 2003). They have drastically improved diagnostic accuracy for NCC by providing objective evidence concerning the topography of the lesions, their stage of involution and the degree of the host's inflammatory response against the parasite.

In the present scenario, CT scans have been found to have very good sensitivity and specificity of over 95 % for the diagnosis of NCC (Wadia et al. 1987). CT scans are especially helpful in detecting calcifications which appear as hyperdense areas. Another diagnostic feature is the presence of perilesional oedema which can be visualized around dying cysts.

MRI has the distinct advantage of helping in differentiating between the various cyst stages of the parasite (Jena et al. 1998). MRI has also been reported to be much more sensitive than CT scan in detecting cysts at locations like the cerebellum and brain stem. (Martinez et al. 1989).Although the sensitivity of MRI for the detection of calcified cysts has been reported to be lower than that of CT scan, MRI has the distinct advantage of recognizing forms of cysticercosis that are not clearly visualized by CT (Suss et al. 1986).

Although diagnosis of NCC can be achieved with a high degree of accuracy by both CT scan and MRI, they often show false-positive results, non-specific findings and also fail in some case with intraventricular cysts. These methods also have the disadvantage of being quite expensive and inaccessible in much of the developing world where NCC is endemic (Chung et al. 1999).

Immunodiagnosis

Immunodiagnostic tests have played a major role in making the diagnosis of NCC easier, accurate and more accessible in the developing world. Immunodiagnostic techniques include detection methods for specific antibodies and for circulating parasite antigen in serum or cerebrospinal fluid. Immunological techniques for the diagnosis of NCC include both antibody detection and antigen detection. The samples which have been used and evaluated for antigen and antibody detection include serum, CSF and urine. (Dorny et al. 2003). Whereas antibody detection is indicative of exposure to the parasite at some point of time, antigen detection is better as it reflects the presence of viable parasites and is positive in recent infection.

Extensive studies have been made on the use and evaluation of serological tests for the diagnosis of NCC, but most of them do not provide univocal proof. There is occurrence of variation in sensitivity and specificity of various immunoassays due to (1) host heterogenous immune response, (2) use of variety of cysticercus antigens from different sources and (3) antigenic drift or variation exhibited by the parasite (Zini et al. 1999).

Detection of Cysticercus Antibodies

Antibodies against the *Taenia solium* larval stage, that is, the *C. cellulosae,* results in antibody response which can be detected in serum, CSF and some other body fluids like urine and saliva. Techniques which have been evaluated for antibody detection include radioimmunoassay, complement fixation test (CFT), dipstick test, latex agglutination, ELISA and immunoblot (Western Blot).Methods like radioimmunoassay and CFT have fallen out of use. ELISA and Electroimmunotransfer blot (EITB) are the two antibody detection methods which are used most widely.

ELISA for the detection of cysticercosis antibodies: *ELISA* for the detection of specific antibodies against *C. cellulosae* has been widely used with variable sensitivity and specificity reported by many authors. It can be used for both field-based surveys as well as laboratory diagnosis where large numbers of samples are to be tested at a time.

ELISA for detecting antibodies using crude antigens of *C. cellulosae* has been reported to have a sensitivity of only 71 % and specificity of 95 % for CSF samples by Proano-Narvaez et al. (2002). Mittal et al. (2001) reported a very low sensitivity of 10.4 % for detecting anti-cysticercus antibodies from CSF samples of clinically suspected cases of NCC by using crude antigens derived from cysts. They also reported considerable rates of cross-reaction seen in cases of hydatid disease when they used crude antigens.

Lee et al. (1993) reported a sensitivity and specificity of 81.7 and 88.5 %, respectively, for detecting antibodies against *C. cellulosae* by using crude antigens. In a similar study from North India, Malla and Mahajan (2001) reported a sensitivity of 68.18 % and specificity of 87.3 % using crude extracts of *Cysticercus cellulosae* cysts for serodiagnosis of NCC. Using crude antigens derived from *C. cellulosae* for designing immunodiagnostic tests for cysticercosis has the disadvantage of cross-reactions with other parasitic diseases like hydatid disease (Neto et al. 2007).

Various components of the cysticercus cysts like the cyst wall, cyst fluid and protoscolex extract have been used and evaluated in the designing of immunodiagnostic tests like ELISAs for detecting antibodies by many workers in the field (Pinto et al. 2000; Sahu et al. 2010). Apart from these, *Taenia crassiceps* antigenic peptides have also been evaluated in serological diagnosis of cysticercosis in pigs (Pinto et al. 2000).

Enzyme immunotransfer blot (EITB) in the diagnosis of NCC: EITB has proved very helpful in the serodiagnosis of parasitic diseases like hydatid disease and NCC in the past few years because of its greater specificity and sensitivity when compared to other serodiagnosis methods like ELISA. Among other parasitic diseases for which EITB has been evaluated are schistosomiasis, amoebiasis and filariasis (Al Sherbiny et al. 1999).

EITB has been evaluated for the detection of both antibodies as well as cysticercus antigens in the diagnosis of NCC, although studies dealing with antibody detection predominate. EITB for the detection of antibodies has been evaluated by using various purified fractions of *C. cellulosae* and has shown varying degrees of sensitivity and specificity. EITB assay using a purified fraction of glycoprotein is highly specific (100 %) for the detection of antibodies both in serum and CSF. The presence of 1–7 glycoprotein bands is considered diagnostic of *T.solium* infection. The test is more than 90 % sensitive in NCC patients with more than 2 cysticerci in the brain (Parija et al. 2006).

Detection of Cysticercus Antigen

Detecting cysticercus antibodies in serum may not necessarily indicate established infection always (Garcia et al. 2001). Besides, detection of antibodies for NCC may even lead to false-positive results in some cases where the parasite has already been eliminated after administration of antiparasitic drug therapy. False-negative results are also possible in some cases where there is no or very low immune response (Parija 2005).

To overcome many of the disadvantages of antibody detection in cysticercosis in general and NCC in particular, detection of cysticercus antigen in various samples like serum, CSF and even other body fluids like saliva has been evaluated by many workers. Antigen detection has especially been found to be valuable in monitoring effective therapy against cysticercosis as the antigen levels fall considerably after antihelminthic therapy in most cases (Garcia et al. 2007).

Among the different methods used for antigen detection are antigen capture ELISA, EITB and the co-agglutination test (Parija et al. 2004; Parija and Rajesh 2006; Rodriguez et al. 2009). The co-agglutination test evaluated by Parija and Rajesh (2006) for antigen detection was found to be moderately sensitive and specific for the diagnosis of cysticercosis. The major advantage of the co-agglutination test for antigen detection is the ease with which this test can be performed and interpreted.

Detection of circulating antigens of cysticercosis by Ag-ELISA was first reported as early as 1989 by Harrison et al. (1989) in animals. Variable sensitivities and specificities have been quoted by various workers in this field since then. Although EITB has been widely used for the detection of anti-cysticercus antibodies in many studies and is one of the major criteria for diagnosis is NCC (Del Brutto et al. 1996; Garcia and Del Brutto 2005), number of studies dealing with EITB for the detection of cysticercus antigens is limited. However, a few workers have evaluated this technique for detecting secreted cysticercus antigens in various samples (Pardini et al. 2001).

Recombinant Antigens in the Diagnosis of Cysticercosis

Currently, routine immunodiagnosis of NCC still relies heavily on the usage of native antigens such as complete somatic homogenate, metacestode cyst fluid, cyst wall antigen, protoscolex antigen and excretory/secretory products in various diagnostic tests. The only source of antigen is *T. solium* cysticerci obtained from parasite extracted from diseased pork meat. The amount of cysticerci obtained from the pork varies widely in relation to the parasite burden. Also, obtaining cysts from pork muscle is a laborious and time-consuming procedure (Greene et al. 2000). Moreover, in many cases, preparation of adequate antigenic extracts in sufficient amount is still linked to the detection of swine infected with *T.solium* larvae which is difficult to obtain (Levine et al. 2004). Parasitized pork is also difficult to obtain in areas where the disease is endemic because farmers tend to hide sick animals, denying the occurrence of infected pork and the animals are eventually killed and sold clandestinely (Zarlenga et al. 1994). Difficulties in obtaining somatic cysticercal antigens in large quantities have prevented the standardization of the immunodiagnostic tests.

As the isolation of purified antigens from cysticerci is limited by paucity of the materials, recombinant DNA technology that would yield sufficient quantities of pure antigen appears to be the best alternative to the native antigen (Ferrer et al. 2005). Recombinant antigens have been evaluated in the recent past for the diagnosis of many parasitic diseases such as cystic echinococcosis (Virgino et al. 2003; Ito et al. 2001), Chagas' disease (Umezawa et al. 1999), fascioliasis (Tantrawatpan et al. 2005), filariasis (Lammie et al. 2004), malaria (Kim et al. 2003), leishmaniasis (Qu et al. 1994) and amoebiasis (Lee et al. 2000; Shenai et al. 1996).

Production of antigens by recombinant DNA technology ensures uniform, specific, pure and continuous supply of antigens which can be used in serodiagnosis of infectious disease (De silva et al. 2000; Levine et al. 2007; Ferrer et al. 2009). Theoretically, a mixture of recombinant proteins with immunodiagnostic potential can help to solve the drawbacks associated with native antigens. Unlimited amounts of antigen can be produced under controlled conditions and it may even be possible to identify and remove the cross-reactive epitopes without losing the diagnostic efficiency. Such a permanent source of diagnostic antigens would also be useful in endemic countries where the collection of *T.solium* cysticerci is difficult. Recombinant proteins which are the single well-defined antigen improve the specificity of the serological tests (Hancock et al. 2004, 2006; Handali et al. 2010). Thus, it eliminates the non-specific reactivity in the serological assays. Since the recombinant antigens are the defined antigens and also the antigenic composition is known, standardization of the serological assays is quite easy. Hence, it can overcome major disadvantages like poor reproducibility of serodiagnostic techniques between different laboratories, due to variability in nature and purity of antigens employed.

Newer Diagnostic Approaches

Serum is the commonest sample obtained for the diagnosis of NCC. Both cysticercus antibodies as well as antigens can be detected in serum. CSF samples are also obtained quite commonly for the diagnosis of NCC. Some workers have found CSF samples more useful than serum samples for the detection of anti-cysticercus antibodies in cases of NCC (Proano-Narvaez et al. 2002).

In a bid to prevent any kind of invasive techniques for obtaining samples for NCC, few workers have evaluated samples like urine, saliva and even tears for detecting antigens or anti-cysticercus antibodies. Sahu et al. (2008) evaluated the detection of IgA antibodies specific for cysticercosis in tear. Parija et al. (2004) evaluated the usefulness of antigen detection in non-invasive sample like urine and found it to be quite useful. Urine specimens, apart from being non-invasive also have the advantage of being readily available. Using urine samples for immunodiagnostic tests also implies that the risk of acquiring blood-borne infections like Hepatitis B and HIV is nil. Malla et al. (2005) have evaluated the usefulness of IgG4 in urine samples for the diagnosis of NCC cases.

Detection of parasite antigens from samples like urine and saliva has also been evaluated for many other parasitic infections like amoebiasis and hydatid disease (Parija and Khairnar 2007). Khairnar et al. have evaluated the detection of *Entamoeba histolytica* DNA in saliva of known patients of amoebic liver abscess who had received metronidazole treatment and found it to be of prognostic importance (Khairnar and Parija 2008). The same approach for the diagnosis of NCC still needs to be evaluated.

Newer Diagnostic Methods

Molecular diagnostic methods like polymerase chain reaction (PCR) have brought about a revolution in the diagnosis of many parasitic diseases. The reason and advantage behind using these techniques is their accuracy and sensitivity. Recent years have seen a spurt in the number of studies which have evaluated molecular methods like PCR detecting cysticercus DNA for the diagnosis of NCC using CSF as the sample. In a study by Almeida et al., as little as 10 fg of *T.solium* DNA was detected by PCR from CSF samples (Almeida et al. 2006). In a recent study, Michelet et al. (2011) compared DNA detection from CSF samples by PCR with other diagnostic tests like antibody detection by ELISA, EITB and cysticercus antigen detection and found PCR to exhibit the highest sensitivity (95 %) and variable specificity (80–100 %).

Conclusions

No single modality of diagnosis can be used in isolation for the diagnosis of NCC. Clinical evidence along with imaging findings and immunodiagnostic tests like ELISA and EITB goes a long way in accurately detecting this condition. Among the immunodiagnostic tests, those detecting cysticercus antigens have been found to be more indicative of active infection as compared to antibody detection methods. Recent years have seen an increase in the use of samples other than serum and CSF for the diagnosis of NCC. Use of samples like saliva and urine has been evaluated for antigen detection in order to avoid any kind of invasive procedures being carried out on patients. Development of recombinant antigens has improved the sensitivity, specificity and accuracy of immunodiagnostic tests for NCC.

References

Al Sherbiny MM, Osman AM, Hancock K, Deelder AM, Tsang VC (1999) Application of immunodiagnostic assays: detection of antibodies and circulating antigens in human schistosomiasis and correlation with clinical findings. Am J Trop Med Hyg 60:960–966

Almeida CR, Ojopi EP, Nunes CM, Machado LR, Takayanagui OM, Livramento JA (2006) *Taenia solium* DNA is present in the cerebrospinal fluid of neurocysticercosis patients and can be used for diagnosis. Eur Arch Psychiatry Clin Neurosci 256:307–310

Chandramukhi A (2000) Neurocysticercosis an emerging and re emerging disease. J Int Med Sci Acad 13:32–39

Chung JY, Bahk YY, Huh S, Kang SY, Kong Y, Cho SY (1999) A recombinant 10-kDa protein of Taenia solium metacestodes specific to active neurocysticercosis. J Infect Dis 180:1307–1315

Da Silva AD, Quagliato EM, Rossi CL (2000) A quantitative enzyme-linked immunosorbent assay (ELISA) for the immunodiagnosis of neurocysticercosis using a purified fraction from Taenia solium cysticerci. Diagn Microbiol Infect Dis 37(2):87–92

Del brutto OH, Sotelo J (1988) Neurocysticercosis: an update. Rev Infect Dis 10:1075–1087

Del Brutto OH, Wadia NH, Dumas M, Cruz M, Tsang VC, Schantz PM (1996) Proposal of diagnostic criteria for human cysticercosis and neurocysticercosis. J Neurol Sci 142:1–6

Dorny P, Brandt J, Zoli A, Geerts S (2003) Immunodiagnostic tools for human and porcine cysticercosis. Acta Trop 87:79–86

Fayer R (2000) Global change and emerging infectious diseases. J Parasitol 86:1174–1181

Ferrer E, Cortez MM, Cabrera Z, Rojas G, D'avila I (2005) Oncospheral peptide based ELISA as potential seroepidemiological tools for *T.solium* NCC. Trans Roy Soc Trop Med Hyg 99:568–576

Ferrer E, Martinez-Escribano JA, Barderas ME, Gonzalez LM, Cortez MM, Davila I et al (2009) Peptide epitopes of the Taenia solium antigen TS8B2 are immunodominant in human and porcine cysticercosis. Mol Biochem Parasitol 168:168–171

Garcia HH, Del Brutto OH (2003) Imaging findings in neurocysticercosis. Acta Trop 87:71–78

Garcia HH, Del Brutto OH (2005) Cysticercosis working group in Peru. Neurocysticercosis: updated concepts about an old disease. Lancet Neurol 4:653–661

Garcia HH, Gonzalez AE, Gilman RH, Palacios LG, Jimenez I, Rodriguez A et al (2001) Cysticercosis working group in Peru. Short report: transient antibody response in Taenia solium infection in field conditions-a major contributor to high seroprevalence. Am J Trop Med Hyg 65:31–32

Garcia HH, Moro PL, Schantz PM (2007) Zoonotic helminth infections of humans: echinococcosis, cysticercosis and fascioliasis. Curr Opin Infect Dis 20:489–494

Greene RM, Hancock K, Wilkins PP, Tsang VC (2000) Taenia solium: molecular cloning and serologic evaluation of 14- and 18-kDa related, diagnostic antigens. J Parasitol 86(5):1001–7

Hancock K, Pattabhi S, Greene RM, Yushak ML, Williams F, Khan A et al (2004) Characterisation and cloning of GP50, a *Taenia solium* antigen diagnostic for cysticercosis. Mol Biochem Parasitol 133:115–124

Hancock K, Pattanhi S, Whitfield FW, Yushak ML, Lane WS, Garcia HH et al (2006) Characterisation and cloning of T24, a *Taenia solium* antigen diagnostic for cysticercosis. Mol Biochem Parasitol 147:109–117

Handali S, Klarman M, Gaspard AM, Noh J, Lee YM et al (2010) Multiantigen print immunoassay for comparision of diagnostic antigens for Taenia solium cysticercosis and taeniasis. Clin Vaccine Immunol 17:68–72

Harrison LJ, Joshua GW, Wright SH, Parkhouse RM (1989) Specific detection of circulating surface glycoproteins of viable cysticerci in Taenia saginata cysticercosis. Parasite Immunol 11:351–370

Ito A, Sako Y, Ishikawa Y, Nakao M, Nakaya K (2001) Differential serodiagnosis of cystic and alveolar echinococcosis using native and recombinant antigens in Japan. Southeast Asian J Trop Med Public Health 32:111–115

Jena A, Sanchetee PC, Gupta RK, Khushu S, Chandra R, Lakshmipati N (1998) Cysticercosis of the brain shown by magnetic resonance imaging. Clin Radiol 39:542–546

Khairnar K, Parija SC (2008) Detection of *Entamoeba histolytica* DNA in the saliva of amebic liver abscess patients who received prior treatment with metronidazole. J Health Popul Nutr 26:292–295

Kim SI, Ahn HJ, Kim TS, Nam HW (2003) ELISA detection of vivax malaria with recombinant multiple stage-specific antigens and its application to survey of residents in endemic areas. Korean J Parasitol 41:203–207

Lammie PJ, Weil G, Noordin R, Kaliraj P, Steel C, Goodman D et al (2004) Recombinant antigen-based antibody assays for the diagnosis and surveillance of lymphatic filariasis-a multicenter trial. Filaria J 3:9–12

Lee JH, Kong Y, Ryu JY, Cho SY (1993) Applicability of ABC-ELISA and protein A-ELISA in serological diagnosis of cysticercosis. Korean J Parasitol 31(1):49–56

Lee J, Park SJ, Yong TS (2000) Serodiagnosis of amoebiasis using a recombinant protein fragment of 29 kDa surface antigen of *E.histolytica*. Int J Parasitol 30:1487–1491

Levine MZ, Calderon JC, Wilkins PP, Lane WS, Asara JM, Hancock K, Gonzalez AZ (2004) Characterisation, cloning and expression of two diagnostic antigens for T.solium tape worm infection. J Parasitol 90:631–638

Levine MZ, Lewis MM, Rodriquez S, Jimenez JA, Khan A, Lin S (2007) Development of an EITB assay using two baculovirus expressed recombinant antigens for diagnosis of *T.solium* taeniasis. J Parasitol 93:409–417

Mahajan RC (1982) Geographical distribution of human cysticercosis: in cysticercosis: present state of knowledge and perspectives. Academic press Inc., New York, pp 39–46

Malla N, Kaur S, Kaur R, Ganguly NK, Mahajan RC (1992) Evaluation of enzyme linked immunosorbent assay for the detection of anti-Cysticercus antibodies in cerebrospinal fluid from patients with neurocysticercosis. J Hyg Epidemiol Microbiol Immunol 36:181–190

Malla N, Kaur R, Ganguly NK, Sawhney IM, Mahajan RC (2005) Utility of specific IgG4 response in saliva and serum samples for the diagnosis and follow-up of human neurocysticercosis. Nepal Med Coll J 7:1–9

Malla N, Mahajan RC (2001) Human neurocysticercosis. In: Sood ML (ed) Helminthology in India. International Book Distributors, Dehradun

Martinez HR, Rangel- Guerra R, Elizondo G, Gonzalez J (1989) MR Imaging in neurocysticercosis: a study of 56 cases. Am J Neuroradiol 10:1011–1019

Michelet L et al (2011) Human neurocysticercosis: comparison of different diagnostic tests using cerebrospinal fluid. J Clin Microbiol 49(1):195–200

Mittal V, Singh VK, Ichhpujani RL (2001) Detection of antibodies to Taenia solium in sera of patient with epilepsy using ELISA. J Commun Dis 33(1):23–7

Neto FI, Pianetti-Filho G, Araujo RN, Nascimento E (2007) Immunodiagnosis of human neurocysticercosis by using semipurifed scolex antigens from *Taenia solium* cysticerci. Rev da Socied Brasil de Medic Trop 40:163–169

Pardini AX, Peralta RH, Vaz AJ (2001) Cysticercus antigens in cerebrospinal fluid samples from patients with neurocysticercosis. J Clin Microbiol 39:3368–3372

Parija M, Biswas R, Harish BN, Parija SC (2004) Detection of specific cysticercus antigen in the urine for diagnosis of neurocysticercosis. Acta Trop 92:253–260

Parija SC (2005) Cysticercus antibodies and antigens in serum of blood donors from Pondicherry. Rev Inst Med Trop Sao Paulo 47:227–230

Parija SC, Khairnar K (2007) Detection of excretory *Entamoeba histolytica* DNA in the urine, and *E. histolytica* DNA and lectin antigen in the liver abscess pus for the diagnosis of amoebic liver abscess. BMC Microbiol 7:41–50

Parija SC, Rajesh RS (2006) Co-agglutination test for cysticercus antigen detection in the serum for the diagnosis of neurocysticercosis. Trop Doct 36:144–147

Parija SC, Rao RS (1987) Prevalence of parasitic infections in Pondicherry. Indian J Parasitol 11:63–65

Pawlowski ZS (2006) cRole of chemotherapy of taeniasis in prevention of neurocysticercosis. Parasitology 34:353–356

Pinto PS, Vaz AJ, Germano PM, Nakamura PM (2000) ELISA test for the diagnosis of cysticercosis in pigs using antigens of Taenia solium and Taenia crassiceps cysticerci. Rev Inst Med Trop Sao Paulo 42:71–79

Pittella JE (1997) Neurocysticercosis. Brain Pathol 7:681–693

Praet N, Kanobana K, Kabwe C, Maketa V, Lukanu P, Lutumba P et al (2010) *Taenia solium* cysticercosis in the democratic republic of Congo: how does pork trade affect the transmission of the parasite? PloS Neglected Trop Dis 4:9, e817

Prasad KN, Chawla S, Jain D, Pandey CM, Pal L, Pradhan S et al (2002) Human and porcine T.solium infection in rural North India. Trans R Soc Trop Med Hyg 96:515–516

Prasad KN, Prasad A, Gupta RK, Pandey CM, Singh U (2007) Prevalence and associated risk factors of Taenia solium taeniasis in a rural pig farming community of north India. Trans R Soc Trop Med Hyg 101(12):1241–7

Prasad KN, Prasad A, Verma A, Singh AK (2008), Human cysticercosis and Indian scenario: a review. J. Biosci. Nov 33(4):571–82

Proano-Narvaez JV, Meza-Lucas A, Mata-Ruiz O, Garcia-jeronimo RC, Correa D (2002) Laboratory diagnosis of human neurocysticercosis :double-blind comparison of enzyme-linked immunosorbent assay and electroimmunotransfer blot assay. J Clin Microbiol 40:2115–2158

Qu JQ, Zhong L, Maoosm Yasinzai M, Abdul-Rab M, Aksu HS, Reed SG et al (1994) Serodiagnosis of Asian leishmaniasis with a recombinant antigen from the repetitive domain of *L.kinesin*. Trans R Soc Trop Med Hyg 88:543–545

Rajashekhar V, Chandy MJ (2000) Clinical manifestation of solitary cysticercus granuloma. In: Rajashekar V, Chandy MJ (eds) Solitary cysticercus granuloma. Orient Longman, Chennai, pp 29–39

Ramesh GN, Malla N, Raju GS, Sehgal R, Ganguly NK, Mahajan RC et al (1991) Epidemiological study of parasitic infestations in lower socio economic group in Chandigarh (North India). Indian J Med Res 93:47–50

Rodriguez S, Dorny P, Tsang VC, Pretell EJ, Brandt J, Lescano AG et al (2009) Detection of Taenia solium antigens and anti-T.solium antibodies in paired serum and cerebrospinal fluid samples from patients with intraparenchymal or extraparenchtmal neurocysticercosis. J Infect Dis 199:1345–1352

Sahu PS, Parija SC, Jayachandran S (2010) Antibody specific to 43 kDa excretory-secretory antigenic peptide of *Taenia solium* metacestode as a potential diagnostic marker in human neurocysticercosis. Acta Trop 115:257–261

Sahu PS, Parija SC, Sahu PK (2008) Tear IgA ELISA: a novel and sensitive method for diagnosis of ophthalmic cysticercosis. Acta Trop 106:168–174

Sawhney IM (1996) Evaluation of epilepsy management in a developing country. Acta Neurologia Scandinavica 94:19–23

Schantz PM, Cruz M, Sarti E, Pawlowski Z (1993) Potential eradicability of taeniasis and cysticercosis. Bull Pan Am Health Organ 27:397–403

Schmidt GD, Roberts LS (2001) Foundations of parasitology, 6th edn. McGraw Hill, New york

Shenai BR, Komalam BL, Arvind AS, Krishnaswamy PR, Rao PV (1996) Recombinant antigen-based avidin-biotin microtiter enzyme-linked immunosorbent assay for serodiagnosis of invasive amebiasis. J Clin Microbiol 34(828):833

Srivastava VK, Pandey GK (1986) Parasitic infestation in sewage farm workers. Indian J Parasitol 10:193–194

Suss RA, Maravilla KR, Thompson J (1986) MR imaging of intracranial cysticercosis:comparision with CT and anatomopathologic features. Am J Neuroradiol 7:235–242

Takayanagui OM, Odashima NS (2006) Clinical aspects of neurocysticercosis. Parasitol Int 55:S111–S115

Tantrawatpan C, Maleewong W, Wongkham S, Wongkham C, Intapan PM, Nakashima K (2005) Serodiagnosis of human fascioliasis by a cystatin capture ELISA with recombinant *F.gigantica* cathepsin L antigen. Am J Trop Med 72:82–86

Umezawa ES, Bastos SF, Camargo ME, Yamauchi LM, Santos MR, Gonzalez A et al (1999) Evaluation of recombinant antigens for serodiagnosis of Chaga's disease in south and central America. J Clin Microbiol 37:1554–1560

Virginio VG, Hernandez A, Rott MB, Monteiro KM, Zandonai AF, Nieto A et al (2003) A set of recombinant antigens from *E.granulosus* with potential for use in the immunodiagnosis of human cystic hydatid disease. Clin Exp Immunol 132:309–315

Wadia RS, Makhale CM, Kelkar AV, Grant KB (1987) Focal epilepsy in India with special reference to lesions showing ring or disc like enhancement on contrast computed tomography. J Neurol Neurosurg Psychiatry 50:1298–1301

WHO (2006) The control of neglected zoonotic diseases. A route to poverty alleviation. In: Organization WH (ed) Report of a joint WHO/DFID-AHP meeting with the participation of FAO and OIE 20 and 21 Sept 2005. WHO Headquarters, Geneva, p 54

Zarlenga DS, Rhodas ML, Al-Yaman FM (1994) A *T.crassiceps* cDNA sequence encoding a putative immunodiagnostic antigen for bovine cysticercosis. Mol Biochem Parasitol 67:215–223

Zini D, Farrell VJ, Wadee AA (1999) The relationship of antibody levels to the clinical spectrum of human neurocysticercosis. J Neurol Neurosurg Psychiatry 53:656–661

Human Cryptosporidiosis and Drinking Water: Looking Beyond HIV 14

Ujjala Ghoshal and Asmita Dey

Abstract

Cryptosporidium is an obligate intracellular apicomplexan parasite, responsible for significant morbidity and mortality in both humans and animals. It causes protracted and life-threatening diarrhea in HIV-infected individuals, along with some extra-intestinal symptoms. It also causes infection in other immunocompromised hosts like organ transplant recipients, patients with hematological malignancies, though there are fewer reports on this issue. Cryptosporidial infection is transmitted through feco-oral route by contaminated food and water. Oocyst of *Cryptosporidium* species is highly resistant to common water disinfection procedures, which has led to waterborne Cryptosporidium outbreaks, in both large and small communities. UV radiation, flocculation, sedimentation, and filtration are efficient against water contamination by cryptosporidial oocysts. Diagnosis of the parasite is based mainly on acid-fast staining, fluorescent stains like auramine-carbolfuchsin stain, ELISA of specific antigens, as well as antibodies, and polymerase chain reaction. The principle defense mechanism against *Cryptosporidium* sp. is cell-mediated immunity, particularly macrophage activation by CD4 + T cells–derived cytokines. Th1 response assists in developing resistance to infection, whereas activation of Th2 cells results in increased parasite survival and exacerbation of lesions, due to macrophage suppressive action of Th2 cytokines, or during recovery stages of the infection. Immune restoration is very important for treatment. Nitazoxanide and Paramomycin are the mainstay of therapy, however complete elimination of the parasite is difficult.

U. Ghoshal (✉) · A. Dey
Department of Microbiology, Sanjay Gandhi Postgraduate Institute of Medical Sciences, Rae Bareli Road, Lucknow, 226014, India
e-mail: ujjalaghoshal@yahoo.co.in

P. P. Singh and V. Sharma (eds.), *Water and Health*,
DOI: 10.1007/978-81-322-1029-0_14, © Springer India 2014

Keywords
Cryptosporidium · Diagnosis · Epidemiology · Immunocompromised · Pathogenesis

Introduction

Cryptosporidium species are obligate intracellular apicomplexan parasites that have emerged as an important cause of diarrheal illness in humans and animals. It is widespread among vertebrates causing mainly gastrointestinal disease in mammals, reptiles, and birds (Chalmers and Giles 2010). Infected individuals show a wide spectrum of clinical presentations. Previously, it was thought to be a rare, opportunistic, animal pathogen (Seydel et al. 1998), until the late 1970s that the first case in humans was reported. Globally, Cryptosporidium is responsible for the majority of gastrointestinal parasitic infections, representing a significant cause of morbidity and mortality in its hosts (Tzipori et al. 1980a; Jerrett and Snodgrass 1981). In immunocompetent individuals, it is responsible for self-limiting diarrhea, lasting for only 1–2 weeks (Arora and Arora 2009). In immunocompromised individuals like HIV-infected patients, it causes prolonged, protracted voluminous, non-bloody, and life-threatening diarrhea. Most of the data on cryptosporidiosis is from HIV-positive patients (Tali et al. 2011; Erhabor et al. 2011; Del Chierico et al. 2011; Ojurongbe et al. 2011; Lim et al. 2011; Nel et al. 2011; Le et al. 2008; Lucca et al. 2009; Singla et al. 2010; Kaushik et al. 2008, 2009a; Rao Ajjampur et al. 2007a; Muthusamy et al. 2006; Agarwal et al. 1998a). There are scanty data on patients with other immunocompromised status like renal transplant recipients (Bandin et al. 2009a; Ok et al. 1997) and on patients with hematological malignancies (Nahrevanian and Assmar 2008; Botero et al. 2003; Udgiri et al. 2004). So, data from this subgroup of patients are urgently required. Cryptosporidium infection is transmitted by feco-oral route, mainly through contaminated food and water. Oocyst of Cryptosporidium is the infective stage of the parasite. These oocysts can resist harsh environmental conditions, also surviving through water disinfection procedures, like chlorination (Smith 1989). Thus, water distribution systems and swimming pools are highly vulnerable to contamination with Cryptosporidium, thereby posing a considerable threat to public health. A number of waterborne outbreaks have been associated with cryptosporidiosis, the most important being that of Milwaukee, Wisconsin, in 1993, known as the largest waterborne outbreak in American history (Mac Kenzie et al. 1994).

History

Jackson Clarke, in 1895, noted *Cryptosporidium* sp. in the gastric epithelium of mice (Current and Garcia 1991). However, the discovery of the coccidian parasite is mainly associated with Ernest Edward Tyzzer (1875–1965), who first made his observations on the genus *Cryptosporidium* sp. in 1907, as a cell-associated organism in the gastric mucosa of mice (Vuorio et al. 1991). *Cryptosporidium* sp. was so named because of the absence of sporocysts within the oocysts, a characteristic of other *Coccidia* (Tzipori and Widmer 2008). Slavin, in 1955, was the first to associate cryptosporidiosis with morbidity and mortality. He described severe diarrhea and deaths in turkey poults and attributed the illness to a new species of *Cryptosporidium*, *C. meleagridis* (Slavin 1955). Cryptosporidium (*C. parvum*) aroused interest in veterinarians when this protozoan was first reported to be associated with bovine diarrhea in 1971 (Panciera et al. 1971). Since then, it has been shown to be associated with diarrhea in various animal species, mainly calves (Tzipori et al. 1980b; Jerrett and Snodgrass 1981; Morin et al. 1976), lambs (Angus et al. 1982; Barker and Carbonell 1974), and goat kids (Mason et al. 1981).The first case of human cryptosporidiosis was reported by Nime, in 1976, in a 3-year-old girl from rural

Tennessee who suffered from severe gastroenteritis for 2 weeks (Kaur et al. 2002a). The first case of cryptosporidiosis in a homosexual man with AIDS was reported in 1982 (Ma and Soave 1983a).

Now, it is one of the important causes of diarrhea in AIDS patients and other immunocompromised patients like renal transplant recipients and patients with various hematological malignancies.

Classification and Genetic Characterization

Cryptosporidium belongs to the Phylum: Apicomplexa, Class: Conoidasida, Subclass: Coccidiasina, Family: Cryptosporidiidae. To date, 22 species of Cryptosporidium have been named based on host occurrence and preference, parasite morphology, and site of infection, which are known to cause infection in more than 150 different mammals, reptiles, and birds. However, only 8–10 species are considered valid by some researchers (Levine 1984; Fayer et al. 1997; Sulaiman et al. 2000). They are Cryptosporidium parvum, *C. andersoni*, *C. baileyi*, *C. felis*, *C. meleagridis*, *C. muris*, *C. nasorum*, *C. saurophilum*, *C. serpentis*, and *C. wrairi* (Leav et al. 2003). The two species, Cryptosporidium hominis (previously known as *C. parvum*, genotype 1) and Cryptosporidium parvum (*C. parvum*, genotype 2), are responsible for most cases of cryptosporidiosis in humans (Tzipori and Ward 2002). *C. parvum* is a zoonotic species, whereas *C. hominis* is anthroponotic. Other Cryptosporidium species that have been shown to cause illness in humans include *C. meleagridis*, *C. felis*, *C. canis*, *C. suis*, *C. muris*, and *C. andersoni* (Vuorio et al. 1991; Tzipori and Ward 2002; Molbak et al. 1993; Martins and Guerrent 1995; Current and Reese 1986; Marcial and Madara 1986). *C. hominis*, *C. parvum*, *C. parvum* (mouse/ferret), *C. meleagridis*, *C. felis*, and *C. muris* have been reported from AIDS patients and children in India (Muthusamy et al. 2006; Das et al. 2006; Rao Ajjampur et al. 2007b). However, there are no data on genetic characterization of Cryptosporidium in transplant recipients and those with hematological malignancies and other subgroups of patients.

Epidemiology

In humans, *Cryptosporidium* sp. infection is found in all age groups, particularly in children, less than 5 years of age and in elderly people (Snelling et al. 2007; Chacín-Bonilla 1995). In children, the age group most affected is from 1 to 9 years old (Yoder and Beach 2007; Mathan et al. 1985). Worldwide, the prevalence of *Cryptosporidium* sp. in children varies from 0.3 to 35 % (Sejdini et al. 2011; Ranjbar-Bahadori et al. 2011; Muñoz-Antoli et al. 2011; Al-Mekhlafi et al. 2011; Antonios et al. 2010; Molloy et al. 2011; Davies et al. 2009; Baxby and Hart 1986). In India, the prevalence of *Cryptosporidium* sp. varies from 0 to 9 %, in asymptomatic children, while 1.3–18.9 % (Das et al. 2006; Mathan et al. 1985; Kaur et al. 2002b; Ballal and Shivananda 2002; Shetty et al. 1995; Jindal et al. 1995), in children, with diarrhea. In immunocompetent individuals, cryptosporidiosis has been described in over 26 countries, with a reported prevalence of 0.6–20 % in Western countries and 4–20 % in developing countries (Smith 1990). The elderly population has greater susceptibility to cryptosporidiosis. Prevalence of this infection has been reported to be 18 % in people from 60 to 86 years of age in India (Gambhir et al. 2003), while as high as 46 % in a study, from the USA, on elderly hospitalized patients (Neill et al. 1996).

Persistent diarrhea occurs in up to 75 % of patients with HIV/AIDS in Western populations (Dworkin et al. 1985; Antony et al. 1988; Smith et al. 1988; Rene et al. 1989; Rolston et al. 1989; May et al. 1993) and in up to 100 % of those infected in some developing countries (Manabe et al. 1998) and has a high morbidity and mortality. At least 80 % of cases of chronic diarrhea can be attributed to a specific enteropathogen of which the coccidial parasites are most important. *Cryptosporidium* sp. infection in AIDS patients is 10–20 % in the USA and Western Europe, while as high as 50 % in developing countries (Manabe et al. 1998). Regarding cryptosporidiosis, in India, there have been reports from the mid-1990s on the prevalence of symptomatic disease in HIV-infected adults from different parts of the country, ranging from as low as 0.7 % to as high as 81 % (Rao Ajjampur et al. 2008b). A high prevalence of ~80 % was reported from a study in Imphal (Agarwal et al. 1998b) and another in Maharashtra (Ghorpade et al. 1996) but both had very small sample sizes. At our center the prevalence of Cryptosporidium sp. is 12.7% in HIV infected patients (unpublished report).

Patients with some type of immunocompromised condition and those submitted to immunosuppressive therapy, like transplant recipients, have an increased chance of acquiring opportunistic parasitic infections. Transplant recipients can be at increased risk of parasitic infection. There may be 'Recrudescence' of latent infection, already present in a recipient, occurrence of a *de novo* infection by means of natural infection, or transmission of the parasite, into a näıve recipient, through the transplanted organ (or blood product, either before or after transplantation) (Kottona and Lattes 2009). In renal transplant recipients, prevalence of *Cryptosporidium* sp. varies from 3.5 to 34.8 % (Ok et al. 1997; Bandin et al. 2009b; Chieffi et al. 1998), while, in a study from North India, it is reported to be 20 % (Udgiri et al. 2004). The prevalence of *Cryptosporidium* sp. in renal transplant recipients is 7.3% (unpublished report).

Parasitic infections are controlled in the human body by both Th1 and Th2 cellular responses. Patients with various hematological malignancies have qualitative and/or quantitative alterations in their cellular or humoral responses, which hinder their efficient action against parasites they harbor. Thus, they are also at increased risk of acquiring opportunistic parasitic infections (Botero et al. 2003). Prevalence of cryptosporidiosis in patients with hematological malignancies has been reported to be 1.4–3.6 % (Nahrevanian and Assmar 2008; Botero et al. 2003). In a study from Vellore, cryptosporidiosis has been reported in 2.9 % of adult and 1.7 % of pediatric allogeneic BMT recipients (George et al. 2004, 2006). The prevalence of *Cryptosporidium* sp. is 3% in patients with hematological malignancies, at our center (unpublished report).

Patients with other types of immunosuppression, such as congenital hypogammaglobulinemia, protein-calorie malnutrition, diabetes mellitus, and hematological neoplasias, also show a high predisposition to developing severe forms of cryptosporidiosis (Moura et al. 1989; Costa-Cruz and Ferreira 1996; Pedersen et al. 1996; Sorvillo et al. 1998).

Transmission

Cryptosporidial infection can be transmitted through feco-oral route by contaminated food and water. The majority of the documented outbreaks of waterborne infection in the world have been attributed to contaminated drinking water supplies, although contaminated water used for recreational activities has also been shown to be responsible in transmission of the disease (Leav et al. 2003). Foodborne and person-to-person spread have also been documented. Cryptosporidiosis has been attributed to ingestion of contaminated apple cider, chicken salad, milk, and food prepared by an infected food handler (Leav et al. 2003). *Cryptosporidium* sp. has also been detected in seawater and has even been found in commercially harvested oysters (Leav et al. 2003). Another potential source of infection may be raw vegetables sold in the marketplaces in developing countries (Leav et al. 2003). The probability of transmission from just a small amount of contamination is fairly high, since a recent study has determined that the 50 % infective dose (ID50) of *C. parvum* is only 132 oocysts for healthy persons with no previous serological immunity to cryptosporidiosis (DuPont et al. 1995). Hence, Cryptosporidium has got a low minimum infectious dose (MID) values and is highly resilient to both host immune defense (sporozoite forms) and environmental conditions (oocyst forms), as oocysts are resistant to the commonly known disinfectants like chlorine (Fig. 14.1).

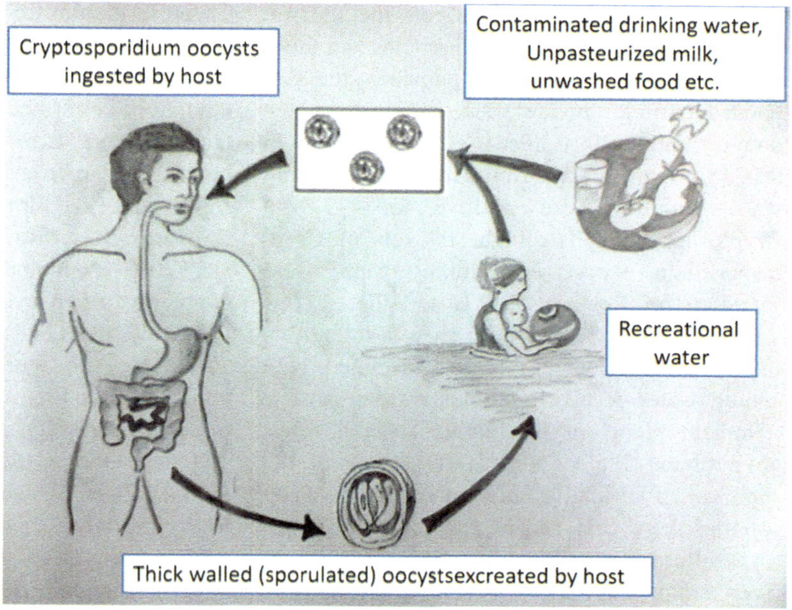

Fig. 14.1 Transmission of *Crytopsporidium* sp

Cryptosporidium and Water Sources

Cryptosporidium sp. has been found to contaminate surface water more than ground water, because surface water is more susceptible to direct contamination from sewage discharges and runoff; 5.6 and 87.1 % of source waters (i.e., surface, spring, and groundwater samples not impacted by domestic and/or agricultural waste) were tested and found to be contaminated with 0.003–4.74 *Cryptosporidium* sp. oocysts/L (Lisle and Rose 1995). In another study, 60.2 % of surface waters tested in the USA and Canada were found to contain Cryptosporidium oocysts (LeChevallier and Norton 1995). Cryptosporidium oocysts were also found to contaminate 50 % of raw sewage samples (Bukhari et al. 1997; Zuckerman et al. 1997), 4.5 % of raw water samples, and 3.5 % of treated water samples (Wallis et al. 1996). *Cryptosporidium* sp. oocysts are found less frequently in ground water than in surface waters. In a study done on 199 ground water samples, 5 % of vertical wells, 20 % of springs, 50 % of infiltration galleries, and 45 % of horizontal wells were tested and found to be contaminated with *Cryptosporidium* sp. oocysts (Hancock et al. 1998). A 10-year study of water supplies in Seoul, South Korea, showed 0.04–1.90 Cryptosporidium oocysts/10 L, and the pollution level was a little higher in winter (Lee et al. 2010). *Cryptosporidium* sp. oocysts were also detected in river water samples from Iran (Mahmoudi et al. 2011). There are scanty data from India. A study on 199 drinking water samples collected from ten zones of Chennai, Southern India, revealed that three zones of the city were highly contaminated with coliforms and parasitic protozoa. The contamination of water samples with *Cryptosporidium* sp. oocysts may present a public health hazard (Anbazhagi et al. 2007). A study of 47 samples collected from a complicated river network system, which is a source of drinking water in Tongxiang, China, showed 37 (78.7 %) to be *Cryptosporidium* sp. positive, with a mean concentration of 0.51 oocysts per liter (Xiao et al. 2012). In a study from Yaoundé, Cameroon, done on waste and surface water, the highest concentrations of *Cryptosporidium* sp. were 1,110 oocysts/L in the stream/surface water and 1,500 oocysts/L in the residual effluent. They found that turbidity had a significant effect on the population dynamics of Cryptosporidium.

Due to the interactions between the electric charges carried by suspended particles and those present at the surface of the parasites, the suspended particles in the wastewater adsorb the oocysts and cysts (Gideon et al. 2007); 6/45 drinking water samples from Hungarian drinking water supplies were positive for (13.3 %) Cryptosporidium. The highest levels of Cryptosporidium oocysts in drinking water were found to be 3 oocysts/100 L and that in raw water were 50 oocysts/100 L. It was found that higher oocyst densities were associated with source water receiving effluents from sewage treatment plants or originating from a forest environment (Plutzer et al. 2007). A total of 284 samples of drinking and recreational water supplies were analyzed in Alava, Spain. Cryptosporidium oocysts were found in 63.5 % of river samples, 33.3 % of reservoirs samples, 15.4 and 22.6 % of raw water samples from conventional and small water treatment facilities (respectively), 30.8 % of treated water from small treatment facilities, and 26.8 % of tap water from municipalities with chlorination treatment only. The samples were contaminated with Cryptosporidium more frequently during the autumn season than during the spring and winter seasons (Carmena et al. 2007).

Cryptosporidium Outbreaks

Oocyst of Cryptosporidium is highly resistant to common water disinfection procedures, which has led to waterborne Cryptosporidium outbreaks, in both large and small communities. In 1989, there was an outbreak at Swindon and Oxfordshire, affecting some 5,000 people (Richardson et al. 1991). In 1991, a cross-connection to a contaminated creek, in Sweden, led to contamination of the community water supply, resulting in 600 infections including cryptosporidiosis (Thulin 1991). The largest Cryptosporidium-associated waterborne outbreak was that of Milwaukee, Wisconsin in 1993 (Mac Kenzie et al. 1994). Another in Dracy Le Fort, Burgundy, France, in 2001, led to severe gastroenteritis in 563 of the 1,100 inhabitants. Oocysts of Cryptosporidium were identified in the tap water supply, used for consumption (Dalle et al. 2003). In Northern Ireland, between 2000 and 2001, 347 laboratory-confirmed cases were linked to contamination of the drinking water supply, by human sewage, from a septic tank and wastewater from a blocked drain, which seeped into the drinking water distribution system (Glaberman et al. 2002). A number of cases of cryptosporidial contamination of river, spring, and ground water due to treatment deficiencies have been reported; 12,960 cases in Georgia, due to river water contamination, in 1987; 551 cases in Pennsylvania, due to ground water contamination, in 1991; and 15,000 in Oregon, due to spring/river water contamination, in 1992, have been reported all due to treatment deficiencies (USEPA 2001). Cryptosporidium was also found to be the causative agent of the outbreak in Nablus, Palestine, which took place in October 2008, in which several people were admitted to hospital showing symptoms such as diarrhea, strong abdominal pain, and periodic vomiting (Hussein 2011).

Life Cycle

Cryptosporidium sp. is taxonomically classified as a Sporozoa, since its oocyst releases four sporozoites (its motile infectious agents) upon excystation. However, it differs from related parasites such as *Toxoplasma* by its monoxenous life cycle—completing its entire cycle within a single host (Flanigan and Soave 1993). The life cycle is complex; there are both sexual and asexual cycles; and there are six distinct developmental stages (Keusch et al. 1995) (Fig. 14.2).

- Excystation of the orally ingested oocyst in the small bowel with release of the four sporozoites.
- Invasion of intestinal epithelial cells via the differentiated apical end of the sporozoite within the parasitophorous vacuole formed of both host and parasite membranes and the initiation of the asexual intracellular multiplication stage. The developed trophozoites undergo asexual proliferation by merogony and form: Type I meronts (form eight

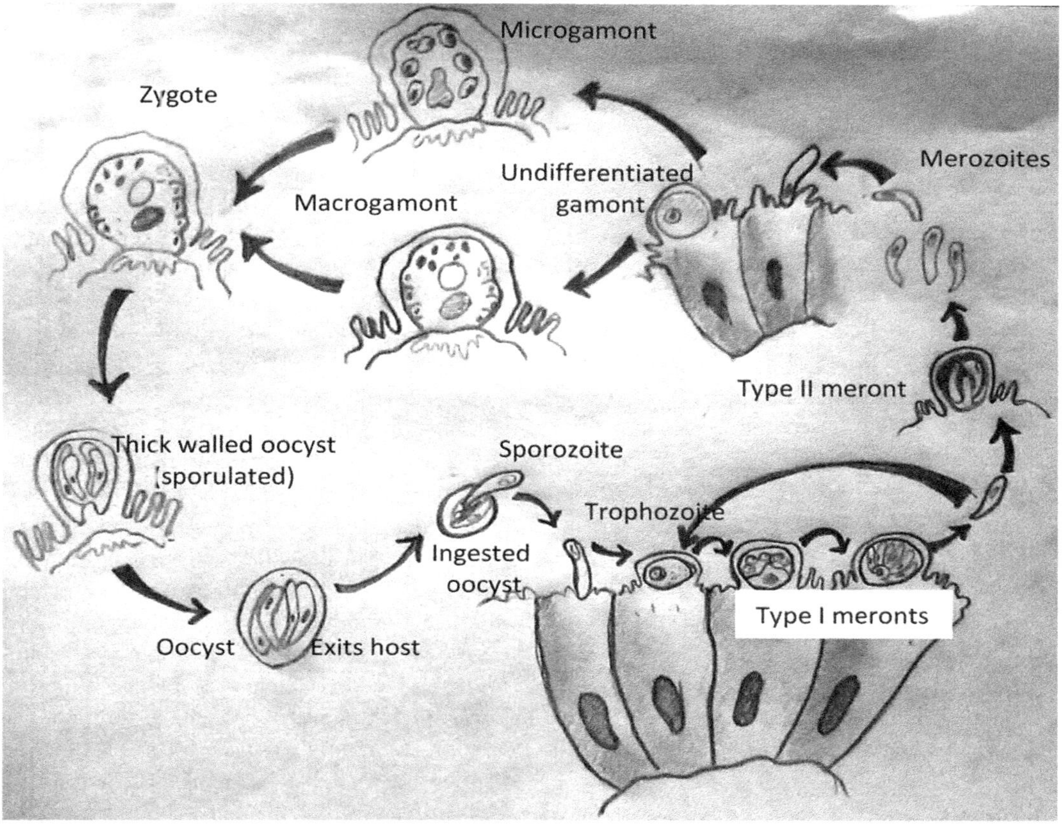

Fig. 14.2 Life cycle of *Cryptosporidium* sp

merozoites that are liberated from the parasitophorous vacuole when mature; the merozoites then invade other epithelial cells where they undergo another cycle of type I merogony or develop into type II meronts) and type II meronts (form four merozoites which do not undergo further merogony but produce sexual reproductive stages called gamonts).

- Sexual reproduction by gametogony: Differentiation of microgametes (male gametes) and macrogametes (female gametes).
- Fertilization of gametes, producing zygotes, which undergo sporogony.
- Development of oocysts, containing four sporozoites (thick-walled oocysts, excreted in feces; thin-walled oocysts, excysting within the hosts causing autoinfection).
- Excretion of the oocysts in the stool of hosts.

Mechanism of *Cryptosporidium* Sporozoite Infecting Host Enterocytes

- Specific ligands on the sporozoite surface attach to receptors on the host cell, inducing host cell membrane protrusion that encapsulates the parasite.
- Galactose-N-acetylgalactosamine Gal/GalNAc epitopes of sporozoite: gp900, gp40/15, and circumsporozoite surface ligand (CSL), thrombospondin-1 (TSP-1) domains like thrombospondin-related adhesive protein (TRAP C1) on the parasite surface recognize specific receptors on the host enterocytes.
- First, the contents of Cryptosporidium micronemes are arranged in an array of 15-nm cubic crystals arranged in a pine-cone-like pattern.

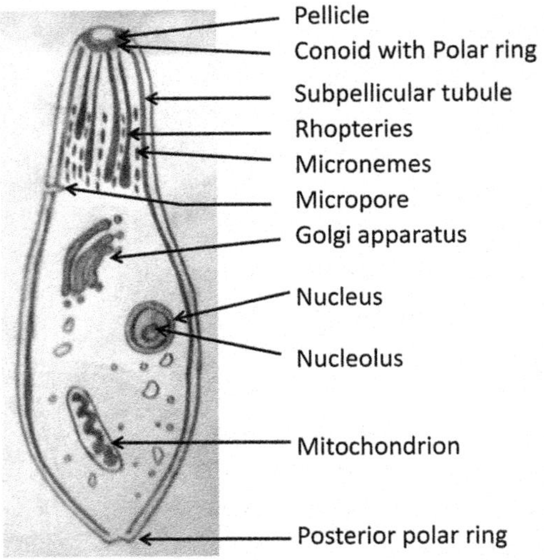

Fig. 14.3 Structure of a typical apicomplexan sporozoite

- A number of micronemes are discharged in rapid succession, and a strong measure of organization is presumably necessary to achieve efficient deployment.
- The internal dimensions of micronemes (~75 × 150 nm) cannot accommodate some of the larger microneme proteins in their fully extended state (see also below), and therefore, these proteins must be packaged in orderly fashion so that they are primed for secretion onto the parasite surface.
- The parasite is retained on the apical surface of the cell, and a unique 'electron-dense band' is established that separates the organism from the host cell cytoplasm and may facilitate the uptake of nutrients from the host through an ABC transporter system (Fig. 14.3).

Pathogenesis

Pathogenic mechanism of this parasite is poorly recognized. Cryptosporidium sp. invades and resides in the epithelial cells, most commonly in the small intestine, without usually invading deeper mucosal layers, thus can be viewed as 'minimally invasive' pathogen, residing in an intracellular, but extracytoplasmic, parasitophorous vacuole. However, in the immunodeficient host, infection may be widespread, involving epithelial cells of the biliary tract, pancreatic duct, stomach, esophagus, and respiratory tract (Hunter and Nichols 2002a). Three types of mechanisms have been proposed regarding the pathophysiology of diarrhea induced by Cryptosporidium. One is parasite induced disruption of the intestinal epithelium, causing villous atrophy, crypt hyperplasia leading to malabsorption and osmotic diarrhea. Second is release of inflammatory and hormonal metabolites, as host immune response to the infection which also induces intestinal secretion, associated with malabsorption. The enterotoxic moiety produced by the parasite, also responsible for secretory diarrhea (Fig. 14.4).

Clinical Features

In immunocompetent individuals, Cryptosporidium causes self-limiting diarrhea lasting for only 1–2 weeks (Erhabor et al. 2011). In immunocompromised individuals, it causes prolonged, protracted voluminous, non-bloody, and life-threatening diarrhea (Hunter and Nichols 2002a; Davies et al. 2007; Casemore et al. 1985a; Verdon

Fig. 14.4 Host cell–parasite interaction

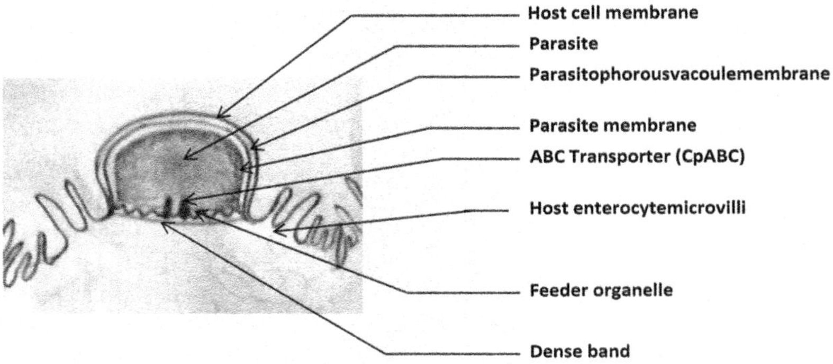

et al. 1998). In immunocompromised patients, it causes prolonged and severe diarrhea. Diarrhea of upto 25 L/day have been reported. Infected patients suffer from dehydration, shock, severe abdominal cramps especially in the upper right quadrant, malaise, low-grade fever, weight loss, and anorexia (Casemore et al. 1985). In developing countries, Cryptosporidium also affects malnourished children, resulting in long-term negative effects on growth, weight gain, physical and cognitive development. Extraintestinal infections, including parasite development on respiratory and biliary epithelial surfaces, have been described in immunosuppressed patients (Current and Garcia 1991). There are numerous reports on respiratory Cryptosporidiosis (Brady et al. 1984; Egger et al. 1990; Forgacs et al. 1983; Goodstein et al. 1989; Hojlyng and Jensen 1988); 40–50 % of healthy children experience respiratory symptoms during intestinal cryptosporidiosis (Egger et al. 1990; Weikel et al. 1985; Shahid et al. 1987). However, in AIDS patients, co-infection with organisms such as cytomegalovirus and P. jiroveci can occur, obscuring the specific contribution of cryptosporidiosis to respiratory disease. Autopsy reports and prospective studies have showed an etiologic role for the organism in biliary syndromes like sclerosing cholangitis and acalculous cholecystitis (Teare et al. 1997; French et al. 1995; Lopez-Velez et al. 1995). Infection of the pancreas has also been reported in cryptosporidiosis, indicated by right upper quadrant pain and elevated serum lipase and amylase levels (Calzetti et al. 1997; Godwin 1991). Chronic infection occurs in patients with CD4 counts below 200 cells/cumm (Shetty et al. 1995). Severe infection has been shown to occur in patients with CD4 count below 50 cells/cumm (Hunter and Nichols 2002b).

Immune Response

The principle defense mechanism against protozoan parasites is cell-mediated immunity, particularly macrophage activation by CD 4 + T cells–derived cytokines. Th1 response assists in developing resistance to infection, whereas activation of Th2 cells results in increased parasite survival and exacerbation of lesions, due to macrophage suppressive action of Th2 cytokines, or during recovery stages of the infection. Parasitic invasion activates expression of a range of pro-inflammatory cytokines and chemokines including IFN-γ, TNF-α, IL-2, TGF-beta. IFN-γ expression has been shown to be associated with resistance to infection as judged by prevention of oocyst shedding (Hu et al. 2010). Role and expression of IL-6 has been studied in only murine models but not in humans (Singh et al. 2005; Lacroix et al. 2001; Maillot et al. 2000). IL-8 expression has been shown to be induced by *C. parvum* infection in cell lines (Alcantara et al. 2003) and fecal samples (Kaushik et al. 2009b). Elevated levels of TNF-α have been detected in mice models and enterocyte cell lines (Lacroix et al. 2001; Lean et al. 2006). However, its mRNA level was low in IFN-γ of knockout mice, and injections of TNF-α adult mice significantly reduced oocyst shedding (Lacroix et al. 2001). Thus, TNF-α may participate in the control of parasite development. It has been shown that interferon (IFN)-gamma is important for early control of the infection and acts directly on enterocytes to inhibit parasite development (Choudhry et al. 2009). IFN-gamma has been shown to inhibit *C. parvum* infection in cell lines (Pollok et al. 2001). Elevated levels of the regulatory cytokine IL-10 have been found in Cryptosporidium-infected mice, calves as well as in humans (in children) (Singh et al. 2005; Wyatt et al. 2002; Kirkpatrick et al. 2002). Most of the studies done so far based on cytokine and chemokine expression have been on cultured epithelial cell lines, or animal models. ICAM-1 is intercellular adhesion molecule-1, a member of the immunoglobulin superfamily, which has been shown to be expressed in human intestinal epithelium in association with active inflammatory response. Up-regulation of ICAM-1, at both message and protein levels in epithelial cells following *C. parvum* infection, has been detected (Parr et al. 2007).

Increase in the levels of IL-2, IL-4, IL-10, IFN-γ, and TNF-α has been reported in Cryptosporidium-infected AIDS patients (Udgiri et al. 2004). There has been no study on systemic cytokine and chemokine profile of Cryptosporidium-infected transplant recipients or in patients with hematological malignancies.

Chemokines are chemotactic cytokines, family of small proteins which stimulate leukocyte motility and directed movement (chemotaxis). Levels of chemokine CXCL10 and cytokine IL-1α concentrations were significantly increased in jejunal biopsies of AIDS patients with cryptosporidiosis, compared to that in normal volunteers without cryptosporidiosis (Wang et al. 2007).

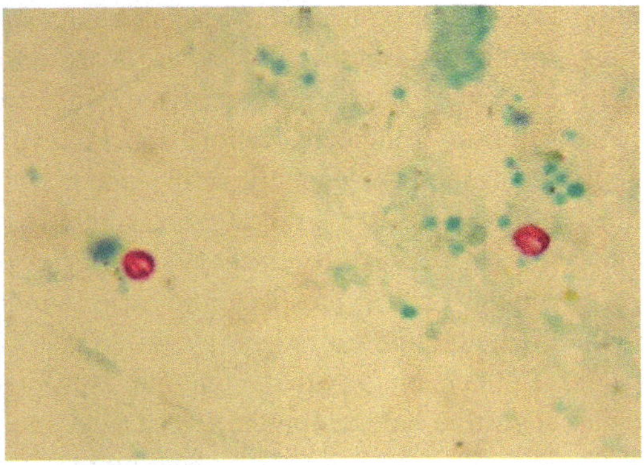

Fig. 14.5 Cryptosporidium sp. oocyst stained by modified (cold) Kinyoun's stain, counterstained by malachite green

Diagnosis

Diagnosis of Cryptosporidium was done previously by histological analysis of biopsy specimen from small intestine and occasionally from rectum, and by observing the different life stages of the parasite at the microvillous region, under both light and electron microscopes (Current and Garcia 1991; Fayer and Ungar 1986; Garcia 2001). However, biopsy specimens may miss the region of intestinal mucosa actually infected by the parasite. So, the recent tools for diagnosis are based on observation of the parasite oocysts from fecal matter and other body fluids, usually by special staining techniques (Ma and Soave 1983b). Specimens, such as stool sample, are most commonly used. Sputum or bronchoalveolar lavage samples or bile may also be used. Collection of multiple stool samples (minimum three specimens) has been recommended, as the number of oocyst in the stool may fluctuate, particularly in case of formed stools, containing fewer oocysts than diarrheic specimens (Current and Garcia 1991). Stool samples can be concentrated by sedimentation with formalin ether (modified Ritchie) or formalin ethyl acetate, or Sheather's sugar floatation techniques, in zinc sulfate (33 % to saturated), or in sodium chloride (36 % to saturated), and then observed, by coverslip preparation on normal saline or iodine. It is very difficult to diagnose Cryptosporidium by routine microscopy. Modified acid-fast staining is the gold standard for the diagnosis of Cryptosporidium (Garcia 2001; Ma and Soave 1983b), in which pink oocysts are seen against the blue/green background. It is important to know that malachite green produces cleaner background and gives better contrast. Safranin-methylene blue (malachite green) compares well with modified acid-fast techniques and stains oocysts red and yeasts and other fecal debris blue (Baxby et al. 1984). A lactophenol cotton blue stain can also be used for the staining of Cryptosporidium oocysts, in direct wet mounts of stool, but it is more useful for the detection of other Coccidian parasites, such as Cyclospora and Isospora oocysts (Parija et al. 2003) (Fig. 14.5).

Fluorescent stains, like auramine-rhodamine and auramine-carbolfuchsin, have been used (Casemore et al. 1985b). Commercial ELISA-based assays for the detection of Cryptosporidium antigens in stool (Kashyap et al. 2010; Tuli et al. 2010; Uppal et al. 2009; Ulçay et al. 2008; Nair et al. 2008; Nagamani et al. 2007; Jayalakshmi et al. 2008), having sensitivity, ranging between 90 and 97 %, and specificity, ranging between 96 and 99 %. At our center, sensitivity of Cryptosporidium ELISA (IVD research, USA) is 84 % and specificity is 96 % (unpublished report). Direct fluorescent antibody (DFA) (Se: 93 %, Sp: 98 %)/immunofluorescent assay (IFA) (Se: 78 %, Sp: 95 %) techniques offer high combination of sensitivity and specificity and is considered the gold-standard by many laboratories by many laboratories (Coutinho et al. 2008; Geurden et al. 2006). Rapid assay techniques like 'ImmunoCard STAT!' (Moyo et al. 2011), 'RIDA (R) Quick Cryptosporidium Strip Test' (Çiçek et al. 2008) have also been used for detection of Cryptosporidium. PCR assay of the small subunit rRNA (18 s rRNA) gene of Cryptosporidium is a highly specific and sensitive technique being used currently for parasite detection and identification (Spano et al. 1997).

Genetic characterization of Cryptosporidium has been done using PCR–RFLP assays of TRAP-C1, COWP, Cpgp40/15, β-tubulin, and HSP70 genes, of the parasite (Hunter and Nichols 2002b; Spano et al. 1997; Coupe et al. 2005; Spano et al. 1998; Akiyoshi et al. 2002; Cohen et al. 2006; Assefa et al. 2009), and real-time PCR assays. Closed-tube, real-time PCR methods, circumvent false-positive results due to cross-contamination, during further processing of the amplified PCR product in PCR–RFLP assays, also obviating the need for time-consuming techniques, like gel electrophoresis (Soliman and Othman 2009; Rasmussen et al. 2007).

The identification of Cryptosporidium oocysts in environmental samples is currently performed by using the US Environmental Protection Agency (USEPA) Method 1623 (US Environmental Protection Agency 2005) and similar testing methods from other countries (Naeini et al. 2010; Weintraub 2006; Yang et al. 2008). USEPA Method 1623 requires concentration of oocysts by filtration, using filters like IDEXX Filta-Max® foam filter, Pall Gelman Envirochek™ HV filter; elution, purification, and isolation of oocysts by immunomagnetic separation (IMS); staining of recovered oocysts with fluorescent antibodies (Waterborne Aqua-Glo™ G/C Direct FL antibody stain, Waterborne Crypt-a-Glo™, and Giardi-a-Glo™ antibody stains, BTF EasyStain™ antibody stain), and fluorescein isothiocyanate (FITC)/4, 6-diamidino-2-phenylindole dihydrochloride (DAPI); and detection and enumeration of the stained oocysts or cysts by differential interference contrast (DIC) microscopy.

Treatment

Severity of Cryptosporidium infection depends on the immune status of the host. Immunocompetent individuals usually recover spontaneously, but supportive therapy, including fluid and electrolyte replacement, should be provided (Pantenburg et al. 2009). Antiparasitic drugs, like nitazoxanide (a nitrothiazole benzamide) and paromomycin (an oligosaccharide aminoglycoside related to kanamycin), are the mainstay drugs for treatment and currently used, as the first-line treatment, for Cryptosporidiosis (Garcia 2001; Pantenburg et al. 2009). Nitazoxanide is US FDA approved for the treatment for Cryptosporidiosis and Giardiasis in immunocompetent individuals. However, treatment with nitazoxanide can be considered in immunocompromised patients due to the seriousness of cryptosporidiosis (Abubakar et al. 2007). Adult dosage in immunocompetent individuals is 500 mg, twice daily, for 3 days, while its dosage in children of the age group 1–3 years is 100 mg, twice daily, for 3 days, of 4–11 years is 200 mg, twice daily, for 3 days, and it is 500 mg, quarterly, for 3 days in children >12 years of age. In immunocompromised patients, treatment for longer durations may be required. Treatment with paromomycin has shown significantly decreased parasite shedding in stool and improved diarrheal symptoms (Pantenburg et al. 2009). Dosage is 500 mg, thrice daily, for 7–14 days, often followed by a maintenance therapy with 500 mg twice daily. Other drugs which have showed a positive clinical response, as well as parasite eradication, are spiramycin and azithromycin (Garcia 2001). According to various case reports, there has been successful elimination of symptoms and parasites in AIDS and bone marrow transplant recipients, after prolonged oral azithromycin and paromomycin combination therapy (Meamar et al. 2006; Palmieri et al. 2005; Nachbaur et al. 1997).

Water Treatment

Exposure to 1.05 and 3 % chlorine, as sodium hypochlorite for up to 18 h even, could not affect its viability. Long-term exposure to 10 % formalin, 5–10 % ammonia, and 70–100 % bleach was declared necessary to completely eliminate parasite infectivity (Campbell et al. 1982; Reduker and Speer 1985). It can also be inactivated by thirty minutes of exposure to temperatures above 65 °C or below freezing (Tzipori 1983). It has been reported in mice inoculated intragastrically with 1,000 treated oocysts that treating drinking water with 1.11 ppm of ozone (1.11 mg/L) completely eliminates parasite infectivity in 5 min and 15–30 min of exposure to 0.43 ppm of chlorine dioxide significantly reduces oocyst production (Peeters et al. 1989). UV radiation is also very effective to inactivate parasite oocysts. Flocculation (a process by which fine particles are made to clump together) of the negatively charged oocysts, by aluminum sulfate, iron (II) sulfate, or iron (III) chloride, and then sedimentation and proper filtration, can remove Cryptosporidium oocysts. Membrane filtration further improves the quality of drinking water (Semenza and Nichols 2007). There have been amendments and continuous monitoring of the water treatment regulations in the USA and UK, to reduce occurrence of oocysts in drinking and recreational water supply (Tzipori and Widmer 2008b).

Prevention

Hands should be washed properly with soap and water, before preparing or eating food, after using the toilet, before and after tending to someone who is suffering from diarrhea, and after handling an animal or its feces. Care should be taken to avoid swallowing recreational water, or untreated water of lakes, ponds, or other water bodies. Foods like raw fruits and vegetables washed with untreated, contaminated water, uncooked or undercooked food, raw meat or unpasteurized milk should not be taken. Better hygiene and sex practices should be followed, and contact and contamination with feces during sex should be prevented (CDC 2010).

Acknowledgments We sincerely acknowledge Department of Biotechnology, India, for its financial aid (Project: 102/IFD/SAN 1839/2008-09), and Asmita Dey would like to acknowledge UGC, India, for fellowship regarding PhD program. We would also like to thank all the members of our laboratory, especially our Technical officer Mr. Dinesh Gangwar and laboratory attendant Raghubeer for their consistent hard work and support.

References

Abubakar I, Aliyu SH, Arumugam C, Hunter PR, Usman NK (2007) Prevention and treatment of cryptosporidiosis in immunocompromised patients. Cochrane Database Syst Rev 1:CD004932

Agarwal A, Ningthouja S, Sharma D, Mohen Y, Singh NB (1998a) Cryptosporidium and HIV. Indian Med Assoc 96(9):276–277

Agarwal A, Ningthouja S, Sharma D, Mohen Y, Singh NB (1998b) Cryptosporidium an HIV. J Indian Med Assoc 96:276–277

Akiyoshi DE, Feng X, Buckholt MA, Widmer G, Tzipori S (2002) Genetic analysis of a Cryptosporidium parvum human genotype 1 isolate passaged through different host species. Infect Immun 70(10): 5670–5675

Alcantara CS, Yang CH, Steiner TS, Barrett LJ, Lima AA, Chappell CL, Okhuysen PC, White AC Jr, Guerrant RL (2003) Interleukin-8, tumor necrosis factor-alpha, and lactoferrin in immunocompetent hosts with experimental and Brazilian children with acquired cryptosporidiosis. Am J Trop Med Hyg 68(3):325–328

Al-Mekhlafi HM, Mahdy MA, 'Azlin MY, Fatmah MS, Norhayati M (2011) Childhood Cryptosporidium infection among aboriginal communities in peninsular Malaysia. Ann Trop Med Parasitol 105(2): 135–143

Anbazhagi M, Loganathan D, Tamilselvan S, Jayabalou R, Kamatchiammal SK, Kumar R (2007) Cryptosporidium Oocysts in drinking water supply of Chennai City, Southern India. Clean—Soil, Air. Water 35(2):167–171

Angus KW, Tzipori S, Gray EW (1982) Intestinal lesions in specific-pathogen-free lambs associated with a cryptosporidium from calves with diarrhea. Vet Pathol 19:67–78

Antonios SN, Tolba OA, Othman AA, Saad MA (2010) A preliminary study on the prevalence of parasitic infections in immunocompromised children. J Egypt Soc Parasitol 40(3):617–630

Antony MA, Brandt LJ, Klein RS, Bernstein LS (1988) Infectious diarrhoea in patients with AIDS. Dig Dis Sci 33:1141–1146

Arora DR, Arora B (2009) AIDS-associated parasitic diarrhea. Indian J Med Microbiol 27(3):185–190
Assefa S, Erko B, Medhin G, Assefa Z, Shimelis T (2009) Intestinal parasitic infections in relation to HIV/AIDS status, diarrhea and CD4 T-cell count. BMC Infect Dis 9:155
Ballal M, Shivananda PG (2002) Rotavirus and enteric pathogens in infantile diarrhoea in Manipal, south India. Indian J Pediatr 69:393–396
Bandin F, Kwon T, Linas MD, Guigonis V, Valentin A, Cassaing S, Carol A, Garnier A, Baudouin V, Decramer S (2009a) Cryptosporidiosis in paediatric renal transplantation. Pediatric Nephrology 24(11):2245–2255
Bandin F, Kwon T, Linas MD, Guigonis V, Valentin A, Cassaing S, Carol A, Garnier A, Baudouin V, Decramer S (2009b) Cryptosporidiosis in paediatric renal transplantation. Pediatr Nephrol 24(11): 2245–2255
Barker IK, Carbonell PL (1974) Cryptosporidium agni sp. n. from lambs, and Cryptosporidium bovis sp. from a calf, with observations on the oocysts. Z Parasitenkd 44:289–298
Baxby D, Hart CA (1986) The incidence of cryptosporidiosis: a two-year prospective survey in a children's hospital. J Hygiene (London) 96(1):107–111
Baxby D, Blundell N, Hart CA (1984) The development and performance of a simple, sensitive method for the detection of Cryptosporidium oocysts in feces. J Hygiene 92:317–323
Botero JH, Castaño A, Montoya MN, Ocampo NE, Hurtado MI, Lopera MM (2003) A preliminary study of the prevalence of intestinal parasites in immunocompromised patients with and without gastrointestinal manifestations. Revista do Instituto de Medicina Tropical de São Paulo 45(4):197–200
Brady E, Margolis ML, Korzeniowski M (1984) Pulmonary cryptosporidiosis in acquired immune deficiency syndrome. J Am Med Assoc 252:89–90
Bukhari Z, Smith HV, Sykes N, Humphreys SW, Paton CA, Girdwood RWA, Fricker CR (1997) Occurrence of Cryptosporidium spp. oocysts and Giardia spp. cysts in sewage influents and effluents from treatment plants in England. Water Sci Technol 35:385–390
Calzetti C, Magnani G, Confalonieri D, Capelli A, Moneta S, Scognamiglio P et al (1997) Pancreatite da Cryptosporidium parvum in pazienti con grave deficit immunitario correlato ad infezione da HIV. Annali Italiani di Medicina Interna 12:63–66
Campbell IS, Tzipori S, Hutchinson G, Angus KW (1982) Effect of disinfectants on survival of Cryptosporidium oocysts. Vet Rec 11:414–415
Carmena D, Aguinagalde X, Zigorraga C, Fernández-Crespo JC, Ocio JA (2007) Presence of Giardia cysts and Cryptosporidium oocysts in drinking water supplies in northern Spain. J Appl Microbiol 102(3):619–629
Casemore DP, Sands RL, Curry A (1985a) Review article, Cryptosporidium species a new human pathogen. J Clin Pathol 38:1321–1336
Casemore DP, Armstrong M, Sands RL (1985b) Laboratory diagnosis of cryptosporidiosis. J Clin Pathol 38:1337–1341
CDC (2010) Morbidity and mortality weekly report: surveillance summaries. 59(SS-6):11
Chacín-Bonilla L (1995) Cryptosporidiosis in humans review. J Clin Inv 36(4):207–250
Chalmers RM, Giles M (2010) Zoonotic cryptosporidiosis in the UK—challenges for control. J Appl Microbiol 109:1487–1497
Chieffi PP, Sens YA, Paschoalotti MA, Miorin LA, Silva HG, Jabur P (1998) Infection by Cryptosporidium parvum in renal patients submitted to renal transplant or hemodialysis. J Brazilian Soc Trop Med 31(4):333–337
Choudhry N, Korbel DS, Edwards LA, Bajaj-Elliott M, McDonald V (2009) Dysregulation of interferon-gamma-mediated signalling pathway in intestinal epithelial cells by Cryptosporidium parvum infection. Cell Microbiol 11(9):1354–1364
Ciçek M, Körkoca H, Gül A (2008) Investigation of Cryptosporidium sp. in workers of the Van municipality slaughterhouse and in slaughtered animals. Turkiye Parazitol Derg 32(1):8–11
Cohen S, Dalle F, Gallay A, Palma MD, Bonnin A, Ward HD (2006) Identification of Cpgp40/15 Type Ib as the Predominant Allele in isolates of Cryptosporidium spp. from a waterborne outbreak of gastroenteritis in South Burgundy, France. J Clin Microbiol 44(2):589–591
Costa-Cruz JM, Ferreira MS (1996) Intestinal parasites in AIDS and +HIV patients in Uberlândia, Minas Gerais. Brazil Mem Inst Oswaldo Cruz 91:685–686
Coupe S, Sarfati C, Hamane S, Derouin F (2005) Detection of Cryptosporidium and identification to the species level by nested PCR and restriction fragment length polymorphism. J Clin Microbiol 43(3):1017–1023
Coutinho BP, Oriá RB, Vieira CM, Sevilleja JE, Warren CA, Maciel JG et al (2008) Cryptosporidium infection causes undernutrition and, conversely, weanling undernutrition intensifies infection. J Parasitol 94(6):1225–1232
Current William L, Garcia Lynne S (1991) Cryptosporidiosis. Clin Microbiol Rev 4(3):325–358
Current WL, Reese NC (1986) A comparison of endogenous development of three isolates of Cryptosporidium in suckling mice. J Protozoology 33:98–108
Dalle F, Roz P, Dautin G, Di Palma M, Kohli E, Sire-Bidault C et al (2003) Molecular characterization of isolates of waterborne Cryptosporidium spp. Collected during an outbreak of gastroenteritis in South Burgundy, France. J Clin Microbiol 41:2690–2693
Das P, Roy SS, Mitradhar K, Dutta P, Bhattacharya MK, Sen A et al (2006) Molecular characterization of Cryptosporidium spp. from Children in Kolkata, India. J Clin Microbiol 44(11):4246–4249
Davies NM, Grinyo J, Heading R, Maes B, Meier-Kriesche HU, Oellerich M (2007) Gastrointestinal side effects of mycophenolic acid in renal transplant

patients: a reappraisal. Nephrol Dial Transplant 22:2440–2448
Davies AP, Campbell B, Evans MR, Bone A, Roche A, Chalmers RM (2009) Asymptomatic carriage of protozoan parasites in children in day care centers in the United Kingdom. Pediatr Infect Dis J 28(9):838–840
Del Chierico F, Onori M, Di Bella S, Bordi E, Petrosillo N, Menichella D et al (2011) Cases of cryptosporidiosis co-infections in AIDS patients: a correlation between clinical presentation and GP60 subgenotype lineages from aged formalin-fixed stool samples. Ann Trop Med Parasitol 105(5):339–349
DuPont HL, Chappell CL, Sterling CR, Okhuysen PC, Rose JB, Jakubowski W (1995) The infectivity of Cryptosporidium parvum in healthy volunteers. New Engl J Med 332:855-859
Dworkin B, Wormser GP, Rosenthal WS, Heier SK, Braunstein M, Weiss L et al (1985) Gastrointestinal manifestations of the acquired immunodeficiency syndrome. Am J Gastroenterol 80:774–778
Egger M, Mausezahl D, Odermatt P, Marti HP, Tanner M (1990) Symptoms and transmission of intestinal cryptosporidiosis. Arch Dis Child 65:445–447
Erhabor O, Obunge O, Awah I (2011) Cryptosporidiosis among HIV-infected persons in the Niger Delta of Nigeria. Nigerian Med J 20(3):372–375
Fayer R, Ungar BLP (1986) Cryptosporidium spp. and cryptosporidiosis. Microbiol Rev 50:458–483
Fayer R, Spear CA, Dubey JP (1997) The general biology of Cryptosporidium. Cryptosporidium and cryptosporidiosis. CRC Press, Boca Raton, pp 1–41
Flanigan TP, Soave R (1993) Cryptosporidiosis. In: Progress in clinical parasitology, pp 1–20
Forgacs P, Tarchis A, Ma P, Federman L, Mele L, Silverman M, Shea JA (1983) Intestinal and bronchial cryptosporidiosis in an immunodeficient homosexual man. Ann Intern Med 99:793–794
French AL, Beaudet LM, Benator DA, Levy CS, Kass M, Orenstein JM (1995) Cholecystectomy in patients with AIDS: clinicopathologic correlations in 107 cases. Clin Infect Dis 21:852–858
Gambhir IS, Jaiswal JP, Nath G (2003) Significance of Cryptosporidium as an aetiology of acute infectious diarrhoea in elderly Indians. Tropical Med Int Health 8(5):415–419
Garcia LS (2001) Diagnostic medical parasitology. American Society for Microbiology, USA, pp 65–73
George B, Mathews V, Srivastava A, Chandy M (2004) Infections among allogeneic bone marrow transplant recipients in India. Bone Marrow Transplant 33:311–315
George B, Mathews V, Viswabandya A, Srivastava A, Chandy M (2006) Infections in children undergoing allogeneic bone marrow transplantation in India. Pediatric Transplant 10:48–54
Geurden T, Berkvens D, Geldhof P, Vercruysse J, Claerebout E (2006) A Bayesian approach for the evaluation of six diagnostic assays and the estimation of Cryptosporidium prevalence in dairy calves. Vet Res 37(5):671–682
Ghorpade MV, Kulkarni SA, Kulkarni AG (1996) Cryptosporidium, Isospora and Strongyloides in AIDS. Natl Med J India 9:201
Gideon AA, Njiné T, Nola M, Menbohan SF, Ndayo MW (2007) Measuring resistant forms of two pathogenic protozoa (Giardia spp and Cryptosporidium spp) in two aquatic biotopes in Yaoundé (Cameroon). Sante 17(3):167–172
Glaberman S, Moore JE, Lowery CJ, Chalmers RM, Sulaiman I, Elwin K et al (2002) Three drinking-water-associated cryptosporidiosis outbreaks, Northern Ireland. Emerg Infectious Dis 8(6):631–633
Godwin TA (1991) Cryptosporidiosis in the acquired immunodeficiency syndrome: a study of 15 autopsy cases. Hum Pathol 22:1215–1224
Goodstein RF, Colombo CS, Illfelder MA, Skaggs RE (1989) Bronchial and gastrointestinal cryptosporidiosis in AIDS. J Am Osteopath Assoc 89:195–197
Hancock CM, Rose JB, Callahan M (1998) Crypto and Giardia in U.S. groundwater. J AWWA 58(3):58–61
Hojlyng N, Jensen BN (1988) Respiratory cryptosporidiosis in HIV-positive patients. Lancet 2:590–591
Hu G, Gong AY, Liu J, Zhou R, Deng C, Chen XM (2010) miR-221 suppresses ICAM-1 translation and regulates interferon-gamma-induced ICAM-1 expression in human cholangiocytes. Am J Physiol Gastrointest Liver Physiol 298(4):G542–G550
Hunter PR, Nichols G (2002a) Epidemiology and clinical features of Cryptosporidium infection in immunocompromised patients. Clin Microbiol Rev 15(1):145–154
Hunter PR, Nichols G (2002b) Epidemiology and clinical features of Cryptosporidium infection in immunocompromised patients. Clin Microbiol Rev 15(1):145–154
Hussein AS (2011) Cryptosporidium parvum causes gastroenteritis epidemics in the Nablus region of Palestine. Tropical Med Int Health 16(1):12–17
Jayalakshmi J, Appalaraju B, Mahadevan K (2008) Evaluation of an enzyme-linked immunoassay for the detection of Cryptosporidium antigen in fecal specimens of HIV/AIDS patients. Indian J Pathol Microbiol 51(1):137–138
Jerrett IY, Snodgrass DR (1981) Cryptosporidia associated with outbreaks of neonatal calf diarrhea. Aust Vet J 57:434–435
Jindal N, Arora R, Bhushan B, Arora S (1995) A study of infective aetiology of chronic diarrhoea in children in Amritsar. J Indian Med Assoc 93:169–170
Kashyap B, Sinha S, Das S, Rustagi N, Jhamb R (2010) Efficiency of diagnostic methods for correlation between prevalence of enteric protozoan parasites and HIV/AIDS status–an experience of a tertiary care hospital in East Delhi. J Parasitic Dis 34(2):63–67
Kaur R, Rawat D, Kakkar M, Uppal B, Sharma VK (2002a) Intestinal parasites in children with diarrhea in Delhi, India. Southeast Asian J Trop Med Publ Health 33:725–729
Kaur R, Rawat D, Kakkar M, Uppal B, Sharma VK (2002b) Intestinal parasites in children with diarrhea in Delhi, India. Southeast Asian J Trop Med Public Health 33:725–729

Kaushik K, Khurana S, Wanchu A, Malla N (2008) Evaluation of staining techniques, antigen detection and nested PCR for the diagnosis of cryptosporidiosis in HIV seropositive and seronegative patients. Acta Trop 107(1):1–7

Kaushik K, Khurana S, Wanchu A, Malla N (2009a) Serum immunoglobulin G, M and A response to Cryptosporidium parvum in Cryptosporidium-HIV co-infected patients. BioMed Central Infectious Dis 9:179

Kaushik K, Khurana S, Wanchu A, Malla N (2009) Lymphoproliferative and cytokine responses to Cryptosporidium parvum in patients coinfected with C. parvum and human immunodeficiency virus. Clin Vaccine Immunol 16(1):116–21

Keusch GT, Hamer D, Joe A, Kelley M, Griffiths J, Ward H (1995) Cryptosporidia-who is at risk? Schweiz Med Wochenschr 125(18):899–908

Kirkpatrick BD, Daniels MM, Jean SS, Pape JW, Karp C, Littenberg B et al (2002) Cryptosporidiosis stimulates an inflammatory intestinal response in malnourished Haitian children. J Infectious Dis 1; 186(1):94–101

Kottona CN, Lattes R (2009) Parasitic infections in solid organ transplant recipients. Am J Transplant 9(Suppl 4):S234–S251

Lacroix S, Mancassola R, Naciri M, Laurent F (2001) Cryptosporidium parvum-specific mucosal immune response in C57BL/6 Neonatal and gamma interferon-deficient mice: role of tumor necrosis factor alpha in protection. Infect Immun 69(3):1635–1642

Le XH, Wang H, Gou JZ, Chen XC, Yang GL, Yang QT et al (2008) Detection of cryptosporidium infection among AIDS patients in Guangdong and Yunnan. Zhonghua Shi Yan He Lin Chuang Bing Du Xue Za Zhi 22(5):339–341

Lean IS, Lacroix-Lamandé S, Laurent F, McDonald V (2006) Role of tumor necrosis factor alpha in development of immunity against Cryptosporidium parvum infection. Infect Immun 74(7):4379

Leav BA, Mackay M, Ward HD (2003) Cryptosporidium species: new insights and old challenges. Clin Infect Dis 36:903–908

LeChevallier MW, Norton WD (1995) Giardia and Cryptosporidium in raw and finished water. J AWWA 87(9):54–68

Lee MY, Cho EJ, Lee JH, Han SH, Park YS (2010) A survey of Cryptosporidium Oocysts in water supplies during a 10-year period (2000–2009) in Seoul. Korean J Parasitol 48(3):219–224

Levine ND (1984) Taxonomy and review of the coccidian genus Cryptosporidium (Protozoa, Apicomplexa). J Protozoology 31:94–98

Lim YA, Iqbal A, Surin J, Sim BL, Jex AR, Nolan MJ, Smith HV, Gasser RB (2011) First genetic classification of Cryptosporidium and Giardia from HIV/AIDS patients in Malaysia. Infection Genet Evol 11(5):968–974

Lisle J, Rose J (1995) Cryptosporidium contamination of water in the USA and UK: a mini-review. J Water SRT-Aqua 44(3):103–117

Lopez-Velez R, Tarazona R, Garcia Camacho A, Gomez-Mampaso E, Guerrero A, Moreira V et al (1995) Intestinal and extraintestinal cryptosporidiosis in AIDS patients. Eur J Clin Microbiol Infect Dis 14:677–681

Lucca P, De Gaspari EN, Bozzoli LM, Funada MR, Silva SO, Iuliano W, Soares RM (2009) Molecular characterization of Cryptosporidium spp. from HIV infected patients from an urban area of Brazil. Revista do Instituto de Medicina Tropical de São Paulo 51(6):341–343

Ma P, Soave R (1983a) Three-step stool examination for cryptosporidiosis in 10 homosexual men with protracted watery diarrhea. J Infectious Dis 147:824–828

Ma P, Soave R (1983b) Three-step stool examination for cryptosporidiosis in 10 homosexual men with protracted watery diarrhoea. J Infect Dis 147:824–828

Mac Kenzie WR et al (1994) A massive outbreak in Milwaukee of Cryptosporidium infection transmitted through the public water supply. N Engl J Med 331:161–167

Mahmoudi MR, Ashrafi K, Abedinzadeh H, Tahvildar-Bideruni F, Haghighi A, Bandehpour M et al (2011) Development of sensitive detection of Cryptosporidium and Giardia from surface water in Iran. Iran J Parasitol 6(3):43–51

Maillot C, Gargala G, Delaunay A, Ducrotte P, Brasseur P, Ballet JJ, Favennec L (2000) Cryptosporidium parvum infection stimulates the secretion of TGF-beta, IL-8 and RANTES by Caco-2 cell line. Parasitol Res 86(12):947–949

Manabe YC, Clark DP, Moore RD, Lumadue JA, Dahlman HR, Belitsos PC et al (1998) Cryptosporidiosis in patients with AIDS: correlates of disease and survival. Clin Infect Dis 27:536–542

Marcial MA, Madara JL (1986) Cryptosporidium: cellular localization, structural analysis of absorptive cell-parasite membrane–membrane interactions in guinea pigs, and suggestion of protozoan transport by M cells. Gastroenterology 90:583–594

Martins CAP, Guerrent RL (1995) Cryptosporidium and cryptosporidiosis. Parasitology Today 11:434–435

Mason RW, Hartley WJ, Tilt L (1981) Intestinal cryptosporidiosis in a kid goat. Aust Vet J 57:386–388

Mathan MM, Venkatesan S, George R, Mathew M, Mathan VI (1985) Cryptosporidium and diarrhoea in southern Indian children. Lancet 2:1172–1175

May GR, Gill MJ, Church DL, Sutherland LR (1993) Gastrointestinal symptoms in ambulatory HIV-infected patients. Dig Dis Sci 38:1388–1394

Meamar AR, Rezaian M, Rezaie S, Mohraz M, Kia EB, Houpt ER et al (2006) Cryptosporidium parvum bovine genotype oocysts in the respiratory samples of an AIDS patient: efficacy of treatment with a combination of azithromycin and paromomycin. Parasitol Res 98(6):593–595

Molbak K, Hojlyng N, Gottschau A, Sa JC, Ingholt L, Da Silva AP, Aaby P (1993) Cryptosporidiosis in infancy and childhood mortality in Guinea Bissau, West Africa. BioMed J 307:417–420

Molloy SF, Tanner CJ, Kirwan P, Asaolu SO, Smith HV, Nichols RA et al (2011) Sporadic Cryptosporidium infection in Nigerian children: risk factors with species identification. Epidemiol Infect 139(6):946–954

Morin M, Lariviere S, Lallier R (1976) Pathological and microbiological observations made on spontaneous cases of acute neonatal calf diarrhea. Can J Comp Med 40:228–240

Moura H, Fernandes O, Viola JPB (1989) Enteric parasites and HIV infection: occurrence in AIDS patients in Rio de Janeiro. Brazil Mem Inst Oswaldo Cruz 84:527–533

Moyo SJ, Gro N, Matee MI, Kitundu J, Myrmel H, Mylvaganam H et al (2011) Age specific aetiological agents of diarrhoea in hospitalized children aged less than five years in Dar es Salaam, Tanzania. BMC Pediatrics 11:19

Muñoz-Antoli C, Pavón A, Marcilla A, Toledo R, Esteban JG (2011) Prevalence and molecular characterization of Cryptosporidium in schoolchildren from department of Rio San Juan (Nicaragua). Trop Biomed 28(1):40–47

Muthusamy D, Rao SS, Ramani S, Monica B, Banerjee I, Abraham OC et al (2006) Multilocus genotyping of Cryptosporidium sp. isolates from human immunodeficiency virus-infected individuals in South India. J Clin Microbiol 44(2):632–634

Nachbaur D, Kropshofer G, Feichtinger H, Allerberger F, Niederwieser D (1997) Cryptosporidiosis after CD34-selected autologous peripheral blood stem cell transplantation (PBSCT). Treatment with paromomycin, azithromycin and recombinant human interleukin-2. Bone Marrow Transplant 19(12):1261–1263

Naeini KM, Asadi M, Chaleshtori MH (2010) Detection and molecular characterization of Cryptosporidium species in recreational waters of Chaharmahal va Bakhtiyari Province of Iran using nested-PCR-RFLP. Iran J Parasitol 6(1):20–27

Nagamani K, Pavuluri PR, Gyaneshwari M, Prasanthi K, Rao MI, Saxena NK (2007) Molecular characterisation of Cryptosporidium: an emerging parasite. Indian J Med Microbiol 25(2):133–136

Nahrevanian H, Assmar M (2008) Cryptosporidiosis in immunocompromised patients in the Islamic Republic of Iran. J Microbiol Immunol Infect 41(1):74–77

Nair P, Mohamed JA, DuPont HL, Figueroa JF, Carlin LG, Jiang ZD, Belkind-Gerson J, Martinez-Sandoval FG, Okhuysen PC (2008) Epidemiology of cryptosporidiosis in North American travelers to Mexico. Am J Trop Med Hyg 79(2):210–214

Neill MA, Rice SK, Ahmad NV, Flanigan TP (1996) Cryptosporidiosis: an unrecognized cause of diarrhea in elderly hospitalized patients. Clin Infectious Dis 22(1):168–170

Nel ED, Rabie H, Goodway J, Cotton MF (2011) A retrospective study of cryptosporidial diarrhea in a region with high HIV prevalence. J Trop Pediatr 57(4):289–292

Ojurongbe O, Raji OA, Akindele AA, Kareem MI, Adefioye OA, Adeyeba AO (2011) Cryptosporidium and other enteric parasitic infections in HIV-seropositive individuals with and without diarrhoea in Osogbo. Nigeria. Br J Biomed Sci 68(2):75–78

Ok UZ, Cirit M, Uner A, Ok E, Akçiçek F, Başçi A, Ozcel MA (1997) Cryptosporidiosis and blastocystosis in renal transplant recipients. Nephron 75(2):171–174

Palmieri F, Cicalini S, Froio N, Rizzi EB, Goletti D, Festa A et al (2005) Pulmonary cryptosporidiosis in an AIDS patient: successful treatment with paromomycin plus azithromycin. Int J STD AIDS 16(7):515–517

Panciera RJ, Thomassen RW, Gardner FM (1971) Cryptosporidial infection in a calf. Vet Pathol 8:479–484

Pantenburg B, Cabada MM, Clinton White JRA (2009) Treatment of Cryptosporidiosis. Expert Rev AntiInfect Ther 7(4):385–391

Parija SC, Shivaprakash MR, Jayakeerthi SR (2003) Evaluation of lacto-phenol cotton blue (LPCB) for detection of Cryptosporidium, Cyclospora and Isospora in the wet mount preparation of stool. Acta Trop 85(3):349–354

Parr Jonathan B, Sevilleja JE, Samie A, Alcantara C, Stroup SE, Kohli A et al (2007) Detection and quantification of cryptosporidium in Hct-8 cells and human fecal specimens using real-time Polymerase chain reaction. Am J Trop Med Hyg 76(5):938–942

Pedersen C, Danner S, Lazzarin A (1996) Incidence of cryptosporidiosis among European AIDS patients. Genitourin Med 72:128–131

Peeters JE, Mazas EA, Masschelein WJ, Martinez de Maturana IV, Debacker E (1989) Effect of disinfection of drinking water with ozone or chlorine dioxide on survival of Cryptosporidium parvum oocysts. Appl Environ Microbiol 55:1519–1522

Plutzer J, Takó MH, Márialigeti K, Törökné A, Karanis P (2007) First investigations into the prevalence of Cryptosporidium and Giardia spp. in Hungarian drinking water. J Water Health 5(4):573–584

Pollok RC, Farthing MJ, Bajaj-Elliott M, Sanderson IR, McDonald V (2001) Interferon gamma induces enterocyte resistance against infection by the intracellular pathogen Cryptosporidium parvum. Gastroenterology 120(1):99–107

Ranjbar-Bahadori Sh, Sangsefidi H, Shemshadi B, Kashefinejad M (2011) Cryptosporidiosis and its potential risk factors in children and calves in Babol, north of Iran. Trop Biomed 1:125–131

Rao Ajjampur SS, Asirvatham JR, Muthusamy D, Gladstone BP, Abraham OC, Mathai D et al (2007a) Clinical features and risk factors associated with cryptosporidiosis in HIV infected adults in India. Indian J Med Res 126(6):553–557

Rao Ajjampur SS, Gladstone BP, Selvapan D (2007b) Molecular and spatial epidemiology of Cryptosporidiosis in Children in a semiurban community in South India. J Clin Microbiol 45(3):915–920

Rao Ajjampur SS, Sankaran P, Kang G (2008) Cryptosporidium species in HIV-infected individuals in India: an overview. Natl Med J India 21:178–184

Rasmussen JP, Saint CP, Monis PT (2007) Use of DNA melting simulation software for in silico diagnostic assay design: targeting regions with complex melting curves and confirmation by real-time PCR using intercalating dyes. BMC Bioinform 8:107

Reduker DW, Speer CA (1985) Factors influencing excystation of Cryptosporidium oocysts from cattle. J Parasitol 71:112–115

Rene E, Marche C, Regnier B, Sainot AG, Vilde JL et al (1989) Intestinal infections in patients with acquired immunodeficiency syndrome. A prospective study in 132 patients. Dig Dis Sci 34:773–780

Richardson AJ et al (1991) An outbreak of waterborne cryptosporidiosis in Swindon and Oxfordshire. Epidemiol Infect 107:485–495

Rolston KV, Rodriguez S, M Hernandez, Bodey GP (1989) Diarrhoea in patients infected with the human immunodeficiency virus. Am J Med 86:137–138

Sejdini A, Mahmud R, Lim YA, Mahdy M, Sejdini F, Gjoni V et al (2011) Intestinal parasitic infections among children in central Albania. Ann Trop Med Parasitol 105(3):241–250

Semenza JC, Nichols G (2007) Cryptosporidiosis, surveillance and water-borne outbreaks in Europe. Eurosurveillance 12:3–6

Seydel KB, Zhang T, Champion GA, Fichtenbaum C, Swanson PE, Tzipori S et al (1998) Cryptosporidium parvum Infection of human intestinal xenografts in SCID mice induces production of human tumor necrosis factor alpha and interleukin-8. Infect Immun 66(5):2379–2382

Shahid NS, Rahman AS, Sanyal SC (1987) Cryptosporidium as a pathogen for diarrhoea in Bangladesh. Trop Geogr Med 39:265–270

Shetty M, Brown TA, Kotian M, Shivananda PG (1995) Viral diarrhea in a rural coastal region of Karnataka India. J Trop Pediatr 41:301–303

Singh I, Theodos C, Li W, Tzipori S (2005) Kinetics of Cryptosporidium parvum-specific cytokine responses in healing and nonhealing murine models of C. parvum infection. Parasitol Res 97(4):309–317

Singla R, Hari S, Sharma SK (2010) Diffuse small bowel thickening in AIDS patient–a case report. BioMed Central Infectious Dis 10:310

Slavin D (1955) Cryptosporidium meleagridis. J Comp Pathol 65:262–266

Smith HV (1989) The effect of free chlorine on the viability of cryptosporidium spp. oocysts. Medmenham, Water Research Centre (Great Britain), report PRU 2023-M

Smith HV (1990) Environmental aspects of Cryptosporidium species: a review. J Royal Soc Med 83(10):629

Smith PD, Lane CL, Gill VJ, Manischewitz JF, Quinnan GV, Fauci AS et al (1988) Intestinal infections in patients with the acquired immunodeficiency syndrome (AIDS). Etiology and response to therapy. Ann Intern Med 108:328–333

Snelling WJ, Xiao L, Ortega-Pierres G, Lowery CJ, Moore JE, Rao JR et al (2007) Cryptosporidiosis in developing countries. J Infect Developing Countries 1(3):242–256

Soliman Rasha H, Othman Ahmad A (2009) Evaluation of DNA Melting Curve Analysis Real-Time PCR for Detection and Differentiation of Cryptosporidium Species. PUJ 2(1):47–54

Sorvillo F, Beall G, Turner PA, Beer VL, Kovacs AA, Kraus P (1998) Seasonality and factors associated with cryptosporidiosis among individuals with HIV infection. Epidemiol Infect 121:197–204

Spano F, Sputignani L, Mc Lauchlin J, Casemore DP, Crisanti A (1997) PCR-RFLP analysis of the Cryptosporidium oocyst wall protein (COWP) gene discriminates between C.wrairi and C.parvum and between C.parvum isolates of human and animal origin. FEMS Microbiol Lett 150:209–217

Spano F, Putignani L, Guida S, Crisanti A (1998) Cryptosporidium Parvum: PCR-RFLP analysis of the Trap-C1 (Thrombospondin-Related Adhesive Protein of Cryptosporidium-1). Gene discriminates between two alleles differentially associated with parasite isolates of animal and human origin. Exp Parasitol 90:195–198

Sulaiman IM, Morgan UM, Thompson RCA, Lal AA, Xiao L (2000) Phylogenetic Relationships of Cryptosporidium Parasites Based on the 70-Kilodalton Heat Shock Protein (HSP70) Gene. Appl Environ Microbiol 66(6):2385–2391

Tali A, Addebbous A, Asmama S, Chabaa L, Zougaghi L (2011) Respiratory cryptosporidiosis in two patients with HIV infection in a tertiary care hospital in Morocco. Ann Biol Clin 69(5):605–608

Teare JP, Daly CA, Rodgers C, Padley SP, Coker RJ, Main J et al (1997) Pancreatic abnormalities and AIDS related sclerosing cholangitis. Genitourin Med 73:271–273

Thulin R (1991) Contamination of tap water in Jongkoping in summer 1991. Rapport Jongkopings Community, Sweden

Tuli L, Singh DK, Gulati AK, Sundar S, Mohapatra TM (2010) A multiattribute utility evaluation of different methods for the detection of enteric protozoa causing diarrhea in AIDS patients. BMC Microbiol 15(10):11

Tzipori S (1983) Cryptosporidiosis in animals and humans. Microbiol Rev 47:84–96

Tzipori S, Ward H (2002) Cryptosporidiosis: biology, pathogenesis and disease. Microbes Infect 4:1047–1058

Tzipori S, Widmer G (2008a) A hundred-year retrospective on cryptosporidiosis. Trends Parasitology 24(4):184–189

Tzipori S, Widmer G (2008b) A hundred-year retrospective on cryptosporidiosis. Trends Parasitol 24(4):184–189

Tzipori S, Campbell L, Sherwood D, Snodgrass DR (1980) An outbreak of calf diarrhea attributed to cryptosporidial infection. Veterinary Record 107:579–580

Udgiri N, Minz M, Kashyap R, Heer M, Gupta CS, Mohandas K et al (2004) Intestinal cryptosporidiasis in living related renal transplant recipients. Transplant Proc 36(7):2128–2129

Ulçay A, Görenek L, Coşkun O, Araz E, Acar A, Eyigün CP (2008) Diagnosis of intestinal-protozoa in patients with immune deficiency. Turkiye Parazitol Derg 32(4):328–333

Uppal B, Kashyap B, Bhalla P (2009) Enteric Pathogens in HIV/AIDS from a Tertiary Care Hospital. Indian J Community Med 34(3):237–242

U.S. Environmental Protection Agency (2005) Method 1623: Cryptosporidium and Giardia by filtration/IMS/FA. U.S. Environmental Protection Agency publication no. 815-R-05-002. U.S. Environmental Protection Agency, Washington, DC

USEPA (2001) Drinking Water Criteria Document Addendum: *Cryptosporidium*. United States environmental protection agency, Office of Water, Washington, DC

Verdon R, Polianski J, Grodets A, Garryt L, Carbon C (1998) Cryptosporidium parvum biliary tract infection in adult immunocompetent and immunosuppressed mice. J Med Microbiol 47:71–77

Vuorio AF, Jokipii AM, Jokipii L (1991) Cryptosporidium in asymptomatic children. Rev Infect Dis 13(2):261–264

Wallis PM, Erlandsen SL, Isaac-Renton JL, Olson ME, Robertson WJ, Van Keulen H (1996) Prevalence of Giardia cysts and Cryptosporidium oocysts and characterization of Giardia spp. isolated from drinking water in Canada. Appl Environ Microbiol 62(8):2789–2797

Wang HC, Dann SM, Okhuysen PC, Lewis DE, Chappell CL, Adler DGet al. 2007. Clinton White, Jr. High Levels of CXCL10 Are Produced by Intestinal Epithelial Cells in AIDS Patients with Active Cryptosporidiosis but Not after Reconstitution of Immunity. Infect Immun 75(1):481–487

Weikel CS, Johnston LI, De Sousa MA, Guerrant RL (1985) Cryptosporidiosis in northeastern Brazil: association with sporadic diarrhea. J Infect Dis 151:963–965

Weintraub JM (2006) Improving Cryptosporidium testing methods: a public health perspective. J Water Health 4(1):23–26

Wyatt CR, Barrett WJ, Brackett EJ, Schaefer DA, Riggs MW (2002) Association of IL-10 expression by mucosal lymphocytes with increased expression of Cryptosporidium parvum epitopes in infected epithelium. J Parasitol 88(2):281–286

Xiao S, An W, Chen Z, Zhang D, Yu J, Yang M (2012) Occurrences and genotypes of Cryptosporidium oocysts in river network of southern-eastern China. Parasitol Res: 110

Yang W, Chen P, Villegas EN, Landy RB, Kanetsky C, Cama V et al (2008) Cryptosporidium source tracking in the Potomac River watershed. Appl Environ Microbiol 74:6495–6504

Yoder JS, Beach MJ (2007) Cryptosporidiosis Surveillance—United States, 2003–2005. Surveillance Summaries 56(SS07):1–10

Zuckerman U, Gold D, Ghelef G, Yuditsky A, Armon R (1997) Microbial degradation of Cryptosporidium parvum by Serratia marcescens with high chitinolytic activity. In: Fricker et al. (eds) Proceedings 1997 International Symposium waterborne Cryptosporidium. Newport Beach, California

Water- and Food-Borne Trematodiases in Humans

15

Sumeeta Khurana and Nancy Malla

Abstract

Water and aquatic organisms such as fish, crustaceans, molluscs and aquatic plants serve as important source of various pathogens especially trematodes. It is estimated that more than 10 % of world population is potentially at a risk of food-borne trematodiases (FBT), and globally, more than 100 million people are infected with one or more of flukes. Most of the infections occur in South-east Asia and Asian–Pacific regions where aquatic animals are eaten undercooked or raw. The trematodes may infect gastrointestinal tract (*Fasciolopsis*, *Echinostoma,* Heterophyes), liver (*Clonorchis, Opisthorchis, Fasciola)*, lungs (*Paragonimus*) and less commonly other organs. *Opisthorchis viverrini* has also been associated with cholangiocarcinoma. The laboratory diagnosis relies mainly on the demonstration of parasite in clinical samples and serological techniques. Recently, molecular techniques have been applied, but sensitivity and specificity vary according to the technique used. Most of these FBTs are preventable, and thus, there is an urgent need for improvement in available diagnostics and initiation of control programmes and health education in endemic communities.

Keywords

Fish · Flukes · Intestinal · Liver · Lung · Snails · Trematodes · Water

Introduction

Water and aquatic organisms such as fish, crustaceans, molluscs and aquatic plants serve as important source of various pathogens. Aquaculture is one of the most important of the world's food producing activities, and Asia is by far the world's chief aquaculture producer; approximately, 90 % of the world aquaculture

S. Khurana · N. Malla (✉)
Department of Parasitology, Postgraduate Institute of Medical Education and Research, Chandigarh 160012, India
e-mail: drmallanancy@gmail.com

P. P. Singh and V. Sharma (eds.), *Water and Health*,
DOI: 10.1007/978-81-322-1029-0_15,

production is in Asia (Bostock et.al. 2010; FAO 2010). Coincidently, food-borne trematodiases (FBT) transmitted by fish are the most prevalent in Asia. The trematode infections, for which aquatic animals serve as hosts, are generally referred to as FBT. The burden of disease and at risk population is ill-defined because of paucity of readily available, accurate assays at point of contact. Most of these FBTs are preventable and occur in world's poor countries. However, despite increasing international awareness, control programmes are not fully successful.

The trematodes belong to phylum Platyhelminthes. Although humans can be infected with >70 species of flukes, the most important species are liver flukes (*Clonorchis sinensis, Opisthorchis viverrini, Opisthorchis felineus, Fasciola hepatica),* lung flukes (*Paragonimus westermani, P. heterotremus, P. philippinensis, P. pulmonalis* etc.) and intestinal flukes (*Fasciolopsis buski*, *Echinostoma,* Heterophyes).

Global Burden of the Disease

It is estimated that more than 10 % of world population is potentially at a risk of FBT, and globally, more than 100 million people are infected with FBTs. Most of the infections occur in areas where aquatic animals are eaten undercooked or raw, for example, Kung Plah, Kung Ten (raw crayfish salad), Nam Prik Poo (crab sauce) and koi-pla (raw fish) are age old popular and widely consumed dishes in Asia pacific regions. Moreover, fish farming of grass carp and other susceptible species in ponds that are routinely contaminated by untreated sewage help in the establishment of infection in fish populations, which makes control of liver fluke infection even more difficult (Lun et al. 2005; Sripa et al. 2007). The most common FBT is *C. sinensis* responsible for about 35 million infections followed by *Paragonimus*, which accounts for 20 million infections. *F. hepatica* infection ranges from 2.4 to 17 million infections, and *O. viverrini* accounts for 10 million infections. Almost half of *Clonorchis* infections are reported from China, while *O. viverrini* infections are almost confined to Thailand and Laos PDR. The FBTs have almost tripled in China in the last decade due to rising aquaculture, while in Thailand, prevalence rate has come down from 63 % in 1984 to 7 % in 2000 due to stringent control measures (Keiser and Utzinger 2005, 2009; Sripa et al. 2010). The infection sources, disease burden and endemic areas are summarized in Table 15.1.

The food-borne trematode most prevalent in India is Paragonimus and *F. buski*. *Paragonimus* is reported mainly from the north-eastern states of India. Most of the cases are reported from Manipur and are due to infection with *P. westermani* (Singh et.al. 1986), but lately cases of *P. heterotremous* have been reported from Arunachal Pradesh and Nagaland (Narain et al. 2003; Singh et al. 2009). *F. buski* is reported mainly from Uttar Pradesh, Bihar, Bengal and Assam (Bhatti et al. 2000; Bhattacharyya et al. 2010; Chandra 1984; Keiser and Utzinger 2009; Muralidhar et al. 2000). *Metagonimus* has been reported from Assam and Delhi (Mahanta et al. 1995; Uppal and Wadhwa 2005), and one case of *F. hepatica* from spine has been reported from Lucknow (Vatsal et al. 2006).

Liver fluke infections (*C. sinensis* and *O. viverrini*) occur more frequently in males than in females, and prevalence increases with age, often reaching a plateau in early adulthood. However, *F. hepatica* infections are more common in females and among children. Intestinal and lung fluke infections are most prevalent in school-aged children. Reinfections occur rapidly after treatment, particularly in communities where a high prevalence of these infections is reported (Sripa et al. 2010). It must be emphasized here that most of trematode infections are multispecies infections as areas of endemicity of various parasites often overlap. In a study in Laos, more than 80 % of patients investigated were coinfected with *O. viverrini* and one or multiple species of intestinal flukes (Sayasone et al. 2009).

Table 15.1 Water- and food-borne trematodes of human medical importance

FBT	Species	Infection source	No. of globally reported infections	Geographical distribution
Liver flukes	*Clonorchis sinensis*	Small freshwater fish	35 million	People's Republic of China, Korea, North Vietnam
	Opisthorchis viverrini	Small freshwater fish	10 million	South-east Asia including Thailand, Lao PDR, Cambodia and Central Vietnam
	Opisthorchis felineus	Small freshwater fish	1.2 million	Russia
	F. hepatica *F. gigantica*	Freshwater vegetables, infected water, infected raw liver	2.4–17 million	Vietnam, Hawaii, Cuba, Egypt, Western European countries
Intestinal fluke	*F. buski*	Freshwater plants, water caltrop, water lily, watercress	40–50 million all intestinal flukes combined	China, Korea, India, Bangladesh, Thailand, Malaysia, Myanmar, Sumatra, Vietnam, Laos, recently in Bolivia, South and Central America
	Echinostoma *Metagonimus* *Heterophyes* *Gastrodiscoides*	Freshwater fish, frogs, tadpoles, molluscs (snails, clams, oysters), crustaceans (crabs, shrimp), etc.		Korea, Indonesia, India, the Philippines, Malaysia, Taiwan, Thailand, Russia
Lung flukes	*Paragonimus westermani* *P.heterotremus*	Freshwater crabs, crayfish, wild boar meat	20.7 million	Asia (People's Republic of China, Japan, Korea, Laos PDR, the Philippines, Vietnam, Taiwan, Thailand, India), South and Central America (Ecuador, Peru, Costa Rica, Columbia) and Africa (Cameroon, Gambia and Nigeria)
	P.africanus *P.kellicotti*			North America

The Parasites: Morphology and Their Life Cycle

The trematodes have a dorsoventrally flattened bilaterally symmetrical body, ranging in size from few millimetres (Heterophyes) to 10 cm (*F. gigantica*), depending on the species. They typically have an oral and a ventral sucker. The reproductive system is hermaphroditic.

Their life cycle is complex, but a common feature is that aquatic snails act as intermediate hosts. These are 4–6 larval stages with alternating sexual and asexual stages. Infection in humans is acquired by drinking contaminated water, ingestion of raw, pickled or incompletely cooked aquatic products (fish, crustaceans, snails, tadpoles and frogs) containing metacercariae or eating contaminated aquatic vegetation like water chestnut, lily, caltrop. Metacercariae encyst in stomach and juvenile worm is released which migrates to target organ. They are typically found in major viscera such as bile ducts, lungs and gut. Infection with Paragonimus spp. may also occur through consumption of undercooked meat of wild boars. Adult worms formed may live up to 25 years in human hosts. They produce eggs, which are released via faeces except in case of Paragonimus where eggs are released in sputum. Egg-laying capacity ranges from 1,000 to 4,000 eggs/day for *C. sinensis* and 16,000 eggs/day for *F. buski*. Eggs embryonate under appropriate environmental condition in water and hatch releasing miracidium which in turn is ingested by snails or penetrates the molluskan intermediate host. A number of snail species are susceptible intermediate host. For liver flukes, important snail species are Alocinma, Bithynia and Parafossarulus; for lung flukes, Melania and Bithynia while for intestinal flukes Segmentina, Hippeutis and Gyraulus. Within the snail, asexual reproduction occurs for several weeks. Miracidia after converting to sporocyst stage multiply and produce daughter sporocysts or radiae which then develop into cercariae which leave the snail and either encyst on aquatic vegetations such as water lotus, water caltrop, chestnut, water lily (*F. hepatica, F. buski*) or penetrate the skin of 2nd intermediate host (fish, frogs, tadpoles, snails). These 2nd intermediate hosts usually breed in stagnant or slow flowing water. In the body of 2nd intermediate host, they are converted into metacercariae which is infective stage for humans/ other mammals which act as definitive hosts (Garcia 2007).

Clinical Manifestations

Morbidity due to trematode infections depends on the worm burden, type of worm, target organ and immune status of the host. Characteristically, there is an acute stage of infection, which is followed by chronic disease, and complications may occur depending on the worm load. Generally, a mild infection with about 100 flukes is usually asymptomatic, while severe infections with >25,000 flukes have severe morbidity.

In intestinal fluke infection (*F. buski*, Heterophyes, echinostomes), acute stage consists of diarrhoea, constipation, headache, flatulence, abdominal pain and fever, while chronic stage comprises of oedema, anaemia, anorexia, malnutrition and extreme prostration (Robinson and Dalton 2009). The attachment of parasites on the duodenal or jejunal wall by suckers leads to mechanical damage and inflammation, and ultimately, intestinal ulcers are formed that may bleed and lead to abscess formation.

Lung fluke *(Paragonimus)* infection leads to symptoms comprising of cough, fever, bloody sputum, loss of appetite, chest pain, headache, night sweats, etc., and clinical picture closely mimics that of pulmonary tuberculosis. Occasionally, lung flukes may migrate to ectopic locations, and the central nervous system is the most common locus of involvement, accounting for approximately 50 % of extra-thoracic disease. When lodged in brain, worms may cause epilepsy, headache, motor and sensory disturbances, aphasia, blindness, etc. In spinal paragonimiasis, paravertebral pain, motor disturbances in lower extremities and urinary disturbances may manifest. Other extrapulmonary sites of spread are skin, liver, eye, etc. (Gadkowski and Jason 2008; Keiser and Utzinger 2009). The presence of the flukes in the lung causes haemorrhage, inflammatory reaction and necrosis of lung parenchyma, ultimately leading to fibrosis.

Liver flukes (*C. sinensis, O. viverrini, O. felineus, F. hepatica*)—In liver fluke infections, acute stage consists of fever, diarrhoea, abdominal pain, loss of appetite, rash, oedema. In chronic-stage cholangitis, cholecystitis, cholelithiasis, pancreatitis, hepatitis and biliary and liver abscesses may manifest. There is definite association of *O. viverrini* with bile duct cancer (cholangiocarcinoma) with an extremely poor prognosis. Thus, Thailand and Laos PDR have the highest rates of cholangiocarcinoma. *C. sinensis* and *O. felineus* are probable carcinogens and are associated with cholangiocarcinoma.

Pathogenesis of disease is either due to direct damage caused by attachment of parasite and its sucking action, or due to chemical injury or immune reactions to parasitic metabolites. Some excretory–secretory products produced by the parasites cause chemical damage, are mitogenic, and lead to inflammation, which becomes chronic. Chronic inflammation leads to DNA damage and makes the tissue more susceptible to endogenously or exogenously produced mitogens. Increased endogenous production of

N-nitroso compounds and enhanced hepatic activation of carcinogens have been described (Satarug et al. 1996, 1998) and may establish highly mutagenic conditions for the chronically inflamed and proliferating bile duct epithelium. Cholangiocarcinoma usually affects more than 40 years old and has poor prognosis (Keiser and Utzinger 2009; Robinson and Dalton 2009; Sripa et al. 2010).

Immune Responses

Most of the trematodes have a good equilibrium between the host and parasite as evidenced by asymptomatic or mild disease in majority of infected individuals. This is because of the long-term coevolution of the two partners and a balance between host defence and immune evasion of the parasite. The immune response of host to most of helminthic infections is characterized by a predominant Th-2 response and typically associated with eosinophilia and significant IgE production, mucous mastocytosis and goblet cell hyperplasia (Anthony et al. 2007). IL-4 is involved in IgE isotype switch, and IL-5 is involved ion eosinophil production. IL-13 has similar functions to IL-4 and involved mainly in the effector phase of inflammation and the development of fibrosis. Antibody-dependent cell-mediated toxicity is dependent on eosinophils, neutrophils, macrophages and mast cells as effector cells and IgE, IgG and IgA as antibodies. The parasites covered with antibodies are destroyed by cells, carrying receptors for Fc fragment by release of products that are toxic to worm. These products include major basic protein, eosinophilic cationic protein, reactive nitrogen intermediates; on the other hand, parasites have developed certain mechanism to evade the effector response of the host, for example, Fasciola (1) produces superoxide dismutase which neutralizes superoxide radicals and (2) releases cathepsin L protease that cleaves IgE and IgG (Anthony et al. 2007; MacDonald et al. 2002).

Diagnosis

The importance of accurate diagnosis cannot be overemphasized for adequate patient management and for monitoring of helminth control programmes.

Liver Fluke Infections

The clinical diagnosis is presumptive, and definite diagnosis primarily depends on detection of trematode eggs in faeces. The most widely used techniques for detection of eggs are Kato-Katz and formalin ethyl acetate technique because these two methods have a reasonable high sensitivity at the level of moderate and heavy infection and also allow the quantification. However, a major problem of microscopy is its low sensitivity, particularly in light infections and low specificity as many trematode eggs are morphologically indistinguishable, leading to underestimation of problem and misidentification. The adult worms, if available, can lead to definite diagnosis and can be obtained following expulsion chemotherapy, but this is not normally practised (Keiser and Utzinger 2009; Maria et al. 2010).

Several serological tests have been developed for serodiagnosis, but their sensitivity and specificity vary because of complexity of antigens. Cross-reactivity with other trematodes is a major problem. There have been attempts to develop ELISA using excretory–secretory proteins, somatic extracts or specific recombinant antigens such as eggshell protein (rOVESP) for diagnosis of *O. viverrini* and recombinant cysteine protease for *C. sinensis* with sensitives ranging from 76 to 100 % and specificity ranging from 53 to 94 % (Hong and Fang 2011; Kaewpitoon et al. 2008; Nagano et al. 2004; Wongratanacheewin et al. 2003). Most of the recombinant proteins were found sensitive and specific for serodiagnosis of clonorchiasis but not specific enough to replace the crude extract which is now used. Moreover, another major problem of serology is the lack of ability to differentiate between past and current infection.

As opposed to antibody detection, antigen detection can indicate the current state of infection. However, no such kit is commercially available. Current tests for immunodiagnosis of human *Fasciola* infection include ELISAs with excretory–secretory antigens or purified/recombinant proteins in serum (Dixit et al. 2008; Espinoza et al. 2005; O'Neill et al. 1999). Detection of Fasciola antigens in faeces from infected humans has been reported with almost 100 % sensitivity and specificity (Ubeira et al. 2009).

In past few years, PCR-based methods have been developed for diagnosis of trematode infections with high sensitivity and specificity, but it is unlikely to become a tool for routine diagnosis due to high cost and expertise required and is thus suitable only for research laboratories. Different PCR protocols have been developed for detecting various species of *Opisthorchis*, *Clonorchis*, etc., but in low worm burdens, most of these techniques lack sensitivity and specificity (Hong and Fang 2011; Keiser and Utzinger 2009; Maria et al. 2010). Moreover, a further simplification and cost reduction would be important for their use in the developing countries.

Imaging techniques like ultrasound and CT scan have been used extensively for diagnosis. Ultrasound scanning is easy and practised commonly for routine health screening and early detection of hidden diseases. The ultrasound diagnosis of clonorchiasis depends on detecting dilatation and thickening of the intrahepatic bile duct, peri ductal echo and floating materials in the gallbladder. However, its diagnostic sensitivity and specificity are not so high (Choi et al. 2004). Also, it has a limitation to differentiate active disease and past history after cure (Choi et al. 2009).

Intestinal Fluke Infections

Direct demonstration of the eggs is the gold standard. However, eggs of many of the minute intestinal flukes like Heterophyes and echinostomes are indistinguishable, and very often, multiparasitism exists. Reports of molecular diagnosis for intestinal flukes are scarce (Keiser and Utzinger 2009; Nagano et al. 2004; Vatsal et al. 2006).

Lung Fluke Infection

For diagnosis of paragonimiasis, sputum examination is the method of choice. The sputum is rust coloured, classically described as 'iron fillings' not only because of hemosiderin released from RBCs but also because of brown eggs. The eggs of *Paragonimus* species are birefringent and will polarize when exposed to plane-polarized light. Demonstration of eggs in the sputum has low sensitivity (30–40 % only for single respiratory sample) but improves to 54–89 % with use of multiple samples. Moreover, eggs are not demonstrable until 2–3 months after infection. Sensitivity of stool samples for diagnosis of paragonimiasis is even poorer (Procop 2009).

Serologic tests have emerged as important tools that aid clinicians in establishing the diagnosis of paragonimiasis. Uchiyama et al. (1999) studied 104 patients with paragonimiasis and found that although only 50 % of patients had eggs in sputum or bronchoalveolar lavage specimens, 99 % had serologic evidence of disease with positive IgG in serum and/or pleural effusion fluid. Immunodiagnostic techniques are useful in the diagnosis of patients with cerebral paragonimiasis, particularly early in the course of disease, as antibodies decline with chronicity (Nakamura-Uchiyama et al. 2002). A variety of serologic assays like immunodiffusion assays, immunoelectrophoresis assays and monoclonal antibody-based ELISAs have been developed, and most of these are highly sensitive, often exceeding 95 % sensitivity. Immunoassays with specificities less than 100 % usually have cross-reactivity with sera from patients with infections caused by certain other helminths, such as Clonorchis. The degree of cross-reactivity depends on the particular antigen used in the assay for antibody capture. Several assays with 99–100 % specificity have been developed after extensive refinements;

these include a monoclonal antibody-based IgG ELISA and an immunoblot assay. The development and incorporation of monoclonal antibodies into some of these assays have contributed to an increase in assay specificity (Yong et al. 1993). Centre for Disease Control, USA, has approved an immunoblot assay which targets antibodies directed against an 8-kDa antigen of *P. westermani.* It has a sensitivity of 96 % and specificity of 99 % (Slemenda et al. 1988). However, none of these can differentiate between various species.

Molecular tools for detection of egg DNA have been developed for Paragonimus. Broad-range PCR assays that target conserved regions within the Paragonimus genome and species-specific PCR assays have been constructed for the detection of some Paragonimus species. The commonly used genetic targets for species differentiation following sequencing include the internal transcribed spacer (ITS) regions of ribosomal genetic complexes and the mitochondrial cytochrome c oxidase gene (Cui et al. 2003; Procop 2009; Iwagami et al. 2003; Ryu et al. 2000).

Complementary tools for diagnosis are ultrasound, CT and MRI techniques. CT is used to detect peripheral density, bronchial wall thickening and centrilobular nodules. However, the differentiation between the pulmonary paragonimiasis and other infections cannot be made by CT. CT and MRI findings of clustered ring-enhancing lesions in the brain are very suggestive of cerebral paragonimiasis (Yao et al. 2004, 2006). Supportive tests include total leucocyte counts, and eosinophilia is a common finding in patients with paragonimiasis, with approximately two-thirds having 500×10^9 eosinophils/mm^3. Eosinophils and Charcot-Leyden crystals may be found in the respiratory secretions and stools of patients with paragonimiasis (Procop 2009).

Chemotherapy

Only 2 drugs, that is, praziquantel and triclabendazole, are available for treatment for trematode infections. Former has broad-spectrum activity against trematodes and an excellent safety profile and hence the drug of choice. It is given as a single oral dose of 25 mg/kg body weight for intestinal infections and 25 mg/kg TDS for 2 days for liver and lung fluke infections. In mass drug administration programmes, a single 40 mg/kg dose is used. Though at present there is no drug resistance, a low cure rate (29 %) was documented for patients infected with *C. sinensis* who were treated with praziquantel (25 mg/kg × 3 days) in Vietnam. Side effects are minor and include abdominal pain, headache, nausea, urticaria (Keiser and Utzinger 2004).

Triclabendazole, developed initially as veterinary drug used for treatment for fascioliasis and Paragonimus infections, is registered in only 4 countries, that is, Egypt, France, Equador and Venezuela. A single postprandial dose of 10 mg/kg is recommended. Abdominal pain, fever, nausea, pruritus, vomiting and liver enlargement following treatment have been reported. Preliminary studies on rodent models have shown the antihelminthic properties of artemisinins (artemether and artesunate) (Keiser et al. 2005). However, those need to be evaluated further before they can be used for humans.

Prevention and Control

All food-borne trematode infections can be controlled by interruption of life cycle that can be achieved through proper diagnosis, treatment and prevention of reinfection. Since most of infections occur from eating raw molluscs, fish, crustaceans and amphibians, these can be prevented by educating people, implementation of control programmes, including mass antihelminthics, and sanitary control of human excreta to break the cycle of flukes. Control of snail intermediate hosts by molluscicides is not considered feasible because of widespread distribution of snails and low prevalence of fluke infection of snails even in endemic communities and economic considerations. Health education includes avoidance of eating raw molluscs, fish, crustaceans, amphibians, water plants and untreated water; crab meat should never be eaten raw. Crabs pickled in alcohol as a method of preservation are insufficient to kill the parasite. Whenever fresh crustaceans are cut or processed, all the utensils and cutlery boards that have been used should be thoroughly cleaned prior to using these to prepare any other foods, particularly those foods that will be consumed without cooking (e.g., salads). Heating is the most effective method for eliminating the risk of parasitic disease from seafood. The internal temperature of the thickest part of the product must reach a minimum of 63 °C (145 °F) for 15 s or longer. Fish can be frozen at −20 °C for 7 days or −35 °C or below for 15 h prior to consumption to reduce the risk (Butt et al. 2004). Other control measures for FBT include restraining pigs from having access to ponds and canals, prohibiting use of aquatic green fodders to feed them and unsterilized night soil as a fertilizer. There have been some vaccine trials in animals for *F. hepatica* with protection rates ranging from 19 to 72 %, but no such trials have been carried out in humans (Spithill and Dalton 1998). Currently, no vaccine is available for prevention of FBTs and should be the focus of research.

Conclusions

It is estimated that more than 10 % of world population is potentially at a risk of FBT, and globally, more than 100 million people are infected with one or more of flukes. Most of the infections occur in areas where aquatic animals are eaten undercooked or raw. In spite of increasing awareness about water- and food-borne infections and association of some flukes with cholangiocarcinoma, scarce reports are available for control of these infections. There is an urgent need for improvement in diagnostics and initiation of control programmes and health education in endemic communities.

References

Anthony RM, Rutitzky LI, Urban JF Jr, Stadecker MJ, Gause WC (2007) Protective immune mechanisms in helminth infection. Nat Rev Immunol 7(12):975–987

Bhattacharyya S, Khurana S, Bhatti HS, Singhi S, Malla N (2010) Fasciolopsis buski, Ascaris lumbricoides and Hookworm coinfection in a child: a case report. Trop Gastroenterol 31:126–127

Bhatti HS, Malla N, Mahajan RC, Sehgal R (2000) Fasciolopsiasis–a re-emerging infection in Azamgarh (Uttar Pradesh). Indian J Pathol Microbiol 43:73–76

Bostock J, McAndrew B, Richards R, Jauncey K, Telfer T, Lorenzen K, Little D, Ross L, Handisyde N, Gatward I, Corner R (2010) Aquaculture: global status and trends. Philos Trans R Soc Lond B Biol Sci 365:2897–2912

Butt AA, Aldridge KE, Charles VS (2004) Infections related to the ingestion of seafood. Part II: parasitic infections and food safety. Lancet Infect Dis 4:294–300

Chandra SS (1984) Epidemiology of fasciolopsis buski in Uttar Pradesh. Indian J Med Res 79:55–59

Choi D, Hong ST, Lim JH, Cho SY, Rim HJ, Ji Z et al (2004) Sonographic findings of active Clonorchis sinensis infection. J Clin Ultrasound 32:17–23

Choi D, Jeon YH, Lee GC, Choi MH, Hong ST (2009) Changes in sonographic findings after treatment of patients with clonorchiasis in a heavy endemic area. Korean J Parasitol 47:19–23

Cui AL, Chang ZS, Chen MG, Blair D, Chen SH, Zhang YN, Feng Z (2003) Taxonomic status in DNA sequences of 5 species of genus paragonimus. Zhongguo Ji Sheng Chong Xue Yu Ji Sheng Chong Bing Za Zhi 21:27–30

Dixit AK, Dixit P, Sharma RL (2008) Immunodiagnostic/protective role of cathepsin L cysteine proteinases secreted by Fasciola species. Vet Parasitol 154:177–184

Espinoza JR, Timoteo O, Herrera-Velit P (2005) Fas2-ELISA in the detection of human infection by Fasciola hepatica. J Helminthol 79:235–240

FAO (2010) The state of world fisheries and aquaculture 2010. Food and Agriculture Organization of the United Nations, Rome

Gadkowski LB, Jason ES (2008) Cavitary pulmonary disease. Clin Microbiol Rev 21:305–333

Garcia LS (ed) (2007) Diagnostic medical parasitology, 5th edn. ASM Press, Washington DC

Hong S-T, Fang Y (2011) Clonorchis sinensis and clonorchiasis, an update. Parasitol Int. doi: 10.1016/j.parint.2011.06.007

Iwagami M, Monroy C, Rosas MA, Pinto MR, Guevara AG, Vieira JC, Agatsuma T (2003) A molecular phylogeographic study based on DNA sequences from individual metacercariae of Paragonimus mexicanus from Guatemala and Ecuador. J Helminthol 77:33–43

Kaewpitoon N, Kaewpitoon SJ, Pengsaa P, Sripa B (2008) Opisthorchis viverrini: the carcinogenic human liver fluke. World J Gastroenterol 14:666–674

Keiser J, Utzinger J (2004) Chemotherapy for major food-borne trematodes: a review. Expert Opin Pharmacother 5:1711–1726

Keiser J, Utzinger J (2005) Emerging foodborne trematodiasis. Emerg Infect Dis 11:1507–1514

Keiser J, Utzinger J (2009) Food-borne trematodiases. Clin Microbiol Rev 22:466–483

Keiser J, Engels D, Büscher G, Utzinger J (2005) Triclabendazole for the treatment of fascioliasis and paragonimiasis. Expert Opin Investig Drugs 14:1513–1526

Lun ZR, Gasser RB, Lai DH, Li AX, Zhu XQ, Yu XB et al (2005) Clonorchiasis: a key foodborne zoonosis in China. Lancet Infect. Dis. 5:31–34

MacDonald AS, Araujo MI, Pearce EJ (2002) Immunology of parasitic helminth infections. Infect Immun 70:427–443

Mahanta J, Narain K, Srivastava VK (1995) Heterophyid eggs in human stool samples in Assam: first report for India. Indian J Commun Dis 27:142–145

Maria VJ, Sithithaworn P, Bergquist R, Utzinger J (2010) Towards improved diagnosis of zoonotic trematode infections in Southeast Asia. Adv Parasitol 73:172–195

Muralidhar S, Srivastava L, Aggarwal P, Jain N, Sharma DK (2000) Fasciolopsiasis–a persisting problem in eastern UP: a case report. Indian J Pathol Microbiol 43:69–71

Nagano I, Pei F, Wu Z, Wu J, Cui H, Boonmars T, Takahashi Y (2004) Molecular expression of a cysteine proteinase of Clonorchis sinensis and its application to an enzyme-linked immunosorbent assay for immunodiagnosis of clonorchiasis. Clin Diagn Lab Immunol 11:411–416

Nakamura-Uchiyama F, Mukae H, Nawa Y (2002) Paragonimiasis: a Japanese perspective. Clin Chest Med 23:409–420

Narain K, Rekha Devi K, Mahanta J (2003) Paragonimus and Paragonimiasis: a new focus in Arunachal Pradesh, India. Curr Sci 84:985–987

O'Neill SM, Parkinson M, Dowd AJ, Strauss W, Angles R, Dalton JP (1999) Immunodiagnosis of human fascioliasis using recombinant Fasciola hepatica cathepsin L1 cysteine proteinase. Am J Trop Med Hyg 60:749–751

Procop GW (2009) North American Paragonimiasis (caused by paragonimus kellicotti) in the context of global paragonimiasis. Clin Microbiol Rev 22:415–446

Robinson MW, Dalton JP (2009) Zoonotic helminth infections with particular emphasis on fasciolosis and other trematodiases. Trans R Soc B 364:2763–2776

Ryu JS, Hwang UW, Min DY, Shin KS, Nam SJ, Lee OR (2000) Molecular identification of Paragonimus ohirai and P. westermani from Anhui Province, China. Parasite 7:305–309

Satarug S, Lang MA, Yongvanit P, Sithithaworn P, Mairiang E, Mairiang P et al (1996) Induction of cytochrome P450 2A6 expression in humans by the carcinogenic parasite infection, Opisthorchiasis viverrini. Cancer Epidemiol Biomark Prev 5:795–800

Satarug S, Haswell-Elkins MR, Sithithaworn P, Bartsch H, Ohshima H, Tsuda M et al (1998) Relationships between the synthesis of N-nitrosodimethylamine and immune responses to chronic infection with the carcinogenic parasite, Opisthorchis viverrini, in men. Carcinogenesis 19:485–491

Sayasone S, Vonghajack Y, Vanmany M, Rasphone O, Tesana S, Utzinger J, Akkhavong K, Odermatt P (2009) Diversity of human intestinal helminthiasis in Lao PDR. Trans R Soc Trop Med Hyg 103:247–254

Singh TS, Mutum SS, Razaque MA (1986) Pulmonary paragonimiasis: clinical features, diagnosis and treatment of 39 cases in Manipur. Trans R Soc Trop Med Hyg 80:967–971

Singh TS, Sugiyama H, Umehara A, Hiese S, Khalo K (2009) Paragonimus heterotremus infection in Nagaland: a new focus of paragonimiasis in India. Indian J Med Microbiol 27:123–127

Slemenda SB, Maddison SE, Jong EC, Moore DD (1988) Diagnosis of paragonimiasis by immunoblot. Am J Trop Med Hyg 39:469–471

Spithill TW, Dalton JP (1998) Progress in development of liver fluke vaccines. Parasitol Today 14:224–228

Sripa B, Kaewkes S, Sithithaworn P, Mairiang E, Laha T, Smout M et al (2007) Liver fluke induces cholangiocarcinoma. PLoS Med 4:e201

Sripa B, Kaewkes S, Intapan PM, Maleewong W, Brindley PJ (2010) Food-borne trematodiases in Southeast Asia: epidemiology, pathology, clinical manifestation and control. Adv Parasitol 72:305–350

Ubeira FM, Muino L, Valero MA, Periago MV, Perez-Crespo I, Mezo M et al (2009) MM3-ELISA detection of Fasciola hepatica coproantigens in preserved human stool samples. Am J Trop Med Hyg 81:156–162

Uchiyama F, Morimoto Y, Nawa Y (1999) Re-emergence of paragonimiasis in Kyushu, Japan. Southeast Asian J Trop Med Public Health 30:686–691

Uppal B, Wadhwa V (2005) Rare case of Metagonimus yokogawai. Indian J Med Microbiol 23:61–62

Vatsal DK, Kapoor S, Venkatesh V, Vatsal P, Husain N (2006) Ectopic fascioliasis in the dorsal spine: case report. Neurosurgery 59:E706–E707

Wongratanacheewin S, Sermswan RW, Sirisinha S (2003) Immunology and molecular biology of Opisthorchis viverrini infection. Acta Trop 88:195–207

Yao LX, Yao CY, Qian WK et al (2004) MRI appearance of cerebral paragonimiasis. Radiol Pract 19:274–276

Yao A, Hammond N, Alsadi R, Nikolaidis P (2006) Central hepatic involvement in paragonimiasis: appearance on CT and MRI. AJR Am J Roentgenol 187:w236–w237

Yong TS, Seo JH, Yeo IS (1993) Serodiagnosis of human paragonimiasis by ELISA-inhibition test using monoclonal antibodies. Korean J Parasitol 31:141–147

Water and Lymphatic Filariasis

16

Shailja Misra-Bhattacharya, Susheela Kushwaha, and Preeti Bajpai

Abstract

Lymphatic filariasis (LF) is one of the most prevalent diseases of tropical and subtropical countries that is transmitted by mosquito and accompanied by a number of pathological conditions. It has been recognized as second leading cause of permanent and long-term disability. Research within the last decade has provided newer and better diagnostics methods for LF, improved disease management and treatment control strategies. World Health Organization (WHO) recognized LF as a potentially eradicable disease and launched Global Programme to Eliminate Lymphatic filariasis (GPELF) in 2000 to eliminate this disease by the year 2020. The release of sequenced and annotated genome of *Brugia malayi* and its endosymbiont *Wolbachia* has provided new insight into the pathogenesis of filarial disease, chemotherapy, and the mechanism of host–parasite interaction. The genome information is useful to identify novel antifilarial drug targets or design potent inhibitors for the existing or new targets. This chapter briefly gives an overview of LF including the diagnostic methods, symptoms, treatment strategies, and molecular biology of filarial parasite.

S. Misra-Bhattacharya (✉)
Division of Parasitology, CSIR—Central Drug Research Institute, Chattar Manzil Palace, Mahatma Gandhi Marg, 173 Lucknow 226001, India
e-mail: shailjacdri@yahoo.com

S. Kushwaha
Division of Parasitology, CSIR—Central Drug Research Institute, 10/1, Sector 10, Jankipuram Extension, Sitapur Road, Lucknow, Uttar Pradesh 226031, India

P. Bajpai
Department of Biotechnology, Integral University, Dasauli, P.O. Bas-ha Kursi Road, Lucknow, 226026, India

P. P. Singh and V. Sharma (eds.), *Water and Health*,
DOI: 10.1007/978-81-322-1029-0_16, © Springer India 2014

Keywords
Brugia malayi · Chyluria · Diethylcarbamazine · Genome · Lymphatic filariasis · Lymphedema · *Wolbachia*

Introduction

Neglected tropical diseases (NTDs), a group of parasitic and bacterial diseases, are the source of tremendous suffering because of their disfiguring, debilitating, and sometimes deadly impact. More than 1 billion people (about one-sixth of the world's population) are infected with one or many of these NTDs. LF commonly known as elephantiasis is one of the 17 WHO-listed NTDs. LF is the second leading cause of permanent and long-term disability imposing a strong impediment to socioeconomic development and causes mental and physical trauma as a result of serious disabilities caused by this disease. It is caused by the thread-like nematodes *Wuchereria bancrofti, Brugia malayi, and B. timori*. More than 95 % of the LF infection is caused by *W. bancrofti* and remaining by *B. malayi* and *B. timori*. These tissue-dwelling worms belong to the phylum Nematoda, class Secernentea, order Spirurida, and family Filariidae.

Filarial Parasite: Life Cycle and Transmission

Filarial nematodes are transmitted by female mosquitoes of *Culex, Anopheles, Aedes,* and *Mansonia* species. *Culex quinquefasciatus* is the most common mosquito found in human habitations in the tropics and subtropics which transmits the nocturnally periodic, *W. bancrofti*. *Aedes polynesiensis* incriminated in the transmission of the nonperiodic, *W. bancrofti* mainly in endemic islands in the Pacific, and Mansonoides spp. (*M. annulifera, M. uniformis, M. indiana* and *M. tongipalanis*) transmit *B. malayi*. The infection is transmitted by introduction of third-stage infective larvae (L3) of the parasite into the host by the bites of L3-bearing mosquitoes. Following deposition on the skin L3 migrates into local lymphatic vessels (Maizels et al. 2001). After 8 ± 1 day of entry, L3 molt and fourth-stage larvae (L4) appear. Subsequent development of L4 to adult worms occurs over a period of 3–12 months. Sexually mature male (13–23 mm length) and female adult worms (43–55 mm length) reside in afferent lymphatic vessels, copulate and fecund females viviparously produce microfilariae (L1-stage larvae). These microfilariae are sheathed, 177–230 μm in length, exhibit either nocturnal or diurnal periodicity, coinciding with peak feeding time of the mosquito vector (Moulia-Pelat et al. 1993). Microfilariae migrate into the lymph and enter the blood stream reaching the peripheral blood and are taken up by the mosquito vector. Within the mosquito, ingested microfilariae exsheath, penetrate the insect gut wall and migrate to the thoracic muscles where they mature into the third-stage larvae (L3) after two molts (Fig. 16.1).

Global Prevalence of LF

W. bancrofti is distributed throughout the tropical regions of Asia, Africa, Americas, and the Pacific, and is particularly prevalent in areas with hot and humid climate, while *B. malayi* is found in Southeast Asia and in areas of Southwest India, whereas *B. timori* occurs only on some islands in Indonesia. Currently, an estimated 129 million people are infected worldwide and about 1.39 billion live in filariasis endemic areas (Fig. 16.2). Approximately 40 million people suffer from the stigmatizing and disabling clinical manifestations of the disease, including 15 million who have lymphedema (elephantiasis), and 25 million men who have urogenital swelling, principally scrotal hydrocele. Southeast Asia region contributes to almost two-third of global cases with India having a total of 590 million people at risk and 10 % of

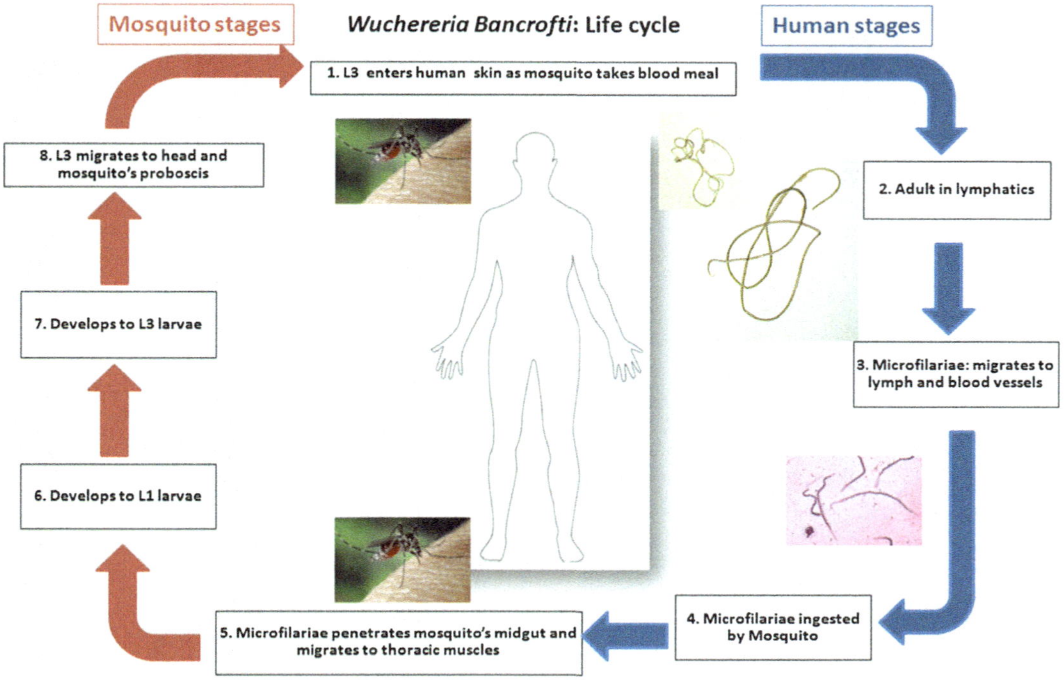

Fig. 16.1 Life cycle of lymphatic filariasis: *Wuchereria bancrofti*

them already suffering from LF. According to WHO 2010 recent report, nine countries, such as Burundi, Cape Verde, Costa Rica, Mauritius, Rwanda, Seychelles, Solomon Islands, Suriname, and Trinidad and Tobago, earlier categorized as endemic for LF are now classified as nonendemic while China and the Republic of Korea have been officially recognized as having eliminated LF as a public health problem.

Historical Background

The first written documents of LF come from the ancient Greek and Roman writers who could differentiate between the similar symptoms of leprosy and LF, describing leprosy as "elephantiasis graecorum" and LF as "elephantiasis arabum." Jan Huygen Linschoten during his trip to Goa between 1588 and 1592 first documented the disease symptoms and wrote that inhabitants "all born with one of their legs and one foot from the knee downwards as thick as an elephant's leg." In 1849, William Prout in his book "*On the Nature and Treatment of Stomach and Renal Diseases*" documented chyluria. In 1863, French surgeon Jean-Nicolas Demarquay first reported microfilaria in hydrocoele fluid, and Otto Henry Wucherer in 1866 observed microfilaria in urine. Adult worms were recovered by Joseph Bancroft in Australia in 1876 and named *Filaria bancrofti*. In 1921, this species was included in the genus *Wuchereria*. The most important and significant contribution in discovery of LF was made by Patrick Manson in 1877 (McGregor 1995) who observed the development of the microfilaria in mosquito fed on the blood of his microfilaremic gardener and speculated its transmission by mosquito.

Clinical Manifestation of LF

Clinical Manifestations of LF are influenced by number of cofactors, including patient's age and gender, the species and strain of parasite,

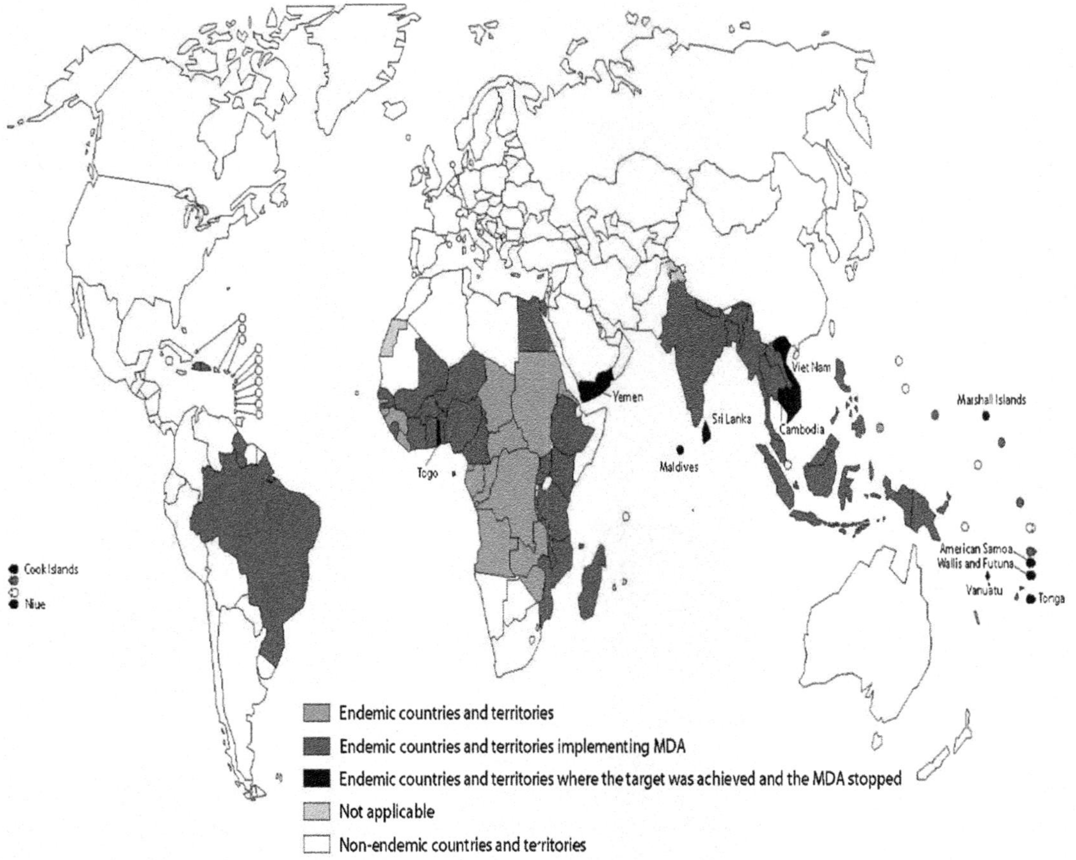

Fig. 16.2 World map showing countries where lymphatic filariasis is endemic and status of mass drug administration in those countries, WHO (2010)

anatomical location of the adult worms, immune response to the parasite, and secondary bacterial infections (Friedman and Kalisher 2002). The clinical manifestations of the disease are due to the inflammation and damage of lymphatic vessels. Most infections do not produce symptoms, but in people where the lymphatic drainage is badly impaired, the common symptoms include *hydrocele* (swelling of the scrotum), swelling of the legs and feet, and thickening of the skin into folds. Lymphedema is a common clinical manifestation of LF. Filariasis due to *W. bancrofti* involves the entire limb, the genitals, or breasts, whereas *B. malayi* infection differs in that the lymphedema involves only the legs below the knee and upper limbs below the elbow, without any genital or breast involvement. The spectrum of LF infection in filaria-endemic area ranges from individuals who despite constant exposure to infection remain free from microfilariae or symptoms (endemic normal), to asymptomatic microfilaremic (carriers), to subjects with acute or chronic clinical manifestations (symptomatics) (Fig. 16.3).

Asymptomatic Microfilaremic Carriers

"Asymptomatic microfilaremia" is most common manifestation of filariasis in endemic population and includes individuals having no clue of presence of microfilariae circulating in blood for decades without any progression to overt clinical

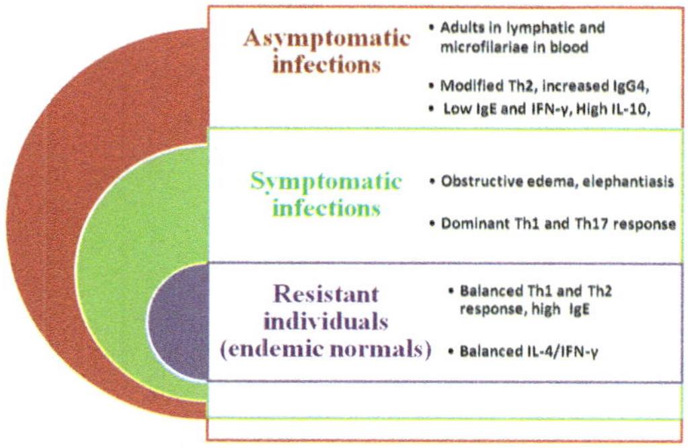

Fig. 16.3 Clinical spectrum of human filarial infection

disease (Ottesen 1992). These carriers are likely to harbor fecund adults in the lymphatics and appear to be immunologically tolerant to parasite antigens demonstrating reduced cellular immunity to filarial antigens (Maizels et al. 1991). Earlier considered to be totally symptom free, profound lymphatic tissue changes with collateral channeling and/or glomerulonephritis has been observed by noninvasive diagnostic methods (Langhammer et al. 1997; Dissanayake 2001).

Acute Clinical Manifestation

The most common acute clinical manifestations in LF are adenitis and lymphangitis that are usually associated with chills, fever, and pain in the involved region. Though occasionally seen during the early stages of the disease, these episodes are more frequent in higher grades of lymphedema. Several factors have been associated with these acute episodes such as secondary bacterial infection, immunological responses to a variety of filarial antigens, excreted or secreted products by the parasite or exposure to fresh infection with L3, release of substances from death of the adult worm (Ottesen and Poindexter 1980). Acute filarial adenolymphangitis (AFL) is supposed to be caused by death of adult worm either spontaneously or as a result of treatment with a macrofilaricidal drug. Dilatation of the lymphatic vessels induced by the presence of the adult worm eventually leads to lymphatic dysfunction and accumulation of protein-rich fluid in the tissues. Latter may result into recurrent bacterial infections which further aggravates the disease. Trauma, interdigital fungal infections, and onychomycosis provide entry sites for these bacteria, which multiply rapidly and cause a reticular lymphangitis of the small collecting vessels; the condition is known as acute dermatolymphangioadenitis (ADLA) (Dreyer et al. 1999) which is a major risk factor for the development of elephantiasis (Addiss and Brady 2007). The lymphatics of the male genitalia are frequently affected, leading to funiculitis, epididymitis, and orchitis. Repeated episodes of acute attacks lead to the formation of fibrous and calcified tissues in and around the lymphatic vessels (Olszewski et al. 1997).

Chronic Manifestation of LF

Common chronic manifestation of LF is lymphedema of extremities which on progression result into elephantiasis. The upper limbs, male genitalia, and rarely breasts in the females may also be affected. The lymphedema of the limbs is commonly graded as grade I, grade II, and grade III (Shenoy 2008). Grade I is mostly pitting edema; reversible on elevation, grade II lymphedema is mostly nonpitting edema, not spontaneously reversible on elevation while grade III is called elephantiasis which is a gross increase in volume in grade II lymphedema accompanied with dermatosclerosis and papillomatous lesions. The most conspicuous symptoms occur due to blockage of lymphatics in the chronic stage leading to the intermittent discharge of intestinal lymph (chyle) into the renal pelvis and subsequently into the urine resulting into an incurable stage termed as chyluria (Hashim et al. 1964). The urine may be milky white in color, particularly after a fatty meal. Chyluria is often intermittent, associated with increased high endothelin-1 (ET-1) level and vascular endothelial growth factor (VEGF) (Esterre et al. 2005; Debrah et al. 2007). Hydrocoele is the most frequent chronic manifestation of bancroftian filariasis. Chronic epididymitis, funiculitis (inflammatory swelling of the spermatic cord), and lymphedematous

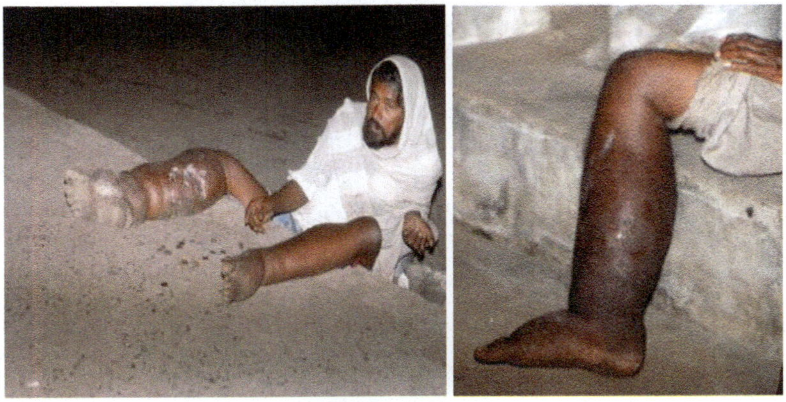

Fig. 16.4 Elephantiasis stage of lymphatic filariasis (pictures were taken from Andhra Pradesh, India)

thickening of the scrotal skin are also genital manifestations of chronic bancroftian filariasis (Shenoy et al. 1998). Secondary bacterial infections, inflammatory reactions, and immune response to parasite antigens add to the severity of morbidity (Fig. 16.4).

Tropical Pulmonary Eosinophilia (TPE) or Weingarten's Syndrome

TPE is a syndrome characterized by symptoms of bronchial asthma with paroxysmal nocturnal cough and anorexia. It is a relatively unusual manifestation of infection with the filariae *Wuchereria bancrofti* or *Brugia malayi*. TPE affects more males than females (4:1 ratio). Characteristic features of TPE include very high blood eosinophilia of >3,000/mm3, absence of microfilariae in the blood, and high antifilarial IgG4 and IgE antibodies. Diethylcarbamazine (DEC) is effective in the treatment of TPE (Rom et al. 1990). If left untreated, TPE tends to relapse and progress to a condition of chronic pulmonary fibrosis (Magnaval and Berry 2005; Vijayan et al. 2007).

Diagnosis

The precise diagnosis of filariasis is important for successful treatment. During the last few years, new diagnostics methods have been developed that are useful in both initial diagnosis of LF and also in post-treatment follow-up.

Parasitological Diagnosis

Traditional diagnosis of LF is based on the presence of microfilaria in patient's blood collected around midnight in areas where microfilariae exhibit nocturnal periodicity, and around midday where periodicity is diurnal (Khamboonruang et al. 1987; Sabry 1992). The Giemsa stained thick blood smear method is the most commonly used method that is simple but has low sensitivity owing to small amount of blood examined and loss of microfilariae during the staining procedure (McMahon et al. 1981). Further, microfilariae are not normally seen in peripheral blood in acute and chronic infections as well as occult filarial infections, making it difficult to diagnose clinically for successful treatment. The counting chamber technique is fast, quantitative, cheap, and most suitable for routine hospital diagnosis as well as for field surveys. Another sensitive method used for diagnosis is the Knott concentration technique using formalin, triton x-100 treatment to the diluents to overcome protein precipitation due to formalin (Melrose et al. 2000). The membrane (Nuclepore) filtration techniques are sensitive where staining of filters enables identification of the microfilariae but impractical for field surveys because venous blood is needed. In DEC provocation test, nocturnally periodic microfilariae come into the peripheral blood during the day time by giving DEC (Sabry 1988).

Immunodiagnosis

Immunodiagnostic tests are very sensitive and specific and have the potential to replace traditional nighttime blood surveys where both microfilaremic and amicrofilaremic cases can be detected any time during 24 h.

Antigen Detection

Filarial antigen tests have revolutionized the diagnosis of bancroftian filariasis since their introduction in the 1980s and are now regarded as the "gold standard." Two monoclonal antibody–based enzyme-linked immunosorbent assays (ELISAs) that detect circulating *W. bancrofti* antigens have been developed. The first assay, based on the monoclonal antibody AD12, recognizes a 200-kDa antigen and other utilizes the monoclonal antibody Og4C3 (Weil et al. 1987; Turner et al. 1993). Since circulating antigen remains available 24 h, blood can be collected during the day. The Og4C3 ELISA was commercialized and is available as Trop-Ag *W. bancrofti*, manufactured by JCU Tropical Biotechnology Pvt. Ltd., Townsville, Queensland, Australia. Its sensitivity ranges from 72 to 100 % and specificity 98.6–100 % (Lammie et al. 1993; Chanteau et al. 1994; Rocha et al. 1996). The monoclonal antifilarial antibody AD12 has recently been incorporated into a commercially available rapid format card test by ICT Diagnostics (Balgowlah, New South Wales, Australia). The assay has a reported sensitivity of 96–100 and 100 % specificity. Sensitivity appears to be lower in infected persons who are microfilaria negative or those with ultralow microfilarial densities. The test is simple, very quick, and easy to perform in field, and cards can be marked and stored for records.

Antibody detection assay: Filarial worms induce a wide range of immune responses in the host, and several immunodiagnostic techniques for LF based on the detection of specific antibodies in the patients' serum have been devised in the past. These have generally been of limited value in endemic areas because, (1) most individuals are positive for antibodies to crude filarial antigens as a result of constant exposure, (2) cross-reaction with other nematode infections, (3) unable to distinguish past and present infection. It may be of some value in diagnosing visitors to endemic areas who develop symptoms of LF without microfilaremia. Cross-reactions have been reduced by detecting specific IgG4 antibody which has been reported to be a good marker of active infection. Such tests may be of particular value in brugian filariasis, for which progress in development of circulating antigen detection-based diagnosis has been limited (Ottesen et al. 1985; Lal and Ottesen 1988; Kwan-Lim et al. 1990; Lammie et al. 1993; Estambale et al. 1994). A rapid format dipstick test (Brugia Rapid) detecting IgG4 to a *B. malayi* recombinant antigen in whole blood is currently undergoing field evaluation. Other recombinant antigen–based antibody assays are also being developed and tested for their usefulness in monitoring human exposure and infection in control programs. For *W. bancrofti*, recombinant protein Bm14 has given promising results for detection of antifilarial IgG4 antibodies and has been employed in longitudinal studies in Egypt. An ELISA that detects filarial specific IgG4 antibodies in urine has also been devised and reported to have high sensitivity (96 %) and specificity (99 %). However, ELISA-based diagnostic tests require a well-equipped laboratory and equally well-trained staff to conduct the assays.

DNA Detection-Based Assays

Polymerase chain reaction (PCR) assays for detection of microfilarial infections in humans have been developed for both *W. bancrofti* and

B. malayi. Sensitivity is high, although these assays appear to be positive only when circulating microfilariae are detectable (Lizotte et al. 1994; Williams et al. 1996). The techniques need at least one microfilaria in the volume of blood used for DNA extraction and therefore are more sensitive than microscopic blood examination for microfilariae. Detection of DNA appears to be a powerful tool for detection of infection in vectors (Lainy et al. 2008).

Ultrasonography and Lymphoscintigraphy

Adult *W. bancrofti* can be detected by ultrasonography in lymphatic vessels of the scrotal area of infected males. They are more dispersed and more difficult to detect within the lymphatic system of infected females (Amaral et al. 1994). The live worms wriggle continuously inside the dilated vessel ("filaria dance sign" or FDS) (Dreyer et al. 1999). Though less useful in *B. malayi* infections, worms have been seen in the breast, thigh, calf, and inguinal lymph node. Lymphoscintigraphy can also be useful in documentation of the diagnosis, evaluation, as a screening procedure to prevent progression, and to enhance management of filarial lymphedema (Shelley et al. 2006).

Control Methods

There are several approaches to control vector-borne parasitic diseases such as vaccination, vector control, breaking of vector-host contact by use of repellents and bed nets and chemotherapy. Successful programs for the control of LF are based on distribution and dynamics of the disease. In endemic communities, the disfiguring and debilitating clinical manifestations result in much suffering and have severe socioeconomic and psychological consequences for those affected. The objective of control is to reduce transmission and morbidity, thereby eliminating LF as a public health problem. The diverse characteristics of communities in endemic foci, as well as differences in vector, parasite, and disease parameters, emphasize the importance of having multiple measures and approaches for control.

Chemotherapy

Chemotherapeutic control of LF is generally based on mass treatment, that is, administration of drug to total population in a community (except individuals in whom it is contraindicated). The strategy of using mass drug administration programs for 4–6 years is based on the assumption that reduction in the mf to very low levels would slowdown transmission and re-emergence of disease. The drugs used in LF control programs are diethylcarbamazine (DEC), ivermectin alone or in combination with albendazole.

Diethylcarbamazine (*DEC*): DEC was first discovered against *L. sigmodontis* in cotton rat in 1944 and has been in use for the treatment and control of LF since 1947 (Hewitt et al. 1947). The drug is inactive *in vitro* and is known to act through the host immune system. A single dose of DEC (6 mg/kg) can reduce microfilaria production from adult females by 67–87 % and blood mf density by 57–52 % at 1–2 years post-treatment in humans (Stolk et al. 2005). DEC is also known to have partial macrofilaricidal (adulticidal) effects (Chen 1964; Noroes et al. 1997) and is not recommended in onchocerciasis or areas with *O. volvulus* and *Loa loa* coendemicity due to severe side effects known as the "Mazzotti reaction." DEC is also given as DEC mixed in the cooking salt (0.1–0.6 %) that caused long-lasting reduction in microfilariae compared with DEC tablets. Fortifying salt with DEC is a safe and cost-effective method for controlling LF transmission.

Ivermectin (*IVM*): Ivermectin was introduced in 1981 in worm control program. It was found to be safe and effective for treating river blindness in Africa, reducing clinical symptoms, and halting progression to blindness (Aziz et al. 1982b). IVM exhibits potent microfilaricidal activity against major human filarial parasites

(Aziz et al. 1982a; Diallo et al. 1986; Ismail et al. 1996; Addiss et al. 1997) without affecting adult parasites, the effect lasting up to 12 months (Thylefors 2004). IVM targets the glutamate-gated Cl and K+ channels in nematodes, insects, and ticks (Arena et al. 1992; Cully et al. 1994) and paralyses nematode body wall muscle, including the pharynx (Feng et al. 2002; Sheriff et al. 2005). Severe adverse neurological reactions have been reported in people with high *L. loa* infection intensity, which hindered the IVM treatment program in areas potentially coendemic for *L. loa* (Chippaux et al. 1996). The genetic evidence of IVM resistance has recently been demonstrated by various studies (Lustigman and McCarter 2007) and the picture is complicated by the fact that the molecular basis of resistance to IVM is not understood in any parasite (Wolstenholme et al. 2004).

Albendazole (*ALBZ*): A single dose of ALBZ (600 mg), especially in combination with either IVM (400 mg/kg) or DEC (6 mg/kg) (Horton 1997), proved to have both long-term effectiveness and safety in decreasing microfilariae in *W. bancrofti* infections (Ismail et al. 1996). Macrofilaricidal properties of ALBZ apparently prolonged and reinforced the microfilaricidal suppression produced by the other two drugs (Rajendran et al. 2004, 2006). Apart from these drugs, several new drugs (moxidectin, emodepside, polycorpol), organic compounds and their derivatives, plant products, synthetic compounds, and marine products have also been investigated (Singh et al. 2010). The impact of combination therapy was first described in 1987 against LF (Molyneux et al. 2003). The recommended drug combinations are ALB and DEC or ALB and IVM for LF and IVM and ALB or IVM alone for onchocerciasis (Fox 2006; Caffrey 2007; Udall 2007).

Mass drug administration (*MDA*): In 1997, the WHO began an initiative "The Global Programme to Eliminate Lymphatic Filariasis (GPELF)" by organizing alliance between governments, health organizations, and pharmaceutical companies (GlaxoSmithKline, Pfizer, and Merck and Co., Inc.) under the Global Programme to Eliminate Lymphatic Filariasis. The program was supported by a large-scale donation of albendazole for worldwide elimination of LF (from Glaxo SmithKline), and by an expansion of the Mectizan® Donation Program for onchocerciasis to include free ivermectin for elimination of LF in African countries where the two infections co-exist.

LF MDA program uses repeated annual single-dose mass treatment of the filaria-endemic population; DEC and IVM in a single-dose combination were found to have better microfilaricidal efficacy than either of the single drugs. Most MDA programs appear to be less effective than originally planned and, depending upon the initial disease prevalence, need to be continued for a much longer period, a situation that in view of the risk of the development of drug resistance is also highly undesirable.

Antiwolbachial therapy: Another novel approach to control filariasis is to target the endosymbiont bacteria, *Wolbachia* which live intracellularly within filariids, and appear to be essential to healthy growth and development of the filarial parasite (Hoerauf et al. 2001). In 2003, the common antibiotic doxycycline was suggested for treating elephantiasis (Hoerauf et al. 2003). Taylor et al. (2005) reported that an 8-week course of doxycycline was safe and well-tolerated treatment for LF with significant activity against adult worms and microfilaremia. Long-term treatment of IVM (6 μg/kg) combined with an antiwolbachia antibiotic doxycycline (DOXY; 10 mg/kg) against *Dirofilaria immitis* in experimentally infected dogs resulted in a significantly faster decrease in circulating microfilariae and higher adulticidal activity as compared with either drug individually (Bazzocchi et al. 2008). The crucial point in the coadministration of drugs in MDA programs is that the drug combinations do not show pharmacokinetic interactions with no increased adverse effects. The existing data provide valuable estimates of drug effect; however, they also advocate a need for more comprehensive comparative drug studies. Filarial parasites have symbiotic bacteria in the genus *Wolbachia*, which live inside the worm and which seem to play a major role in both its reproduction and the development of the disease. Clinical trials in June 2005 by the Liverpool School of Tropical Medicine reported that an 8-week course almost completely eliminated microfilaremia. In an area of Cameroon, coendemic for onchocerciasis and loiasis, Turner et al. (2010) conducted a trial comparing doxycycline with or without ivermectin treatment to ivermectin treatment alone. A six-week course of doxycycline delivered macrofilaricidal and sterilizing activities, which is not dependent upon coadministration of ivermectin. Doxycycline was well tolerated in patients coinfected with moderate intensities of *L. loa* microfilariae. The trial indicates that antiwolbachial therapy is a feasible alternative to ivermectin in communities coendemic for onchocerciasis and loiasis.

Vector Control

Vector control can provide a useful supplement to chemotherapy in reducing LF transmission. The feasibility of vector control programs depends upon the local epidemiological conditions, including the species of vectors, their biting, resting, and breeding habits, and the type of environment (e.g., rural or urban). The main antivector measures are environmental control of breeding sites, use of insecticides against adult mosquitoes. The usefulness of *Bacillus sphaericus*, a toxin-producing bacterium, as a biological agent for control of *C. quinquefasciatus* is also being assessed. In areas where malaria and filariasis transmission depends on the same *Anopheles* species or vectors with similar bionomics, filariasis control may benefit from malaria vector control programs. Vector control programs along with chemotherapeutic efforts provided better filariasis control results.

Morbidity Control

Morbidity control focuses on improved hygiene, treatment of secondary infections, proper limb care and increased access to hydrocelectomy, to alleviate suffering and decrease disability among those already affected.

Vaccine Development

Development of a vaccine against filarial infection is an effective adjunct to existing control methods. Successful protection using irradiated L3 as a vaccine has been reported in various animal models of varied filarial sp (Oothuman et al. 1979; Chusattayanond and Denham 1986; Abraham et al. 1989; Devaney et al. 1993). Crude, semi-purified, purified, and recombinant filarial antigens derived from various life cycle stages offered varying degree of protection against parasite challenges (Tanner and Weiss 1981; Kazura and Davis 1982; Vedi et al. 2008; Shakya et al. 2009a, b). Despite this significant research, the development of vaccines for the prevention of filarial nematode infections is still remains in a state of relative infancy.

Brugia malayi Genomics

The WHO and the United Nations Development Program/World Bank/TDR launched Filarial genome project (FGP) in 1994 to study the genomes of lymphatic filarial parasites and complete draft-assembled and annotated genome of the TRS strain of *B. malayi* was released in 2007. Most filarial nematodes, including *B. malayi*, carry three genomes: nuclear, mitochondrial (available at GenBank, accession no. AF538716), and that of an alphaproteobacterial endosymbiont, *Wolbachia*. According to released genome, the *B. malayi* nuclear genome was found to be organized as five chromosomes, including an XY sex determination pair, has been estimated to be 80–100 mega bases (Mb), and has 14,500–17,800 protein-coding genes (Ghedin et al. 2007). From the sequenced genome, we can predict and identify several processes and proteins such as molting, nuclear receptors, collagens and collagen processing, neuronal signaling, reliance on host and endosymbiont metabolism, and various kinases to be potential drug targets that can accelerate the drug discovery programs. A total of 1,771 predicted gene products were used in the data snapshot of the *B. malay,i and* 7,435 was identified to have an ortholog in *C. elegans,* and 3,059 of these were mapped to the RNAi positive set, constituting a predicted "essential" *B. malayi* genome. The majority of these essential genes having close human homologs were removed, and finally, 589 genes were identified as a potential drug targets (Kumar et al. 2007).

Endosymbiont Bacteria *Wolbachia*

Wolbachia is an alpha proteobacterium, belonging to the order Rickettsiales and closely related to the genera *Ehrlichia*, *Cowdria,* and *Anaplasma*. In filarial nematodes, it was first observed in hypodermal tissues of lateral chords, uterine wall and in embryos of filarial nematodes (Fig. 16.5), embedded as single or multiple organisms in host-derived vacuoles (McLaren et al. 1975; Vincent et al. 1975; Kozek 1977). Ultrastructural studies showed that they are 0.6–1.5 µm in size, covered with a double membrane enclosing the cytoplasm which is rich in dense ribosomes and can attain different shapes oval, round, or rod-shaped (Martin and Gavotte 2010). In filarioid nematodes, *Wolbachia* have only been identified in two subfamilies of onchocercidae: onchocercinae and Dirofilariinae (Casiraghi et al. 2004; Bain et al. 2008). *Wolbachia* are present in *O. volvulus*, the filaria responsible for human onchocerciasis, in *W. bancrofti*, *B. malayi,* and *B. timori*, agents of LF, and in *Dirofilaria immitis*, known as heart worm in dogs. However, in both subfamilies, some filarial species do not contain *Wolbachia* (Casiraghi et al. 2004; Bain et al. 2008). The phylogeny of the symbionts was found congruent with the host phylogeny and indicated stable and specific association of bacteria and the filarial host (Bandi et al. 1999). The obligate dependency of *Wolbachia*-positive filariae on their symbiont was demonstrated by prolonged antibiotic treatment of filarial infections in animal models, which led to stunting and sterilization or death of adult worms.

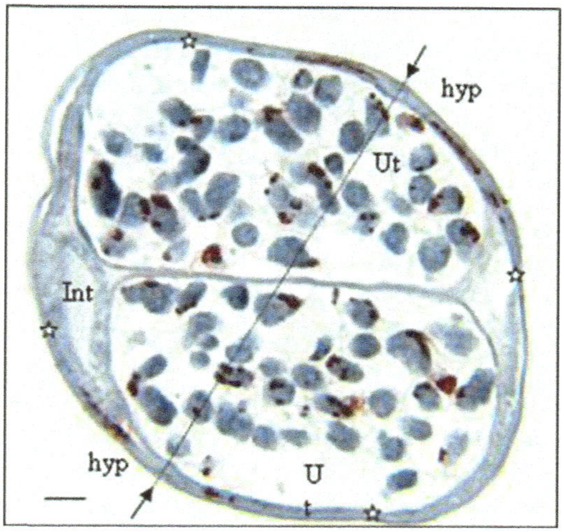

Fig. 16.5 *Wolbachia* in filaria *Litomosoides sigmodontis* visualized by immunostaining (taken from Martin and Gavotte 2010)

Wolbachia Genome

The *Wolbachia* Genome Consortium was held in 1999 (Slatko et al. 1999) to investigate and sequence the representative *Wolbachia* genomes. The *w*Mel genome is the first sequenced genome of the *Wolbachia* of *D. melanogaster* (*w*Mel) published in 2004 (Wu et al. 2004). Three new species of *Wolbachia* and their genomes were discovered within the sequenced genome of several *Drosophila* species: *Wolbachia wAna*, *Wolbachia wSim*, and *Wolbachia wMoj*. Complete genome DNA sequence and analysis for *B. malayi Wolbachia* (*w*Bm, super group D *Wolbachia*) was presented in 2005 (Foster et al. 2005). The genome of *w*Bm is represented by a single circular chromosome 1.1 Mb of size and is 34 % G + C, smaller than *w*Mel and *Rickettsia prowazekii*. The smaller genome size of wBm relative to wMel reflects loss of genes possibly, required for infecting host cells and avoiding host defense systems. Analysis of wBm genome revealed loss of transcriptional regulators, genes required for DNA repair, and RNA modification. While most signal transduction systems are also depleted, stress response and heat shock proteins are present. *Wolbachia* has incomplete pathways for biosynthesis of certain vitamins and cofactors such as NAD, biotin, lipoic acid, ubiquinone, folate, pyridoxal phosphate, and Coenzyme A, making them dependent on external sources, such as from the host. To date, there is no evidence of genes for riboflavin and heme synthesis in the *B. malayi* genome (Ghedin et al. 2004, 2007). All but one of the C4-type heme biosynthetic genes is readily identified from the *w*Bm genome. The filarial worms are incapable to synthesize heme de novo due to absence of ferrochelatase thus these worms presumably salvage heme/intermediates from their surroundings and/or acquire them from their *Wolbachia* endosymbionts. Heme from *Wolbachia* could be vital to worm embryogenesis, as there is evidence that molting and reproduction are controlled by ecdysteroid-like hormones (Warbrick et al. 1993) whose synthesis requires heme. Heme is also an essential cofactor for many proteins including cytochromes, hemoglobins, peroxidases, and catalases that are involved in a wide range of critical biological processes, including oxidative metabolism and electron transport (Wu et al. 2009). Depletion of *Wolbachia* would therefore halt production of these hormones and block embryogenesis. *Wolbachia* could be a source of glutathione which aside from its role in the detoxification of methylglyoxal is required for host protection against oxidative stress from oxygen-reactive species secreted by mammalian immune cells (Selkirk et al. 1998). *Wolbachia* as a source of the above metabolites would explain the block in embryogenesis and the sterility seen in worms after depletion of the endobacteria (Fenn et al. 2006). Examining the genome of wBm will help us understand the molecular basis for the endosymbiosis between filarial nematodes and *Wolbachia* (Pfarr and Hoerauf 2005).

Role of *Wolbachia* in Filarial Immunobiology

Wolbachia are speculated to play an important role in nutrition, metabolism and in subverting the host immune responses (Hosokawa et al. 2010; Brownlie et al. 2009). The presence of large number of bacteria in the parasitized tissues suggests the exposure of mammalian host to *Wolbachia* and their products as a consequence of degeneration of infective larvae, adult worms, or mf and/or as a result of their destruction by the host's immune mechanisms (Taylor et al. 2001; Brattig et al. 2004; Gillette-Ferguson et al. 2006), study conducted in filaria-infected individuals demonstrated contribution of *Wolbachia* in the pathogenesis of filarial disease. *Wolbachia* DNA and even *Wolbachia* bacteria can be detected within 1 day of DEC or ivermectin administration, in the plasma of onchocerciasis, or LF patients (Cross et al. 2001; Keiser et al. 2002; Hise et al. 2007). *Wolbachia* released into the bloodstream by degenerating or dead microfilaria cause the acute adenolymphangitis and fever, and these acute post-treatment reactions are accompanied by increase in plasma TNF-α, IL-6, and LPS-binding protein (Njoo et al. 1994; Turner et al. 1994). Role of *Wolbachia* in initiating or perpetuating local or systemic inflammatory reactions is proposed by several observations such as (1) filaria-induced innate immune cellular activation is impaired when parasite or infected animal hosts are pretreated with tetracycline that partially eliminate intracellular *Wolbachia* (Brattig et al. 2000; Taylor et al. 2005; Turner et al. 2006, 2010); (2) *Wolbachia* isolated from filaria or from insect cell lines elicit inflammatory responses similar to those of filarial extracts containing *Wolbachia* (Hise et al. 2007); (3) *O. volvulus* extract (containing *Wolbachia*) when injected into the corneal stroma using a murine model of ocular onchocerciasis was observed that the neutrophil infiltration and loss of corneal clarity were markedly reduced when Wolbachia-free *Acanthocheilonema viteae* or *Wolbachia*-depleted *O. volvulus* extracts were used (Saint Andre et al.

2002). *Wolbachia*-containing filarial extracts induce activation and tolerance in murine macrophages (Taylor et al. 2000; Turner et al. 2006), activate human monocytes (Brattig et al. 2000), and activate human and murine neutrophils (Gillette-Ferguson et al. 2006, 2007). Inflammatory responses to *Wolbachia* have been shown to be mediated primarily by engagement of TLR2 and the coreceptor TLR6 and are dependent on the adaptor molecules MyD88 and Toll/IL-1R domain-containing adaptor protein (TIRAP) 3/MyD88 adaptor-like. Presence of IgG antibodies against the *Wolbachia* surface protein (WSP) has been observed in infected humans and experimental animals as also WSP-mediated production of TNF-alpha, IL-12, and interferon (IFN-γ) (Brattig et al. 2004) and neutrophil chemotaxis via IL-8 production, a strong chemoattractant for neutrophils (Bazzocchi et al. 2003; Brattig et al. 2004). Recombinant HSP60 of *Wolbachia* decreased T-cell activation and lymphoproliferation in filarial patients that has been linked to increased T regulatory cell activity through increased expression of CTLA4 and CD25 on CD4+ T cells. Recently, *Wolbachia* diacyl lipoproteins were identified as being responsible for TLR 2 and TLR6 activation and attributed in pro-inflammatory cytokine production (Turner et al. 2009). In Wolbachia, lipoproteins also induce anti-inflammatory mediators as a feedback loop to limit the inflammatory burst (Haarbrink et al. 1999, 2000). Early TNF-α production by monocytes exposed to *Onchocerca* extracts was followed by the production of IL-10 and a reduced expression of HLA-DR and costimulatory molecules. The persistent low-level release of bacterial molecules may induce down-regulation of inflammatory mediators, thus contributing to the anergic state or hyporesponsiveness in chronic infections. Alternatively, *Wolbachia* may play an important role in protecting the nematodes against the host immune responses. For example, catalase, important in neutralizing hydrogen peroxide, is produced by the *Wolbachia* of *O. volvulus* (Henkle-Duhrsen et al. 1998) and confers longevity to *O. ochengi* through a defensive mutualism, by diverging a potentially lethal effector cell response. *O. ochengi*-infected cattle showed increased eosinophil degranulation with reduction in nodular bovine neutrophilic chemokines gene expression after treatment with oxytetracycline. Bacterial products released from both living (Landmann et al. 2010) and dead (Keiser et al. 2002) worms activate the innate immune system triggering the release of pro-inflammatory mediator; thus, *Wolbachia* also contributes to the clinical presentation of filarial disease.

Wolbachia: A Novel Drug Target

The obligate and mutualistic association of *Wolbachia* with filarial host offers novel target for antibiotic-based therapy in filariasis. The effect of antirickettsial antibiotic, tetracycline, against filarial nematodes was first observed in studies on *B. pahangi* by J. McCall showing that tetracycline could inhibit the development of third-stage larvae (L3) to adults and the establishment of microfilaremia. Several studies in human and animal models of disease have demonstrated that targeting of *Wolbachia* with tetracycline had detrimental consequences on filariae such as (1) complete and long-term inhibition of embryogenesis with cytotoxicity of developing embryos and inhibition of transovarial transmission, (Hoerauf et al. 2000b) (2) interference in molting process from larval stage 4 to adults (Casiraghi et al. 2002) (3) impaired microfilariae development into L3 (Arumugam et al. 2008), and (4) eventual loss of viability, including adulticidal activity (Hoerauf et al. 2001, 2008 ; Volkmann et al. 2001; Mand et al. 2008; Specht et al. 2008; Mand et al. 2009). Evidence for *Wolbachia* as antibiotic targets emerged from studies on *Acanthocheilonema viteae, and* filarial parasites free of *Wolbachia* infection where antibiotic treatment fails to have any effect on parasites (Taylor 2000). Also, use of antibiotics ineffective against rickettsial bacteria fails to clear *Wolbachia* and has no effect on filarial parasites (Hoerauf et al. 2000a). Several antibiotics such as tetracycline, doxycycline, or rifampicin using typically three- to eight-week courses of antibiotic on worms maintained *in vitro*, or *in vivo*, including several clinical trials in humans (Taylor et al. 2005; Bazzocchi et al. 2008; Hoerauf et al. 2008; Wanji et al. 2009), had antifilarial effects. Coadministration of antibiotics tetracycline and doxycycline along with conventional antifilarial drugs has also revealed promising filaricidal efficacy in both experimental and clinical trials (Johnston and Taylor 2007; Hoerauf et al. 2008). Improvement in antifilarial efficacy of DEC and reduction in inflammation and pathogenesis have also been noticed when doxycycline was given prior to DEC in rodent model of B. *malayi* (Shakya et al. 2008). Antibiotic therapy is of particular interest in individual treatment for onchocerciasis where loiasis is also endemic (Hoerauf 2008). Main difficulties associated with antibiotic treatment are treatment duration (daily drug administration for 4–6 weeks) and contraindications in pregnant women and children under eight years of age making doxycycline unsuitable for mass drug administration (MDA). Discovery of newer safe molecules against *Wolbachia* is needed. Recent studies regarding the targeted delivery approaches viz. liposomized tetracycline, doxycycline, or rifampicin reduced the dose and treatment schedule to a great extent and need more investigation (Bajpai et al. 2005; Dangi et al. 2010). Proteomics approach is also being utilized to identify the protein families and domains related to stage-specific expression of functionally important proteins of the parasite and its endobacterium, *Wolbachia* (Bennuru et al. 2011). A recent study has shown that regulation of several filarial proteins is affected after killing *Wolbachia* by tetracycline (Dangi et al. 2009). Proteomic study of *B. malayi* excretory and secretory stage-specific proteins reflects different strategies of immune evasion by the parasite in host–parasite immunobiology (Hewitson et al. 2008; Moreno and Geary 2008).

New Developments

RNA interference (RNAi) is a natural post-transcriptional process that cells use to turn down, or silence, the activity of specific genes. Double-stranded RNAs or dsRNA are used to silence the expression of target genes via RNA interference. The first evidence that dsRNA could achieve efficient gene silencing through RNAi came from studies on the nematode *Caenorhabditis elegans* (Fire et al. 1998). RNAi studies in parasitic nematodes have been initiated, and a mixed success has been achieved. In filarial nematodes, gene silencing by dsRNA of the size of 300 bp targeting beta-tubulin, RNA polymerase II large subunit, *B. malayi* mf sheath protein, and ATP-dependent RNA helicase caused parasite mortality (Aboobaker and Blaxter 2003; Singh et al. 2011), ubiquitin and tropomyosin genes were more efficiently silenced in *Trichostrongylus colubriformis* (Issa et al. 2005) using still smaller (22 bp) siRNA and the technique could reduce target gene expression in several other parasitic helminths including *Onchocerca volvulus* (Ford et al. 2005; Visser et al. 2006). Genome-wide expression profiling of filarial nematodes along with robust bioinformatics analysis would enhance our understanding of the molecular biology of reproduction (Li et al. 2011) or immune responses (Kariuki et al. 2010). Key molecules and pathways associated with reproductive and other biological processes can be identified using such approaches, and results obtained could be utilized to identify novel antifilarial drug targets or to design potent inhibitors for the existing targets (Strubing et al. 2010).

Conclusions

There have been several advancements in the diagnosis, chemotherapeutic control, immunobiology, and molecular biology of filarial parasites over the last two decades. MDA programs have been quite effective in field settings; however, in some geographical areas, in spite of several annual repeated treatments with antifilarial drugs, the microfilarial levels could not be significantly decreased. There is no drug which can specifically kill the adult filarial parasites which reside deep into the lymphatic tissues.

Targeting of *Wolbachia* by the use of antibiotics though have shown promising results in experimental settings as also in limited clinical trials; their practical use in field is still debatable. The threat of development of drug resistance in case of ivermectin and albendazole is a serious issue, and newer molecules with newer mode of action need to be urgently discovered. The genomic information may be helpful in designing new drugs and or developing an efficient vaccine candidate. The chronic pathological lesions are normally irreversible and cause great misery to the people suffering from this disease.

References

Aboobaker AA, Blaxter ML (2003) Use of RNA interference to investigate gene function in the human filarial nematode parasite Brugia malayi. Mol Biochem Parasitol 129:41–51

Abraham D, Grieve RB, Holy JM, Christensen BM (1989) Immunity to larval Brugia malayi in BALB/c mice: protective immunity and inhibition of larval development. Am J Trop Med Hyg 40:598–604

Addiss DG, Brady MA (2007) Morbidity management in the global programme to eliminate Lymphatic Filariasis: a review of the scientific literature. Filaria J 6:2

Addiss DG, Beach MJ, Streit TG, Lutwick S, LeConte FH, Lafontant JG, Hightower AW, Lammie PJ (1997) Randomised placebo-controlled comparison of ivermectin and albendazole alone and in combination for Wuchereria bancrofti microfilaraemia in Haitian children. Lancet 350:480–484

Amaral F, Dreyer G, Figueredo-Silva J, Noroes J, Cavalcanti A, Samico SC, Santos A, Coutinho A (1994) Live adult worms detected by ultrasonography in human Bancroftian filariasis. Am J Trop Med Hyg 50:753–757

Arena JP, Liu KK, Paress PS, Schaeffer JM, Cully DF (1992) Expression of a glutamate-activated chloride current in Xenopus oocytes injected with Caenorhabditis elegans RNA: evidence for modulation by avermectin. Brain Res Mol Brain Res 15:339–348

Arumugam S, Pfarr KM, Hoerauf A (2008) Infection of the intermediate mite host with Wolbachia-depleted Litomosoides sigmodontis microfilariae: impaired L1 to L3 development and subsequent sex-ratio distortion in adult worms. Int J Parasitol 38:981–987

Aziz MA, Diallo S, Diop IM, Lariviere M, Porta M (1982a) Efficacy and tolerance of ivermectin in human onchocerciasis. Lancet 2:171–173

Aziz MA, Diallo S, Lariviere M, Diop IM, Porta IM, Gaxotte P (1982b) Ivermectin in onchocerciasis. Lancet 2:1456–1457

Bain O, Casiraghi M, Martin C, Uni S (2008) The nematoda Filarioidea: critical analysis linking molecular and traditional approaches. Parasite 15:342–348

Bajpai P, Vedi S, Owais M, Sharma SK, Saxena PN, Misra-Bhattacharya S (2005) Use of liposomized tetracycline in elimination of Wolbachia endobacterium of human lymphatic filariid Brugia malayi in a rodent model. J Drug Target 13:375–381

Bandi C, Slatko B, O'Neill SL (1999) Wolbachia genomes and the many faces of symbiosis. Parasitol Today 15:428–429

Bazzocchi C, Genchi C, Paltrinieri S, Lecchi C, Mortarino M, Bandi C (2003) Immunological role of the endosymbionts of Dirofilaria immitis: the Wolbachia surface protein activates canine neutrophils with production of IL-8. Vet Parasitol 117:73–83

Bazzocchi C, Mortarino M, Grandi G, Kramer LH, Genchi C, Bandi C, Genchi M, Sacchi L, McCall JW (2008) Combined ivermectin and doxycycline treatment has microfilaricidal and adulticidal activity against Dirofilaria immitis in experimentally infected dogs. Int J Parasitol 38:1401–1410

Bennuru S, Meng Z, Ribeiro JM, Semnani RT, Ghedin E, Chan K, Lucas DA, Veenstra TD, Nutman TB (2011) Stage-specific proteomic expression patterns of the human filarial parasite Brugia malayi and its endosymbiont Wolbachia. Proc Natl Acad Sci USA 108:9649–9654

Brattig NW, Rathjens U, Ernst M, Geisinger F, Renz A, Tischendorf FW (2000) Lipopolysaccharide-like molecules derived from Wolbachia endobacteria of the filaria Onchocerca volvulus are candidate mediators in the sequence of inflammatory and antiinflammatory responses of human monocytes. Microbes Infect 2:1147–1157

Brattig NW, Bazzocchi C, Kirschning CJ, Reiling N, Buttner DW, Ceciliani F, Geisinger F, Hochrein H, Ernst M, Wagner H, Bandi C, Hoerauf A (2004) The major surface protein of Wolbachia endosymbionts in filarial nematodes elicits immune responses through TLR2 and TLR4. J Immunol 173:437–445

Brownlie JC, Cass BN, Riegler M, Witsenburg JJ, Iturbe-Ormaetxe I, McGraw EA, O'Neill SL (2009) Evidence for metabolic provisioning by a common invertebrate endosymbiont, Wolbachia pipientis, during periods of nutritional stress. PLoS Pathog 5:e1000368

Caffrey CR (2007) Chemotherapy of schistosomiasis: present and future. Curr Opin Chem Biol 11:433–439

Casiraghi M, McCall JW, Simoncini L, Kramer LH, Sacchi L, Genchi C, Werren JH, Bandi C (2002) Tetracycline treatment and sex-ratio distortion: a role for Wolbachia in the moulting of filarial nematodes? Int J Parasitol 32:1457–1468

Casiraghi M, Bain O, Guerrero R, Martin C, Pocacqua V, Gardner SL, Franceschi A, Bandi C (2004) Mapping the presence of Wolbachia pipientis on the phylogeny of filarial nematodes: evidence for symbiont loss during evolution. Int J Parasitol 34:191–203

Chanteau S, Moulia-Pelat JP, Glaziou P, Nguyen NL, Luquiaud P, Plichart C, Martin PM, Cartel JL (1994) Og4C3 circulating antigen: a marker of infection and adult worm burden in Wuchereria bancrofti filariasis. J Infect Dis 170:247–250

Chen TT (1964) Demonstration of Macrofilaricidal action of Hetrazan, Antimony and Arsenic preparations in man. Chin Med J (Engl) 83:625–640

Chippaux JP, Boussinesq M, Gardon J, Gardon-Wendel N, Ernould JC (1996) Severe adverse reaction risks during mass treatment with ivermectin in loiasis-endemic areas. Parasitol Today 12:448–450

Chusattayanond W, Denham DA (1986) Attempted vaccination of jirds (Meriones unguiculatus) against Brugia pahangi with radiation attenuated infective larvae. J Helminthol 60:149–155

Cross HF, Haarbrink M, Egerton G, Yazdanbakhsh M, Taylor MJ (2001) Severe reactions to filarial chemotherapy and release of Wolbachia endosymbionts into blood. Lancet 358:1873–1875

Cully DF, Vassilatis DK, Liu KK, Paress PS, Van der Ploeg LH, Schaeffer JM, Arena JP (1994) Cloning of an avermectin-sensitive glutamate-gated chloride channel from Caenorhabditis elegans. Nature 371:707–711

Dangi A, Vedi S, Nag JK, Paithankar S, Singh MP, Kar SK, Dube A, Misra-Bhattacharya S (2009) Tetracycline treatment targeting Wolbachia affects expression of an array of proteins in Brugia malayi parasite. Proteomics 9:4192–4208

Dangi A, Dwivedi V, Vedi S, Owais M, Misra-Bhattacharya S (2010) Improvement in the antifilarial efficacy of doxycycline and rifampicin by combination therapy and drug delivery approach. J Drug Target 18:343–350

Debrah AY, Mand S, Toliat MR, Marfo-Debrekyei Y, Batsa L, Nurnberg P, Lawson B, Adjei O, Hoerauf A, Pfarr K (2007) Plasma vascular endothelial growth factor-A (VEGF-A) and VEGF-A gene polymorphism are associated with hydrocele development in lymphatic filariasis. Am J Trop Med Hyg 77:601–608

Devaney E, Bancroft A, Egan A (1993) The effect of irradiation on the third stage larvae of Brugia pahangi. Parasite Immunol 15:423–427

Diallo S, Aziz MA, Lariviere M, Diallo JS, Diop-Mar I, N'Dir O, Badiane S, Py D, Schulz-Key H, Gaxotte P et al (1986) A double-blind comparison of the efficacy and safety of ivermectin and diethylcarbamazine in a placebo controlled study of Senegalese patients with onchocerciasis. Trans R Soc Trop Med Hyg 80:927–934

Dissanayake S (2001) In Wuchereria bancrofti filariasis, asymptomatic microfilaraemia does not progress to amicrofilaraemic lymphatic disease. Int J Epidemiol 30:394–399

Dreyer G, Noroes J, Addiss D, Santos A, Medeiros Z, Figueredo-Silva J (1999) Bancroftian filariasis in a paediatric population: an ultrasonographic study. Trans R Soc Trop Med Hyg 93:633–636

Estambale BB, Simonsen PE, Vennervald BJ, Knight R, Bwayo JJ (1994) Bancroftian filariasis in Kwale district of Kenya. II. Humoral immune responses to filarial antigens in selected individuals from an endemic community. Ann Trop Med Parasitol 88:153–161

Esterre P, Plichart C, Huin-Blondey MO, Nguyen LN (2005) Soluble cellular adhesion molecules, selectins, VEGF and endothelin-1 in patients with Wuchereria bancrofti infection and association with clinical status. Parasite Immunol 27:9–16

Feng XP, Hayashi J, Beech RN, Prichard RK (2002) Study of the nematode putative GABA type-A receptor subunits: evidence for modulation by ivermectin. J Neurochem 83:870–878

Fenn K, Conlon C, Jones M, Quail MA, Holroyd NE, Parkhill J, Blaxter M (2006) Phylogenetic relationships of the Wolbachia of nematodes and arthropods. PLoS Pathog 2:e94

Fire A, Xu S, Montgomery MK, Kostas SA, Driver SE, Mello CC (1998) Potent and specific genetic interference by double-stranded RNA in Caenorhabditis elegans. Nature 391:806–811

Ford L, Guiliano DB, Oksov Y, Debnath AK, Liu J, Williams SA, Blaxter ML, Lustigman S (2005) Characterization of a novel filarial serine protease inhibitor, Ov-SPI-1, from Onchocerca volvulus, with potential multifunctional roles during development of the parasite. J Biol Chem 280:40845–40856

Foster J, Ganatra M, Kamal I, Ware J, Makarova K, Ivanova N, Bhattacharyya A, Kapatral V, Kumar S, Posfai J, Vincze T, Ingram J, Moran L, Lapidus A, Omelchenko M, Kyrpides N, Ghedin E, Wang S, Goltsman E, Joukov V, Ostrovskaya O, Tsukerman K, Mazur M, Comb D, Koonin E, Slatko B (2005) The Wolbachia genome of Brugia malayi: endosymbiont evolution within a human pathogenic nematode. PLoS Biol 3:e121

Fox LM (2006) Ivermectin: uses and impact 20 years on. Curr Opin Infect Dis 19:588–593

Friedman PD, Kalisher L (2002) Case 43: filariasis. Radiology 222:515–517

Ghedin E, Wang S, Foster JM, Slatko BE (2004) First sequenced genome of a parasitic nematode. Trends Parasitol 20:151–153

Ghedin E, Wang S, Spiro D, Caler E, Zhao Q, Crabtree J, Allen JE, Delcher AL, Guiliano DB, Miranda-Saavedra D, Angiuoli SV, Creasy T, Amedeo P, Haas B, El-Sayed NM, Wortman JR, Feldblyum T, Tallon L, Schatz M, Shumway M, Koo H, Salzberg SL, Schobel S, Pertea M, Pop M, White O, Barton GJ, Carlow CK, Crawford MJ, Daub J, Dimmic MW, Estes CF, Foster JM, Ganatra M, Gregory WF, Johnson NM, Jin J, Komuniecki R, Korf I, Kumar S, Laney S, Li BW, Li W, Lindblom TH, Lustigman S, Ma D, Maina CV, Martin DM, McCarter JP, McReynolds L, Mitreva M, Nutman TB, Parkinson J, Peregrin-Alvarez JM, Poole C, Ren Q, Saunders L, Sluder AE, Smith K, Stanke M, Unnasch TR, Ware J, Wei AD, Weil G, Williams

DJ, Zhang Y, Williams SA, Fraser-Liggett C, Slatko B, Blaxter ML, Scott AL (2007) Draft genome of the filarial nematode parasite Brugia malayi. Science 317:1756–1760

Gillette-Ferguson I, Hise AG, Sun Y, Diaconu E, McGarry HF, Taylor MJ, Pearlman E (2006) Wolbachia- and Onchocerca volvulus-induced keratitis (river blindness) is dependent on myeloid differentiation factor 88. Infect Immun 74:2442–2445

Gillette-Ferguson I, Daehnel K, Hise AG, Sun Y, Carlson E, Diaconu E, McGarry HF, Taylor MJ, Pearlman E (2007) Toll-like receptor 2 regulates CXC chemokine production and neutrophil recruitment to the cornea in Onchocerca volvulus/Wolbachia-induced keratitis. Infect Immun 75:5908–5915

Haarbrink M, Terhell AJ, Abadi GK, Mitsui Y, Yazdanbakhsh M (1999) Inflammatory cytokines following diethylcarbamazine (DEC) treatment of different clinical groups in lymphatic filariasis. Trans R Soc Trop Med Hyg 93:665–672

Haarbrink M, Abadi GK, Buurman WA, Dentener MA, Terhell AJ, Yazdanbakhsh M (2000) Strong association of interleukin-6 and lipopolysaccharide-binding protein with severity of adverse reactions after diethylcarbamazine treatment of microfilaremic patients. J Infect Dis 182:564–569

Hashim SA, Roholt HB, Babayan VK, Vanitallie TB (1964) Treatment of Chyluria and Chylothorax with medium-chain Triglyceride. N Engl J Med 270:756–761

Henkle-Duhrsen K, Eckelt VH, Wildenburg G, Blaxter M, Walter RD (1998) Gene structure, activity and localization of a catalase from intracellular bacteria in Onchocerca volvulus. Mol Biochem Parasitol 96:69–81

Hewitson JP, Harcus YM, Curwen RS, Dowle AA, Atmadja AK, Ashton PD, Wilson A, Maizels RM (2008) The secretome of the filarial parasite, Brugia malayi: proteomic profile of adult excretory-secretory products. Mol Biochem Parasitol 160:8–21

Hewitt RI, Kushner S et al (1947) Experimental chemotherapy of filariasis; effect of 1-diethyl-carbamyl-4-methylpiperazine hydrochloride against naturally acquired filarial infections in cotton rats and dogs. J Lab Clin Med 32:1314–1329

Hise AG, Daehnel K, Gillette-Ferguson I, Cho E, McGarry HF, Taylor MJ, Golenbock DT, Fitzgerald KA, Kazura JW, Pearlman E (2007) Innate immune responses to endosymbiotic Wolbachia bacteria in Brugia malayi and Onchocerca volvulus are dependent on TLR2, TLR6, MyD88, and Mal, but not TLR4, TRIF, or TRAM. J Immunol 178:1068–1076

Hoerauf A (2008) Filariasis: new drugs and new opportunities for lymphatic filariasis and onchocerciasis. Curr Opin Infect Dis 21:673–681

Hoerauf A, Volkmann L, Hamelmann C, Adjei O, Autenrieth IB, Fleischer B, Buttner DW (2000a) Endosymbiotic bacteria in worms as targets for a novel chemotherapy in filariasis. Lancet 355:1242–1243

Hoerauf A, Volkmann L, Nissen-Paehle K, Schmetz C, Autenrieth I, Buttner DW, Fleischer B (2000b) Targeting of Wolbachia endobacteria in Litomosoides sigmodontis: comparison of tetracyclines with chloramphenicol, macrolides and ciprofloxacin. Trop Med Int Health 5:275–279

Hoerauf A, Walter RD, Remme H, Lazdins J, Fleischer B (2001) Call to consolidate achievements for onchocerciasis and lymphatic filariasis control. Trends Parasitol 17:566–567

Hoerauf A, Mand S, Fischer K, Kruppa T, Marfo-Debrekyei Y, Debrah AY, Pfarr KM, Adjei O, Buttner DW (2003) Doxycycline as a novel strategy against bancroftian filariasis-depletion of Wolbachia endosymbionts from Wuchereria bancrofti and stop of microfilaria production. Med Microbiol Immunol (Berl) 192(4):211–216

Hoerauf A, Specht S, Buttner M, Pfarr K, Mand S, Fimmers R, Marfo-Debrekyei Y, Konadu P, Debrah AY, Bandi C, Brattig N, Albers A, Larbi J, Batsa L, Taylor MJ, Adjei O, Buttner DW (2008) Wolbachia endobacteria depletion by doxycycline as antifilarial therapy has macrofilaricidal activity in onchocerciasis: a randomized placebo-controlled study. Med Microbiol Immunol 197:295–311

Horton RJ (1997) Albendazole in treatment of human cystic echinococcosis: 12 years of experience. Acta Trop 64:79–93

Hosokawa T, Koga R, Kikuchi Y, Meng XY, Fukatsu T (2010) Wolbachia as a bacteriocyte-associated nutritional mutualist. Proc Natl Acad Sci USA 107, pp 769–774

Ismail MM, Weil GJ, Jayasinghe KS, Premaratne UN, Abeyewickreme W, Rajaratnam HN, Sheriff MH, Perera CS, Dissanaike AS (1996) Prolonged clearance of microfilaraemia in patients with bancroftian filariasis after multiple high doses of ivermectin or diethylcarbamazine. Trans R Soc Trop Med Hyg 90:684–688

Issa Z, Grant WN, Stasiuk S, Shoemaker CB (2005) Development of methods for RNA interference in the sheep gastrointestinal parasite, Trichostrongylus colubriformis. Int J Parasitol 35:935–940

Johnston KL, Taylor MJ (2007) Wolbachia in filarial parasites: targets for filarial infection and disease control. Curr Infect Dis Rep 9:55–59

Kariuki MM, Hearne LB, Beerntsen BT (2010) Differential transcript expression between the microfilariae of the filarial nematodes, Brugia malayi and B. pahangi. BMC Genomics 11:225

Kazura JW, Davis RS (1982) Soluble Brugia malayi microfilarial antigens protect mice against challenge by an antibody-dependent mechanism. J Immunol 128:1792–1796

Keiser PB, Reynolds SM, Awadzi K, Ottesen EA, Taylor MJ, Nutman TB (2002) Bacterial endosymbionts of Onchocerca volvulus in the pathogenesis of posttreatment reactions. J Infect Dis 185:805–811

Khamboonruang C, Thitasut P, Pan-In S, Morakote N, Choochote W, Somboon P, Keha P (1987) Filariasis in Tak province, northwest Thailand: the presence of subperiodic variant Wuchereria bancrofti. Southeast Asian J Trop Med Public Health 18:218–222

Kozek WJ (1977) Transovarially-transmitted intracellular microorganisms in adult and larval stages of Brugia malayi. J Parasitol 63:992–1000

Kumar S, Chaudhary K, Foster JM, Novelli JF, Zhang Y, Wang S, Spiro D, Ghedin E, Carlow CK (2007) Mining predicted essential genes of Brugia malayi for nematode drug targets. PLoS One 2:e1189

Kwan-Lim GE, Forsyth KP, Maizels RM (1990) Filarial-specific IgG4 response correlates with active Wuchereria bancrofti infection. J Immunol 145:4298–4305

Lal RB, Ottesen EA (1988) Enhanced diagnostic specificity in human filariasis by IgG4 antibody assessment. J Infect Dis 158:1034–1037

Lammie PJ, Addiss DG, Leonard G, Hightower AW, Eberhard ML (1993) Heterogeneity in filarial-specific immune responsiveness among patients with lymphatic obstruction. J Infect Dis 167:1178–1183

Landmann F, Foster JM, Slatko B, Sullivan W (2010) A symmetric Wolbachia segregation during early Brugia malayi embryogenesis determines its distribution in adult host tissues. PLoS Negl Trop Dis 4 e758

Laney SJ, Buttaro CJ, Visconti S, Pilotte N, Ramzy RMR, Weil GJ, Williams SA (2008) A reverse transcriptase-PCR assay for detecting Filarial infective Larvae in mosquitoes. PLoS Negl Trop Dis 2(6):e251

Langhammer J, Birk HW, Zahner H (1997) Renal disease in Lymphatic filariasis: evidence for tubular and glomerular disorders at various stages of the infection. Trop Med Int Health 2:875–884

Li BW, Rush AC, Jiang DJ, Mitreva M, Abubucker S, Weil GJ (2011) Gender-associated genes in filarial nematodes are important for reproduction and potential intervention targets. PLoS Negl Trop Dis 5:e947

Lizotte MR, Supali T, Partono F, Williams SA (1994) A polymerase chain reaction assay for the detection of Brugia malayi in blood. Am J Trop Med Hyg 51:314–321

Lustigman S, McCarter JP (2007) Ivermectin resistance in Onchocerca volvulus: toward a genetic basis. PLoS Negl Trop Dis 1:e76

Magnaval JF, Berry A (2005) Tropical pulmonary eosinophilia. Clin Infect Dis 40:635–636

Maizels RM, Kurniawan A, Selkirk ME, Yazdanbakhsh M (1991) Immune responses to filarial parasites. Immunol Lett 30:249–254

Maizels RM, Gomez-Escobar N, Gregory WF, Murray J, Zang X (2001) Immune evasion genes from filarial nematodes. Int J Parasitol 31:889–898

Mand S, Buttner DW, Hoerauf A (2008) Bancroftian filariasis–absence of Wolbachia after doxycycline treatment. Am J Trop Med Hyg 78:854–855

Mand S, Pfarr K, Sahoo PK, Satapathy AK, Specht S, Klarmann U, Debrah AY, Ravindran B, Hoerauf A (2009) Macrofilaricidal activity and amelioration of lymphatic pathology in bancroftian filariasis after 3 weeks of doxycycline followed by single-dose diethylcarbamazine. Am J Trop Med Hyg 81:702–711

Martin C, Gavotte L (2010) The bacteria Wolbachia in filariae, a biological Russian dolls' system: new trends in antifilarial treatments. Parasite 17:79–89

McGregor IA (1995) Royal Society of tropical medicine and hygiene meeting at Manson House, London, 3 October 1994. The third Manson oration. Patrick Manson 1844–1922: the birth of the science of tropical medicine. Trans R Soc Trop Med Hyg 89:1–8

McLaren DJ, Worms MJ, Laurence BR, Simpson MG (1975) Micro-organisms in filarial larvae (Nematoda). Trans R Soc Trop Med Hyg 69:509–514

McMahon JE, Magayauka SA, Kolstrup N, Mosha FW, Bushrod FM, Abaru DE, Bryan JH (1981) Studies on the transmission and prevalence of Bancroftian filariasis in four coastal villages of Tanzania. Ann Trop Med Parasitol 75:415–431

Melrose WD, Turner PF, Pisters P, Turner B (2000) An improved Knott's concentration test for the detection of microfilariae. Trans R Soc Trop Med Hyg 94:176

Molyneux DH, Bradley M, Hoerauf A, Kyelem D, Taylor MJ (2003) Mass drug treatment for lymphatic filariasis and onchocerciasis. Trends Parasitol 19: 516–522

Moreno Y, Geary TG (2008) Stage- and gender-specific proteomic analysis of Brugia malayi excretory-secretory products. PLoS Negl Trop Dis 2:e326

Moulia-Pelat JP, Nguyen LN, Glaziou P, Chanteau S, Gay VM, Martin PM, Cartel JL (1993) Safety trial of single-dose treatments with a combination of ivermectin and diethylcarbamazine in bancroftian filariasis. Trop Med Parasitol 44:79–82

Njoo FL, Hack CE, Oosting J, Luyendijk L, Stilma JS, Kijlstra A (1994) C-reactive protein and interleukin-6 are elevated in onchocerciasis patients after ivermectin treatment. J Infect Dis 170:663–668

Noroes J, Dreyer G, Santos A, Mendes VG, Medeiros Z, Addiss D (1997) Assessment of the efficacy of diethylcarbamazine on adult Wuchereria bancrofti in vivo. Trans R Soc Trop Med Hyg 91:78–81

Olszewski WL, Jamal S, Manokaran G, Pani S, Kumaraswami V, Kubicka U, Lukomska B, Dworczynski A, Swoboda E, Meisel-Mikolajczyk F (1997) Bacteriologic studies of skin, tissue fluid, lymph, and lymph nodes in patients with filarial lymphedema. Am J Trop Med Hyg 57:7–15

Oothuman P, Denham DA, McGreevy PB, Nelson GS, Rogers R (1979) Successful vaccination of cats against Brugia pahangi with larvae attenuated by irradiation with 10 krad cobalt 60. Parasite Immunol 1:209–216

Ottesen EA (1992) The wellcome trust lecture. infection and disease in lymphatic filariasis: an immunological perspective. Parasitology 104:S71–79

Ottesen EA, Poindexter RW (1980) Modulation of the host response in human schistosomiasis. II. Humoral factors which inhibit lymphocyte proliferative responses to parasite antigens. Am J Trop Med Hyg 29:592–597

Ottesen EA, Skvaril F, Tripathy SP, Poindexter RW, Hussain R (1985) Prominence of IgG4 in the IgG antibody response to human filariasis. J Immunol 134:2707–2712

Pfarr K, Hoerauf A (2005) The annotated genome of Wolbachia from the filarial nematode Brugia malayi: what it means for progress in antifilarial medicine. PLoS Med 2:e110

Rajendran R, Sunish IP, Mani TR, Munirathinam A, Abdullah SM, Arunachalam N, Satyanarayana K (2004) Impact of two annual single-dose mass drug administrations with diethylcarbamazine alone or in combination with albendazole on Wuchereria bancrofti microfilaraemia and antigenaemia in south India. Trans R Soc Trop Med Hyg 98:174–181

Rajendran R, Sunish IP, Mani TR, Munirathinam A, Arunachalam N, Satyanarayana K, Dash AP (2006) Community-based study to assess the efficacy of DEC plus ALB against DEC alone on bancroftian filarial infection in endemic areas in Tamil Nadu, south India. Trop Med Int Health 11:851–861

Rocha A, Addiss D, Ribeiro ME, Noroes J, Baliza M, Medeiros Z, Dreyer G (1996) Evaluation of the Og4C3 ELISA in Wuchereria bancrofti infection: infected persons with undetectable or ultra-low microfilarial densities. Trop Med Int Health 1:859–864

Rom WN, Vijayan VK, Cornelius MJ, Kumaraswami V, Prabhakar R, Ottesen EA, Crystal RG (1990) Persistent lower respiratory tract inflammation associated with interstitial lung disease in patients with tropical pulmonary eosinophilia following conventional treatment with diethylcarbamazine. Am Rev Respir Dis 142:1088–1092

Sabry M (1988) A quantitative analysis of the diagnostic value of diethylcarbamazine provocation in endemic Wuchereria bancrofti infection. Trans R Soc Trop Med Hyg 82:117–121

Sabry M (1992) Critical analysis of an epidemetrons model for the assessment of bancroftian filariasis endemicity in some areas in Egypt. J Trop Med Hyg 95:260–265

Saint Andre A, Blackwell NM, Hall LR, Hoerauf A, Brattig NW, Volkmann L, Taylor MJ, Ford L, Hise AG, Lass JH, Diaconu E, Pearlman E (2002) The role of endosymbiotic Wolbachia bacteria in the pathogenesis of river blindness. Science 295:1892–1895

Selkirk ME, Smith VP, Thomas GR, Gounaris K (1998) Resistance of filarial nematode parasites to oxidative stress. Int J Parasitol 28:1315–1332

Shakya S, Bajpai P, Sharma S, Misra-Bhattacharya S (2008) Prior killing of intracellular bacteria Wolbachia reduces inflammatory reactions and improves antifilarial efficacy of diethylcarbamazine in rodent model of Brugia malayi. Parasitol Res 102:963–972

Shakya S, Singh PK, Kushwaha S, Misra-Bhattacharya S (2009a) Adult Brugia malayi approximately 34 kDa (BMT-5) antigen offers Th1 mediated significant protection against infective larval challenge in Mastomys coucha. Parasitol Int 58:346–353

Shakya S, Srivastava AK, Misra-Bhattacharya S (2009b) Adult Brugia malayi mitochondrial and nuclear fractions impart Th1-associated sizeable protection against infective larval challenges in Mastomys coucha. J Helminthol 83:83–95

Shelley S, Manokaran G, Indirani M, Gokhale S, Anirudhan N (2006) Lymphoscintigraphy as a diagnostic tool in patients with lymphedema of filarial origin—an Indian study. Lymphology 39(2):69–75

Shenoy RK (2008) Clinical and pathological aspects of filarial lymphedema and its management. Korean J Parasitol 46:119–125

Shenoy RK, Suma TK, Rajan K, Kumaraswami V (1998) Prevention of acute adenolymphangitis in brugian filariasis: comparison of the efficacy of ivermectin and diethylcarbamazine, each combined with local treatment of the affected limb. Ann Trop Med Parasitol 92:587–594

Sheriff JC, Kotze AC, Sangster NC, Hennessy DR (2005) Effect of ivermectin on feeding by Haemonchus contortus in vivo. Vet Parasitol 128:341–346

Singh PK, Ajay A, Kushwaha S, Tripathi RP, Misra-Bhattacharya S (2010) Towards novel antifilarial drugs: challenges and recent developments. Future Med Chem 2:251–283

Singh M, Singh PK, Misra-Bhattacharya S (2011) RNAi mediated silencing of ATPase RNA helicase gene in adult filarial parasite Brugia malayi impairs in vitro microfilaria release and adult parasite viability. J Biotechnol

Slatko BE, O'Neill SL, Scott AL, Werren JL, Blaxter ML (1999) The Wolbachia Genome Consortium. Microb Comp Genomics 4:161–165

Specht S, Mand S, Marfo-Debrekyei Y, Debrah AY, Konadu P, Adjei O, Buttner DW, Hoerauf A (2008) Efficacy of 2- and 4-week rifampicin treatment on the Wolbachia of Onchocerca volvulus. Parasitol Res 103:1303–1309

Stolk WA, Van Oortmarssen GJ, Pani SP, De Vlas SJ, Subramanian S, Das PK, Habbema JD (2005) Effects of ivermectin and diethylcarbamazine on microfilariae and overall microfilaria production in bancroftian filariasis. Am J Trop Med Hyg 73:881–887

Strubing U, Lucius R, Hoerauf A, Pfarr KM (2010) Mitochondrial genes for heme-dependent respiratory chain complexes are up-regulated after depletion of Wolbachia from filarial nematodes. Int J Parasitol 40:1193–1202

Tanner M, Weiss N (1981) Dipetalonema viteae (Filarioidea): evidence for a serum-dependent cytotoxicity against developing third and fourth stage larvae in vitro. Short commun Acta Trop 38:325–328

Taylor MJ (2000) Wolbachia bacteria of filarial nematodes in the pathogenesis of disease and as a target for control. Trans R Soc Trop Med Hyg 94:596–598

Taylor MJ, Cross HF, Bilo K (2000) Inflammatory responses induced by the filarial nematode Brugia malayi are mediated by lipopolysaccharide-like activity from endosymbiotic Wolbachia bacteria. J Exp Med 191:1429–1436

Taylor MJ, Cross HF, Ford L, Makunde WH, Prasad GB, Bilo K (2001) Wolbachia bacteria in filarial immunity and disease. Parasite Immunol 23:401–409

Taylor MJ, Makunde WH, McGarry HF, Turner JD, Mand S, Hoerauf A (2005) Macrofilaricidal activity after doxycycline treatment of Wuchereria bancrofti: a double-blind, randomised placebo-controlled trial. Lancet 365:2116–2121

Thylefors B (2004) Eliminating onchocerciasis as a public health problem. Trop Med Int Health 9:A1–A3

Turner P, Copeman B, Gerisi D, Speare R (1993) A comparison of the Og4C3 antigen capture ELISA, the Knott test, an IgG4 assay and clinical signs, in the diagnosis of Bancroftian filariasis. Trop Med Parasitol 44:45–48

Turner PF, Rockett KA, Ottesen EA, Francis H, Awadzi K, Clark IA (1994) Interleukin-6 and tumor necrosis factor in the pathogenesis of adverse reactions after treatment of lymphatic filariasis and onchocerciasis. J Infect Dis 169:1071–1075

Turner JD, Langley RS, Johnston KL, Egerton G, Wanji S, Taylor MJ (2006) Wolbachia endosymbiotic bacteria of Brugia malayi mediate macrophage tolerance to TLR- and CD40-specific stimuli in a MyD88/TLR2-dependent manner. J Immunol 177:1240–1249

Turner JD, Langley RS, Johnston KL, Gentil K, Ford L, Wu B, Graham M, Sharpley F, Slatko B, Pearlman E, Taylor MJ (2009) Wolbachia lipoprotein stimulates innate and adaptive immunity through Toll-like receptors 2 and 6 to induce disease manifestations of filariasis. J Biol Chem 284:22364–22378

Turner JD, Tendongfor N, Esum M, Johnston KL, Langley RS, Ford L, Faragher B, Specht S, Mand S, Hoerauf A, Enyong P, Wanji S, Taylor MJ (2010) Macrofilaricidal activity after doxycycline only treatment of Onchocerca volvulus in an area of Loa loa co-endemicity: a randomized controlled trial. PLoS Negl Trop Dis 4:e660

Udall DN (2007) Recent updates on onchocerciasis: diagnosis and treatment. Clin Infect Dis 44:53–60

Vedi S, Dangi A, Hajela K, Misra-Bhattacharya S (2008) Vaccination with 73 kDa recombinant heavy chain myosin generates high level of protection against Brugia malayi challenge in jird and mastomys models. Vaccine 26:5997–6005

Vijayan VA, Sathish Kumar BY, Ganesh KN, Urmila J, Fakoorziba MR, Makkapati AK (2007) Efficacy of piperonyl butoxide (PBO) as a synergist with deltamethrin on five species of mosquitoes. J Commun Dis 39:159–163

Vincent AL, Ash LR, Frommes SP (1975) The ultrastructure of adult Brugia malayi (Brug 1927) (Nematoda: Filarioidea). J Parasitol 61:499–512

Visser A, Geldhof P, de Maere V, Knox DP, Vercruysse J, Claerebout E (2006) Efficacy and specificity of RNA interference in larval life-stages of Ostertagia ostertagi. Parasitology 133:777–783

Volkmann L, Saeftel M, Bain O, Fischer K, Fleischer B, Hoerauf A (2001) Interleukin-4 is essential for the control of microfilariae in murine infection with the filaria Litomosoides sigmodontis. Infect Immun 69:2950–2956

Wanji S, Mafo FF, Tendongfor N, Tanga MC, Tchuente E, Bilong Bilong CE, Njine T (2009) Spatial distribution, environmental and physicochemical characterization of Anopheles breeding sites in the Mount Cameroon region. J Vector Borne Dis 46:75–80

Warbrick EV, Barker GC, Rees HH, Howells RE (1993) The effect of invertebrate hormones and potential hormone inhibitors on the third larval moult of the filarial nematode, Dirofilaria immitis, in vitro. Parasitology 107(Pt 4):459–463

Weil GJ, Jain DC, Santhanam S, Malhotra A, Kumar H, Sethumadhavan KV, Liftis F, Ghosh TK (1987) A monoclonal antibody-based enzyme immunoassay for detecting parasite antigenemia in bancroftian filariasis. J Infect Dis 156:350–355

Williams SA, Nicolas L, Lizotte-Waniewski M, Plichart C, Luquiaud P, Nguyen LN, Moulia-Pelat JP (1996) A polymerase chain reaction assay for the detection of Wuchereria bancrofti in blood samples from French Polynesia. Trans R Soc Trop Med Hyg 90:384–387

Wolstenholme AJ, Fairweather I, Prichard R, von Samson-Himmelstjerna G, Sangster NC (2004) Drug resistance in veterinary helminths. Trends Parasitol 20:469–476

Wu M, Sun LV, Vamathevan J, Riegler M, Deboy R, Brownlie JC, McGraw EA, Martin W, Esser C, Ahmadinejad N, Wiegand C, Madupu R, Beanan MJ, Brinkac LM, Daugherty SC, Durkin AS, Kolonay JF, Nelson WC, Mohamoud Y, Lee P, Berry K, Young MB, Utterback T, Weidman J, Nierman WC, Paulsen IT, Nelson KE, Tettelin H, O'Neill SL, Eisen JA (2004) Phylogenomics of the reproductive parasite Wolbachia pipientis wMel: a streamlined genome overrun by mobile genetic elements. PLoS Biol 2:E69

Wu B, Novelli J, Foster J, Vaisvila R, Conway L, Ingram J, Ganatra M, Rao AU, Hamza I, Slatko B (2009) The heme biosynthetic pathway of the obligate Wolbachia endosymbiont of Brugia malayi as a potential antifilarial drug target. PLoS Negl Trop Dis 3:e475

Recent Update on Echinococcosis and Water

17

Abhiruchi Galhotra and Priya Datta

Abstract

Echinococcosis in humans occurs as a result of infection by the larval stages of taenid cestodes of genus Echinococcus. In this chapter, we discuss the aspects of life cycle, etiology, distribution, transmission, control, and prevention of echinococcosis. The four species of public interest include *Echinococcus granulosus, Echinococcus multilocularis, Echinococcus vogeli,* and *Echinococcus oligarthus*. Two new species have recently been identified: *Echinococcus shiquicus* and *Echinococcus felidis*. There is emergence or re-emergence of human cystic echinococcosis in parts of China, Central Asia, Eastern Europe, and Israel. Increasing trends in the seroprevalence of human hydatidosis have been observed in north India, and there are numerous reports and surveys that reveal the occurrence of human hydatidosis in most states. The mainstay of diagnosis remains serology and radiological methods. In spite of efforts to control of Echinococcosis, this zoonosis continues to be a major public health problem in several countries.

Keywords

Echinococcosis · *E. granulosus* · Epidemiology · Diagnosis · Prevention

Recent Update on Echinococcosis

Echinococcosis is one of the parasitic diseases that have been recognized since times immemorial. It is a near cosmopolitan zoonosis caused by adult or larval stages of tapeworms (cestodes) belonging to the genus *Echinococcus* (family *Taeniidae*). Larval infection (hydatid disease; hydatidosis) is characterized by long-term growth of metacestode (hydatid) cysts in the intermediate host.

A. Galhotra (✉)
Department of Community Medicine and Family Medicine, All India Institute of Medical Sciences, Raipur, Chattisgarh, India
e-mail: abhiruchigalhotra@yahoo.com

P. Datta
Department of Microbiology, Government Medical College Hospital, Sector 32-A, Chandigarh, India

P. P. Singh and V. Sharma (eds.), *Water and Health*,
DOI: 10.1007/978-81-322-1029-0_17, © Springer India 2014

Six species have been recognized, but four are of public health concern: *Echinococcus granulosus* (which causes cystic echinococcosis), *Echinococcus multilocularis* (which causes alveolar echinococcosis), and *Echinococcus vogeli* and *Echinococcus oligarthus* (which causes polycystic echinococcosis). Two new species have recently been identified: *Echinococcus shiquicus* in small mammals from the Tibetan plateau and *Echinococcus felidis* in African lions, but their zoonotic transmission potential is unknown (Pedro and Peter 2009).

In this chapter, we discuss the aspects of life cycle, etiology, distribution, transmission, control, and prevention of echinococcosis.

Distribution

Echinococcosis continues to be a major public health problem in several countries. *E granulosus* has a worldwide geographical distribution. It is found on all continents, with highest prevalence in parts of Eurasia (especially Mediterranean countries, the Russian Federation and adjacent independent states, and China), north and East Africa, Australia, and South America. There is clear evidence for the emergence or re-emergence of human cystic echinococcosis in parts of China, Central Asia, Eastern Europe, and Israel (McManus 2003; Eckert et al. 2001).

Communities involved in sheep farming harbor the highest rates of infection, showing the public health importance of the sheep–dog cycle and the sheep strain of *E. granulosus* in transmission to people (Thompson and McManus 2001).

Wild animals are also involved in sylvatic cycles in different parts of the world, although, generally, their zoonotic importance is small compared with the domestic cycles (Thompson and McManus 2001). A significant sylvatic cycle involving the common sheep strain operates on the mainland of Australia between dingoes (and feral dogs) and macropod marsupials (such as wallabies). This cycle overlaps and interacts with the domestic sheep–dog cycle, impeding control efforts (Thompson and McManus 2001). Another sylvatic life cycle, involving wolves or sled dogs and cervids, such as moose and reindeer, occurs in the higher latitudes in Northern North America and Eurasia (McManus 2003).

The distribution of *E. multilocularis* is limited to the northern hemisphere. In North America, the parasite is present in subarctic regions of Alaska and Canada and in a few northern states of the US. In Europe, it is present in the central and eastern countries and in Asia in the former USSR, Turkey, Iraq, northern India, Japan, and central China. In some regions of central Europe, approximately 40–75 % of the red fox populations are infected with *E. multilocularis*. On St. Lawrence Island, Alaska, up to 100 % of the arctic foxes are infected. In Gansu, a province of China, 8.8 % of the human population was found seropositive (http://www.who.int/zoonoses/diseases/echinococcosis/en/).

Disability-adjusted life years (DALYs) and monetary losses resulting from human and livestock cystic echinococcosis have been calculated at the global level assuming substantial under-reporting. The estimated global human burden of echinococcosis may be as high as 1,009,662 DALYs—or an annual loss of US$ 763,980,979. A maximum annual livestock production loss of US$ 2,190,132,464 is also estimated (http://www.who.int/neglected_diseases/diseases/zoonoses_figures/en/).

Echinococcosis in India

In several regions of India, there are alarming indications of increasing human health risks associated with Echinococcosis. Increasing trends in the seroprevalence of human hydatidosis have been observed in north India, and there are numerous reports and surveys that reveal the occurrence of human hydatidosis in most states. The conditions in the country are ideal for the establishment and transmission of hydatidosis in both livestock and humans. Cultural, educational, socioeconomic, agricultural, and environmental factors contribute to the transmission of the disease. The presence of stray dogs and fallen carcasses plays an important role in the transmission of the disease in the country. *Echinococcus granulosus* has a wide geographical distribution in livestock, and it is prevalent throughout the country. The overall prevalence of *E. granulosus* cysts in southern India was 7.0 % (106/1,519), 7.1 % (31/439), 9.4 % (46/489), and 11.5 % (10/87) in sheep, cattle, buffalo, and pigs, respectively. Another study carried out from 1995 to 1997 in Puducherry showed higher infection rates: 37.8 % in 325 sheep and 47.6 % in 680 goats. In Uttar Pradesh, the prevalence of *E. granulosus* was found to be 2.9 % (9/312), 1.4 % (39/2,710), and 0.9 % (27/2,980) in sheep, goats, and pigs, respectively. (Singh et al. 2010).

Epidemiological Transmission

Exposure to *Echinococcus* eggs may be affected by occupational and behavioral factors. In the case of *E. multilocularis*, hunters, trappers, and mushroom pickers would be expected to be more highly exposed than the general population, but there is little evidence that these groups are at increased risk of infection. The wide distribution and generally high frequency of *E. multilocularis* in foxes are not reflected in rates of infection in man which, for reasons not fully understood, are low in most endemic areas. Immunogenetic factors might play a part in this situation (McManus 2003; Eckert et al. 2001; Gottstein et al. 2001).

The dynamics of *E. multilocularis* transmission are complex, being affected by many factors that include seasonal fluctuations in the size of fox populations, the dispersal of foxes, involvement of other wild carnivores in the life cycle, the susceptibility and immunity of definitive hosts, worm burdens, prepatent period and egg definitive hosts, dispersal of eggs, and resistance of eggs to environmental factors. The contributions of intermediate hosts and the ecology of small mammalian hosts to the transmission dynamics of *E. multilocularis* are also likely to be very important but are less well defined. The sylvatic cycle can persist with low (<2 %) or high (>80 %) rates of *E. multilocularis* in foxes and with variable infection rates (<1 to >80 %) in rodents (Eckert et al. 2001).

With *E. granulosus*, acquired immunity in intermediate hosts represents an important density-dependent constraint for transmission, but parasite-induced mortality in livestock does not seem to play a role in the regulation of the cycle. In *E. multilocularis*, the natural intermediate rodent hosts are short-lived, the parasite evades the host immune responses, and, generally, but not always, large numbers of protoscolices are produced in a short period after infection. In contrast, the intermediate hosts of *E. granulosus* are long-lived and infection by eggs provokes a high degree of protective immunity, a characteristic that has been used for the development of a highly effective vaccine. Although knowledge of the epidemiology of *E. multilocularis* and the life histories of its hosts are limited, mathematical models of the life cycles of both *E. multilocularis* and *E. granulosus* have been formulated, but not yet rigorously tested and verified. The recent application of satellite remote sensing, geographical information systems, and landscape approaches in mammalian ecology to the study of *E. multilocularis* provides a new approach for exploring spatial relationships between landscape composition and transmission that will allow predictive models of alveolar echinococcosis risk to be formulated (McManus 2003; Eckert et al. 2001).

Echinococcus Granulosus

Morphology

Adult Worm—It is a small tapeworm, measuring 3–6 mm in length. It comprises of a scolex (head), neck, and strobila consisting of 3 segments. The first segment is immature, the second one is mature, and the last one is gravid. The terminal segment is by far the biggest, measuring 2–3 mm in length by 0.6 mm in breadth. The scolex bears four suckers and a protrusible rostellum with two circular rows of hooklets.

Egg—It is ovoid in shape and resembles eggs of *Taenia* species. It measures 32–36 μm in length by 25–32 μm in breadth and contains a hexacanth embryo with 3 pairs of hooklets. The egg is infective to man, cattle, sheep, and other herbivorous animals.

Larval form—This is found within the hydatid cyst developing inside the intermediate hosts. It represents the structure of the scolex of the future adult worm and remains invaginated within a vesicular body. On entering the definitive host, the scolex with four suckers and rostellar hooklets becomes evaginated and develops into an adult worm (Chatterjee 2009).

Life Cycle

MODE OF INFECTION. The eggs in the dog's feces are ingested by man. This occurs in the following ways:

1. By direct contact (handling and fondling) with infected dogs,
2. By allowing the dog to feed from the same dish, and
3. By eating/consuming uncooked vegetables contaminated with infected canine feces. Infection through contaminated water is not common as the eggs being heavier sink to the bottom.

Infection is generally acquired in childhood (due to intimate association with dogs) though the disease does not manifest before adult life (Chatterjee 2009).

Infecting Agent—Eggs in dog's faeces.

Portal of Entry—Alimentary tract.

Sites of Localization—Viscera (liver, lungs, and other organs).

The worm passes its life cycle in two hosts.

1. Definitive hosts. dog, wolf, fox, and jackal. The adult worm resides in the small intestine of these animals that discharge a large number of eggs in their feces. The dog is the optimum definitive host
2. Intermediate hosts. sheep, pig, cattle, horse, goat, and man. The larval stage is passed in these animals and man, known as hydatid cyst. The sheep appears to be the optimum intermediate host.

The eggs are discharged with the feces of the definitive hosts. These are swallowed by the intermediate hosts, sheep, and other domestic animals while grazing in the field and also by man (particularly children) due to intimate handling of infected dogs. In the duodenum, the hexacanth embryos are hatched out. After ingestion, the embryos bore their way through the intestinal wall and enter the radicles of the portal vein. The embryos are carried to the liver to be arrested in the sinusoidal capillaries (the liver acts as the first filter). Some of the embryos may pass through the hepatic capillaries, enter the pulmonary circulation, and filter out in the lungs (lungs act as the second filter). A few of the embryos may pass the pulmonary capillaries, enter the general blood stream, and lodge in the various organs like brain, heart, kidney, muscle, bone, and urogenital tract.

Wherever the embryo settles, it forms a hydatid cyst, the young larva being transformed into a hollow bladder. From the inner side of the cyst, brood capsules with a number of scolices are developed. A hydatid cyst developing from a single egg (oncosphere) may contain thousand of scolices. These fertile hydatid cysts, when ingested by the dog, are capable of growing into adult worms in about 6–7 weeks' time in the intestine. Thus, the cycle is repeated.

As the dogs have no access to the hydatid cyst developed in the viscera of man, the life cycle of the parasite comes to a dead-end. The natural cycle is thus maintained by dog and sheep. Life span of the larval worm is considerable, and it may continue to develop for many years. (Chatterjee 2009).

Pathogenicity

The adult worms of *E. granulosus* in dogs do not cause much inconvenience. They are found in large numbers (by hundreds or even thousands) in the small intestine of an infected dog where they lie embedded in the mucous membrane and appear on postmortem as small white specks on the reddish mucous surfaces (owing to the minuteness of size, they are often overlooked).

The larval worm of *E. granulosus* in man causes unilocular hydatid disease (Thompson 1995).

HYDATID CYST. The cyst wall secreted by the embryo consists of 2 layers:

1. Outer Cuticular Layer (Ectocyst). It is a laminated hyaline membrane having a thickness up to 1 mm. To the naked eye, the ectocyst has the appearance of the white of a hard-boiled egg. It is elastic and when incised or ruptured curls on itself exposing the inner layer containing the brood capsules and daughter cysts
2. Inner or Germinal Layer (Endocyst). It is cellular and consists of a number of nuclei embedded in a protoplasmic mass. It is very thin and measures about 22–25 μm in thickness. It is the vital layer of the cyst and (a) gives rise to brood capsules with scolices, (b) secretes the specific hydatid fluid, and (c) forms the outer layer.

Composition and Character of Hydatid Fluid:

1. Clear colorless fluid (may be pale yellow in color)
2. Specific gravity low, 1.005–1.010
3. Reaction slightly acid, pH 6.7
4. Contains sodium chloride, sodium sulfate, sodium phosphate, and sodium and calcium salts of succinic acid (a Fehling reducing substance)
5. Antigenic, being used for immunological tests
6. Highly toxic, when absorbed gives rise to anaphylactic symptoms
7. Hydatid sand—A granular deposit found to settle at the bottom. It consists of liberated brood capsules, free scolices, and loose hooklets.

Acephalocysts. Sometimes brood capsules are not developed and, if developed, are without any scolices; these cysts are sterile and called acephalocysts. These sterile hydatid cysts are found in large number in cattle.

Endogenous daughter cyst. formation in hydatid cysts is the result of growth over many years and is therefore particularly seen in man. The daughter cysts develop inside the mother cyst and may arise from the detached fragment of the germinal layer or from regressive changes of the young brood capsule and scolex bud. The daughter cyst also consists of an outer protective layer and an inner germinative layer from which brood capsules and scolices arise and even grand-daughter cysts may develop (Chatterjee 2009).

Development of brood capsules and scolices. Brood capsules sprout from the germinal layer. It is at first spherical, but soon becomes vacuolated and transformed into vesicle. The scolices, numbering 5–20 or more, develop within these brood capsules. A fully developed scolex represents the future "head" of the adult worm with suckers, and a circle of hooklets invaginated inside the scolex.

In a growing hydatid cyst, all the stages of development of a scolex may be found, beginning from the undifferentiated cellular bud to the fully developed stage with suckers and hooklets. The scolices may remain attached to the wall by means of pedicles or may remain free inside the cavity of the cyst and form the grain of "hydatid sand" previously referred to (Chatterjee 2009).

RATE OF GROWTH OF HYDATID CYST. The development of hydatid cyst in man is very slow. At the end of a year, it is approximately 4 cm in diameter and the brood capsules and scolices begin to appear.

DISTRIBUTION OF HYDATID CYST. The organ most commonly involved is the liver, because it acts as the first filter and the next organ involved is the lung which forms the second filter. After it enters the systemic circulation, it may be distributed in various organs.

Clinical Features. The chief clinical manifestations are entirely dependent upon site, size, and location of the cyst. Most primary infections in humans consist of a single cyst; however, 20–40 % has multiple cysts or multiple organ involvement. The liver is the most common site of the echinococcal cyst followed by lung (25 %), the cyst is seen less frequently in the spleen, kidneys, heart, bone, and CNS. In the majority of cases, the disease remains latent (symptomless) for many years and its presence is only detected at autopsy or by its pressure effects on the surrounding tissues or when the cyst ruptures or suppurates. The rate of growth of cyst is variable, ranging from 1 to 5 cms in diameter per year. Only 10–20 % of the cases are diagnosed in patients younger than 16 years.

However, cyst located in the brain or an eye can cause clinical symptoms even when small. The signs and symptoms of hepatic echinococcosis include hepatic enlargement, right epigastric pain, nausea, and vomiting. Rupture of a hydatid cyst is associated with anaphylactic symptoms and formation of localized or generalized secondary echinococcosis. In the lung, ruptured cyst membranes can be evacuated entirely through the bronchus or can be retained to serve as a nidus for bacterial or fungal infections (Pawlowski et al. 2001).

Laboratory Diagnosis. The presence of cyst in a person with a history of exposure to dogs in areas where *E. granulosus* is endemic supports the diagnosis of cystic echinococcosis. However, echinococcal cyst must be differentiated from benign cysts, cavitary tuberculosis, mycoses, abscesses, and neoplasm. This consists of the following (Chatterjee 2009).

1. *Casoni's Reaction.* An immediate hypersensitivity skin test introduced by Casoni in 1911. Intradermal injection of 0.2 ml of fresh sterile hydatid fluid (sterilized by Seitz filter) produces within half an hour, in all positive cases, a large wheal (5 cm in diameter) with multiple pseudopodia; it fades in an hour. Sterile normal saline, 0.2 ml, is injected in the other arm for control. Hydatid fluid from human cases (removed by operation) or from animals (obtained from slaughterhouse) is used as antigen (Chatterjee 2009)
2. *Blood examination.* may reveal a generalized eosinophilia of 20–25 %
3. *Serological tests.*—antibody assays are useful to confirm presumptive radiologic diagnoses, although some patients with cystic echinococcosis do not demonstrate a detectable immune response. Hepatic cysts are more likely to elicit an immune response than pulmonary cysts. The sensitivity of serological test is inversely related to degree of sequestration of the antigen inside the cysts; ruptured or leaking cysts are associated with stronger response as compared to healthy intact cysts. Various serological tests for the diagnosis of echinococcosis include (Chatterjee 2009);

(a) *IHAT, ELISA* using *E. granulosus.* hydatid fluid antigen are diagnostically 25–98 % sensitive for hepatic cases, 50–60 % sensitive for pulmonary cases and 50–60 % sensitive for other multiple organs localization cases
(b) *Haemagglutination test.* using fresh or formalinised sheep red cells sensitized with tannic acid and coated with echinococcal antigen
(c) *Flocculation tests.* using bentonite particles coated with hydatid cyst fluid as antigen. Bentonite flocculation test. The antigen is first adsorbed on to standardized bentonite particles (1–2 μm.). On a wax-ringed slide 0.1 ml of bentonite-antigen suspension is added to 0.1 ml of serial dilution of serum dilution of the serum in which more than half of the particles are aggregated into floccules is considered positive. In a positive case, the serum is titrated to an end point
(d) *Latex test*—The antigen used is polystyrene latex suspension of particles 0.81 μm diameter, sensitized with a pooled hydatid fluid. The diluted serum (double dilutions) is mixed with the antigen, and the mixture is incubated, in a water bath, at 37 °C for 90 min and then kept overnight at 4 °C, after which the reading is taken. A positive result is recorded, if there is a heavy precipitate
(e) *Indirect fluorescent antibody test.* (*IFA*) has also been found to be of great promise
(f) *Immunoelectrophoresis* (Arc 5). It is now well accepted

Immunoblotting for a relatively specific 8 kDa/12 kDa hydatid fluid polypeptide antigens is available now for diagnosis of disease.

(g) *Molecular methods.* such as DNA probe and PCR have been developed, but their application is of limited value because of their technical complexity. (Zhang and McManus 2006).

4. *Exploratory Cyst Puncture.* Though an accurate diagnosis may be made by withdrawing a few milliliters of the hydatid fluid and examining it under the microscope for scolices or hooklets, yet it is often attended with serious results and

is therefore not advised. Ultrasonographic guidance of the puncture, antihelminthic coverage, and anticipation of the possible need to treat an allergic reaction minimizes risks. Protoscolices can sometimes be demonstrated in sputum or bronchial washings; identification of hooklets is facilitated by acid fast stains (Hira et al. 1988)

5. *Radiological.* This is often helpful in the diagnosis of hydatid cysts of lungs and liver. Radiography permits detection of echinococcal cysts in the lungs. Computed tomography, magnetic resonance imaging, and ultrasonography are useful for diagnosis of deep-seated lesions in all organs and also for determination of the extent and condition of the avascular fluid-filled cysts. Abdominal ultrasonography has emerged as the most widely used imaging technique for echinococcosis because of its widespread availability and usefulness for defining number, site, dimensions, and vitality of cysts. World Health Organization (WHO) has given a standardized classification system for hepatic cysts detected by ultrasonography. This classification includes

(a) Type CL—unilocular cystic lesions, with uniform anechoic content
(b) Type CE 1—unilocular cystic lesions, with uniform anechoic content and with pathognomic signs that include visible cyst wall and snowflake sign
(c) Type CE 2—multivesicular, multiseptated cysts
(d) Type CE 3—anechoic content with detachment of laminated membrane from the cyst wall visible as floating membrane or as water-lily sign
(e) Type CE 4—heterogeneous hypoechoic or hyperechoic degenerative contents, no daughter cysts present
(f) Type CE 5—cysts characterized by thick calcified wall which is arch shaped, producing a cone-shaped shadow, the degree of calcification may vary from partial to complete (Zhang and McManus 2006; World Health Organization 2003).

6. *IV pyelogram.* is helpful for detection of renal hydatid cyst.

Echinococcus Multilocularis (Alveolar Echinococcosis)

Alveolar echinococcosis results from infection by larval stage of *E. multilocularis*. The natural intermediate host is rodent; the larval stage proliferates rapidly by exogenous budding of germinative tissue and produces an alveolar—like pattern of microvesicles filled with protoscolices. In humans, the larval mass resembles a malignancy in appearance and behavior, because it proliferates indefinitely by exogenous budding and invades the surrounding tissues. Protoscolices are rarely observed in human infections (Wilson and Rausch 1980).

Life cycle

It is same as in *E. granulosus* with the following exceptions.

1. Definitive host—fox, dogs, cats, coyotes, wolves
2. Intermediate host—small rodents.

Clinical Manifestation

Liver is the primary location of the larval stage in humans as well as in rodents. Local extension of the lesions and metastases to lung and brain may follow. In chronic alveolar echinococcosis infections, the lesions consist of a central necrotic cavity filled with a white amorphous material that is covered with a thin peripheral layer of dense fibrous tissue. The initial symptoms of alveolar hydatid disease are usually vague. Mild upper quadrant and epigastric pain with hepatomegaly may progress to obstructive jaundice. Patients die due to hepatic failure, invasion of contagious structures, and metastases to the brain (Wilson and Rausch 1980).

Labortory Findings and Diagnosis

Alveolar echinococcosis closely mimics hepatic carcinoma or cirrhosis. Usually, CT scan shows indistinct solid tumor with central necrosis and perinecrotic plaque-like calcification. Serological tests using *E. multilocularis* antigen are very specific and permit discrimination between *E. granulosus* and *E. multilocularis* (Ito et al. 1999).

Echinococcus Vogeli (Polycystic Echinococcosis)

The characteristics of polycystic echinococcosis are intermediate between those of cystic and alveolar forms.

Life cycle

It is same as in *E. granulosus* with the following exceptions.

1. Definitive host—bush dogs and dogs
2. Intermediate host—pacas and spiny rats (Moro et al. 2008).

The characteristics of polycystic echinococcosis are intermediate between those of the cystic and alveolar forms. Large cysts are filled with liquid and contain brood capsule with numerous protoscolices. Primary lesion is in liver, but cyst may spread to contiguous sites. Currently available immunodiagnostic tests do not permit species diagnosis as *E. vogeli* shares antigens with other *Echinococcus* spp (Moro et al. 2008; D'Allesandro 1997).

Echinococcus Shiquicus

DNA sequencing has identified E. *shiquicus* as a new species. The larval stage is found in the plateau pika, western Sichuan, China. Tibetan fox, *Vulpes ferrilata,* harbors the adult worm. There is no data available on the pathogenicity of *E. schiquicus* to humans (Xiao et al. 2006).

Echinococcus Felidis

Another new species *E. felidis* has been identified by DNA studies. The adult stage has been isolated in African lions, and the larval stage is believed to occur in wild ungulates. The zoonotic potential is presently unknown (Hüttner et al. 2008).

Prevention and Control

Controlling Echinococcus spp. that occurs in domesticated animal cycles, such as *E. granulosus* in the sheep/dog cycle, reduces human exposure. Dogs should not be fed the entrails from livestock at slaughter. Because dogs and cats can also be infected from parasites in wildlife cycles, they should not be allowed to hunt wild animals, or be fed any tissues from these species. Regular testing and/or treatment are advisable in animals allowed outside in endemic areas.

It is difficult to completely prevent exposure to *Echinococcus spp.* eggs from wild animals; however, food safety precautions, combined with good hygiene, can be helpful. All fruits and vegetables, particularly those picked in the wild, should be washed thoroughly to remove any eggs. Fences should be placed around vegetable gardens to keep animals, especially dogs and other canids, away. The hands should always be washed after handling pets, farming, gardening or preparing food, and before eating. Untreated water from sources such as lakes may also contain *Echinococcus* eggs and should be avoided. Wild carnivores, especially canids and felids, should be discouraged from coming close to homes. If these animals or their carcasses must be handled, gloves should be used. In some areas, foxes have been treated with antiparasitic drugs in bait, to decrease the prevalence of *E. multilocularis*. Meat, particularly canine intestines, should be thoroughly cooked before eating.

Anyone who handles the definitive hosts or material that may be contaminated with eggs should use appropriate personal protective equipment. In some countries, *Echinococcus* spp. must be handled in a BSL-3 laboratory. Regular surveillance with serological tests can be helpful in high-risk populations such as laboratory personnel working with eggs, or children who have been exposed to the feces of infected foxes. The purpose of testing is to detect cysts in the early stages, when they are most treatable. Vaccines are not available for people.

In view of the high potential pathogenicity of the Echinococcus infection to humans safety, precautions in laboratories and for field workers are of special importance. Heat remains the most reliable method for killing of Echinococcus eggs. They may also be inactivated by deep freezing, but only at temperatures of −70 to −80 °C and minimum exposure times of 96 h and 48 h, respectively. The high cold resistance of the eggs of *E.multilocularis* is well documented. On the other hand, it is still unclear whether strains of E. *granulosus* may differ in various regions with regard to cold resistance of their eggs. Chemical disinfection is difficult as most of the commercial disinfectants are ineffective against Echinococcus eggs.

Health education (HE) is a basic component of any program for control of *Echinococcus granulosus* and cystic echinococcosis (CE) and should be closely linked to and coordinated with all phases of the campaign.

Health education requires the motivation and participation of various population groups that are described. The educational material should address local problems in order to be effective and have the needed impact on governmental officials, managers, farmers, health professionals, etc. Health education has to take into consideration the beliefs, perceptions, behaviors, expectations, and needs of the people; it should not be a passive, but a dynamic procedure, adjusted to the changing demands and progress of the control campaign. Educational materials include audiovisual aids (video films, television programs), posters, pamphlets, text books, and others. Educational programs at schools and personal visits of dog owners, farmers, and other involved groups are of special significance. Continuing evaluation of the impact and the limitations of HE should be undertaken, and modifications should be made as and when indicated (Gemmell et al. 2001; Eckert et al. 2001).

References

Chatterjee KD (2009) CESTODES. Parasitology (Protozoology and Helminthology) in relation to clinical medicine, 13th edn. In: Chatterjee D (ed) CBS Publishers and Distributors Pvt. Ltd, New Delhi, pp 159–166

D'Allesandro A (1997) Polycystic echinococcosis in tropical America: Echinococcus vogeli and E. oligarthrus. Acta Trop 67:43–65

Eckert J, Schantz P, Gasser R et al (2001a) Geographic distribution and prevalence. In: Eckert J, Gemmell MA, Meslin F-X, Pawlowski ZS (eds) WHO/OIE manual on echinococcosis in humans and animals: a public health problem of global concern. World Organization for Animal Health, Paris, pp 100–141

Eckert J, Rausch R, Gemmell M et al (2001b) Epidemiology of Echinococcus multilocularis, Echinococcus vogeli and Echinococcus oligarthus. In: Eckert J, Gemmell MA, Meslin F-X, Pawlowski ZS (eds) WHO/OIE manual on echinococcosis in humans and animals: a public health problem of global concern. World Organisation for Animal Health, Paris, pp 164–182

Eckert J, Gottstein B, Heath D, Liu FJ (2001c) Prevention of Echinococcosis in humans and safety precautions. In: Eckert J, Gemmell MA, Meslin F-X, Pawlowski ZS (eds) WHO/OIE manual on echinococcosis in humans and animals: a public health problem of global concern. World Organization for Animal Health, Paris, pp 238–246

Gemmell MA, Roberts MG, Beard TC et al (2001) Control of Echinococcosis. In: Eckert J, Gemmell MA, Meslin F-X, Pawlowski ZS (eds) WHO/OIE manual on echinococcosis in humans and animals: a public health problem of global concern. World Organization for Animal Health, Paris, pp 195–236

Gottstein B, Saucy F, Deplazes P et al (2001) Is high prevalence of Echinococcus multilocularis in wild and domestic animals associated with disease incidence in humans? Emerg Infect Dis 7:408–412

Hira PR, Shweiki H, Lindberg LG, Shaheen Y, Francis I, Leven H et al (1988) Diagnosis of cystic hydatid disease: role of aspiration cytology. Lancet 17:655–657

Hüttner M, Nakao M, Wassermann T, Siefert L, Boomker JD, Dinkel A et al (2008) Genetic characterization and phylogenetic position of Echinococcus felidis (Cestoda: Taeniidae) from the African lion. Int J Parasitol 38:861–868

Ito A, Ma L, Schantz PM, Gottstein B, Liu YH, Chai JJ et al (1999) Differential serodiagnosis for cystic and alveolar echinococcosis using fractions of Echinococcus granulosus cyst fluid (antigen B) and E. multilocularis protoscolex (EM18). Am J Trop Med Hyg 60:188–192

McManus DP, Zhang W, Jun LI, Bartley PB (2003) Echinococcosis. Lancet 362:1295–1304

Moro P, Schantz PM (2009) Echinococcosis: a review. Int J Infect Dis 13:125–133

Moro PL, Cavero CA, Tambini M, Bricenõ Y, Jiménez R, Cabrera L (2008) Identification of risk factors for cystic echinococcosis in a periurban population of Peru. Trans R Soc Trop Med Hyg 102:75–78

Pawlowski Z, Eckert J, Vuitton DA, Ammann RW, Kern P, Craig PS et al (2001) Echinococcosis in humans: clinical aspects, diagnosis and treatment. In: Eckert J, Gemmell MA, Meslin FX, Pawlowski Z (eds) WHO/OIE Manual on Echinococcosis in humans and animals. Office International des Épizooties, Paris, pp 20–71

Singh BB, Sharma R, Sharma JK, Juyal PD (2010) Parasitic Zoonoses in India: an overview. Rev Sci Tech Off Int Epiz 29(3):629–637

Thompson RC (1995) Biology and systematics of Echinococcus. In: Thompson RC, Lymbery AJ (eds) Echinococcus and hydatid disease. CAB International, London, pp 1–37

Thompson RCA, McManus DP (2001) A etiology: parasites and life-cycles. In: Eckert J, Gemmell MA, Meslin F-X, Pawlowski ZS (eds) WHO/OIE manual on echinococcosis in humans and animals: a public health problem of global concern. World Organization for Animal Health, Paris, pp 1–19

Wilson JF, Rausch RL (1980) Alveolar hydatid disease: a review of clinical features of 33 indigenous cases of Echinococcus multilocularis infection in Alaskan Eskimos. Am J Trop Med Hyg 29:340–349

World Health Organization (2003) International classification of ultrasound images in cystic echinococcosis for application in clinical and field epidemiological settings. Acta Trop 85:253–261

Xiao N, Qiu J, Nakao M, Li T, Yang W, Chen X et al (2006) Echinococcus shiquicus, a new species from the Qinghai-Tibet plateau region of China: discovery and epidemiological implications. Parasitol Int 55(Suppl):S233–S236

Zhang W, McManus DP (2006) Recent advances in the immunology and diagnosis of echinococcosis. FEMS Immunol Med Microbiol 47:24–41

Ascariasis and Water: An Overview

18

D. K. Chhina, R. S. Chhina, and P. Suri

Abstract

Ascaris lumbricoides is globally the most widespread of all human intestinal roundworms. It infects more than one-fourth of the world's population. Infection occurs with greatest frequency in tropical and subtropical regions. Poor personal hygiene, poverty, overcrowding and inadequate sanitation are important predisposing factors. Children are more prone to pulmonary and intestinal ascariasis while hepatobiliary ascariasis is seen with more frequency in adults. Various radioimaging techniques are available for the diagnosis of ascariasis. Ultrasonography (US) is an inexpensive, rapid, accurate and safe modality for the diagnosis. Endoscopic retrograde cholangiopancreatography (ERCP) has both diagnostic and therapeutic role. Management is mainly by benzimidazoles, and prevention remains the best strategy at community level. New control methods are under study to decrease emergence of drug resistance to the common antihelminthic compounds.

Keywords

Ascaris · Ascariasis · Benzimidazoles · Geohelminth

Introduction

Intestinal parasitic infections are more common in tropical and subtropical countries but lately seen in almost all parts of the world due to increased frequency of international travel, immigration and consumption of exotic foods. Generally, the intestinal helminths have a long lifespan in hosts and cause no symptoms or illness. It takes years for the worm to cause serious disease. Thorough history from patient including the recent or remote travel to an endemic area in the past is important for diagnosing helminthic infections.

D. K. Chhina (✉) · P. Suri
Department of Microbiology, Dayanand Medical College and Hospital, Ludhiana, India
e-mail: deepinder.chhina@rediffmail.com

R. S. Chhina
Department of Gastroenterology, Dayanand Medical College and Hospital, Ludhiana, India

P. P. Singh and V. Sharma (eds.), *Water and Health*,
DOI: 10.1007/978-81-322-1029-0_18, © Springer India 2014

Ascaris lumbricoides, the causative agent of ascariasis, is the largest and globally the most widespread of all human intestinal roundworms, infecting more than one-fourth of the world's population. It is assumed that in areas where ascariasis is hyperendemic, untreated wastewater has an *Ascaris* egg concentration of 1,000 per litre (Mara and Sleigh 2010). Children in the age group of 3–8 years are infected more than adults. Poor nutrition predisposes to severe infection. Children are often infected by putting their hands in their mouth after playing in contaminated soil. Eating raw vegetables grown in contaminated soil is another frequent cause of infection. Ascariasis presents without specific signs and symptoms; thus, the clinician needs a high index of suspicion.

A. lumbricoides is also known as geohelminth because its egg or larvae must develop on soil before becoming infective to humans. Because of this requirement, it cannot be transmitted directly from one person to another and cannot multiply in the host unlike another geohelminth *Strongyloides stercoralis* which can complete its life cycle within the same host as well as in soil. Therefore, *Ascaris* does not act as an opportunistic parasite in persons who have acquired immunodeficiency syndrome (AIDS) or are receiving immunosuppressive drugs. Although it cannot act as an opportunistic parasite in patients with human immunodeficiency virus (HIV) infection, it elicits T-helper type 2 (Th2) cytokine responses that lead to more rapid progression to AIDS.

Taxonomy (Crompton 1998)

Phylum	Nematoda
Class	Secernentea
Order	Ascaridida
Superfamily	Ascaridoidea
Family	Ascarididae
Genus	*Ascaris*
Species	*lumbricoides*
Trivial name	Roundworm

Ascaridida comprises a large number of nematodes usually living in the gastrointestinal tract of the definitive host. Superfamily Ascaridoidea comprises stout worms of large size. The mouth is provided with 3 lips but lacks a buccal capsule. They have a simple digestive tract. Male has one or two copulatory spicules, and the female is somewhat longer than the male.

History

A. lumbricoides is one of the six worms listed and named by Linnaeus. The name *lumbricoides* is derived from the Latin word *Lumbricus* (meaning earthworm) due to its resemblance to earthworms. It has been known to humans since 1,500 BC as described in Egyptian medical papyri; the works of Hippocrates in the fifth century BC; Chinese writings of the second and third century BC (Hoeppli 1959); and the texts of Roman and Arabic physicians (Grove 1990). *A. lumbricoides* eggs have been found in human coprolites from Peru dating from 2,277 BC (Patrucco et al. 1983; Horne 1985) and Brazil from about 1,660 to 1,420 BC (Ferreira et al. 1980; Ferreira et al. 1983). The detailed anatomy of the worm was described by Edward Tyson (Tyson 1683), Francesco Redi (1684) independently in the late seventeenth century. The life cycle in humans was discovered in 1922 by Shimesu Koino, who infected himself and a volunteer and found large numbers of larvae in sputum (Koino 1922; Kean et al. 1978).

Epidemiology

A. lumbricoides is a remarkably infectious and persistent parasite (Crompton 2001). Infection occurs with greatest frequency in tropical and subtropical regions and in areas with inadequate sanitation. Global disability-adjusted life years (DALYs) lost due to diseases caused by soil-transmitted helminths (STH) are shown in Table 18.1, and estimated global number of STH infections and related morbidity and mortality (WHO 2002a) is shown in Table 18.2.

Table 18.1 Global DALYs lost due to diseases caused by soil-transmitted helminths (STH) (WHO 2002a)

Infection	DALYs lost (millions)
Hookworm disease	22.1
Ascariasis	10.5
Trichuriasis	6.4

The highest prevalence of *A. lumbricoides* occurs in China (39 %), East Asia and the Pacific islands (36 %), South Asia (27 %) and sub-Saharan Africa (25 %). In India, the prevalence of ascariasis and the number of cases are 14 % and 140 million, respectively (de Silva et al. 2003). In the Indian subcontinent, ascariasis is highly endemic in Kashmir (70 %), Bangladesh (82 %) and central and south-west India (20–49 %) (Khuroo 1996).

The high prevalence of ascariasis worldwide is a consequence of the tremendous egg output from the female worms and the remarkable ability of ova to resist unfavourable external environments. Chronic ascariasis is known to contribute to morbidity, including growth retardation and effects on cognitive development, particularly in growing children (O'Lorcain and Holland 2000). There are two species of *Ascaris*—*Ascaris lumbricoides* (in humans) and *Ascaris suum* (in swine). Recently, *A. suum* has emerged as an important zoonotic parasite. An outbreak of visceral larva migrans due to *A. suum* in Kyushu, Japan, has been reported (Maruyama et al. 1996).

Children acquire the infection by playing in soil contaminated with eggs, whereas adults most often are infected by farming or eating raw vegetables fertilized with untreated sewage or human excreta "night soil". The most intense infections are seen in children aged between 5 and 15 years. In India, Indonesia, Kenya, Myanmar and the United Republic of Tanzania, it has been repeatedly shown that nutritional status improves (weight and height gain, increased skin fold thickness) following deworming treatment. Treatment probably has an impact on impaired fat digestion, reduced vitamin absorption and temporary lactose intolerance, which are known consequences of ascariasis (WHO 2002a).

Climate is an important determinant of transmission of these infections, with adequate moisture and warm temperature essential for larval development in the soil (Brooker and Michael 2000; Brooker et al. 2006). Equally important determinants are poverty and inadequate water supplies and sanitation (de Silva et al. 2003). Because morbidity from these infections and the rate of transmission are directly related to the number of worms harboured in the host (Anderson and May 1991), intensity of infection is measured by the number of eggs per gram of faeces, generally by the Kato-Katz faecal thick-smear technique (Katz et al. 1972). Genetics and host immunity determine individual susceptibility to infections. Immigrants and travellers account for most of the cases in non-endemic areas. Many studies involving the provision of antihelminthic treatment and subsequent observation of intensity of reinfection have shown that individuals tend to reacquire similar worm burdens as that harboured before treatment. This phenomenon is known as predisposition (Elkins et al. 1986; Holland et al. 1989) and can be detected over multiple rounds of chemotherapy (Holland et al. 1989).

Infection caused by *A. lumbricoides* is often referred to as being "overdispersed" in endemic communities, such that most worms are harboured by a few individuals in an endemic area and they are at highest risk of severe disease (Anderson and May 1985).

Table 18.2 Estimated global number of STH infections and related morbidity and mortality (WHO 2002a)

Helminth	Number of infections (millions)	Morbidity (cases, millions)	Mortality (deaths per year, thousands)
Ascaris lumbricoides	1,450	350	60
Hookworms	1,300	150	65
Trichuris trichiura	1,050	220	10

Habitat

Adult worm resides in the small intestine (jejunum and ileum) of the infected persons.

Morphology and Life Cycle

Freshly recovered adult male and female *A. lumbricoides* are cylindrical worms, with tapering ends, the anterior end being more pointed than the posterior end (Fig. 18.1). They are pinkish in colour and measure 200–300 mm in length. The mouth is surrounded by three well-developed lips—one dorsal and two ventral—rows of small "teeth" on the lips and a cuticle with transverse striations. Collagen, stabilized by disulphide links, forms the main structural protein of the adult cuticle, which becomes more complex with each moult (Bird and Bird 1991). The morphology of denticles is a character widely used to distinguish *A. lumbricoides* from *A. suum*. A male, which is usually smaller than a female of the same age, has a curved posterior tail with two copulatory spicules (Fig. 18.2). The female *Ascaris* is larger with straight and conical posterior end. It has a didelphic uterus, and the vulva is situated near the junction of the anterior and middle third of the body surrounded by a distinct groove known as vulvar waist/genital girdle. Figure 18.3 shows bunch of *Ascaris* worms.

A. lumbricoides passes two types of eggs—fertilized eggs, which are embryonated and develop into infective eggs, and unfertilized eggs, which are non-embryonated and non-infective. Both fertilized and unfertilized *Ascaris* eggs are bile stained. They may or may not possess a cortical layer. Such eggs which lack the cortical layer are known as decorticated eggs.

The fertilized egg is spherical/ovoid (50–70 × 40–50 μm) (Fig. 18.4) and consists of three layers—an inner lipid layer responsible for selective permeability, a chitin-protein layer responsible for structural strength and an outer vitelline layer. The inner layer contains a lipoprotein, known as ascaroside, which explains how the eggs with enclosed infective larvae can survive formaldehyde, disinfectants and other destructive chemicals. The fertilized egg is frequently observed to have an uneven deposit of mucopolysaccharide (the cortical layer) on its surface. This deposit is obtained when the egg passes through the uterus of the female worm (Foor 1967) and is responsible for its adhesive

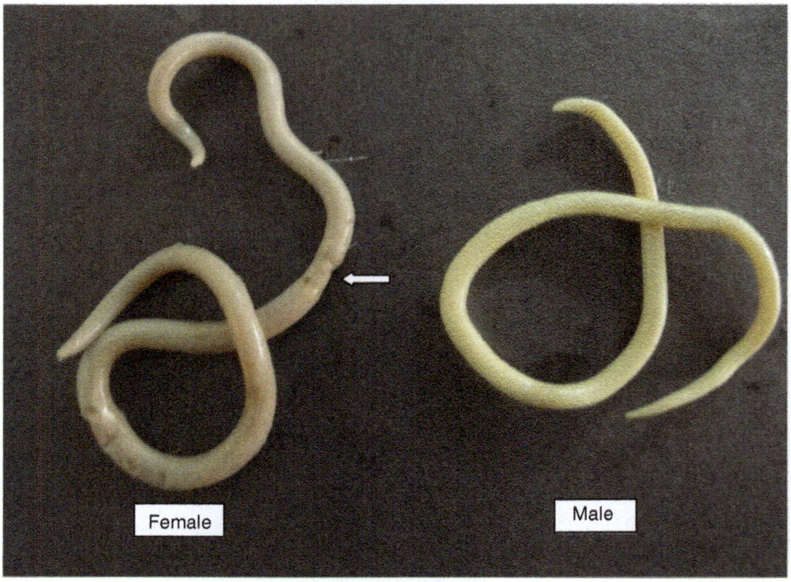

Fig. 18.1 Adult female and male *Ascaris lumbricoides*. Arrow marks, the vulvar waist in female worm

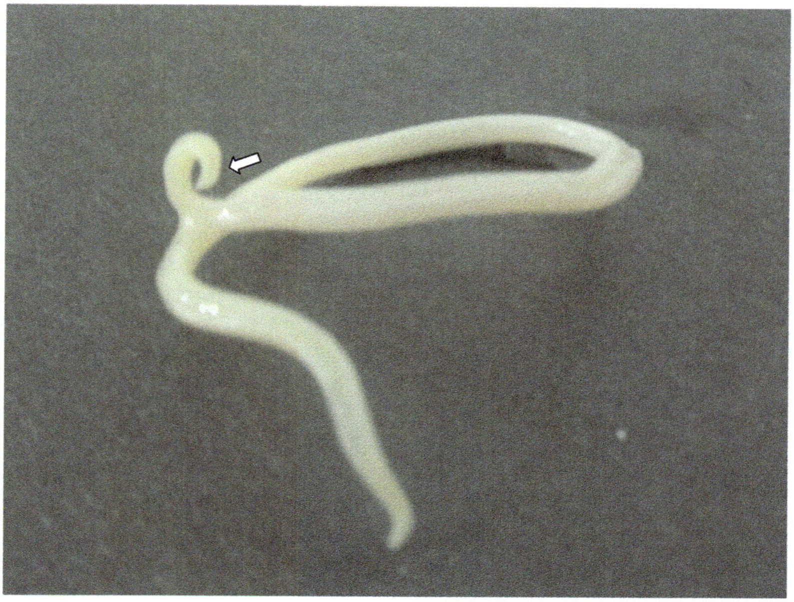

Fig. 18.2 Adult male worm with curved posterior end

Fig. 18.3 Bunch of *Ascaris* worms

properties (Kagei 1983). The middle of the egg contains a large unsegmented ovum with clear crescentic space at either pole.

The unfertilized egg is longer up to 90 μm. Its shell is thinner with the outer mamillary coat, scanty and irregular. It contains an atrophic ovum and numerous highly refractile granules of various sizes. The unfertilized egg is heavy and does not float in saturated salt solution.

The first-stage larva of *A. lumbricoides* develops inside the eggshell, moults there and forms the second-stage larva. Soon after hatching, second-stage larvae have a typical filariform appearance and measure about 250 × 14 μm.

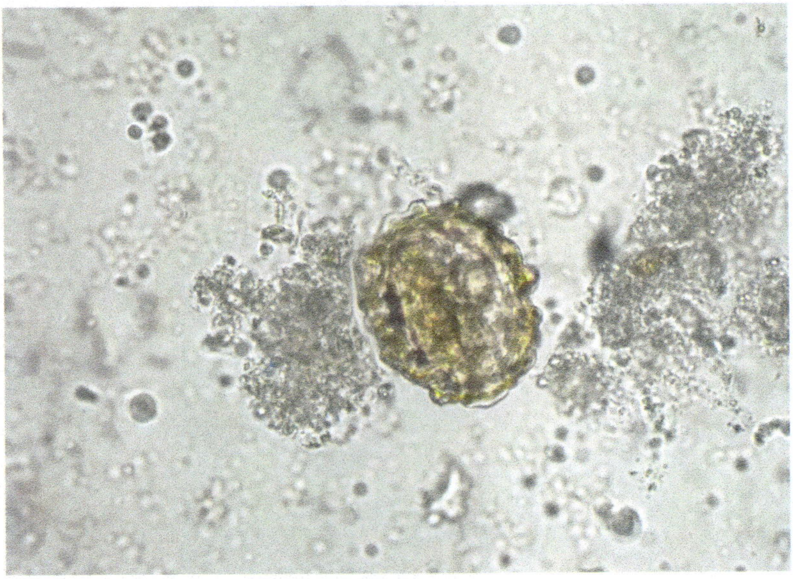

Fig. 18.4 Fertilized egg of *Ascaris lumbricoides* (x400)

Just before the second moult in the lungs, they measure about 560 × 28 μm in length. After the fourth (final) moult in the small intestine, growth is rapid.

Life cycle of *A. lumbricoides* (shown in Fig. 18.5) is completed in only one host, human. No intermediate host is required. Infection occurs when the eggs containing infective rhabditiform larva are swallowed by ingestion of contaminated food, drinks, vegetables and salads and by dirty fingers of children playing in mud transmitting embryonated eggs to mouths. In highly endemic areas, an airborne infection is also possible. Occasionally, migrating larva can also pass through the placenta and result in congenital ascariasis (daCosta-Macedo and Rey 1990). The ingested eggs hatch to liberate larva in small intestine. These larvae are actively motile and penetrate the intestinal mucosa and are carried by the portal circulation to the liver, where they stay for 3–4 days. Then they pass via the hepatic vein, inferior vena cava and the right heart and reach the lungs in about 4 days. In lungs the larvae grow in size and moult twice (first on 5th day and second on 10th day). Then the larvae enter alveoli from where it crawls up the bronchi, trachea and larynx and are swallowed. The larvae moult and develop into adults in the upper part of small intestine. Moulting takes 50–70 days. They become sexually mature in 6–10 weeks, and by 12 weeks the gravid females start laying eggs, which are passed in stool, and the cycle is repeated. The adult worm has a lifespan of 12–20 months.

Immunity

Immunity acquired due to past infections of *A. lumbricoides* is only partial. Immunoglobulins of all classes—IgM, IgG1-4, IgA and immunoglobulin E (IgE)—are raised but have no protective role. However, Hagel et al. have demonstrated negative associations between specific IgE and reinfection, and McSherry et al. found an inverse association between IgE specific to *A. lumbricoides* recombinant allergen ABA-1 and persistent susceptibility (Hagel et al. 1993; McSherry et al. 1999). IgE induces immediate hypersensitivity reaction by competitive binding of FcεR1 receptors (Hagel et al. 1993).

A. lumbricoides induces Th2 immune response which includes the production of cytokines such as interleukin (IL)-4, IL-5, IL-10 and IL-13 (Turner et al. 2003; Jackson et al. 2004), parasite-specific immunoglobulin and

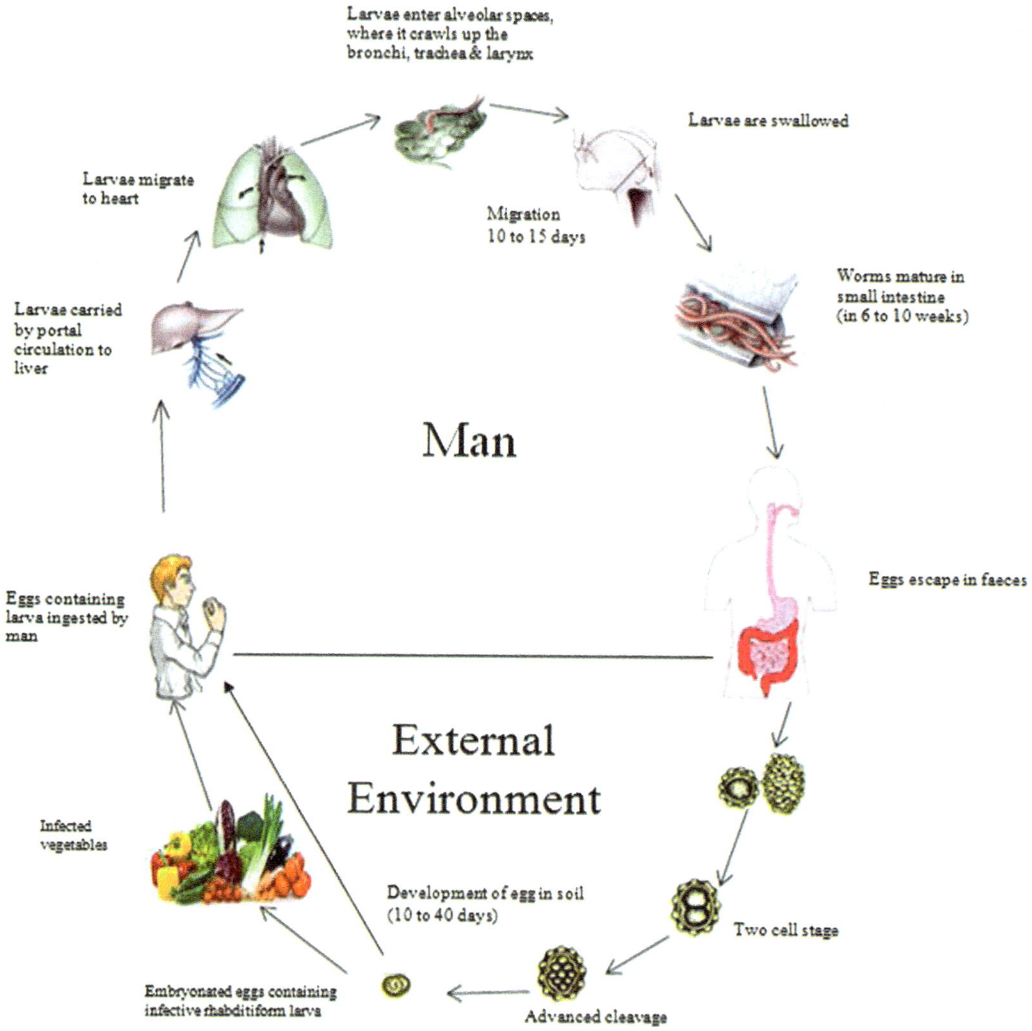

Fig. 18.5 Life cycle of *Ascaris lumbricoides*

non-specific IgE, and expansion and mobilization of mast cells, eosinophils and basophils (Maizel et al. 1993). The functional effectors at the base of the Th2 cascade are likely to include eosinophilia, IgA, IgE, mastocytosis and increased mucus secretion (Bradley and Jackson 2004). Turner et al. in 2003 demonstrated an inverse relationship between Th2 cytokines and intensity of *A. lumbricoides* infection in children aged more than 11 years (Turner et al. 2003).

Excretory-secretory (ES) products, produced from parasitic nematodes, have immunomodulatory effects. One such ES product produced by *Ascaris* is phosphorylcholine which inhibits lymphoid proliferation (Deehan et al. 2002).

Pathogenicity and Clinical Features

Most patients with *Ascaris* infections are asymptomatic. The pathogenicity and clinical features of *Ascaris* infections can be classified into acute manifestations associated with larval migration through the skin and viscera, and the acute and chronic manifestations resulting from parasitism of the gastrointestinal tract by adult worms.

Early Larval Migration

The mechanical trauma due to migrating larvae is one of the factors responsible for the pathophysiologic consequences of *Ascaris* infection. Pulmonary ascariasis develops 4–16 days after ingesting infective eggs. When larvae of *Ascaris* invade lung tissue, the larval antigens cause an intense inflammatory response consisting of eosinophilic infiltrates, resulting in an immune-mediated hypersensitivity response. Children are more susceptible to *Ascaris* pneumonia, and the disease is more severe on reinfection. The verminous pneumonia is commonly accompanied by wheezing, dyspnoea, a non-productive cough, chest discomfort and fever produced during heavy infections. Sputum may be blood tinged and may contain *Ascaris* larvae and Charcot–Leyden crystals. In severe cases the patients develop dyspnoea and an eosinophilic pneumonia (Loeffler's syndrome) with transient patchy infiltrates seen on chest radiographs. The pneumonia usually is self-limited but can be life-threatening.

Allergic inflammatory reaction to migrating larvae may involve other organs such as liver and kidneys. *Ascaris* larvae that die during migration through the liver can induce eosinophilic granulomas (Kaplan et al. 2001).

Intestinal Parasitism

Ascaris is pathogenic to man. It secretes from its body wall a pepsin inhibitor that protects maturing worms from digestive enzymes in the stomach before they reach the small intestine. It also produces phosphorylcholine linked to secreted glycoconjugates which suppress lymphocyte proliferation. Toxic body fluid (Ascaron) released by *Ascaris* may lead to various allergic manifestations.

Spoliation action. Adult worms in the lumen of the small bowel provoke no symptoms or produce only mild abdominal discomfort, dyspepsia or nausea. Moderate and heavy infections, however, can impair the nutritional status by robbing the host of its nutrition, which may lead to protein energy malnutrition, lactose intolerance, malabsorption of vitamin A and possibly other nutrients and depress appetite and food intake among children, which might partly cause the nutritional and growth failure. The impact on nutrition, cognitive performance and growth are likely the most important health-related consequences of ascariasis worldwide.

Mechanical obstruction. The presence of adult worms in intestinal tract may lead to abdominal discomfort, distension and acute colic pains. In young children, adult worms can aggregate in the ileum and cause partial obstruction because the lumen is small (Khuroo 1996). In case of heavy worm burden, total obstruction can occur. Several hundred thousands of worms are responsible for such blockage which can be fatal. Other complications of chronic ascariasis such as obstruction of bile and pancreatic ducts, appendicitis and intestinal perforation are mostly mechanical. Obstruction may be partial or complete, and occasionally, it is complicated by intussusception, volvulus, haemorrhagic infarction of the bowel, perforation or death (Khuroo 1996; Wasadikar and Kulkarni 1997).

Biliary Ascariasis

Ascaris worms are highly motile. Mature worms can enter the ampulla of Vater and migrate into the bile or pancreatic ducts, resulting in hepatobiliary ascariasis. The commonest presentation is biliary colic, and other presentations can be obstructive jaundice, ascending cholangitis, a-calculous cholecystitis, acute pancreatitis, choledocholelithiasis and liver abscess. The presence of a dead worm forms the nidus for a common bile duct (CBD) or hepatic stone. These worms have high glucuronidase activity which deconjugates bilirubin and forms pigment stones (Maki 1966). Biliary ascariasis is predominantly a disease of adults as compared to children, presumably because the adult biliary tree is large enough to accommodate an adult worm. Among adults it is more common in women (Khuroo and Zargar 1985) possibly because of the reason that the hormone progesterone in females leads to relaxation of the smooth muscles of the sphincter of Oddi (Tiemey et al. 1999), allowing the *Ascaris* worms easy access to the bile duct.

Ectopic Ascariasis

Increased migration of worms occurs with general anaesthesia, high fever, fasting or treatment with certain antihelminthic drugs such as mebendazole or albendazole. Wandering worms may invade appendix, pulmonary artery, trachea, bronchioles, maxillary sinus, bulbar conjunctiva, auditory canal and urinary tract causing ectopic symptoms. Haematemesis in gastric ascariasis was reported by Bhasin and Chinna in 1989.

Diagnosis

Clinical diagnosis of ascariasis is difficult because the signs and symptoms are quite vague and are indistinguishable from those of other intestinal nematode infections or from non-parasitic infections.

Direct Detection

(a) Demonstration of adult worm

The adult worm may pass out spontaneously in the stool or be vomited or escape through nares and are easily identified because of their large size and unsegmented body.

(b) Demonstration of larvae

Larvae can be found in sputum and gastric aspirate during their migration through the lungs even before eggs are detected in the stools.

(c) Demonstration of eggs

Microscopic identification of eggs in the stool is the most common method for diagnosing intestinal ascariasis. Eggs begin to appear in the stool about 2 months after initial exposure. As the *Ascaris* eggs are passed in the stool in enormous numbers (>200,000 eggs/day), they can be easily detected by the microscopic examination of stool sample. Concentration by floatation method may be used for the detection of eggs in the stool in case of light infections. It is to be noted that unfertilized eggs do not float in salt solution and also if the patient harbours a solitary male, *Ascaris* eggs are not found in the stool. For quantitative assessment of infection, various methods such as the Kato-Katz can be used. It is also a WHO-recommended technique for assessing the prevalence and infection intensity of major STH like *A. lumbricoides*, hookworms and *T. trichura.* Categories of intensity of infection according to egg counts (eggs per gram) are depicted in Table 18.3 (WHO 2002a). But the limitation of this technique is low sensitivity if only a single stool sample is examined especially in areas with low-intensity infections. For mass examination, cellophane thick-smear technique can be used which is simple, time saving, economic as well as highly sensitive for *Ascaris* egg detection.

Serodiagnosis

The use of serological techniques in detection of ascariasis is limited because of extensive cross-reactivity between the antigenic epitopes of different nematodes infective to humans. Ascariasis is associated with elevation of IgE and IgG antibodies. Serological tests based on IgG detection may overestimate the prevalence of

Table 18.3 Egg counts (eggs per gram) used to describe intensity of infection (WHO 2002a)

Causative pathogen	Intensity of infection (egg count per gram)		
	Light	Moderate	Heavy
A. lumbricoides	1–4,999	5,000–49,999	≥50,000
T. trichiura	1–999	1,000–99,999	≥1,00,00
Hookworm	1–1,999	2,000–3,999	≥4,000

infection, due to the persistence of antibodies for a long time after deworming of the patients. Among the IgG subclasses, IgG4 antibodies are highly elevated and do not cross-react with the sera of patients infected with other nematodes. This enhanced specificity of IgG4-based assays can be used for the serodiagnosis especially for epidemiological survey of *Ascaris* infection in large populations (Chatterjee et al. 1996). It is a marker of active infection, and its levels are found to be independent of worm burden (Bhattacharya et al. 2001).

Radiodiagnosis

Most patients suffering from intestinal ascariasis are asymptomatic or present with non-specific complaints. Radiological findings may sometimes be useful for diagnosis of *Ascaris* infection.

Plain X-ray abdomen will show segments of dilated small intestines packed with round worms well contrasted against gas in intestines, which can help in the early diagnosis of intestinal obstruction caused by *Ascaris*.

During *barium studies Ascaris* worm(s) may appear as single or multiple filling defects in the small bowel or they may retain barium after it has cleared from the patient's gastrointestinal tract, producing linear opacities (Fig. 18.6).

Ultrasonography (US)

Abdominal US is the best tool for the diagnosis of biliary ascariasis, but the US findings depend on the worm's site of infestation, orientation to and resolution of the transducer probe, the presence or absence of fluid around the worm, the part of the worm imaged and whether the worm is dead or alive (Mahmood et al. 2001). Though not sensitive, it is an inexpensive, rapid, accurate and safe modality for the diagnosis of *Ascaris*. US examination of biliary tree and pancreas will show long, straight, tubular, non-shadowing, echogenic structures showing characteristic erratic zig-zag movements suggestive of *A. lumbricoides*. The worms sometimes can be seen as one or more non-shadowing, tubular, echogenic structures with brighter parallel walls showing characteristic "Stripe" or "Railway track" sign and with central anechoic tube ("Inner tube sign"). If multiple, they may completely fill the bile duct, producing either the "Spaghetti" or "Whirlpool" sign. In cross section the worm appears like a "Ring", "Target" or "Bulls eye" sign.

Endoscopic retrograde cholangiopancreatography (ERCP)

ERCP can help in diagnosing biliary and pancreatic ascariasis. Worms in the biliary tree and pancreatic duct can be easily visualized and extracted with balloon, Dormia basket or forceps.

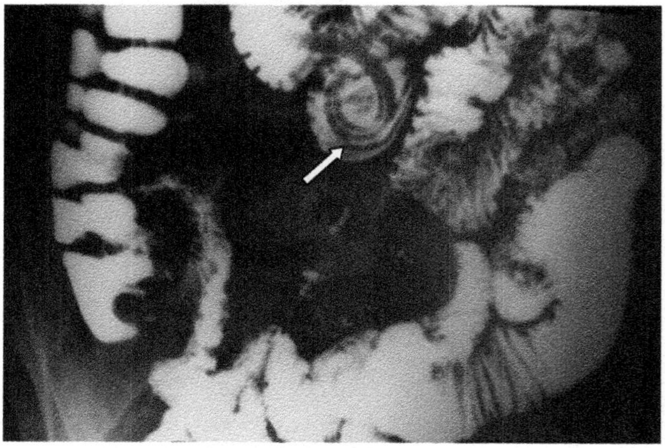

Fig. 18.6 Barium meal follow-through showing long tubular filling defect in a jejuna loop. A thin streak of barium seen within this—barium ingested by *Ascaris*

Magnetic resonance cholangiopancreatography (MRCP)

MRCP is a non-invasive diagnostic tool used to provide more details of biliary tree and has replaced ERCP in cases where papilla is inaccessible (Adaletli et al. 2005). It is not operator dependent. Intraductal worms are seen as linear hypo intense filling defects in the biliary tract. Hwang et al. 2001 have described two cases of biliary ascariasis undetected by both US and computed tomography (CT) but correctly diagnosed after magnetic resonance (MR) cholangiography (Fig. 18.7a, b).

Computer Tomography (CT)

In recent years CT has become increasingly popular. Contrast-enhanced CT findings have been described in several cases of intestinal ascariasis (Rodrigues et al. 2003; Hendi et al. 2006; Pratt and Blachar 2007). The worms may be seen as linear or cylindrical filling defects in the intestinal lumen. They may also be organized in masses or demonstrate a whirled pattern.

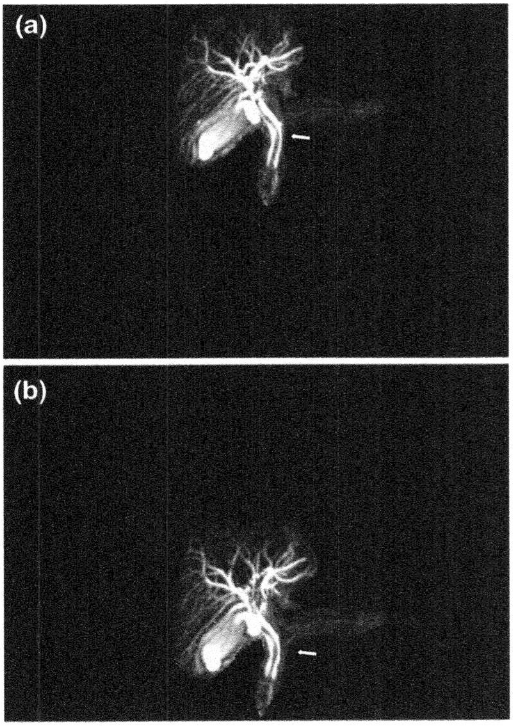

Fig. 18.7 a and b MRCP pictures showing *Ascaris* worm in the common bile duct

Indirect Evidences

In an endemic region pulmonary ascariasis can be diagnosed in a patient who presents with dyspnoea, dry cough, fever and peripheral eosinophilia. Sputum may show Charcot–Leyden crystals, and the chest radiograph may reveal fleeting pulmonary infiltrates. Stool samples may be negative in pulmonary ascariasis till two to three months after respiratory symptoms have appeared, unless the patient was previously infected. However, larvae can sometimes be demonstrated in respiratory samples or gastric aspirates.

Dermal reaction: "Scratch Test" with *Ascaris* antigen has often been found to be positive, but the results are variable and unreliable.

Management

Asymptomatic colonization with *A. lumbricoides* is treated easily with a single 400 mg oral dose of albendazole. All infections, including those that are mild, should be treated preferably with oral albendazole (400 mg single dose for adults and children more than 2 years of age), mebendazole (100 mg twice daily for 3 days irrespective of patients' age) or pyrantel pamoate (10 mg/kg of body weight, maximum 1 gm). Levamisole (2.5 mg/kg body weight, maximum 150 mg) is now the drug of choice, preferred by many. The benzimidazoles (BZ), that is, albendazole and mebendazole, act by inhibiting microtubule formation by binding to parasite beta tubulin causing death of the worm. Other biochemical changes seen are inhibition of mitochondrial fumarate reductase, reduced glucose transport and uncoupling of oxidative phosphorylation. Albendazole is teratogenic, and therefore, it should not be used in pregnant women. Pyrantel pamoate has been considered safe for use during pregnancy.

Pulmonary Ascariasis

Pulmonary ascariasis usually does not require any treatment, as it is a self-limiting disease. Two 400 mg doses of albendazole are given one month apart. The mature worms that finished migration to intestine get killed with the first dose, and the worms which were in transit are killed by the second dose. Symptomatic treatment may be necessary with bronchodilators and systemic steroids to reduce pneumonitis or antibiotics for bacterial complications.

Intestinal Ascariasis

Intestinal obstruction is managed conservatively with nasogastric suction, repletion of fluids and electrolytes, and antibiotics, and once the bowel motility is restored, antihelminthic therapy is given (Wasadikar and Kulkarni 1997). Surgery (resection/enterotomy) is indicated if the signs of complete obstruction with inadequate decompression, acute appendicitis, intussusception, volvulus or perforation develop in the patient.

Hepatobiliary Ascariasis

Hepatobiliary ascariasis can be treated conservatively with nasogastric suction, antispasmodics, analgesics and intravenous fluids. Albendazole is given each day for several days because the worms become susceptible only when they migrate out of the bile duct. Ascending cholangitis, acute obstructive jaundice or acute pancreatitis requires ERCP in which worm is extracted from the ducts by balloon, basket or forceps—with or without sphincterotomy, depending on the location of the worm. Worms that die or are trapped in ducts or invade the liver are removed by endoscopy or surgery.

New Control Methods

Emergence of drug resistance to the common antihelminth compounds used for targeted and mass treatment of intestinal parasites in humans and livestock has become a matter of concern. To deal with this problem, new control methods are under study.

1. Nitazoxanide is a nitroimidazole compound with broad-spectrum activity against numerous intestinal protozoa, helminths and anaerobic bacteria. It is presently approved to treat infections due to *Giardia intestinalis* in children and adults and infections due to *Cryptosporidium* species in children. It interferes with pyruvate ferredoxin oxidoreductase (PFOR) enzyme–dependent electron transfer reaction which is important for anaerobic glucose energy metabolism. Nitazoxanide appears very promising for the treatment of ascariasis (Diaz et al. 2003). The results are closely comparable to broad-spectrum anthelminthic drugs like albendazole, mebendazole and praziquantel (Ortiz et al. 2002).
2. Tribendimidine is a symmetrical diamidine derivative of amidantel. It is safe and efficacious against *A. lumbricoides* and hookworm (Xiao et al. 2005; Steinmann et al. 2008). In the study done by Steinmann et al. in 2008, promising results were obtained with tribendimidine against *S. stercoralis* and *Taenia* spp which need further investigations.
3. The use of combinations of antihelminthic drugs in control programmes is also expected to reduce or delay selection for resistance. Combined treatment with mebendazole and levamisole; pyrantel and oxantel; and albendazole and ivermectin has been tested.
4. Vaccination—The development of effective vaccines against STH may aid in control of these parasites, but extensive research is still

needed in this area. Tsuji et al. in 2004 have tested recombinant *Ascaris* 16-kilodalton protective antigen (rAS16) coupled with cholera toxin in pigs and demonstrated significantly elevated levels of rAS16-specific serum IgG and mucosal-associated IgA antibodies, suggesting a promising mucosal vaccine candidate for pig and human ascariasis.

Prevention and Control

Three main strategies for prevention and control of ascariasis in human hosts are as follows:

1. Long-term control and eradication of *A. lumbricoides* infection by improvement in sanitation for safe disposal of human faeces. It interrupts transmission, prevents reinfection and slowly decreases worm burden.
2. Health education—Providing messages as follows results in change in behaviour related to environment sanitation and family and personal hygiene:
   - Avoid contact with soil that may be contaminated with human faeces;
   - Hand washing with soap and water before handling food;
   - Raw vegetables should be washed, peeled or cooked.

WHO 2006 guidelines based on tolerable additional burden of disease of $\leq 10^{-6}$ DALY/ per person per year (pppy) for safe use of wastewater in agriculture recommend ≤ 1 human intestinal nematode egg/L of treated wastewater (WHO 2006). However, in WHO 2007 guidelines for quality of drinking water in areas with high endemicity, less stringent level of acceptable risk of $10^{-5}/10^{-4}$ DALY/pppy has been suggested (WHO 2007). Esrey et al. 1991 noted 29 % reduction in morbidity by ascariasis with improved water supply and sanitation.

3. Antihelminthic treatment—Infected individuals and domestic animals should be treated with medication to reduce morbidity and disease transmission. Children tend to acquire heavier worm burdens and so are considered a suitable group for targeted treatment. According to WHO, targeted treatment of school-age children in areas with a prevalence exceeding 70 % is two to three times a year and in areas with prevalence between 50 and 70 % is once a year (WHO 2002a). Studies among younger children have shown improvement in physical fitness, growth and appetite (Stephenson et al. 1993) and reduction in school absenteeism (Miguel and Kremer 2003).

Results from an intervention, which was based on health education alone, showed a decrease of 26 % in the prevalence of *A. lumbricoides*, while health education in combination with chemotherapy reduced the prevalence and intensity of infection by 42–75 % and 73–85 %, respectively (Albonico et al. 1996; Hadidjaja et al. 1998).

Conclusion

Hence, ascariasis is one of the commonest helminthic infections worldwide. Rough estimates state that up to one-fourth of the world's population may be infected with *Ascaris*. Ascariasis can cause varied signs and symptoms. Management is mainly by benzimidazoles, and prevention remains the best strategy at community level. Vaccination is still in its infancy. Proper understanding of the varied manifestations of the disease is essential for its proper management and eradication.

Acknowledgments Authors would like to thank Department of Radiology, Dayanand Medical College and Hospital, Ludhiana (Punjab), for providing the image of barium meal follow-through.

References

Adaletli I, Selcuk D, Gulsen M, Savas C, Korman U (2005) MRCP findings of biliary ascariasis before and after medical treatment. Turk J Gastroenterol 16:98–101

Albonico M, Shamlaye N, Shamlaye C, Savioli L (1996) Control of intestinal parasitic infections in Seychelles: a comprehensive and sustainable approach. Bull World Health Organ 74:577–586

Anderson RM, May RM (1985) Helminth infections of humans: mathematical models, population dynamics, and control. Adv Parasitol 24:1–101

Anderson RM, May RM (1991) Infectious diseases of humans: dynamics and control. Oxford University Press, Oxford

Bhasin DK, Chinna RS (1989) Hematemesis in gastric ascariasis. Am J Gastroenterol 84:1585–1586

Bhattacharyya T, Santra A, Majumder DNG, Chatterjee BP (2001) Possible approach for serodiagnosis of ascariasis by evaluation of immunoglobulin G4 response using *Ascaris lumbricoides* somatic antigen. J Clin Microbiol 39:2991–2994

Bird AF, Bird J (1991) The structure of nematodes, 2nd edn. Academic Press, London

Bradley JE, Jackson JA (2004) Immunity, immunoregulation and the ecology of trichuriasis and ascariasis. Parasite Immunol 26:429–441

Brooker S, Michael E (2000) The potential of geographical information systems and remote sensing in the epidemiology and control of human helminth infections. Adv Parasitol 47:245–288

Brooker S, Clements A, Bundy DAP (2006) Global epidemiology, ecology and control of soil-transmitted helminth infections. Adv Parasitol 62:223–265

Chatterjee BP, Santra A, Karmaker PR, Mazumder DNG (1996) Evaluation of IgG, response in ascariasis by ELISA for serodiagnosis. Tropical Med Int Health 1:633–639

Crompton DWT (1998) Gastrointestinal nematodes—*Ascaris*, hookworm, *Trichuris* and *Enterobius*. In: Kreier JP, Wakelin D (eds) Topley and Wilson's microbiology and microbial infections, vol 5. FEG Cox, Parasitology, 9th edn. Arnold, London, Sydney, Auckland, p 562

Crompton DWT (2001) *Ascaris* and ascariasis. Adv Parasitol 48:285–375

da Costa-Macedo LM, Rey L (1990) *Ascaris lumbricoides* in neonate: evidence of congenital transmission of intestinal nematodes. Rev Inst Med Trop Sao Paulo 32:351–354

de Silva NR, Brooker S, Hotez PJ, Montresor A, Engels D, Savioli L (2003) Soil-transmitted helminth infections: updating the global picture. Trends Parasitol 19:547–551

Deehan MR, Goodridge HS, Blair D, Lochnit G, Dennis RD, Geyer R et al (2002) Immunomodulatory properties of *Ascaris suum* glycosphingolipids: phosphorylcholine and non-phosphorylcholine-dependent effects. Parasite Immunol 24:463–469

Diaz E, Mondragon J, Ramirez E, Bernal R (2003) Epidemiology and control of intestinal parasites with nitazoxanide in children in Mexico. Am J Trop Med Hyg 68:384–385

Elkins DB, Haswell-Elkins M, Anderson RM (1986) The epidemiology and control of intestinal helminths in the Pulicat Lake region of Southern India. I. Study design and pre- and post-treatment observations on *Ascaris lumbricoides* infection. Trans R Soc Trop Med Hyg 80:774–792

Esrey SA, Potash JB, Roberts L, Shiff C (1991) Effects of improved water supply and sanitation on ascariasis, diarrhoea, dracunculiasis, hookworm infection, schistosomiasis, and trachoma. Bull World Health Organ 69:609–621

Ferreira LF, Araujo A, Confalonieri U (1980) The finding of eggs and larvae of parasitic helminths in archaeological material from Unai, Minas Gerais, Brazil. Trans R Soc Trop Med Hyg 174:798–800

Ferreira LF, Araujo A, Confalonieri U (1983) The findings of helminth eggs in a Brazilian mummy. Trans R Soc Trop Med Hyg 77:65–67

Foor WE (1967) Ultrastructural aspects of oocyte development and shell formation in *Ascaris lumbricoides*. J Parasitol 53:1245–1261

Grove DI (1990) A history of human helminthology. CAB International, Wallingford

Hadidjaja P, Bonang E, Suyardi MA, Abidin SA, Ismid IS, Margono SS (1998) The effect of intervention methods on nutritional status and cognitive function of primary school children infected with *Ascaris lumbricoides*. Am J Trop Med Hyg 59:791–795

Hagel I, Lynch NR, di Prisco MC, Rojas E, Perez M, Alvarez N (1993) *Ascaris* reinfection of slum children: relation with the IgE response. Clin Exp Immunol 94:80–83

Hendi JM, Horton KM, Fishman EK (2006) Ascariasis infection of the colon: MDCT evaluation. Emerg Radiol 12:180–181

Hoeppli R (1959) Parasites and parasitic infections in early science and medicine. University of Malaya Press, Singapore

Holland CV, Asaolu SO, Crompton DWT, Stoddart RC, MacDonald R, Torimiro SEA (1989) The epidemiology of *Ascaris lumbricoides* and other soil-transmitted helminths in primary school children from Ile-Ife. Niger Parasitolo 99:275–285

Horne PD (1985) A review of the evidence for human endoparasitism in the pre-Columbian new world through the study of coprolites. J Archaeol Sci 12:299–310

Hwang CM, Kim TK, Ha HK, Kim PN, Lee MG (2001) Biliary ascariasis: MR cholangiography findings in two cases. Korean J Radiol 2:175–178

Jackson JA, Turner JD, Rentoul L, Faulkner H, Behnke JM, Hoyle M et al (2004) T helper cell type 2 responsiveness predicts future susceptibility to gastrointestinal nematodes in humans. J Infect Dis 190:1804–1811

Kagei N (1983) Techniques for the measurement of environmental pollution by infective stages of soil-transmitted helminths. Collected Pap Control Soil-Transm Helminthiasis 2:24–46

Kaplan KJ, Goodman ZD, Ishak KG (2001) Eosinophilic granuloma of the liver: a characteristic lesion with relationship to visceral larva migrans. Am J Surg Pathol 25:1316–1321

Katz N, Chaves A, Pellegrino J (1972) A simple device for quantitative stool thick-smear technique in *Schistosomiasis mansoni*. Rev Inst Med Trop Sao Paulo 14:397–400

Kean BH, Mott KE, Russell AJ (eds) (1978) Tropical medicine and parasitology: classic investigations. Cornell University Press, Ithaca

Khuroo MS (1996) Ascariasis. Gastroenterol Clin North Am 25:553–577

Khuroo MS, Zargar SA (1985) Biliary ascariasis: a common cause of biliary and pancreatic disease in an endemic area. Gastroenterology 88:418–423

Koino S (1922) Experimental infection of the human body with ascarides. Jpn Med World 15:317–320

Mahmood T, Mansoor N, Quraishy S, Ilyas M, Hussain S (2001) Ultrasonographic appearance of *Ascaris lumbricoides* in the small bowel. J Ultrasound Med 20:269–274

Maizels RM, Bundy DA, Selkirk ME, Smith DF, Anderson RM (1993) Immunological modulation and evasion by helminth parasites in human populations. Nature 365:797–805

Maki T (1966) Pathogenesis of calcium bilirubinate gallstone: role of *E. coli*, p-glucuronidase and coagulation by inorganic ions polyelectrolytes and agitation. Ann Surg 164:90–100

Mara D, Sleigh A (2010) Estimation of *Ascaris* infection risks in children under 15 from the consumption of wastewater-irrigated carrots. J Water Health 8:35–38

Maruyama H, Nawa Y, Noda S, Mimori T, Choi WY (1996) An outbreak of visceral larva migrans due to *Ascaris suum* in Kyushu. Jpn Lancet 347:1766–1767

McSharry C, Xia Y, Holland CV, Kennedy MW (1999) Natural immunity to *Ascaris lumbricoides* associated with immunoglobulin E antibody to ABA-1 allergen and inflammation indicators in children. Infect Immun 67:484–489

Miguel EA, Kremer M (2003) Worms: identifying impacts on education and health in the presence of treatment externalities. Econometrica 72:159–217

O'Lorcain P, Holland CV (2000) The public health significance of *Ascaris* lumbricoides. Parasitology 121:S51–S71

Ortiz JJ, Lopez Chegne N, Gargala G, Favennec L (2002) Comparative clinical studies of nitazoxanide, albendazole and praziquantel in the treatment of ascariasis, trichuriasis and hymenolepiasis in children from Peru. Trans R Soc Trop Med Hyg 96:193–196

Patrucco R, Tello R, Bonavia D (1983) Parasitological studies of coprolites of pre-Hispanic Peruvian populations. Curr Anthropol 24:393–394

Pratt LT, Blachar A (2007) Computed tomography diagnosis of intestinal ascariasis: incidental finding in a trauma patient. Isr Med Assoc J 9:688–689

Redi F (1684) Osservazioni intorno agli animali viventi che si trovano negli animali viventi. Pietro Martini, Florence

Rodrigues EJ, Gama MA, Ornstein SM, Anderson WD (2003) Ascariasis causing small bowel volvulus. Radiographics 23:1291–1293

Steinmann P, Zhou XN, Du ZW, Jiang JY, Xiao SH, Wu ZX et al (2008) Tribendimidine and albendazole for treating soil-transmitted helminths, *Strongyloides stercoralis* and *Taenia* spp.: open-label randomized trial. PLoS Neglected Trop Dis 2:e322

Stephenson LS, Latham MC, Adams EJ, Kinoti SN, Pertet A (1993) Physical fitness, growth and appetite of Kenyan school boys with hookworm, *Trichuris trichiura* and *Ascaris lumbricoides* infections are improved four months after a single dose of albendazole. J Nutr 123:1036–1046

Tiemey S, Nakeeb A, Wong O, Lipsett PA, Sostre S, Pitt HA et al (1999) Progesterone alters biliary flow dynamics. Ann Surg 229:205–209

Tsuji N, Miyoshi T, Islam MK, Isobe T, Yoshihara S, Arakawa T et al (2004) Recombinant *Ascaris* 16-kilodalton protein–induced protection against *Ascaris suum* larval migration after intranasal vaccination in pigs. J Infect Dis 190:1812–1820

Turner JD, Faulkner H, Kamgno J, Cormont F, Van Snick J, Else KJ et al (2003) Th2 cytokines are associated with reduced worm burdens in a human intestinal helminth infection. J Infect Dis 188:1768–1775

Tyson E (1683) Lumbricus teres, or some anatomical observations on the round worm bred in human bodies. Philos Trans Royal Soc Lond 13:153–161

Wasadikar PP, Kulkarni AB (1997) Intestinal obstruction due to ascariasis. Br J Surg 84:410–412

World Health Organization (2002a) Prevention and control of schistosomiasis and soil-transmitted helminthiasis: report of a WHO expert committee. Geneva, Switzerland

World Health Organization (2007) Levels of protection. Geneva, Switzerland. Available at: http://www.who.int/water_sanitation_health/gdwqrevision/levels ofprotection/en/index.html

World Health Organization (2006) Guidelines for the safe use of wastewater, excreta and grey water: Wastewater Use in Agriculture, vol 2. Geneva, Switzerland

Xiao SH, Wu HM, Tanner M, Utzinger J, Wang C (2005) Tribendimidine: a promising, safe and broad-spectrum anthelminthic agent from China. Acta Trop 94:1–14

Soil-Transmitted Helminths and Water 19

Qudsia Tahseen

Abstract

The number of species of soil-transmitted helminths (STHs) infecting humans in the tropical and subtropical parts of the world may be up to a dozen or more. However, the most common STHs include the large roundworm, *Ascaris lumbricoides*, the whipworm, *Trichuris trichiura*, and two species of hookworm, *Necator americanus* and *Ancylostoma duodenale*. Besides, three other helminths, *Enterobius vermicularis*, *Strongyloides stercoralis* and *Oesophagostomum bifurcum*, show considerable morbidity of disease though at a smaller scale than the four common STHs. In addition to their health effects in the form of iron deficiency anaemia, rectal prolapse and chronic dysentery, helminth infections also impair physical and mental growth in childhood, impede educational advancement and hinder economic development. The greatest numbers of soil-transmitted helminth infections occur in Asia, especially the sub-Saharan Africa, India, China, and Southeast Asia. Periodic deworming launched by government and non-government organizations has shown to improve growth, micronutrient status (iron and vitamin A) and motor and language development in preschool children. The present chapter describes and summarizes, through a systematic review of the published literature and notifications of World Health Organization (WHO), the prevalence and intensity of infection caused by the common STHs, the morbidity caused by them, the risk factors and the methods of intervention.

Q. Tahseen (✉)
Department of Zoology, Aligarh Muslim University,
Aligarh, 202002, India
e-mail: qtahseen@yahoo.com

P. P. Singh and V. Sharma (eds.), *Water and Health*,
DOI: 10.1007/978-81-322-1029-0_19, © Springer India 2014

Keywords
Epidemiology · Etiology · Morphology · Soil-transmitted helminths · Treatment

Introduction

Soil-transmitted helminthiasis remains an important reason for morbidity or mortality in developing tropical countries and predominantly the cause of physical and mental growth retardation, particularly among children (World Health Organization 2006). Estimates of annual deaths from soil-transmitted helminth infection vary widely, from 12,000 (WHO 2002) to as many as 135,000 (WHO 2004). Soil-transmitted helminths (STHs) are commonly known as a group of nematodes that infect more than a billion people worldwide (Bethony et al. 2006) without any intermediate host but via the soil-borne contamination. The mode of infection is through their eggs or infective larvae that survive in the warm and moist soil contaminated with human faeces. The parasites cause health problems while living in the human gastrointestinal tract for years. In developing countries, where control measures are often difficult to implement, STHs remain a grave problem. According to an early report, together with schistosomiasis, the STH infections accounted for 40 % of the global morbidity caused by all infectious diseases, exclusive of malaria (WHO 2000). Diseases caused by parasitic worms inflict an enormous public health burden and the poor people in the world bear the brunt of these infectious diseases (Bethony et al. 2006; Hotez et al. 2008). Preschool as well as school going children and pregnant women are the groups at highest risk of morbidity due to these infections (Montresor et al. 2002a, b; Goodman et al. 2007). According to World Health Organization (WHO) out of approximately two billion affected people worldwide, 400 million infected individuals are children of school age and 300 million suffer from heavy worm burdens sufficient to result in poor growth, reduced physical activity, impaired cognitive function and learning ability (Nokes et al. 1992; Adams and Shaw 1994; Stoltzfus et al. 1996; Stephenson et al. 1998) and the major cause of human misery in the tropics. The high STH prevalence may not directly result in poor economic growth, but is mostly one of contributing factors. As a result, repeated drug administration is considered the best solution to reduce the prevalence and intensity of worm infections. Although, in some regions, there has been considerable decline in STHs prevalence primarily because of economic development and specific control; however, in many cases, the prevalence rates are still close to those estimated more than 50 years ago by Norman Stoll (1962, 1999) who termed the STH infections to be "The Great Infection of Mankind" second only to malaria. Despite their educational, economic and public health importance, these infections largely remain neglected by the medical and international community due to three reasons: First, the mostly affected people are from the world's most unprivileged class; second, the infections have insidious clinical presentation; and third, quantification is a difficult task and often not very precise.

The STHs of worldwide importance that pose significant threat to human health are the roundworm/giant intestinal worm (*Ascaris lumbricoides*), whipworm (*Trichuris trichiura*), hookworms (*Necator americanus* and *Ancylostoma duodenale*), pin worm (*Enterobius vermicularis*), thread worm (*Strongyloides stercoralis*) and nodular worm (*Oesophagstomum bifurcum*). Some other species including *Toxocara canis* also qualify to the category of STHs according to some scientists, but owing to their low impact, they have not been included in the present review. The prevalence of STH infections has been reported from the USA, China, East Asia and sub-Saharan Africa. The life cycles of most STHs follow a general

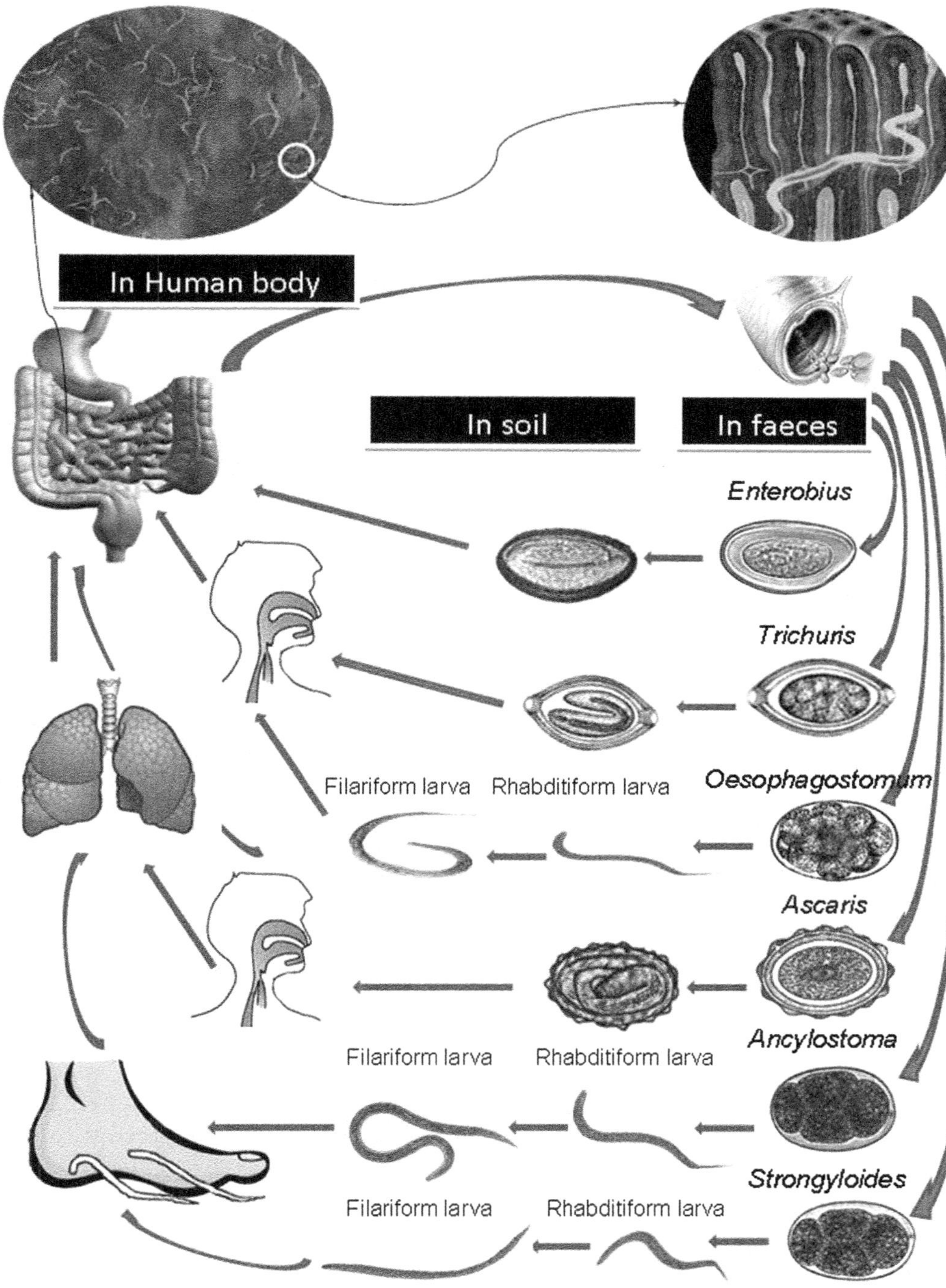

Fig. 19.1 General life cycle trends in some common soil-transmitted helminths

pattern (Fig. 19.1). The adult parasite stages inhabit the gastrointestinal tract, reproduce sexually and produce eggs, which are passed in human faeces and deposited in the external environment/soil that again forms the source of further infections.

Roundworm (*A. lumbricoides*)

Ascaris was probably the first STH known with its scientific description dating back to 1683 (Crompton 1989). *A. lumbricoides* is a human parasite, while *Ascaris suum* parasitizes pigs. Both have demonstrated reproductive isolation (Peng et al. 1998) with zoonotic transmission not occurring frequently. However, occasional instances of the visceral *larva migrans* from *A. suum* infections have been reported (Sakakibara et al. 2002). According to an earlier estimate, approximately 10,500 deaths occur due to acute illness caused by *A. lumbricoides* (de Silva et al. 1997a).

Morphology and Taxonomic Status

A. lumbricoides, Linnaeus 1758, belongs to the family Ascarididae of the superfamily Ascaridoidea. It is the largest of the intestinal nematodes found in human. The adults are cylindrical and pink or cream-coloured with finely striated cuticle. Females can attain 20–49 cm length and 5 μm diameter with a straight pointed posterior end, while males measure 12–31 cm long and 3–4 μm in cross section with a curved posterior region bearing the copulatory apparatus (Fig. 19.3a–f). The oral aperture is guarded by one dorsal and two ventral lips. The vulva in female is located at about one-third of body length. The uteri may accommodate up to 27 million eggs at a time. Unfertilized eggs are longer measuring 75–94 μm × 35–50 μm, whereas the fertilized eggs measure 45–75 μm × 35–50 μm in dimension and are oval to round with a thick shell showing an irregular coat of mucopolysaccharide. The shell layers of the egg provide a very resistant structure, while its lipid layer protects it from acids and alkalis and many other chemicals.

Life Cycle

A. lumbricoides infecting a human lays its eggs in the intestine that are passed out in faeces. Eggs excreted in faeces require a period of maturation in soil. Under favourable conditions, in warm and moist soil, the newly developed larva moults within the eggshell. The embryonic development in the soil is temperature-dependent and may take from 2 weeks to several months, while the infectious egg having second-stage larva can persist in soil for a considerable period. The infective larva does not hatch in soil or invade through skin. When the eggs reach a healthy person's gut through contaminated food, water or other unhygienic practices, the infective larvae hatch out and penetrate the wall of the duodenum to enter the blood stream. The larvae are carried to the liver and heart and then move in pulmonary circulation to reach the alveoli. In the alveoli, the second-stage larvae grow and moult two times to finally develop into 1.5-mm-long fourth-stage larvae. From alveoli, the larvae move to the trachea and are coughed up, swallowed to finally return back to the intestine. The larvae undergo final moult to form adult males and females. After fertilization, each female produces as many as 200,000 eggs per day. From ingestion of infective eggs to the production of mature adults, the development takes about 10–12 weeks. The adult worm has a life span of about a year in the intestine; therefore, persistent infection requires frequent re-exposure and re-infection.

Ascariasis: Causes, Symptoms

Ascariasis is the infection of the small intestine caused by the parasite and considered the most common STH disease. Presumably, one quarter of the world's population receives parasite infection through uncooked food/vegetables contaminated with soil/faecal matter having *Ascaris* eggs. It is common in populations observing low personal hygiene, poor sanitation and in areas where night soil is used as fertilizer.

The symptoms of ascariasis are often not very conspicuous. Occasionally, the transfer of larvae via the lungs can lead to coughing, wheezing, rales, nausea, vomiting, pulmonary eosinophilia and tachypnea while urticaria followed by fever may result late in the migratory phase. Heavy infections also lead to expulsion of adult worms through the mouth or nose besides anus. There are reports of abdominal cramps, malabsorption and partial or complete intestinal blockage (Villamizar et al. 1996) in children due to heavy worm burden. Various grave consequences may include intussusception, volvulus and complete obstruction (Khuroo et al. 1990; Khuroo 1996) leading to bowel infarction and intestinal perforation. The peritonitis caused as a result can be fatal, although if the child survives, the wandering adult worms can die and cause a chronic granulomatous peritonitis. Mature worms sometimes block the appendix, biliary tract or pancreatic duct, producing severe abdominal pain. Ascariasis also causes deficiency of vitamins A and C, as well as of protein (Bethony et al. 2006).

Epidemiology

The prevalence of ascariasis is highest in children of 2–10 years age; however, children aged 5–15 years often show simultaneous infections with other helminths such as *T. trichiura* and hookworm; especially in Vietnam, women of rural areas living in households without a latrine demonstrated high prevalence for ascariasis (Do et al. 2007). The worldwide ascariasis (Fig. 19.2) rates in 2005 were 86 million cases in China, 204 million elsewhere in East Asia and the Pacific, 173 million in sub-Saharan Africa, 140 million in India, 97 million elsewhere in South Asia, 84 million in Latin America and the Caribbean, and 23 million in the Middle East and North Africa. An estimated 2 per 1,000 infected children develop intestinal obstruction per year (Gangopadhyay et al. 2007). The prevalence of ascariasis in Japan in 1949 was 63 %; however, the disease was essentially eliminated by 1973.

Hook worms (*A. duodenale* and *N. americanus*)

The human hookworms include the nematode species, *A. duodenale* and *N. americanus*. Different species of hookworms infecting animals can invade and parasitize humans (*A. ceylanicum*) or can cause cutaneous *larva migrans* but do not develop (*A. braziliense*, *A. caninum* and *Uncinaria stenocephala*) any further. Occasionally, *A. caninum* larvae may migrate to the human intestine, causing eosinophilic enteritis or diffuse unilateral subacute neuroretinitis. The two species of hookworm (*A. duodenale* and *N. americanus*), which infect humans, exhibit differences in their pathogenicity, mode of transmission, geographic distribution, thus causing different degrees of morbidity of disease.

Morphology and Taxonomic Status

Hookworms belong to the family Ancylostomatidae of superfamily Ancylostomatoidea. The adults *A. duodenale*, Dubini 1843, range in length from 8–13 mm and are whitish to reddish brown in colour with a hooked anterior end. The females are slightly larger than males. The anterior end of a hookworm is conspicuous with a spacious buccal cavity armed with powerful teeth or cutting plates. The male's posterior end is expanded to form a copulatory bursa. The eggs are 55–68 μm × 35–40 μm in dimension, oval-shaped with thin and colourless shell. Morphologically, *N. americanus* is most widely distributed in tropical and subtropical climatic regions and can be differentiated from *A. duodenale* (Fig. 19.3g–l) in having slightly smaller, S-shaped body (*vs* C-shaped body); one pair of cutting plates (vs two pairs of teeth); oval (vs circular) bursa and fused (vs free spicules in males and females with preequatorial vulva (vs post-equatorial vulva in *A. duodenale*).

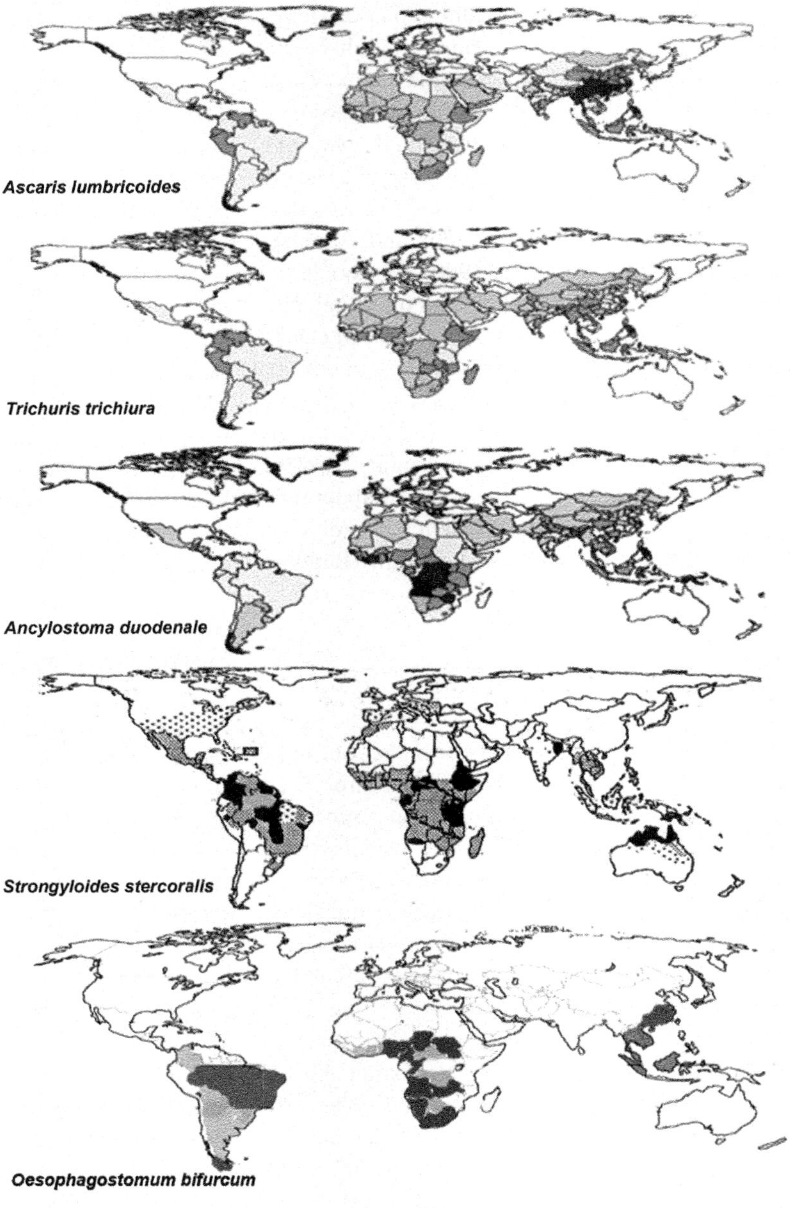

Fig. 19.2 Common soil-transmitted helminths (darker areas showing higher incidence of disease)

Life Cycle

The female worm lays thousands of eggs that pass out in faeces. In soil under favourable conditions of moisture, warmth and shade, the rhabditiform larvae hatch in 1–2 days and feed on soil microbes. Within 5–10 days, the larvae moult twice to develop into 500–700 μm long, infective filariform (third-stage) larvae. The larvae can survive 3–4 weeks in favourable environmental conditions (24–32 °C favourable temperature for survival) and seek transportation by showing nictation behaviour. On contact with human feet, the larvae penetrate through the skin in about 5 min or more. Ground itch at the site of penetration is more common with

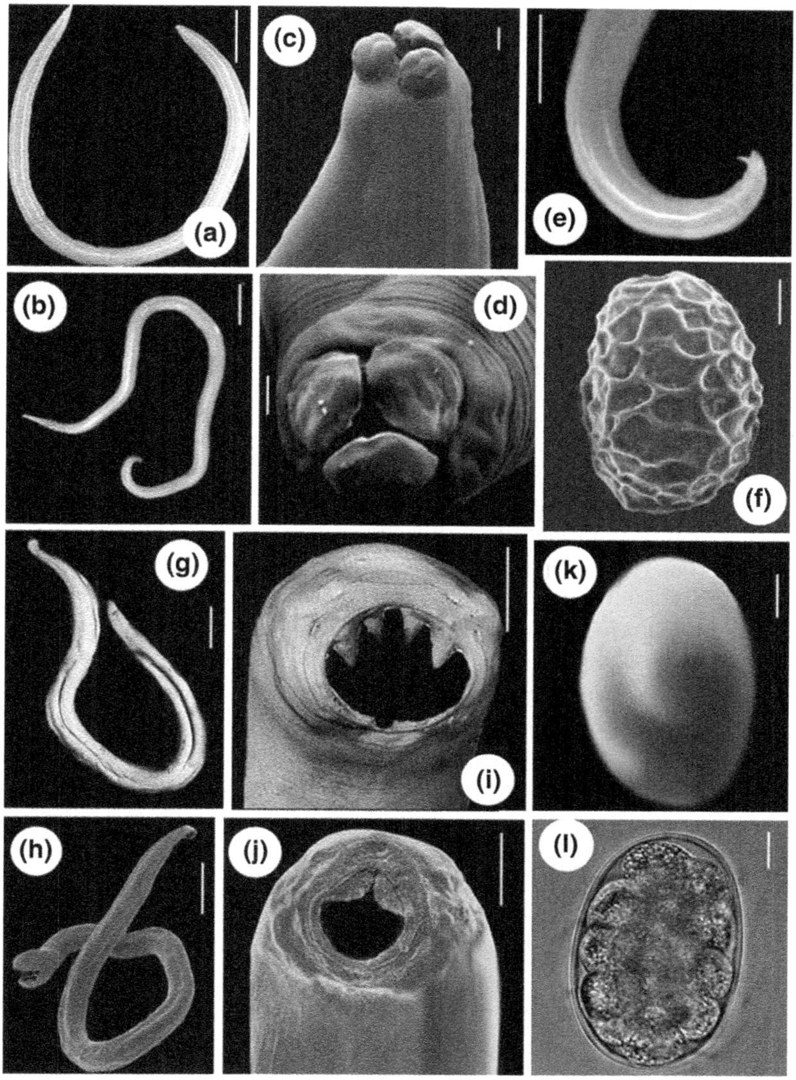

Fig. 19.3 *Ascaris lumbricoides* **a–f**. **a**, **b** Adult female, male; **c** anterior body region; **d** *en face* view; **e** posterior body region; **f** egg. *Ancylostoma duodenale* **g–i**,**k**. **g**, **h** Adult female, male; **i** *en face* view; **k** an egg; *Necator americanus* **j**, **l**. **j** *En face* view; **l** embryonating egg (*Scale bar* **a**, **b**, **g**, **h**, **e** = 1 mm; **i**, **j** = 500 μm; **c**, **d** = 100 μm; **f**, **k**, **l** = 10 μm)

Ancylostoma sp. than with *Necator* sp. After penetration, the larvae reach heart and then lungs through the blood vessels, enter the pulmonary alveoli and finally ascend the bronchial tract to the pharynx. Coughing brings the larvae to the mouth, where swallowing transports them further to the intestine (Fig. 19.1). The larvae reach the gut and puncture mucosal capillaries in the jejunum, small intestine, and moult to finally mature into adults. Approximately 5 weeks after entering the body, adult females start laying eggs. Adult worms embed their buccal capsules in the intestinal mucosa and cause blood loss of the host. *Necator* has 5-year lifespan, whereas *Ancylostoma* is generally eliminated in 1–2 years though the longevity in occasional cases has been reported to be 5 years. If the host is not re-exposed, the infection disappears after the worm dies as it cannot multiply within the host. If *Ancylostoma* larvae enter via ingestion, they do not migrate into the body but lie dormant in tissues instead.

Hookworm Disease: Causes and Symptoms

Hookworms are called the "vampires" of the gut, sucking blood from the capillaries of the intestinal mucosa. Thus, the clinical manifestations of disease resemble those of iron deficiency anaemia from other causes (Hotez 1989). The adult hookworms pierce the mucosa of the duodenum and intestine with hooked teeth (*Ancylostoma*) or with two cutting buccal plates (*Necator*), ingest the blood, rupture the erythrocytes and degrade the haemoglobin (Williamson et al. 2003). Therefore, the disease is attributed to silent blood loss leading to iron deficiency anaemia and protein malnutrition. The extent of intestinal haemorrhage is proportionate to parasite intensity/worm burden (Hotez et al. 2004; Stoltzfus et al. 1997) although heavy infections can be fatal, particularly in infants. The threshold worm loads for anaemia differ internationally, and 40 worms may be enough to produce anaemia (haemoglobin concentrations below 11 g/dL) in countries with low iron consumption. However, the severity depends on the species of hookworm, viz. *A. duodenale*, ingests 5–10 times more blood (260 μl blood/day) than *N. americanus* (30 μl blood/day). Chronic hookworm disease is associated with fatigue, anasarca, pallor, tachycardia and dyspnoea on exertion. Hypoproteinemia may cause oedema with signs of malabsorption and malnutrition. It is estimated that 56 % of all pregnant women in developing countries suffer from anaemia and 20 % of all maternal deaths are either directly or indirectly related to anaemia (Gyorkos et al. 2006). Thus, the hookworms are associated with adverse maternal–foetal outcomes (Bundy et al. 1995; Brooker et al. 1999, 2004). Severe anaemia affects intellectual, physical and cognitive development in children (Smilie and Augustine 1926; Hotez 1989, 2000, 2002; Sakti et al. 1999; Lwambo et al. 2000; Stoltzfus et al. 2001; Beasley et al. 2002; Hotez et al. 2005) and the cardiovascular functions in adults. The movement of third-stage larvae through the vascular tissue to lungs results in eosinophilia and pneumonitis of less severity than *Ascaris* infection. Oral ingestion of *A. duodenale* larvae can result in *Wakana* syndrome, characterised by nausea, vomiting, pharyngeal irritation, cough, dyspnoea and hoarseness.

Epidemiology

A. duodenale is geographically more restricted to higher elevation and occurs in more extreme climates hence prevalent in Europe and the Mediterranean, whereas *N. americanus* is the cosmopolitan hookworm found in the USA, sub-Saharan Africa and Asia (Hotez et al. 2005), responsible for 90 % of human infections that occur in tropical and subtropical regions of the world (Fig. 19.2).

Hookworm infections are endemic; however, despite its global importance, hookworm disease was not listed as one of the six major tropical infections in WHO report (1999) because of insidiousness that is not directly associated with mortality. Moreover, because of the misconception that it could be prevented by increased use of shoes in tropical countries, hookworm was often neglected in scientific studies. Yet, more than a billion people are infected with at least one species, and it has been found extremely difficult to be eliminated in areas of poverty and poor sanitation. Likewise, high transmission rates occur in poor rural areas in the tropics, including southern China, the Indian subcontinent and the USA (Hotez 2002; Yadla et al. 2003). Consequently, hookworm has been associated with decreased future economic productivity (Bleakley 2003) and was reported to cause a 23 % drop in the probability of school attendance (Brooker et al. 1999; Sakti et al. 1999; Bleakley 2003). Brooker et al. (2006) reported 576–740 million individuals to be infected with hookworm and about 80 million to be severely affected (Gasser et al. 2009). Heavy infections are concentrated in sub-Saharan Africa and East Asia/the Pacific Islands. An estimate was given by Brooker et al. (2006) of the region-wise infections including Latin

America and the Caribbean (50 million), South Asia (59 million), Middle East/North Africa (10 million). Co-infection of hookworm with *Plasmodium falciparum* makes the conditions worse in such populations (Brooker et al. 2006).

Whip Worm (*T. trichiura*)

Whip worms (*Trichuris* spp.) are most commonly seen in tropical climates and in areas where sanitation is poor. They seem to occur in areas particularly where *Ascaris* and hookworms are found due to the eggs requiring similar conditions for development. There are several species within this genus each infecting specific host, but only *T. trichiura* infects human causing trichuriasis. It is a parasite that infects many more people than is generally considered, up to 800 million people throughout the tropics and temperate regions.

Morphology and Taxonomic Status

T. trichiura, Linnaeus 1771, is a member of the superfamily Trichocephaloidea of family Trichuridae. Other species of the genus are not significant pathogens of humans. However, a related species, *Trichuris suis*, may occasionally infect humans as a zoonosis transmitted from the reservoir host, pig (Beer 1976). The worm looks like a whip, the anterior 3/5 of body is slender, and the posterior 2/5 is considerably thicker (Fig. 19.4a–e). It is pinkish grey in colour. Adult female measures 35–50 mm in length and the male 30–45 mm long. Its wider posterior part contains the reproductive organs and intestine, whereas it is long, finely attenuated anterior part shows the stichosome along with elongated capillary-like pharynx (Cooper 1995). Lips are not discernible, and vulva is located at the junction of the thread-like and thickened parts of the body. The thin anterior portion of the worm is found embedded in the mucosa. The fertilized females lay several thousands of eggs, which are passed out in the faeces. Eggs are brownish, barrel or spindle shaped with translucent polar plug at either ends and are 49–65 μm × 20–29 μm in dimension. In the soil (a warm, moist environment with plenty of oxygen), the eggs undergo embryonation after which they become infective.

Life Cycle

Unembryonated eggs (unsegmented) laid by females are passed in the faeces of an infected host. In the soil contaminated with faeces, these eggs embryonate to become infective in about 15–30 days. The infective eggs are ingested by human through soil-/ faecal-contaminated hands or food and hatch inside the small intestine (Fig. 19.1). The hatched larvae initially burrow into a villus and then migrate into the caecum (or upper part of the colon), thus reaching their final attachment site. The larvae embed their anterior portion (fine whip-like end) into the tissue mucosa and obtain nutrition from the host tissues. The larvae reach maturity within 3 weeks to several months after infection, during which they undergo four moults. About 60–70 days after infection, female adults begin to lay eggs into the caecum at a rate of 3,000–20,000 eggs per day. The average life span of the parasite is about 1 year.

Trichuriasis: Causes, Symptoms

T. trichiura, as with *A. lumbricoides*, is spread via faecal-/soil-contaminated food or drink. The infective whipworm larva after reaching the host buries its anterior half and feeds on intestinal mucosal tissue/secretions. This invasion causes occasional peripheral eosinophilia. Light infections are usually asymptomatic. Adult whipworms lead both an intracellular and an extracellular existence, with the anterior end embedded in epithelial tunnels within the tissue mucosa and the posterior end located in the lumen of gut. The preferential site is caecum, but in heavy infections, the worms can be seen throughout the colon and rectum. The clinical

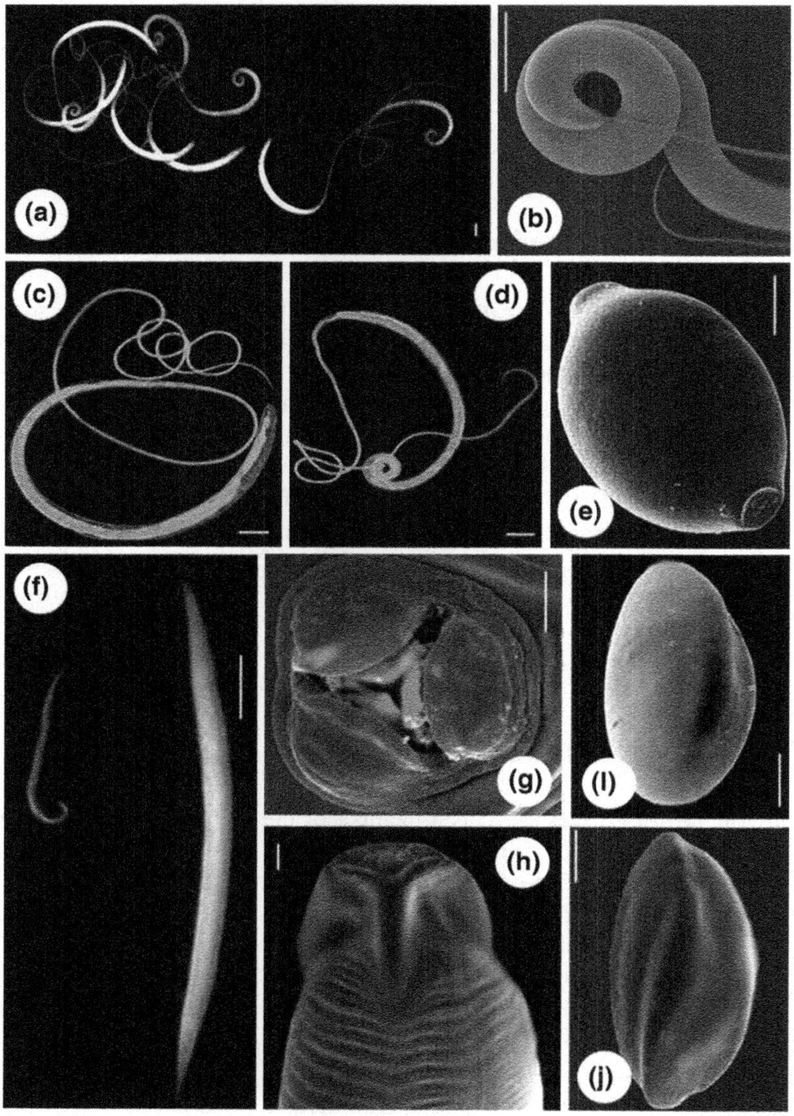

Fig. 19.4 *Trichuris trichiura* **a–e**. **a** Group of individuals; **b** part of body with thin anterior region and thick posterior region; **c** adult female; **d** adult male; **e** egg. *Enterobius vermicularis* **f–j**. **f** Male (small) and female (long); **g** *en face* view; **h** anterior region; **i**, **j** eggs (*Scale bar* **a**, **d**, **f** = 1 mm; **g–j** = 10 μm)

manifestations are abdominal pain, diarrhoea and constipation. Inflammation at the attachment site due to large numbers of whipworms results in colitis resembling inflammatory bowel disease and a series of complications such as anorexia, impaired growth, anaemia and finger clubbing. Heavy whipworm infections result in chronic dysentery, bloody diarrhoea, emaciation and rectal/anal prolapse. In small children with worm burden, there may be intestinal blockage. Rare instances of pneumonitis occur when larvae penetrate into the intestinal tissues and by way of the lymph and blood vessels reach the lungs. The colitis caused by *Campylobacter jejuni* can be exacerbated by whipworm infection. Vitamin A deficiency has also been reported in patients with *T. trichiura* infection (Albonico et al. 2008).

Epidemiology

Globally, trichuriasis is a very common STH infection with about one quarter of the world's population infected with the parasite (Fig. 19.2). The problem is of serious nature in tropical Asia, Africa and South America where the patients with heavy parasite burden become symptomatic. Poor hygiene is associated with *T. trichiura* transmission, and children are vulnerable due to high exposure risk, especially in developing countries, where poor sanitary conditions correlate with heavy disease burden.

Pin Worm (*E. vermicularis*)

Pin worm, *E. vermicularis* is one of the most common nematodes in the world. The earliest known record is evidenced by the eggs found in fossilized animal faeces dating back to 7837 B.C. (Cook 1994). There is no intermediate host. Infection is via ingestion of eggs through contaminated food/water.

Morphology and Taxonomic Status

The pinworm *E. vermicularis* belongs to family Oxyuridae of the superfamily Oxyuroidea. Adults are white, small and delicate nematodes with females 8–13 mm long and 0.5 mm thick having sharply pointed posterior ends; the males are considerably smaller, measuring 2–5 mm in length and 0.2 mm in thickness with curved posterior ends (Fig. 19.4f–j). The oral aperture is guarded by three lips. The eggs are translucent and have a surface that adheres to environmental objects (Cook and Alimuddin 2009). The eggs are thick-shelled, ovoid and asymmetrical in shape (one side being more convex than the other) and measure 50–60 μm × 20–30 μm in dimension.

Life Cycle

The entire life cycle from egg to adult takes place in the gut of a human host. The eggs are passed in the faeces and often deposited in soil where they remain viable in a moist environment for up to 3 weeks (Burkhart and Burkhart 2005) at low temperatures. Caldwell (1982) found about two-thirds of the eggs to be viable after 18 h at 8 °C. Pinworms are common among people living in close contact and are common in people within a household (Gutiérrez 2000). Infections occur by ingesting/swallowing infectious pinworm eggs (Fig. 19.1) through contaminated water or food and/or by anal sex. The eggs hatch in the duodenum to release the first-stage larvae that quickly attain a length of 140–150 μm and migrate through the small intestine towards the colon. During this migration, they moult four times and become adults. The males usually die after mating and are passed out in faeces. The life span of females ranges from 5 to 13 weeks, while the males survive for about 7 weeks. The gravid females may settle in the ileum, caecum, appendix or ascending colon, carrying a load of eggs ranging from 11,000 to 16,000 and attaching themselves to the mucosal wall. The egg-laying process begins approximately 5 weeks after ingestion of eggs. The gravid females migrate through the colon towards the rectum and often crawl out of anus at night for laying the eggs in the peri-anal area and to obtain the oxygen necessary for their maturation. This causes the nightly itching in the peri-anal area. After depositing the eggs, the female occasionally dies. Sometimes the worms migrate to the adjacent orifice and reach the female vagina/urinogenital tract (Burkhart and Burkhart 2005; Dundas et al. 1999; Hong et al. 2002; Smolyakov et al. 2003), or other unusual sites (McDonald and Hourihane 1972; Gargano et al. 2003). The females may disintegrate or rupture due to scratching of area by the host. The scratching also facilitates the transfer of eggs to the fingers/nails, then to the mouth, ensuring autoinfection. Finger sucking (Burkhart and Burkhart 2005) and nail biting (Cook 1994) increase both incidence and relapse rates. The worms occasionally migrate to other organs such as kidneys.

Enterobiasis: Causes, Symptoms

The infective eggs can be transmitted through clothing and bed linen (Cook et al. 2009) or through contaminated food, water, furniture, toys, bathroom utilities, etc. (Cook 1994; Burkhart and Burkhart 2005). Household pets often carry the eggs in their fur though not actually being infected (Caldwell 1982). Eggs often get widely dispersed along with dust while being airborne, with reports of their entry into the mouth and nose (Garcia et al. 1999; Cook et al. 2009). In chronic infections, occasionally due to retroinfection (Caldwell 1982), some pinworm larvae may hatch on the anal mucosa and migrate up the bowel and back into the gastrointestinal tract (Burkhart and Burkhart 2005; Cook et al. 2009). Despite having a 13-week lifespan, pinworms often inhabit the same host indefinitely due to autoinfection or retroinfection.

Pinworm infections are generally asymptomatic but often can be disturbing; however, they rarely cause serious health problems or severe morbidity unless ectopic infections result, for example inflammatory bowel disease (parasites migrate through the bowel wall and are found in extracolonic sites). Intense itching around the anus and/or vagina is the classic symptom. Other symptoms range from intermittent stomach pains, upset stomach to loss of appetite, irritability, restlessness and insomnia. Ectopic enterobiases have been reported to occur in various locations (Gutiérrez 2000; Cateau et al. 2010), including the vagina, salpinx, inguinal area, genital area, pelvic peritoneum, omentum, liver, salivary glands, male genital tract and even the lungs. They have also been associated with acute appendicitis (Cook 1994; Aydin 2007; Da Silva et al. 2007; Ramezani and Dehghani 2007), eosinophilic colitis and eosinophilic gastroenteritis (Hasegawa et al. 1998). *E. vermicularis* granulomas are usually incidental and cause no clinical problems (Smolyakov et al. 2003).

Epidemiology

Pinworm is a cosmopolitan parasite of humans. The parasite shows a high prevalence particularly in countries with a temperate climate and affects approximately 200 million people internationally. The infections form the most common helminth infection in the USA and Western Europe. However, its prevalence is not associated with any gender, social class/community, race or culture (Fig. 19.2). Pinworms are an exception to the belief that intestinal parasites are uncommon in affluent and rich states (Burkhart and Burkhart 2005). In the United States, a study (1992) by the Center of Disease Control revealed an overall incidence rate of 11.4 % among people of all ages. In Western Europe, the prevalence rates in some communities have been reportedly 30–50 % (Burkhart and Burkhart 2005). Pinworms are particularly common in children, with prevalence rates having been reported to be 61 % in India, 50 % in England, 39 % in Thailand, 37 % in Sweden and 29 % in Denmark. Finger sucking and nail biting have been shown to increase both incidence and relapse rates of the infection.

Threadworm (*S. stercoralis*)

The most common and clinically important strongyloid species in humans is *S. stercoralis*. Another species *S. fuelleborni* is found sporadically in Africa and Papua New Guinea. *S. stercoralis* infects 30 million people in 70 countries. It has the ability to persist and replicate within a host while remaining undetected for decades and producing minimal or no symptoms. However, in patients receiving long-term corticosteroid therapy, hyperinfection can lead to high mortality rates up to 87 % (Arsic-Arsenijevic et al. 2005; Asdamongkol et al. 2006; Weller and Leder 2008). Strongyloidiasis is difficult to diagnose because the parasite load is low and the larval output is irregular. Often a single stool examination using conventional techniques fails to detect larvae in up to 70 % of cases (Siddiqi and Berk 2001).

Morphology and Taxonomic Status

Like *Enterobius*, another nematode of relatively lesser importance among the STHs is *S. stercoralis*. It belongs to family Strongyloididae of superfamily Strongyloidea.

The females (2.0–2.7 mm) are found attached to mucosa of the small intestine. Males (0.9 mm) are usually not parasitic. The anterior end of nematode shows tiny buccal capsule connected to a cylindrical pharynx. Oral aperture is guarded by six lips (Fig. 19.5a–g). Eggs are ovoid with smooth shell and measure 40–70 μm × 20–35 μm in dimension.

Life Cycle

The parasite shows heterogonic (alternation of amphimictic and parthenogenetic cycles) mode of reproduction. The complex life cycle is advantageous as it allows reproduction in the absence of a host. The females reproduce by parthenogenesis inside the host body and lay eggs which usually hatch into rhabditiform larvae hence often not found in a faecal sample. The larvae show resemblance with hookworm larvae; therefore, a correct diagnosis is essential. In hyperinfection, larvae may be found in the sputum or in bronchoalveolar fluid. In the free-living cycle, the newly hatched rhabditiform larvae pass in the human faeces and may be deposited in soil. The larvae can further moult twice and become infective filariform larvae (direct development). These larvae may penetrate the human skin usually at the feet (occasionally via the mouth) to initiate the parasitic cycle (Fig. 19.1). The larvae then migrate to the lungs through blood vessels where they penetrate the alveolar spaces, get carried through the bronchial tree to the trachea, are coughed up and then swallowed to reach the small intestine. In the small intestine, they moult twice and become adult worms. The females lie attached to intestinal mucosa, while males are non-parasitic and cannot penetrate the intestinal mucosa hence die after some time. The females by parthenogenesis produce eggs, which yield rhabditiform larvae. The rhabditiform larvae can either be passed in the faeces or can cause autoinfection by transforming into infective filariform larvae, that penetrate the intestinal mucosa (internal autoinfection) or the skin of the perianal area (external autoinfection). Such infections may persist long for many years. In soil, the rhabditiform larvae moult four times and become free-living adult males and females that mate and produce eggs from which next generation of rhabditiform larvae develop to repeat the free-living cycle or develop into infective filariform larvae that may penetrate the human host skin for a parasitic cycle. The free-living males and females of *S. stercoralis* usually die after one generation and do not persist in the soil for long time.

Strongyloidiasis: Causes, Symptoms

Mild infections are generally asymptomatic (Montes et al. 2010). The symptoms of eosinophilia mostly develop, except when immune suppression exists. A history of *larva currens* is indicative of strongyloidiasis. Initially with the entry of worm, a rash appears on the foot followed by subsequent rashes showing the migration of worm beneath the skin. Upon their entry into the lungs, the larvae cause Löffler's syndrome characterized by infiltration of white blood cells, cough, shortness of breath and cloudy chest as evident in X-ray examination. The severity of symptoms may depend on the number of larvae

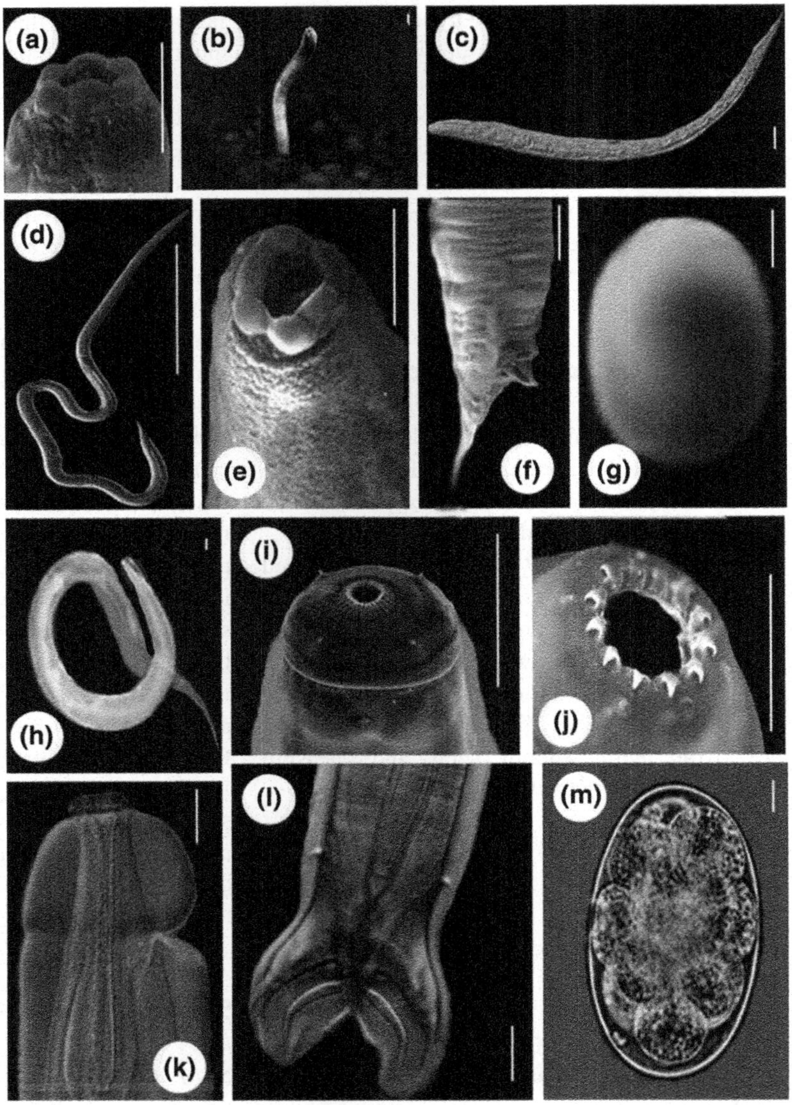

Fig. 19.5 *Strongyloides stercoralis* **a–g**. **a** Anterior body region; **b** nictating filariform larva; **c** filariform larva; **d** adult female; **e** *en face* view; **f** posterior body region (male); **g** egg; *Oesophagostomum dentatum* **h–m**. **h** Adult female; **i**, **j** *en face* view; **k** anterior body region showing pharynx; **l** posterior body region (male); **m** egg (*Scale bar* **d** = 1 mm; **a**, **b**, **c**, **e**–**g** = 10 μm; **h**–**l** = 100 μm; **m** = 10 μm)

and can result in dyspnoea, bronchospasms, bloody sputum and meningoencephalitis (Graeff-Teixeira et al. 2009). The intestinal invasion leads to intestinal discomfort, diarrhoea and stomach aches. Auto re-infection via the skin may give rise to significant itching, red swollen lines due to urticarial reaction. Immune suppression (Kramer et al. 1990; Zygmunt 1990; Gill et al. 2004), especially HTLV-1 infection, achlorhydria, haematological malignancies including lymphoma, cytotoxic medication, nephrosis, burns and especially the long-term use of systemic corticoids, all such conditions lead to hyperinfection (Igra-Siegman et al. 1981; Marcos et al. 2008). The latter may be sometimes accompanied by bacterial septicaemia.

Epidemiology

S. stercoralis can be found in areas with tropical and subtropical climates although low incidence also occurs in temperate areas (Fig. 19.2). The parasite has a high prevalence in populations with faecal contamination of soil or water. Hence, it is rarely encountered in developed countries; while in developing countries, it is less prevalent in urban areas than in rural areas (Zaha et al. 2000). Though strongyloidiasis is endemic in Africa, the prevalence is typically low. The infections are frequently found in rural areas and lower socio-economic groups and countries like India, China, Vietnam, Cambodia and Laos still have endemic strongyloidiasis with the prevalence rates being 10 % or less. Improved sanitation and increased use of shoes in children are likely to decrease the prevalence, although surveys of rural Kentucky schoolchildren demonstrated 3 % prevalence (Walzer et al. 1982). The control programmes have eliminated the disease from endemic areas in Japan although strongyloidiasis still has a high prevalence in some areas of Brazil and Central America.

Nodular Nematode, *Oesophagostomum bifurcum*

The nematode, seldom included among the STHs is the nodular nematode, *Oesophagostomum* (*O. bifurcum*, *O. aculeatum* and *O. stephanostomum*) that usually infects monkeys and animals like goats and pigs Humans are accidental final hosts in some regions although they are increasingly becoming the favourable hosts day by day.

Morphology and Taxonomic Status

Oesophagostomum spp. are usually considered free-living nematodes of the family Strongyloididae. Their developed buccal capsule and club-shaped pharynx are useful to distinguish them from hookworms. Adult females measure 6–24 mm in length and are generally larger than males (6–17 mm). Anterior end of parasite possesses cephalic inflation and is separated from trunk by a deep groove that bears the opening of secretory–excretory system; the oral opening is lined with both internal and external leaf crowns. Males can be distinguished by their bell-like copulatory bursa, located in the tail, and their paired rod-like spicules (Fig. 19.5h–l).

Eggs are ovoid and measure 51–75 μm × 29–40 μm in dimension and are morphologically identical to eggs of *N. americanus* or *A. duodenale* (eggs of other Oesophagostomum species are larger than hookworm eggs). Eggs are often in an advanced stage of cleavage compared to those of hookworm species when shed in faeces.

Life Cycle

When the eggs are deposited in soil, the rhabditiform larvae hatch out and undergo two moults in 5–7 days. Usually, the resulting third-stage filariform larvae can resist long periods of dehydration. The larvae may get swallowed with contaminated food or water and penetrate the human intestinal wall inducing abscesses and nodule formation (Fig. 19.1). The larvae develop further and moult two times to transform into adults, then move to the intestinal lumen and mate. After copulation, the females usually lay around 5,000 eggs per day that are deposited in the faeces. As soon as the worms become adults and reach the intestinal lumen, they do not cause illness. However, many larvae often do not complete development and remain in colon nodules, as humans are generally unsuitable hosts for *Oesophagostomum* spp. (Gasser et al. 2006).

Oesophagostomiasis: Causes, Symptoms

Although the routes of human infection are not well established, the transmission occurs with humans ingesting filariform larvae along with faecal-/soil-contaminated food or water or contaminated meat obtained from infected livestock. Oesophagostomiasis involves nodule formation in the host intestine that can lead to more serious problems such as abscesses with a necrotic content (helminthoma) and dysentery.

The clinical symptoms include abdominal pain in the lower right quadrant, accompanied by the presence of one or several protruding abdominal masses. In advanced stages, larvae can invade the colon wall causing "pimply gut" (multinodular disease) that is characterized by the formation of many nodular lesions containing worms and pus and countless abscesses under the serosa. Epigastric or periumbilical growths may also result. The intestinal lesions including eosinophilic granulomas can lead to further complications, such as bowel obstruction, peritonitis and intestinal volvulus.

Epidemiology

O. bifurcum infections in humans commonly occur in a small part of West Africa (Polderman and Blotkamp 1995) with virtually every village of northern Togo and Ghana and some rural areas exhibiting as much as 90 % prevalence. Oesophagostomiasis is endemic or potentially endemic to 35 countries (Fig. 19.2), and approximately 250,000 people are estimated to be infected worldwide (Polderman et al. 1991) hailing from countries in South America and Southeast Asia, including Brazil, Indonesia and Malaysia (Gideon Infectious Diseases Database 2009). However, the vast majority of clinical cases have been reported from Africa (Ziem 2006).

Distribution and Estimates of Soil-Transmitted Helminths

Among the STH infections, the highest rates of *Ascaris* infection occur in China and Southeast Asia, in the coastal regions of West Africa and Central Africa. *Trichuris* infections reach their highest prevalence in Central Africa, southern India and Southeast Asia. Hookworm infections, however, are common throughout much of sub-Saharan Africa, in addition to South China and Southeast Asia; whereas the pinworm infections are prevalent throughout the world including the temperate regions and are the most common helminthic infection in the United States. The intensity of infection is the key aspect to be studied if transmission, parasite population regulation and morbidity are to be manipulated for the benefit of individuals and communities (Crompton 2001). Some earlier estimates (de Silva et al. 2003) indicate that ascariasis remains common with >1.2 billion infections globally. Almost 50 % of these infections are in China, which still has the highest prevalence. Trichuriasis and hookworm amount to 700–800 million infections each. According to WHO report (2012), there were about 290.1 million, 45.3 million, 371.6 million, 4.3 million people having STH infections and requiring chemotherapy in Africa, the USA, Southeast Asia and Europe, respectively, by 2010 (Figs. 19.6, 19.7).

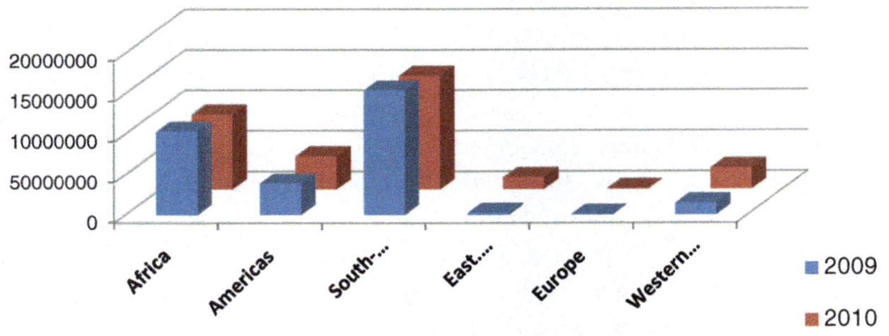

Fig. 19.6 Total number of people given chemotherapy in the year 2009 and 2010. (Data source WHO 2012 report. 87:17–28)

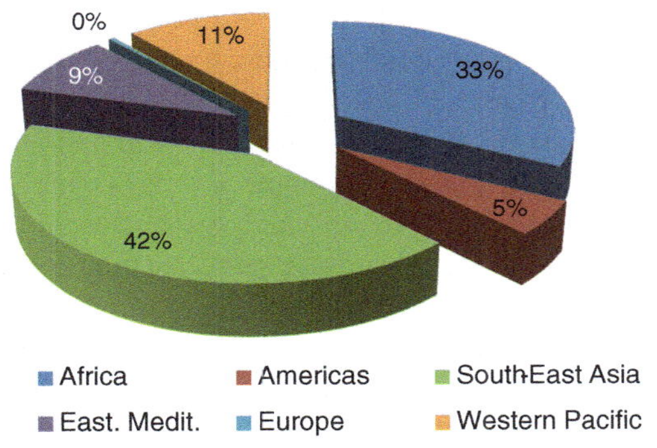

Fig. 19.7 Total number of people requiring chemotherapy in the year 2010. (Data source WHO 2012 report. 87:17–28)

Changes in the Global Situation

There appears to be a marked decline in the prevalence and abundance of all STH infections since 1994 in both the USA and Asia. This dramatic decrease is largely due to launching of national programmes, together with social and economic development. It is still true that the poorest people in the world suffer the greatest burden of infectious diseases due to inadequate water and sanitation, crowded living conditions combined with lack of access to health care and low levels of education/awareness (Sachs and Steele 2001; Gwatkin and Guillot 2000a, b). The reductions in infection rates in some parts of the world can be exemplified by specific control schemes launched by Japan and Korea, and a shift from an agrarian to a suburban economy in Jiangsu Province, China (Xu et al. 1995; Feng-hua et al. 1998). However, there appears to be little change in prevalence rates in sub-Saharan Africa. Anaemia arising from STH infections is often associated with reduced work output, impaired cognitive ability and effects on school attendance among children or with consequences of delaying primary school enrolment and school attainment, thereby affecting future labour market outcomes (Guyatt 2000). It, therefore, indirectly serves as one of the several contributing factors for poor economic growth. Ascariasis and trichuriasis are endemic, and their prevalence rises rapidly once infant age passes and tends to remain high in 5–15 years age group before declining steadily (Crompton 2001). Hookworm infections as earlier reported (Stephenson et al. 1989; Labiano-Abello et al. 1999) are common in children in some tropical areas with peak prevalence and infection intensities in individuals of middle age, or even over the age of 50 (Gandhi et al. 2001; Bethony et al. 2002).

Burden of Disease

The STHs exhibit great variability in the worm burdens as most individuals harbour just a few worms in their intestines, while others harbour disproportionately large number of worms (Bundy et al. 1995). WHO has set the threshold levels for assessing the degree of infections (Table 19.1). As a general rule, approximately 70 % of the worm population is harboured by 15 % of the host population. These heavily infected individuals are at highest risk of disease and a major source of contamination (Bundy et al. 1995).

STH infections rarely cause death, and the severity of disease is dependant on the chronic and insidious effects on the hosts' health and nutritional status (Stoltzfus et al. 1997; Stephenson et al. 2000). Thus, the worldwide

Table 19.1 Thresholds for intensity of infection by three common STH parasites

Parasite	Helminth intensity of infection threshold		
	Light	Moderate	Heavy
A. lumbricoides	1–4,999 epg	5,000–49,999 epg	≥50,000 epg
T. trichiura	1–999 epg	1,000–9,999 epg	≥10,000 epg
Hookworms	1–1,999 epg	2,000–3,999 epg	≥4,000 epg

Source WHO. Helminth control in school-age children: a guide for managers of control programmes, 2002, Geneva

burden is typically assessed by disability-adjusted life years (DALY) that indicates retarded physical and mental development. Since beginning, there has been much variability in the quoted DALY estimates (Chan et al. 1994; Murray and Lopez 1996; WHO 2004) regarding helminthiasis. The global burden of disease caused by these intestinal nematodes is an estimated 22.1 million DALYs lost to hookworm, 10.5 million to *A. lumbricoides*, 6.4 million to *T. trichiura*, giving a combined total of 39 million life years (Chan 1997). Hookworms have long been recognized as an important cause of iron deficiency and protein malnutrition (Hotez et al. 2004) particularly in children and pregnant women (Brooker et al. 2003) with infections causing more DALYs lost than any other helminthiasis except lymphatic filariasis. A majority of such populations live in poverty-stricken areas with poor sanitation and wages generally less than $2 a day (Hotez et al. 2005). Therefore, the infections increase the likelihood of the afflicted population to be further mired in poverty. Severe anaemia in pregnancy leads to premature births, reduced birthweight and impaired lactation (Christian et al. 2004). Ascariasis (and trichuriasis in some cases) is associated with low serum vitamin A (retinol) levels (Mahanalabis et al. 1976; Stephensen 2001). In 1997b, de Silva et al. estimated about 11.5 million individuals (almost all of them children) to be at risk of acute clinical illness and accounted 90 % of 10,500 deaths annually due to serious complications of ascariasis. *Ascaris* and *Trichuris* infections are not only found in tropics but also reported from poor, underserved areas of the developed nations, including the United States (Blumenthal and Schultz 1975; Jones 1983; Kappus et al. 1994) and the United Kingdom (Crompton 1989).

Adaptability of Parasites

STHs are tenacious to host immunity, thus establishing chronic infections during the host's life. These parasites pass through different host tissues (skin, lungs and gut) during their developmental cycles with the life stages carrying stage-specific antigens (Maizels et al. 1999). Adult and larval *Ascaris* release volatile allergens (Coles 1985; Kennedy 1992) and may induce asthma-like symptoms. STHs survive within the host by subverting the host immune response to optimise successful establishment, feeding and reproduction. However, some of the host responses include eosinophil-mediated larval killing, production of specific and polyclonal immunoglobulin E, mast cell degranulation, goblet cell hyperplasia and increased mucus secretion. The effector cells, however, show specificity in their actions (Finkelman et al. 1997; Bradley and Jackson 2004); for example, mast cells demonstrate protective responses against hookworms and *Ascaris* but not against *Trichuris*.

As the survival of host is essential for the survival of parasite, STHs achieve a type of balanced parasitism, in which transmission is maintained and acute morbidity to the host is avoided. Their secretomes interact with host tissues and modulate the host's immune responses to maintain the parasitic existence, for example, anticoagulants secreted by *Ancylostoma* at the site of parasite attachment. Likewise, *Ascaris* larvae release a pepsin inhibitor from the body wall to protect themselves from stomach digestive enzymes before they reach small intestine. *T. trichiura* secretes protein TT47 to create ion-conducting pores in lipid bilayers (Drake et al. 1994) in order to penetrate through the caecal epithelium.

Factors Affecting Soil-Transmitted Helminth Infections

Location/Surrounding

STHs mainly rely on the five f's (fingers, faeces, fomites, flies and food) for their transmission. *Ascaris* and *Trichuris* commonly occur both in urban environments (especially urban slums) and in rural areas. Surprisingly, some reports (Phiri et al. 2000) showed greater prevalence of

Ascaris infection in urban environments. In contrast, high rates of hookworm infection are typically restricted to poor areas (Albonico et al. 1997). The infective eggs have enormous capacity to withstand environmental extremes of urban environments, while the mucopolysaccharide coat of *Ascaris* eggs is adhesive to a wide variety of environmental surfaces including vegetables (Raisanen 1985).

Climate/Season/Topography

Adequate warmth and moisture are crucial for every STH. Thus, total rainfall and its seasonal distribution may determine the patterns of STH prevalence in some endemic regions. Hookworm transmission rates are higher during the rainy season (Mark 1975; Udonsi et al. 1980). Though, *Ascaris* and *Trichuris* eggs can tolerate drier climates better than hookworm infective larvae; nevertheless, a minimum of 1,400 mm annual rainfall is necessary for the prevalence of *A. lumbricoides* to exceed 10 % (Prost 1987; Brooker and Michael 2000), and the rates of infection are low in arid climate. At less than 80 % relative humidity, *Ascaris* eggs fail to embryonate (Brooker and Michael 2000). Likewise, the infective juveniles of thread worm show desiccation tolerance and nictate (stand upright on tail) to get hold of the passing host or any transporting agent. Soil moisture is also critical for vertical migration of hookworm infective larvae (L3) and thus can migrate in presence of vegetation (conserved moisture) to a height of 30–40 cm in 24–48 h (Komiya and Yasuraoka 1966) by utilizing lipid reserves (Hotez 1995). Altitude also affects STH transmission and *Ascaris* occurred at altitudes up to about 1,700 m with decreasing prevalences (Appleton and Gouws 1996; Jemaneh 1998). Prevalence rates of 11–15 % have been reported at altitudes of 3,800–4,200 m (Flores et al. 2001). *A. lumbricoides* and *T. trichiura* do not occur in areas where land surface temperature (LST) exceeds 37 °C (Brooker et al. 2002a, b).

Genetic Risk Factors

Some human populations may have increased genetic susceptibility towards infections. Predisposition has also been described for *Trichuris* (Bundy 1986) and *Ascaris* infections (Haswell-Elkins et al. 1987; Hlaing et al. 1987). In some genetically predisposed individuals, two quantitative trait loci on chromosomes 9 and 18 are found to be responsible for the susceptibility to infection with *T. trichiura* (Ellis et al. 2007a, b). The populations resistant to re-infection show very low worm burdens and were noted to increase parasite-specific IgE and eosinophilic responses (Quinnell et al. 1995; Pritchard et al. 1995; Faulkner et al. 2002). The immunoglobulin levels appear to closely parallel worm burdens (Haswell-Elkins et al. 1989) as also observed in IgG4 host antibody responses (Palmer et al. 1996; Xue et al. 2000).

Behaviour, Occupation and Socio-economic Status

The role of sanitation has a major influence on the prevalence and intensity of STHs; likewise, Kucha houses or those made from wood and bamboo were found to have higher rates of STH infections than concrete houses (Holland et al. 1988). *Ascaris* worm burden was reported not to be influenced by ethnicity and sex but socio-economic factors (Kightlinger et al. 1998) with prevalence significantly higher among children of large families (Prakash et al. 1980; Haswell-Elkins et al. 1989) and clustered households (Forrester et al. 1988). Unexpectedly, shoes were observed to have no significant impact on preventing hookworm transmission (Bethony et al. 2002). Nawalinski et al. (1978) found a greater prevalence of hookworm among Muslims in India compared to Hindus. The prevalence and intensity of STH infections were found to depend on the occupations and behaviour of host. Widespread use of faeces as night soil fertilizer (Humphries et al. 1997; Hotez

2002) in some regions results in heavy infections; however, high rates of infections have also been reported in regions not using night soil. Hookworm infections are also commonly reported in people involved with agricultural pursuits (Hotez et al. 1997) particularly the plantation-style agriculture (Sorensen et al. 1994). Therefore, like malaria, hookworm is also supposed to cause "agriculture-related anaemias" (Fleming 1994; Stoltzfus et al. 1997).

Food Items

Occasionally, infective stages of STHs adhere to vegetables and get distributed in food markets. In a survey of 40 Tokyo shops from Japan, 1178 of 2750 items of vegetables were contaminated with *Ascaris* eggs (Kobayashi 1980). *Ascaris* and *Trichuris* infections were high in people consuming sewage irrigated vegetables (Bouhoulm and Schwartzbrod 1997).

Age Group

The elderly people have rarely been observed having high risk for *Ascaris* or *Trichuris* or other STH infections when compared to younger people or children; however, in some developing countries, due to poor nutritional and overall health, the elderly are vulnerable to the morbidity associated with chronic and heavy hookworm infections (Tucker and Buranapin 2001; Hotez 2002).

Gender

Males are generally more susceptible to infections/diseases than females (Goble and Konopka 1973; Bundy 1988; Zuk and McKean 1996; Klein 2000). However, prevalence of *Ascaris* infection has been considered higher among females compared with males (Crompton 1989) in Africa and Southeast Asia. Hookworm infections although showed preference for adults over children; nevertheless, an estimated 44 million cases of hookworm were reported during pregnancy worldwide (Bundy et al. 1995; Navitsky et al. 1998) that led to adverse foetal outcome.

Diagnosis and Treatment

The aim of treatment for STH infections is to remove adult worms from the gastrointestinal tract. Due to the insidious nature, the clinicians opt for a faecal examination on the basis of some indications such as local epidemiology or country of origin or persistent eosinophilia. Several egg concentration techniques—for example formalin ethyl acetate sedimentation—can detect light infections (Faust and Russell 1964). The number of egg counts per gram of faeces is estimated to measure the intensity of infection using Kato–Katz faecal-thick smear and the McMaster methods (Dunn and Keymer 1986; Santos et al. 2005). Ultrasonography and endoscopy are used for diagnostic imaging of the complications of ascariasis, oesophagostomiasis and hookworm disease resulting in intestinal obstruction, hepatobiliary and pancreatic involvement (Koumanidou et al. 2004). The three main treatments for STH infections are drug therapy, sanitation and health education.

Chemotherapy

WHO urges member countries to ensure availability of good quality anthelmintic drugs in endemic areas. The benzimidazole drugs most commonly used for the STH control are mebendazole (single dose: 500 mg) and albendazole (single adult dose: 400 mg; dose for children between 12 and 24 months: 200 mg). The other drugs levamisole and pyrantel also included in WHO essential drugs list are effective against hookworm and *Ascaris* infections but not for trichuriasis. The anthelmintic drugs bind to nematode tubulin and inhibit microtubule polymerization of adult worms causing death in several days. Both albendazole and mebendazole are effective against *Ascaris, Enterobius* and *Strongyloides* in a single dose. However, in hookworm, a single dose of mebendazole has a low cure rate, while a single dose of albendazole is not effective in many cases of trichuriasis (Adams et al. 2004; Albonico et al. 1994). Albendazole is suitable for larvae as it is better absorbed when ingested with fatty meals and metabolized in the liver into a sulphoxide derivative, which reaches the tissues to kill the migrating larvae (*larva migrans*). The drug mebendazole, on the other hand, is poorly absorbed, and its therapeutic activity is largely confined to adult worms. Usually, the systemic toxic effects of benzimidazole anthelmintic drugs on the liver and bone marrow are rare except some transient abdominal pain, diarrhoea, nausea, dizziness and headache. However, the drugs should be used with caution in children below 12 months and pregnant women due to their embryotoxic and teratogenic effects in pregnant rats (Montresor et al. 2002a, b, 2003). Repeated chemotherapy at regular intervals (periodic deworming) in high-risk groups can be effective with different strategies:

1. Treatment of the entire community.
2. Treatment of community groups based on age, sex or other social characteristics.
3. Treatment of individual on the basis of diagnosis or suspicion.

Some desirable effects of large-scale deworming have been seen in affected areas.

The periodic distribution of anthelmintics in preschool children has resulted in improvement in motor and language development and reduction in malnutrition (Stoltzfus et al. 2004). The schools with a skilled workforce are ideal places for launching such schemes due to their close contact with community. Treatment of school-age children has led to their improved nutritional status (Stoltzfus et al. 2004) and enhanced physical fitness, appetite, growth (Stephenson et al. 2000) and intellectual development (Stephenson et al. 1989; Awasthi et al. 2000; Drake et al. 2000). In endemic areas for hookworm infections, anthelmintic treatment is recommended after 3 months of pregnancy (de Silva et al. 1999; Savioli et al. 2003) and the treated women showed substantial improvements in anaemia and foetal health (Atukorala and Lanerolle 1998; Torlesse and Hodges 2001; Christian et al. 2004).

The complete eradication of the parasites has not been achieved in most of the cases due to certain constraints. It has been disastrous that within 2–3 years of treatment, the hookworm infections have been reportedly reached 80 % of pretreatment rates due to re-infections (Quinnell et al. 1993), while *Ascaris* and *Trichuris* infections touched 55 and 44 % of pretreatment levels within 1–2 years, respectively (Elkins et al. 1988). Hence, in such instances, due to some isolated human reports (Albonico et al. 2003) as well as prevalent reports of livestock, the possibility of drug resistance in these human parasitic nematodes cannot be ruled out.

Sanitation and Hygiene

Sanitation is an effective intervention to control STH infections by reducing soil and water contamination on a large scale. The implementation of this strategy is difficult in poor countries due to the high costs involved (Asaolu and Ofoezie 2003). Furthermore, considerable time is required for sanitation to be effective for controlling the STH infections (Brooker et al. 2004). A programme of latrine construction, health education and anthelmintic treatment decreased the prevalence of ascariasis in Korea from 80 % in 1949 to 55 % by 1971.

Health Education

Health education creates awareness among the masses about the consequences of infection and the practices to be adopted to reduce transmission and re-infection. It also requires a change in attitude and habit of the community and the enthusiasm towards deworming. Health education can be provided to masses and presents no contraindications or risks hence should be an integral part of all helminth control programmes.

In northern Bangladesh, after a 4-year educational campaign and latrine construction programme, the infection prevalence dropped to 64 % in children (Northrop-Clewes et al. 2001).

Research and Development

STH infections cannot be controlled effectively unless the correct data on helminth disease burden and incidence of anthelmintic drug resistance are procured. Vaccination seems to be one of the important methods to control STH infections (Hotez et al. 2003, 2010); however, some limitations in vaccine development include the lack of good animal models and a relatively poor understanding of the basis of long-term survival of parasite without a potent immune response. Tests on dogs and hamsters have shown the success of Na-ASP-2 hookworm vaccine consisting of the recombinant larval antigen ASP2 (Bethony et al. 2005; Goud et al. 2004, 2005; Mendez et al. 2005a, b). Such initiatives cannot be undertaken effectively due to lack of sufficient human clinical trials of the hookworm antigens showing partial amino acid sequences homologous with mammalian proteins. Furthermore, there are no takers in the commercial market for anti-hookworm vaccine due to its large application in poor countries. Also, the infrastructure for developing and manufacturing vaccines is rudimentary and inadequate and requires enormous funding for high levels of technological sophistication (Hotez 2001; Broder et al. 2002).

Brighter Side of the Picture

Helminthic infections, though the source of morbidity, have proved beneficial to the hosts suffering from diseases linked to overactive immune systems (Strachan 1989). Chronic STH infections lead to modulation of the immune response (Reina et al. 2011) resulting in reduced prevalence of allergic diseases such as asthma (Leonardi-Bee et al. 2006), sclerosis, ulcerative colitis (Summers et al. 2005). Helminthic therapy thus involves the inoculation of the patient with *N. americanus*, *T. suis* ova or *T. trichiura* ova for the treatment for Crohn's disease (Croese et al. 2006), ulcerative colitis, inflammatory bowel disease (Summers et al. 2005), multiple sclerosis (Coreale and Farez 2007), asthma, eczema, hay fever (Strachan 1989) and food allergies (Montresor et al. 2002a, b; Goodman et al. 2007).

Conclusions

The STH infections are still considered as the most important and prevalent infections of mankind. The two billion people infected with STH infections reflect a remarkably successful adaptation of STHs to parasitism. Although, in some regions, there has been considerable decline in STH prevalence primarily because of economic development and specific control programmes; however, it appears unlikely that the prevalence of infection will decrease dramatically in near future. To successfully curb the STH diseases, the application of available simple and low-cost interventions on a large scale is absolutely essential. However, the process of deworming has to continue to keep the STH infections at low levels. Nevertheless, the widespread and frequent use of anthelmintics may prove detrimental in the long run, thus leading to drug resistance or decline in effectiveness of the front-line drugs (Albonico et al. 2003, 2004). This instigates us to look for the next generation of drugs that need to be developed to replace the existing ones.

Acknowledgements: The author acknowledges with thanks the people who helped in preparation of this review chapter including those whose figures have been used.

References

Adams D, Shaw S (1994) Leukocyte-endothelial interactions and regulation of leukocyte migration. Lancet 343:831–836

Adams VJ, Lombard CJ, Dhansay MA, Markus MB, Fincham JE (2004) Efficacy of albendazole against the whipworm *Trichuris trichiura*-A randomized, controlled trial. S Afr Med J 94:972–976

Albonico M, Renganathan E, Bosman A, Kisumku UM, Alawi KS, Savioli L (1994) Efficacy of a single dose of mebendazole on prevalence and intensity of soil-transmitted nematodes in Zanzibar. Trop Geogr Med 46:142–146

Albonico M, Chwaya HM, Montresor A, Stolfzfus RJ, Tielsch JM, Alawi KS, Savioli L (1997) Parasitic infections in Pemba Island school children. East Afr Med J 74:294–298

Albonico M, Bickle Q, Ramsan M, Montresor A, Savioli L, Taylor M (2003) Efficacy of mebendazole and levamisole alone or in combination against intestinal nematode infection after repeated targeted mebendazole treatment in Zanzibar. Bull World Health Organ 81:343–352

Albonico M, Engels D, Savioli L (2004) Monitoring drug efficacy and early detection of drug resistance in human soil-transmitted nematodes: a pressing public health agenda for helminth control. Int J Parasitol 34:1205–1210

Albonico M, Allen H, Chitsulo L, Engels D, Gabrielli A-F (2008) Controlling soil-transmitted helminthiasis in pre-school age children through preventive chemotherapy. PLoS Negl Trop Dis 2:126

Appleton CC, Gouws E (1996) The distribution of common intestinal nematodes along an altitudinal transect in Kwa-Zulu Natal, South Africa. Ann Trop Med Parasitol 90:181–188

Arsic-Arsenijevic V, Dzamic A, Dzamic Z, Milobratovic D, Tomic D (2005) Fatal *Strongyloides stercoralis* infection in a young woman with lupus glomerulonephritis. J Nephrol 18(6):787–790

Asaolu SO, Ofoezie IE (2003) The role of health education and sanitation in the control of helminth infections. Acta Trop 86:283–294

Asdamongkol N, Pornsuriyasak P, Sungkanuparph S (2006) Risk factors for strongyloidiasis hyperinfection and clinical outcomes. Southeast Asian J Trop Med Public Health 37:875–884

Atukorala TMS, Lanerolle P (1998) Serum folic acid levels of adolescent school girls in a rural area. Ceylon Med J 43:238–239

Awasthi S, Pande VK, Fletcher RH (2000) Effectiveness and cost effectiveness of albendazole in improving nutritional status of preschool children in urban slums. Indian J Pediatr 37:19–29

Aydin O (2007) Incidental parasitic infestations in surgically removed appendices: a retrospective analysis. Diagn Pathol 2:16

Beasley M, Brooker S, Ndinaromtan M, Madjiouroum EM, Baboguel M, Djenguinabe E, Bundy DA (2002) First nationwide survey of the health of school children in Chad. Trop Med Int Health 7:625–630

Beer RJ (1976) The relationship between *Trichuris trichiura* (Linnaeus 1758) of man and *Trichuris suis* (Schrank 1788) of the pig. Res Vet Sci 20:47–54

Bethony J, Chen JZ, Lin SX, Xiao SH, Zhan B, Li SW, Xue HC, Xing FY, Humphries D, Wang Y, Chen G, Foster V, Hawdon JM, Hotez PJ (2002) Emerging patterns of hookworm infection: influence of aging on the intensity of *Necator* infection in Hainan Province, People's Republic of China. Clin Infect Dis 35:1336–1344

Bethony J, Loukas A, Smout M, Brooker S, Mendez S, Olieskatt J, Goud G, Bottazzi ME, Zhan B, Wang Y, Williamson A, Lustigman S, Correa-oliveira R, Xiao S, Hotez PJ (2005) Antibodies against a secreted protein from hookworm larvae reduce intensity of hookworm infection in humans and vaccinated laboratory animals. Fed Am Soc Exp Biol J 19:1743–1745

Bethony J, Brooker S, Albonico M, Geiger SM, Loukas A (2006) Soil-transmitted helminth infections: ascariasis, trichuriasis and hookworm. Lancet 367:1521–1532

Bleakley H (2003) Disease and development: evidence from the American South. J Eur Econ Assoc 1:376–386

Blumenthal DS, Schultz MG (1975) Incidence of intestinal obstruction in children infected with *Ascaris lumbricoides*. Am J Trop Med Hyg 24:801–805

Bouhoum K, Schwartzbrod J (1997) Epidemiological study of intestinal helminthiasis in a Marrakech raw sewage spreading zone. Zentralbl Hyg Umweltmed 200:553–561

Bradley JE, Jackson JA (2004) Immunity, immunoregulation and the ecology of trichuriasis and ascariasis. Parasite Immunol 26:429–441

Broder S, Hoffman SL, Hotez PJ (2002) Cures for the third world's problems: the application of genomics to the diseases plaguing the developing world may have huge medical and economic benefits for those countries and might even prevent armed conflict. EMBO Rep 3:806–812

Brooker S, Michael E (2000) The potential of geographical information systems and remote sensing in the epidemiology and control of human helminth infections. Adv Parasitol 47:245–288

Brooker S, Peshu N, Warn PA, Mosobo M, Guyatt HL (1999) The epidemiology of hookworm infection and its contribution to anaemia among pre-school children on the Kenyan coast. Trans R Soc Trop Med Hyg 93:240–246

Brooker S, Beasley NMR, Ndinaromtan M, Madjiouroum EM, Baboguel M, Djenguinabe E, Hay SI, Bundy DAP (2002a) Use of remote sensing and a geographical information system in a national helminth control programme in Chad. Bull World Health Organ 80:783–789

Brooker S, Hay SI, Bundy DAP (2002b) Tools from ecology: useful for evaluating infection risk models? Trends Parasitol 18:70–74

Brooker S, Pratap S, Waikagul J, Suvanee S, Kojima S, Takeuchi T, Luong TV, Looareesuwan S (2003) Mapping soil-transmitted helminth infections in Southeast Asia and implications for parasite control. Southeast Asian J Trop Med Public Health 34:24–35

Brooker S, Bethony J, Hotez PJ (2004) Human hookworm infection in the 21st century. Adv Parasitol 58:197–288

Brooker S, Clements AC, Hotez PJ (2006) The co-distribution of *Plasmodium falciparum* and hookworm among African school children. Malaria J 5:99. doi:10.1186/1475-2875-5-99

Bundy DAP (1986) Epidemiological aspects of *Trichuris* and trichuriasis in Carribean communities. Trans R Soc Trop Med Hyg 80:706–718

Bundy DAP (1988) Population ecology of intestinal helminth infections in human communities. Philos Trans R Soc B: Biol Sci 321:405–420

Bundy DAP, Chan MS, Savioli L (1995) Hookworm infection in pregnancy. Trans R Soc Trop Med Hyg 89:521–522

Burkhart CN, Burkhart CG (2005) Assessment of frequency, transmission and genitourinary complications of enterobiasis (pinworms). Int J Dermatol 44:837–840

Caldwell JP (1982) Pinworms (*Enterobius vermicularis*). Can Fam Physician 28:306–309

Cateau E, Yacoub M, Tavilien C, Becq-Giraudon B, Rodier MH (2010) *Enterobius vermicularis* in the kidney: an unusual location. J Med Biol 59:860

Chan MS (1997) The global burden of intestinal nematode infections-fifty years on. Parasitol Today 13:438–443

Chan MS, Medley GF, Jamison D, Bundy DAP (1994) The evaluation of potential global morbidity attributable to intestinal nematode infections. Parasitology 109:373–387

Christian P, Khatry SK, West JP (2004) Antenatal anthelminthic treatment, birthweight, and infant survival in rural Nepal. Lancet 364:981–983

Coles GC (1985) Allergy and immunopathology of ascariasis. In: Crompton DWT, Nesheim MC, Pawlowski ZS (eds) Ascariasis and its public health significance. Taylor & Francis, London and Philadelphia, pp 167–184

Cook GC (1994) *Enterobius vermicularis* infection. Gut 35:1159–1162

Cook GC, Zumla-Alimuddin I (2009) Manson's tropical diseases. In: Cook GC, Manson SP, Zumla-Alimuddin I (eds) Soil-transmitted helminths (geohelminths), 22nd edn. Elsevier Health Sciences, pp 1515–1519

Cook DM, Swanson RC, Eggett DL, Booth GM (2009) A retrospective analysis of prevalence of gastrointestinal parasites among school children in the Palajunoj Valley of Guatemala. J Health, Popul Nutr 27:31–40

Cooper ES (1995) Epidemiology and transmission of intestinal helminths. In: Farthing MJG, Keusch GT, Wakelin D (eds) Enteric infection 2. Intestinal helminths. Chapman & Hall Medical, pp 201–212

Correale J, Farez M (2007) Association between parasite infection and immune responses in multiple sclerosis. Ann Neurol 61:97–108. doi:10.1002/ana.21067

Croese J, O'neil J, Masson J (2006) A proof of concept study establishing *Necator americanus* in Crohn's patients and reservoir donors. Gut 55:136–137. doi: 10.1136/gut.2005.079129

Crompton DWT (1989) Prevalence of ascariasis. In: Crompton DWT, Nesheim MC, Pawlowski ZS (eds) Ascariasis and its prevention and control. Taylor & Francis Ltd., London, pp 45–69

Crompton DWT (2001) *Ascaris* and ascariasis. Adv Parasitol 48:285–375

Da Silva JO, Fernandes RS, Ticli FK, Oliveira CZ, Mazzi MV (2007) Triterpenoid saponins, new metalloprotease snake venom inhibitors isolated from *Pentaclethra macroloba*. Toxicon 50:283–291

de Silva NR, Guyatt HI, Bundy DAP (1997a) Morbidity and morality due to *Ascaris*-induced intestinal obstruction. Trans R Soc Trop Med Hyg 91:31–36

de Silva NR, Chan MS, Bundy DAP (1997b) Morbidity and mortality due to ascariasis: re-estimation and sensitivity analysis of global numbers at risk. Trop Med Int Health 2:519–528

de Silva NR, Sirisena JL, Gunasekera DP, Ismail MM, de Silva HJ (1999) Effect of mebendazole therapy during pregnancy on birth outcome. Lancet 353:1145–1149

de Silva NR, Brooker S, Hotez PJ, Montresor A, Engels D (2003) Soil-transmitted helminth infections: updating the global picture. Trends Parasitol 19:547–551

Do TT, Molbak K, Phung DC (2007) Helminth infections among people using wastewater and human excreta in peri-urban agriculture and aquaculture in Hanoi, Vietnam. Trop Med Int Health 12:82–90

Drake L, Korchev Y, Bashford L (1994) The major secreted product of the whipworm, *Trichuris*, is a pore forming protein. Proc R Soc Lond B 257:255–261

Drake LJ, Jukes MCH, Sternberg RJ, Bundy DAP (2000) Geohelminth infections (ascariasis, trichuriasis and hookworm): cognitive and developmental impacts. Semin Pediatr Infect Dis 11:245–251

Dubini A (1843) Nuovo verme intestinale umano (*Anchylostoma duodenale*), constitutente un sesto genere dei nematoidei proprii dell'uomo. Ann Univ di Med Milano 106:5–13

Dundas KC, Calder AA, Alyusuf R (1999) *Enterobius vermicularis* threadworm infestation of paraovarian tissue in a woman who has had a hysterectomy. Br J Obstet Gynaecol 106:605–607

Dunn A, Keymer A (1986) Factors affecting the reliability of the McMaster technique. J Helminthol 60:260–262

Elkins DB, Haswell-Elkins M, Anderson RM (1988) The importance of host age and sex to patterns of reinfection with *Ascaris lumbricoides* following mass anthelmintic treatment in a South Indian fishing community. Parasitology 96:171–184

Ellis MK, Raso G, Li Y-S, Rong Z, Chen H-G, McManus DP (2007a) Familial aggregation of human susceptibility to co- and multiple helminth infections in a population from the Poyang Lake region. China Int J Parasitol 37:1153–1161

Ellis JS, Smith JD, Braham S, Lock M, Barlow K, Zambon MC (2007b) Design and validation of an H5 TaqMan real-time one-step reverse transcription-PCR and confirmatory assays for detection and verification of influenza A virus H5 infections in humans. J Clin Microbiol 45:1535–1543

Faulkner H, Turner J, Kamgno J, Pion SD, Boussinesq M, Bradley JE (2002) Age- and infection intensity-dependent cytokine and antibody production in human trichuriasis: the importance of IgE. J Infect Dis 185:665–672

Faust EC, Russell PF (1964) Craig and Faust's clinical parasitology, 7th edn. Lea & Febiger, Philadelphia, p 1099

Fenghua S, Zhongxing W, Yixing Q, Hangqun C, Haichoux X, Hainan R, Shuhua X, Bin Z, Hawdon JM, Zheng F, Hotez PJ (1998) Epidemiology of human intestinal nematode infections in Wujiang and Pizhou counties, Jiangsu province, China. Southeast Asian J Trop Med Public Health 29:605–610

Finkelman FD, Shea-Donohue T, Goldhill J (1997) Cytokine regulation of host defense against parasitic gastrointestinal nematodes: lessons from studies with rodent models. Annu Rev Immunol 15:505–533

Fleming AF (1994) Agriculture-related anaemias. Br J Biomed Sci 51:345–357

Flores A, Esteban JG, Angles R, Mas-Coma S (2001) Soil-transmitted helminth infections at very high altitude in Bolivia. Trans R Soc Trop Med Hyg 95:272–277

Forrester JE, Scott ME, Bundy DAP, Golden MHN (1988) Clustering of *Ascaris lumbricoides* and *Trichuris trichiura* infections within households. Trans R Soc Trop Med Hyg 82:282–288

Gandhi NS, Chen JZ, Koshnood K, Xing FY, Li SW, Liu YR, Bin Z, Haechou X, Chong-Jin T, Yan W, Wensen W, Dungxing H, Chong C, Shuhua X, Hawdon JM, Hotez PJ (2001) Epidemiology of *Necator americanus* hookworm infections in Xiulongkan Village, Hainan Province, China: high prevalence and intensity among middle-aged and elderly residents. J Parasitol 87:739–743

Gangopadhyay AN, Upadhyaya VD, Gupta DK (2007) Conservative treatment for round worm intestinal obstruction. Indian J Pediatr 74:1085–1087

Garcia A, Abel L, Cot M, Richard P, Ranque S, Feingold J, Demenais F, Boussinesq M, Chippaux JP (1999) Genetic epidemiology of host predisposition microfilaraemia in human loiasis. Trop Med Int Health 4:565–574

Gargano R, Di Legami R, Maresi E, Restivo S (2003) Chronic sialoadenitis caused by *E. vermicularis*: case report. Acta Otorhinolaryngol Ital 23:319–321

Gasser RB, de Gruijter JM, Polderman AM (2006) Insights into the epidemiology and genetic make-up of *Oesophagostomum bifurcum* from human and non-human primates using molecular tools. Parasitology 132:453–460

Gasser M, Riediker M, Mueller L, Perrenoud A, Blank F, Gehr P, Rothen-Rutishauser B (2009) Toxic effects of brake wear particles on epithelial lung cells in vitro. Part Fibre Toxicol 6:30. doi:10.1186/1743-8977-6-30

GIDEON (2009) GIDEON the world's premier global infectious diseases database

Gill GV, Welch E, Bailey JW, Bell DR, Beeching NJ (2004) Chronic *Strongyloides stercoralis* infection in former British far east prisoners of war. Q J Med 97:789–795

Goble FC, Konopka EA (1973) Sex as a factor in infectious disease. Trans NY Acad Sci 35:325–346

Goodman D, Haji HJ, Bickle QD, Stoltzfus RJ, Tielsch JM (2007) A comparison of methods for detecting the eggs of *Ascaris*, *Trichuris* and hookworm in infant stool and the epidemiology of infection in Zanzibari infants. Am J Trop Med Hyg 76:725–731

Goud GN, Zhan B, Ghosh K, Loukas K, Haedon J, Dobardzic A, Deumic V, Liu S, Dobardzic R, Zook RC, Qun J, Liu YY, Hoffman L, Chung-Debose D, Patel R, Mendez S, Hotez PJ (2004) Cloning, yeast expression, isolation and vaccine testing of recombinant *Ancylostoma* secreted protein 1 (ASP-1) and ASP-2 from *Ancylostoma ceylanicum*. J Infect Dis 189:919–929

Goud GN, Bottazzi ME, Zhan B, Mendez S, Deumic V, Pleiskatt J, Lin S, Wang Y, Bueno L, Fujiwara R, Samuel A, Ahn SY, Solanki M, Asoja O, Wen J, Saul A, Bethony JM, Loukas A, Roy M, Hotez PJ (2005) Expression of the *Necator americanus* hookworm larval antigen Na-ASP-2 in *Pichia pastoris* and purification of the recombinant protein for use in human clinical trials. Vaccine 23:4754–4764

Graeff-Teixeira C, da Silva AC, Yoshimura K (2009) Update on eosinophilic meningoencephalitis and its clinical relevance. Clin Microbiol Rev 22:322–348

Gutiérrez Y (2000) Diagnostic pathology of parasitic infections with clinical correlations, 2nd edn. Oxford University Press, Oxford, pp 354–366

Guyatt H (2000) Do intestinal nematodes affect productivity in adulthood? Parasitol Today 16:153–158

Gwatkin DR, Guillot M (2000a) The burden of disease among the global poor: current situation, future trends and implications for strategy. The World Bank, Washington

Gwatkin DR, Guillot M (2000b) The burden of disease among the global poor: current situation, future trends, and implications for strategy. Global Forum for Health Research Publications, Geneva, p 44

Gyorkos TW, Larocque R, Casapia M, Gotuzzo E (2006) Lack of risk of adverse birth outcomes after deworming in pregnant women. Pediatr Infect Dis J 25: 791–794. doi:10.1097/01.inf.0000234068.25760.97

Hasegawa M, Smith MJ, Goedert M (1998) Tau proteins with FTDP-17 mutations have a reduced ability to promote microtubule assembly. FEBS Lett 437:207–210

Haswell-Elkins MR, Elkins DB, Anderson RM (1987) Evidence for predisposition in humans to infection with *Ascaris*, hookworm, *Enterobius* and *Trichuris* in a South Indian fishing community. Parasitology 95:323–337

Haswell-Elkins MR, Kennedy MW, Maizels RM (1989) The antibody recognition profiles of humans naturally infected with *Ascaris lumbricoides*. Parasite Immunol 11:615–627

Hlaing T, Saw T, Lwin M (1987) Reinfection of people with *Ascaris lumbricoides* following single 6-month and 12-month interval mass chemotherapy in Okpo village, rural Burma. Trans R Soc Trop Med Hyg 81:140–146

Holland CV, Taren DL, Crompton DWT (1988) Intestinal helminthiases in relation to the socioeconomic environment of Panamanian children. Soc Sci Med 26:209–213

Hong ST, Choi MH, Chai JY, Kim YT, Kim MK, Kim KR (2002) A case of ovarian enterobiasis. Korean J Parasitol 40:149–151

Hotez PJ (1989) Hookworm disease in children. Pediatr Infect Dis J 8:516–520

Hotez PJ (1995) Human hookworm infection. In: Farthing MJG, Keusch GT, Wakelin D (eds) Enteric infection 2, intestinal helminths. Chapman & Hall Medical, pp 129–150

Hotez PJ (2000) Pediatric geohelminth infections: trichuriasis, ascariasis and hookworm infections. Semin Pediatr Infect Dis 11:236–244

Hotez PJ (2001) Vaccines as instruments of foreign policy. EMBO Rep 2:862–868

Hotez PJ (2002) China's hookworms. China Q 172:1029–1041

Hotez PJ, Feng Z, Xu LQ, Chen MG, Xiao SH, Liu SX, Blair D, McManus DP, Davis GM (1997) Emerging and reemerging helminthiases and the public health of China. Emerg Infect Dis 3:303–310

Hotez PJ, Zhan B, Bethony JM (2003) Progress in the development of a recombinant vaccine for human hookworm disease: the human hookworm vaccine initiative. Int J Parasitol 33:1245–1258

Hotez PJ, Brooker S, Bethony J, Bottazzi ME, Loukas A (2004) Current concepts: hookworm infection. N Engl J Med 351:799–807

Hotez PJ, Bethony J, Bottazzi ME, Brooker S, Buss P (2005) Hookworm: the great infection of mankind. PLoS Med 2:67. doi:10.1371/journal.pmed.0020067

Hotez PJ, Brindley PJ, Bethony JM, King CH, Pearce EJ (2008) Helminth infections: the great neglected tropical diseases. J Clin Invest 118:1311–1321

Hotez PJ, Engels D, Fenwick A, Savioli L (2010) Africa is desperate for praziquantel. Lancet 376:496–498

Humphries DL, Stephenson LS, Pearce EJ, The PH, Dan HT, Khanh LT (1997) The use of human faeces for fertilizer is associated with increased intensity of hookworm infection in Vietnamese women. Trans R Soc Trop Med Hyg 91:518–520

Igra-Siegman Y, Kapila R, Sen P, Kaminski ZC, Louria DB (1981) Syndrome of hyperinfection with *Strongyloides stercoralis*. Rev Infect Dis 3:397–407

Jemaneh L (1998) Comparative prevalences of some common intestinal helminth infections in different altitudinal regions in Ethiopia. Ethiop Med J 36:1–8

Jones JE (1983) Parasites in Kentucky: the past seven decades. J Ky Med Assoc 81:621–626

Kappus KD Jr, Lundgren RG, Juranek DD, Roberts JM, Spencer HC (1994) Intestinal parasitism in the United States: update on a continuing problem. Am J Trop Med Hyg 50:705–713

Kennedy MW (1992) Genetic control of the antibody response to parasite allergens. In: Moqbel R (ed) Allergy and immunity to helminths. Taylor & Francis, London, pp 63–80

Khuroo MS (1996) Ascariasis. Gastroenterol Clin North Am 25:553–577

Khuroo MS, Zargar SA, Mahajan R (1990) Hepatobiliary and pancreatic ascariasis in India. Lancet 335:1503–1506

Kightlinger LK, Seed JR, Kightlinger MB (1998) *Ascaris lumbricoides* intensity in relation to environmental, socioeconomic and behavioral determinants of exposure to infection in children from Southeast Madagascar. J Parasitol 84:480–484

Klein SL (2000) The effects of hormones on sex differences in infection: from genes to behaviour. Neurosci Biobehav Rev 24:627–638

Kobayashi A (1980) Studies on the mode of *Ascaris* infection in Japan. In: Collected papers on the control of soil-transmitted helminthiases, vol 1. Asian Parasite Control Organization, Tokyo, pp 109–118

Komiya Y, Yasuraoka Y (1966) The biology of hookworms. Prog Med Parasitol Jpn 3:4–114

Koumanidou C, Manoli E, Anagnostara A, Polyviou P, Vakaki M (2004) Sonographic features of intestinal and biliary ascariasis in childhood: case report and review of the literature. Ann Trop Paediatr 24:329–335

Kramer JM, French RP, Park EC, Johnson JJ (1990) The *Caenorhabditis elegans* rol-6 gene, which interacts with the sqt-1 collage gene to determine organismal morphology, encodes a collagen. Mol Cell Biol 10:2081–2089

Labiano-Abello N, Canese J, Velazquez JM, Hawdon JM, Wilson ML, Hotez PJ (1999) Epidemiology of hookworm infection in Itagua, Paraguay: a cross sectional study. Memórias do Instituto Oswaldo Cruz 94:583–586

Leonardi-Bee J, Pritchard D, Britton J (2006) Asthma and current intestinal parasite infection: systematic review and meta-analysis. Am J Respir Crit Care Med 174(5):514–523

Linnaeus C (1758) Systema naturae per regna tria naturae, secundum classes, ordines, genera, species, cum characteribus, differentiis, synonymis, locis. In: Tomus I (ed) Editio decima, reformata. Holmiae. (Laurentii Salvii). pp 1–824

Linnaeus C (1771) Mantissa plantarum altera generum editionis VI & specierum editionis II. Holmiae [=Stockholm]. (Laurentii Salvii).143–587

Lwambo NJ, Brooker S, Siza JE, Bundy DA, Guyatt H (2000) Age patterns in stunting and anaemia in African schoolchildren: a cross-sectional study in Tanzania. Eur J Clin Nutr 54:36–40

Mahanalabis D, Jalan KN, Maitra TK, Agarwal SK (1976) Vitamin A absorption in ascariasis. Am J Clin Nutr 29:1372–1375

Maizels RM, Holland MJ, Falcone FH, Zang XX, Yazdanbakhsh M (1999) Vaccination against helminth parasites: the ultimate challenge for vaccinologists? Immunol Rev 171:125–147

Marcos LA, Terashima A, Dupont HL, Gotuzzo E (2008) Strongyloides hyperinfection syndrome: an emerging global infectious disease. Trans R Soc Trop Med Hyg 102:314–318. doi:10.1016/j.trstmh.2008.01.020

Mark DL (1975) Survival of *Ancylostoma caninum* on bluegrass pasture. J Parasitol 61:484–488

McDonald GSA, Hourihane DOB (1972) Ectopic *Enterobius vermicularis*. Gut 13:621–626

Mendez S, Valenzuela J, Wu WH, Hotez P (2005a) Host cytokine production, lymphoproliferation and antibody responses during the coarse of *Ancylostoma ceylanicum* infection in the golden Syrian hamster. Infect Immun 73:3402–3407

Mendez S, Zhan B, Goud G, Ghosh K, Dobardzic A, Wu WH, Liu S, Deumic V, Dobardzic R, Liu YY, Bethony J, Hotez PJ (2005b) Effect of combining the larval antigens *Ancylostoma* secreted protein 2 (ASP-2) and metalloprotease 1 (MTP-1) in protecting hamsters against hookworm infection and disease caused by *Ancylostoma ceylanicum*. Vaccine 23:3123–3130

Montes M, Sawhney C, Barros N (2010) *Strongyloides stercoralis*: there but not seen. Curr Opin Infect Dis 23:500–504. doi:10.1097/QCO.0b013e32833df718

Montresor A, Crompton D, Gyorkos T, Savioli L (2002a) Helminth control in school-age children: a guide for managers of control programmes. Geneva: World Health Organization

Montresor A, Stoltzfus RJ, Albonico M, Tielsch JM, Rice AL, Chwaya HM, Savioli (2002b) Is the exclusion of children under 24 months from anthelmintic treatment justifiable? Trans R Soc Trop Med Hyg 96:197–199

Montresor A, Awasthi S, Crompton DW (2003) Use of benzimidazoles in children younger than 24 months for the treatment of soil transmitted helminthiasis. Acta Trop 86:223–232

Murray CJ, Lopez AD (1996) The global burden of disease: a comprehensive assessment of mortality and disability from diseases, injuries and risk factors in 1990 and projected to 2020. Harvard School of Public Health, (Global Burden of Disease and Injury Series, vol. I), Cambridge MA, p 142

Navitsky RC, Dreyfuss ML, Shrestha J, Khatry SK, Stoltzfus RJ, Albonico M (1998) *Ancylostoma duodenale* is responsible for hookworm infections among pregnant women in the rural plains of Nepal. J Parasitol 84:647–651

Nawalinski TA, Schad GA, Chowdhury AB (1978) Population biology of hookworms in children in rural West Bengal. I. General parasitological observations. Am J Trop Med Hyg 27:1152–1161

Nokes C, Grantham-McGregor SM, Sawyer AW, Cooper ES, Robinson BA, Bundy DAP (1992) Moderate to heavy infection of *Trichuris trichiura* affect cognitive function in Jamaican school children. Proc R Soc, Lond 247:77–81

Northrop-Clewes CA, Rousham EK, Mascie-Tayor CGN, Lunn PG (2001) Anthelmintic treatment of rural Bangladeshi children: effect on host physiology, growth and biochemical status. Am J Clin Nutr 73:53–60

Palmer DR, Bradley M, Bundy DA (1996) IgG4 responses to antigens of adult *Necator americanus*: potential for use in large-scale epidemiological studies. Bull World Health Organ 74:381–386

Peng W, Zhou X, Crompton DWT (1998) Ascariasis in China. Adv Parasitol 41:110–148

Phiri K, Whitty CJ, Graham SM, Ssembatya-Lule G (2000) Urban/rural differences in prevalence and risk factors for intestinal helminth infection in southern Malawi. Ann Trop Med Parasitol 94:381–387

Polderman AM, Blotkamp J (1995) *Oesophagostomum* infections in humans. Parasitol Today 11:451–456

Polderman AM, Krepel HP, Baeta S, Blotkamp J, Gigase P (1991) Oesophagostomiasis, a common infection of man in northern Togo and Ghana. Am J Trop Med Hyg 44:336–344

Prakash D, Chandra R, Bhatnagar JK (1980) Epidemiological study on ascariasis in children. Indian J Parasitol 4:59–60

Pritchard DI, Quinnell RJ, Walsh EA (1995) Immunity in humans to *Necator americanus*: IgE, parasite weight and fecundity. Parasite Immunol 17:71–75

Prost A (1987) L'ascaridiose en Afrique de l'ouest. Revue épidémiologique. Annales de Parasitoligie Humaine et Comparée 62:434–455

Quinnell RJ, Slater AFG, Tighe PJ, Walsh EA, Keymer AE, Pritchard DI (1993) Reinfection with hookworm after chemotherapy in Papua New Guinea. Parasitology 106:379–385

Quinnell RJ, Woolhouse ME, Walsh EA, Pritchard DI (1995) Immunoepidemiology of human necatoriasis: correlations between antibody responses and parasite burdens. Parasite Immunol 17:313–318

Raisanen M (1985) Studies on the synthesis and excretion of bufotenine and N, N-dimethyltryptamine in man. Academic dissertation, University of Helsinki, Helsinki

Ramezani MA, Dehghani MR (2007) Relationship between *Enterobius vermicularis* and the incidence of acute appendicitis. Southeast Asian J Trop Med Public Health 38:20–23

Reina Ortiz M, Schreiber F, Benitez S, Broncano N, Chico ME (2011) Effects of chronic ascariasis and trichuriasis on cytokine production and gene expression in human blood: a cross-sectional study. PLoS Negl Trop Dis 5:e1157. doi:10.1371/journal.pntd.0001157

Sachs JD, Steele H (2001) (eds) Macroeconomics and health: investing in health for economic development. Report of the commission on macroeconomics and health, WHO. Geneva

Sakakibara S, Nakamura Y, Yoshida T, Shibata S, Koike M, Takano H, Ueda S, Uchiyama Y, Noda T, Okano H (2002) RNA-binding protein Musashi family: roles for CNS stem cells and a subpopulation of ependymal cells revealed by targeted disruption and antisense ablation. Proc Natl Acad Sci USA 99:15194–15199

Sakti H, Nokes C, Hertanto WS, Hendratno S, Hall A (1999) Evidence for an association between hookworm infection and cognitive function in Indonesian school children. Trop Med Int Health 4:322–347

Santos FL, Cerqueira EJ, Soares NM (2005) Comparison of the thick smear and Kato-Katz techniques for diagnosis of intestinal helminth infections. Laboratório de Parasitologia Clínica, Faculdade de Farmácia, Universidade Federal da Bahia, Salvador, BA. The Revista da Sociedade Brasileira de Medicina Tropical 38:196–198

Savioli L, Crompton DW, Neira M (2003) Use of anthelminthic drugs during pregnancy. Am J Obstet Gynecol 188:5–6

Siddiqui AA, Berk SL (2001) Diagnosis of *Strongyloides stercoralis* infection. Clin Infect Dis 33:1040–1047

Smilie WG, Augustine DL (1926) Hookworm infestation: the effect of varying intensities on the physical condition of school children. Am J Dis Child 31:151–168

Smolyakov R, Talalay B, Yanai-Inbar I, Pak I, Alkan M (2003) *Enterobius vermicularis* infection of female genital tract: a report of three cases and review of the literature. Eur J Obstet Gynecol Reprod Biol 107:220–222

Sorensen E, Ismail M, Amarasinghe DK, Hettiarachchi I, Dassenaieke TS (1994) The effect of the availability of latrines on soil-transmitted nematode infections on the plantation sector in Sri Lanka. Am J Trop Med Hyg 51:36–39

Stephensen CB (2001) Vitamin A, infection and immune function. Annu Rev Nutr 21:167–192

Stephenson LS, Latham MC, Kurz KM, Kinoti SM, Brigham H (1989) Treatment with a single dose of albendazole improves growth of Kenyan school children with hookworm, *Trichuris trichiuria* and *Ascaris lumbricoides* infections. Am J Trop Med Hyg 41:78–87

Stephenson MD, Dreher K, Houlihan E, Wu V (1998) Prevention of unexplained recurrent spontaneous abortion using intravenous immunoglobulin: a prospective, randomized, double-blinded, placebo-controlled trial. Am J Reprod Immunol 39:82–88

Stephenson LS, Latham MC, Ottesen EA (2000) Malnutrition and parasitic helminth infections. Parasitology 121(Suppl):S23–S38

Stoll NR (1962) Biology of nematodes parasitic in animals. J Parasitol 48:830–838

Stoll NR (1999) This wormy world, 1947. J Parasitol 85:392–396

Stoltzfus RJ, Albonico M, Chwaya HM, Savioli L, Tielsch J, Schulze K (1996) Hemoquant determination of hookworm-related blood loss and its role in iron deficiency in African children. Am J Trop Med Hyg 55:399–404

Stoltzfus RJ, Dreyfuss ML, Chwaya HM, Albonico M (1997) Hookworm control as a strategy to prevent iron deficiency. Nutr Rev 55:223–232

Stoltzfus R, Kvalsvig J, Chwaya HM (2001) Effects of iron supplementation and anthelmintic treatment on motor and language development of preschool children in Zanzibar: double blind, placebo controlled study. Br Med J 323:1389

Stoltzfus RJ, Chwaya HM, Montresor A, Tielsch JM, Jape JK, Albonico M, Savioli L (2004) Low dose daily iron supplementation improves iron status and appetite but not anemia whereas quarterly anthelminthic treatment improves growth, appetite and anemia in Zanzibari preschool children. J Nutr 134:2

Strachan DP (1989) Hay fever, hygiene, and household size. Br Med J 299:1259–1260. doi:10.1136/bmj.299.6710.1259

Summers RW, Elliott DE, Urban JF, Thompson RA, Weinstock JV (2005) *Trichuris suis* therapy for active ulcerative colitis: a randomized controlled trial. Gastroenterology 128:825–832

Torlesse H, Hodges M (2001) Albendazole therapy and reduced decline in haemoglobin concentration during pregnancy (Sierra Leone). Trans R Soc Trop Med Hyg 95:195–201

Tucker KL, Buranapin S (2001) Nutrition and aging in developing countries. J Nutr 131:2417–2423

Udonsi JK, Nwosu ABC, Anya AO (1980) *Necator americanus*: population structure, distribution, and fluctuations in population densities of infective larvae in contaminated farmlands. Z Parasitenkunde 63:251–259

Villamizar E, Mendez M, Bonilla E, Varon H, de Onatra S (1996) *Ascaris lumbricoides* infestation as a cause of intestinal obstruction in children: experience with 87 cases. J Paediatr Surg 31:201–204

Walzer PD, Milder JE, Banwell JG, Kilgore G, Klein M, Parker R (1982) Epidemiologic features of *Strongyloides stercoralis* infection in an endemic area of the United States. Am J Trop Med Hyg 31:313–319

Weller PF, Leder KL (2008) Strongyloidiasis. Wolters Kluwer: Health. Available at http://www.uptodate.com/contents/strongyloidiasis?source=search_result&selectedTitle=1%7E4

Williamson AL, Brindley PJ, Knox DP, Hotez PJ, Loukas A (2003) Digestive proteases of blood-feeding nematodes. Trends Parasitol 19:417–423

World Health Organization (1999) Global tuberculosis control. WHO Report. Geneva, Switzerland, WHO/TB/99.259

World Health Organization (2000) Communicable diseases: control of schistosomiasis and soil-transmitted helminth infections (report by the Secretariat). Executive Board, 107th Session, 27 October

World Health Organization (2002) World health report: annex table 2: deaths by cause, sex and mortality stratum in WHO regions and annex table 3: burden of disease in DALYs by cause, sex and mortality stratum in WHO regions. World Health Organization, Geneva, pp 186–193

World Health Organization (2004) The global burden of disease: 2004 update. World Health Organization, Geneva

World Health Organization (2006) Schistosomiasis and soil-transmitted helminth infections-preliminary estimates of the number of children treated with albendazole or mebendazole. World Health Organization, Geneva, pp 145–164

World Health Organization (2012) Weekly epidemiological record. 87:17–28

Xu LQ, Yu SH, Jiang ZX, Jyang JL, Lai CQ, Zhang XJ, Zheng CQ (1995) Soil-transmitted helminthiasis: nationwide survey in China. Bull World Health Organ 73:507–513

Xue HC, Wang Y, Xiao SH, Liu S, Wang Y, Shen GJ, Wu WT, Zhan B, Drake L, Feng Z, Hotez PJ (2000) Epidemiology of human ancylostomiasis among rural villagers in Nanlin County (Zhongzhou Village), Anhui Province, China: II. Seroepidemiological studies of the age relationships of serum antibody levels and infection status. Southeast Asian J Trop Med Public Health 31:736–741

Yadla S, Sen HG, Hotez PJ (2003) An epidemiological study of ancylostomiasis in a rural area of Kanpur District Uttar Pradesh, India. Indian J Pub Health 47:53–60

Zaha O, Hirata T, Kinjo F, Saito A (2000) Strongyloidiasis-progress in diagnosis and treatment. Intern Med 39:695–700

Ziem JB (2006) Controlling human oesophagostomiasis in northern Ghana. (Doctoral thesis) Department of Parasitology, Faculty of Medicine/Leiden University Medical Center (LUMC). Leiden University

Zuk M, McKean KA (1996) Sex differences in parasite infections: patterns and processes. Int J Parasitol 26:1009–1024

Zygmunt DJ (1990) *Strongyloides stercoralis*. Infect Control Hosp Epidemiol 11:495–497

Water Quality Issues Needing Technological Interventions: Sustainable Availability for All

20

Khandal Dhriti, G. Mathur, R. Prem, Monica Singh, V. S. Rawat, Geetha Seshadri, A. Manjeet, and R. K. Khandal

Abstract

The fact that existence of life is entirely because of the availability of water and since water cannot be manufactured, it becomes utmost essential to take care of the various resources of water for sustainability. In order to ensure quality water, the management of water resources requires extreme care and caution. Water is the key resource for life support systems. It is only due to the existence of water that agriculture, aquaculture, animal husbandry, marines, etc. is possible. Due to its unique properties, water is an important raw material for various industrial and household products. As per the recent trends, industrial products based on organic solvents are being reformulated using water. For setting up of any industry, one always ensures the sustainable availability of quality water. Water has also been used for transportation. Whichever purposes the water is used for, it is important that water should be of desired quality; use of poor quality water can be worse than the non-availability of water. Water is not only non-degradable but also it can neither be created nor be destroyed. The unique nature of imbibing everything, water gets in contact with, is the concern! The moment water gets in contact with any surface or a substance, it picks up the various constituents from the matrix in contact, resulting in the conversion of quality water into wastewater. The deterioration of quality starts when the measures for protection of water from contamination are inadequate. Therefore, to avoid spoilage of quality

R. K. Khandal (✉)
Gautam Buddh Technical University, IET-Campus,
Sitapur Road, Lucknow, 226021, India
e-mail: drrakeshkhandal@gmail.com

G. Mathur · R. Prem · M. Singh · V. S. Rawat ·
G. Seshadri · A. Manjeet
Shriram Institute for Industrial Research, 19,
University Road, Delhi, 110007, India

K. Dhriti
ECPM-ICPEES, Universite de Strasbourg,
25 rue becquerel, strasbourg, France

P. P. Singh and V. Sharma (eds.), *Water and Health*,
DOI: 10.1007/978-81-322-1029-0_20, © Springer India 2014

water and to recover quality water from wastewater are the key priority tasks for which technological interventions are needed. The present paper deals with various aspects of availability of quality water.

Keywords
Contaminants • Maximum residue limits • Pollutants • Water management • Water quality

Introduction

Water is essential both for the existence of life and for the growth of economy of a country. In fact, water is an elixir of life, not only for the humans but also for the existence of all forms of plant and animal life. Besides the availability of water, it is more important to have the availability of quality water. Water for human consumption needs to be of high quality; the norms for drinking water and that used for the food and beverages should be the most stringent (Annan 2003). Although three-fourth of the land is covered with water, only 1 % of it from different sources, that is, surface water, ground water and rain water, is meeting the growing needs of the increasing population (UNDPI 1997).

With the increasing population, changing lifestyles, shifting rainfall pattern due to global warming and increasing industrialization and urbanization, it is estimated that by the year 2030, there will be an additional 2.5 billion people living in regions already lacking sufficient quantity of clean water. Unpredictable and severe weather systems contribute to flooding which in turn leads to contamination of water supplies. The increased risk of drought conditions may also cause deterioration of ground water. Recovery of water of desired quality from the contaminated water would be the key to ensure sustainability of water (Annan 2003). The present paper describes various quality aspects of water and aims at indicating areas where different R & D interventions are desired for ensuring availability of quality water. The subject is covered with the following outline: (1) Hard facts about water, (2) Sources of water, (3) Quality criteria of water, (4) Quality parameters of water, (5) Regulatory requirements for quality water as per various national and international norms, (6) Issues related to water quality, (7) Major sources of water contamination and (8) Technological interventions.

Certain Hard Facts About Water

With its unique characteristics, appreciation of certain hard facts about water would be necessary here.

Water can neither be created nor be destroyed

Just like energy which can only be transformed from one form to another, water cannot be manufactured. That is why there exists no manufacturing unit producing water even in the case of the technologically most advanced countries; the requirements are met only from natural resources. The amount of water consumed should be retrieved back in spite of the fact that natural cycle of water ensures the availability of water in the form of rains, precipitation, etc. Anthropogenic activities detrimental to the factors responsible for water cycle must be controlled. Even though it is impossible to destroy water, it is very easy to spoil it. Adequate measures are required to restrict spoilage of water, while recovering it from its spoilt form. Spoilage happens, when there is misuse of water or the water gets in contact with pollutants. Water used for irrigation gets easily contaminated with various agrochemicals used during agriculture. Similarly, water is used for dying process in textiles, leather, processing of paper, etc., so the effluent will always be contaminated with toxic substances.

Water is a Carrier of All Nutrients

Nutrients required for growth of any organism are water soluble, and it is through water that they are rendered for metabolic processes of all living beings. All types of contaminants or pollutants if present in water would cause destruction of living cells, and hence, they pose danger to life!

Water is a Food

One can live without carbohydrates, fats, proteins, minerals or nutrients but cannot live without water for long time. Moreover, for the production of all types of basic foods by virtue of agriculture, animal husbandry, etc., water of good quality is essential. Use of polluted water would lead to the production of contaminated food (Annan 2003).

Structural Attributes of Water are Responsible for Its Extraordinary Properties

The most prominent attribute pertains to its polarity based on which all the materials are categorized as polar or non-polar. Water is an ideal solvent and even non-polar substances get mixed with water, in trace amounts. Water, therefore, can be easily used as a cleaning agent, which removes all sorts of dirt from surfaces and gets polluted. The ground water and river water are being continuously contaminated with different types of toxic organic as well as inorganic pollutants. River water pollution has already become a serious concern for the policy makers as well as for the regulators. Many of the products are being designed based on water. Aqueous systems are preferred compared with those based on organic solvents. For example, paints, coatings and pesticide formulations based on organic solvents are being reformulated using water as the base to make them environmental friendly. It is one of those naturally occurring substances, which co-exists in three different phases without decomposition during the phase transfer as well as under extreme conditions of nature. This behaviour should be used for recycling of water.

Unique Thermal Behaviour of Water

High specific heat of water is responsible for its use as a thermic fluid meant for the exchange of heat in the industrial processes involving heating or cooling.

With all the above facts known, one must remember that there exists enough water for all living beings on the earth and the availability of enough water can be ensured forever, even with the current rate of growth of industry and population, provided the water resources are conserved and protected (Momodu and Anyakora 2010).

Sources of Water

Basically, there are two major sources of fresh water: (1) surface water in the form of rivers and streams, lakes, ponds, glaciers, etc. and (2) ground water in the form of reservoirs, accumulated below the earth (Momodu and Anyakora 2010).

The natural cycle of water is the main mechanism for the availability of water for both these sources. Surface water undergoes: (1) percolation to get to the ground water reservoirs, (2) evaporation to get to the atmosphere and (3) run-off to get to the ocean. Since surface water is vulnerable to contamination of all types, that is, physical, chemical or biological, the emphasis has always been on protecting the water resources from getting contaminated by preserving them. Further, technologies are adopted for the treatment of the available water before it is supplied to the consumers. While the urban areas are serviced by the municipal authorities supplying quality water, most of the rural areas remain deprived of this. Since the involvement of the society is as essential as policy intervention, it is important for the consumers to understand the quality characteristics of water available from various sources.

Characteristics of Surface Water

Surface water in fact originates from rains, glaciers, snowfall, hailstorms, etc. It exists mainly in the form of rivers, which flow from one place to another. The static sources, that is, reservoirs such as lakes and ponds, are important for water supply. Rainwater is supposed to be the purest form of water in nature as far as the contamination is concerned. It is a clear, bright and sparkling soft water with only traces (<5 ppm) of dissolved solids and free from pathogenic microorganisms. It tends to pick up different types of gaseous impurities from the atmosphere, for example, oxides of carbon, sulphur and nitrogen. The quality of river water is determined largely by the prevailing conditions of the regions through which the river flows and also by the various streams of wastewater falling into it from the settlements, industries, etc. Flowing across an industrial cluster, it may pick up impurities such as suspended matter, organics, oil and heavy metals, and when flowing through agricultural fields, it would pick up fertilizers, pesticides, salts, radioactivity, heavy metals, etc. Similarly, the quality of stagnant water in lakes and ponds is determined by the man-made activities as well as by the prevailing environmental conditions around the reservoirs. Maintaining the desired quality of water of rivers, lakes, ponds, etc. has always been a challenge especially in the developing countries like India. Surface water is always prone to contamination from human and animal resources and as such is never safe for human consumption unless subjected to sanitary protection and subjected to appropriate treatment before use (Mara and Horan 2003).

River Water

Though some amount of self-purification may occur in river water by natural processes of purification, such as dilution, sedimentation, aeration, oxidation, sunlight, plant and animal life, these are not sufficient to render the water potable. The rivers have a certain capacity called the assimilative capacity due to which the rivers can contain different types of contaminants picked up from the surface without letting the quality adversely affected. For example, it has always been believed that river Ganges has the maximum assimilative capacity amongst all the Indian rivers (Jaiswal 2007). Due to the fact that river Ganges flows at a fast velocity through the Himalayas containing minerals, it is able to retain high amount of dissolved oxygen. However, the assimilative capacity cannot be taken as granted for disposing waste into the river. The present scenario suggests that the quality of major rivers including Ganges has been deteriorating at a fast pace and it cannot be used without proper remediation (Mara and Horan 2003; Helmer and Hespanhol 1997).

Lake Water

Because of the colonization, as also industrialization of the areas around the lakes, it has become a challenge to preserve the quality of lake water. The areas of concern in the case of lake water pertain mainly to eutrophication and chemical contamination.

Sea Water

Sea water is the largest and ultimate reservoir of water on earth. Industrial effluents of different types, regularly being disposed off into the surface water bodies, find ways to the sea and also the incidents of oil spills are the major cause of pollution of the sea water. All this pollution contributes significantly towards the contamination of marine food. In spite of the sea water being plentiful, there are always limitations to its usage. Desalination process has always been a technological challenge besides being a costly preposition and that is why it is the last option.

Ground Water

It is available through shallow, deep wells, springs, pumps, etc. In rural areas, water used by humans comes mainly from the ground water (Annan 2003). Ground water under normal cases should be practically free from contaminants including pathogenic bacteria and as such it can be considered as superior to the surface water; the process of percolation through the soil strata ensures purification of water (Momodu and Anyakora 2010). Water in the shallow wells or the springs is more prone to get contaminated by anthropogenic activities as compared to the water in the deep wells. The spring water is supposed to be clean and safe for drinking but springs are equally vulnerable to contamination. With the water table going down, the water in case of deeper wells in India is being found contaminated with fluorides. Because of the inadequate facilities for sewage treatment, the ground water is also found to be contaminated with nitrates, etc. (CPCB 2007; CPCB 2008a, b, c). In the north-eastern part of India, contamination of arsenic is a major cause of concern for quite some time. In an extensive study conducted by Shriram Institute for Industrial Research on 8,500 sources of drinking water spread in the states of Uttar Pradesh, Bihar and Jharkhand bordering with Nepal, a significant number of water sources were found contaminated with arsenic; many of these sources have been identified as the hot spots where the contamination is beyond the desired limits (SRI 2007). Since, not much of industrial activities exist in the areas, the main cause of contamination may be due to the geogenic factors.

Several kinds of toxic pollutants are being added to the water sources: pesticides, agrochemicals, persistent organic pollutants (POPs), compounds of nitrogen, phosphorus and sulphur, hydrocarbons, PCBs, highly volatile compounds from industries, veterinary and pharmaceutical drugs from wastewater of hospitals, leachates from landfill sites, fluoride, nitrate, phosphates, heavy metals, etc. from geological formations and microbial pathogens including faecal contaminants from sewage (book-water: SRI) (Keast and Johnston 2008).

The Central Pollution Control Board (CPCB), India, has specified guidelines for the quality criteria of surface and ground water (Table 20.1), which form the basis to grade the available water as of high, medium and poor quality and accordingly necessary measures can be taken for the treatment. For making the medium grade and poor grade water suitable for drinking, the treatment methodologies must aim for the reduction of chemical contaminants by chlorination, aeration, precipitation, adsorption, coagulation, flocculation, etc. followed by separation and the elimination of microbial contamination by adopting state-of-the-art techniques. The treatment of poor quality water would require extreme care and caution before declaring it suitable for drinking purposes (CPCB 2008a, b, c).

Quality Criteria of Water

Water quality is determined by the physical, chemical and biological characteristics of water as per the regulatory guidelines. A standard for quality can be designed with qualitative description (narrative in nature) or in a quantitative manner (numeric standard of quality) (Annan 2003).

Narrative criteria are statements that establish water quality goals. Narrative criteria are employed for parameters for which numeric criteria are difficult to specify, for example, colour, taste and odour. Some narrative criteria describe a desirable biological condition, such as related to the existence of healthy population of the native aquatic life of water body or may say that the water should be free from substances which may cause any adverse effect to the aquatic life or the human health. For example, surface water shall be virtually free from floating non-petroleum- or petroleum-derived oils (Obreza et al. 2011; Khandal 2010).

Numeric criteria are measurable benchmarks and are important since they serve as the basis for developing allowed limits for various

Table 20.1 General quality criteria (primary parameters) for surface and ground water as per Central Pollution Control Board, India

S.No.	Parameters	Range/limiting value of water quality graded as		
		High	Medium	Poor
1.	pH	6.5–8.5	6–9	6–9
2.	Colour, Pt scale, Hazen units	<10	<50	<500
3.	Total suspended solids, mg/L	<1,000	<1,500	<2,000
4.	Odour dilution factor	<3	<10	<20
5.	Nitrate, mg/L	<50	<50	<50
6.	Sulphates, mg/L	<150	<250	<250
7.	Chloride, mg/L	<200	<300	<400
8.	Fluoride, mg/L	<1	<1.5	<1.5
9.	Surfactants, mg/L	<0.2	<0.2	<0.2
10.	Phosphates, mg/L	<0.4	<0.7	<0.7
11.	Dissolved oxygen	60–110	80–120	90–140
12.	Biochemical oxygen demand, mg/L	<3	<5	<7
13.	Total Kjeldahl nitrogen, mg/L	<1	<2	<3
14.	Ammonia, mg/L	<0.05	<1	<2
15.	Total coliform MPN/100 ml	<500	<5,000	<50,000
16.	Faecal coliform MPN/100 ml	<200	<2,000	<20,000
17.	Faecal streptococci	200	1,000	10,000

There should not be any visible discharge in the upstream (up to 5 kms) of the water intake point

contaminants for determining water quality problems and establishing goals for various technological interventions. For most of the pollutants, numeric criteria are expressed as maximum acceptable concentrations. Contamination beyond the maximum acceptable concentrations makes water unsafe for humans. In the food safety terminology, the acceptable limits defined for the various types of residual contaminants in foods including water are referred as maximum residue limits or MRLs. The idea of specifying limits for such parameters is mainly to ensure that the presence of various undesirable elements never exceeds the MRL values set for them. On the contrary, the elements that contribute to the health of the consumers are specified always in terms of the minimum desired concentration. Basically, for nutrients, the desired limits are specified as minimum desired concentration. If the values for the desired components fall short of the minimum desired concentration, the water is considered of poor quality. For fixing the limits, comprehensive dossiers for toxic effects as well as for beneficial effects are prepared. Other than these two types, there are parameters for which a range is specified. The observed values for such parameters must fall within the specified range. All of these are discussed in the sections to follow (Helmer and Hespanhol 1997; Obreza et al. 2011; Clausen 2003; Borchardt and Walton 1971).

Parameters for Quality of Water

Water quality parameters can be divided into three categories: physical, chemical and biological (Hammer 2009; Eaton et al. 2005; Khandal 2010).

Physical Parameters

pH Value

pH value is the logarithm to the base 10 of the inverse of hydrogen ion concentration. A pH value of neutral or pure water is 7, and if it is less than 7, it is acidic in nature with a higher concentration of H^+ ion activity, while a pH value of more than 7 indicates a higher concentration of OH^- ion activity. The pH of water normally is between 6.5 and 8.0 or 8.5 for natural water and about 6.5 for distilled water because of the fact that CO_2 in atmosphere gets dissolved in water. Streams generally have a pH value ranging between 6 and 9 depending upon the presence of dissolved substances from the rocks, soils or other materials in the water shed (Hammer 2009; Patnaik 1997; Eaton et al. 2005; Khandal 2010).

Turbidity

Turbidity is the measurement of optical clarity as well as of haziness of water. It is measured by passing the light through the specimen of water. Good quality water should not be hazy and should be free of turbidity. Turbidity is due to the presence of insoluble particles of soil, organics, microorganisms and other materials, and turbidity is expressed in Nephelometric Turbidity Unit (NTU). Turbidity in clear lake water is 25 units, and in muddy water, it may exceed 100 units. For drinking water, it should be <1 NTU (Hammer 2009; Eaton et al. 2005).

Colour

The colour of water is due to the presence of dissolved and particulate matter, certain minerals, for example, iron, manganese, etc., humus material, tannins, coloured wastes discharged from different industries, etc. The colour may also arise due to the formation of metallic sulphides, under anaerobic conditions. Colour of water can be described either as apparent or as true colour. Apparent colour is the colour of the water, received as such without any treatment, whereas the true colour is the colour of the filtered water. The concept of true and apparent colour is important for decision support to plan for the treatment and purification of water. For human consumption, the water should be colourless (<5 Hazen units). Colour is expressed in terms of Hazen units (Hammer 2009; Eaton et al. 2005; Khandal 2010).

Total Dissolved Solids

Total Dissolved Solids (TDS) refer to the presence of minerals, salts, metals and cations or anions dissolved in water. TDS are expressed in terms of mg or gram per litre of water. The acceptable concentration recommended for TDS for drinking water is 500 mg/l. It may be noted that once the TDS are found higher than the acceptable limit of 500 mg/l, the water then needs to be further evaluated for the confirmation of the type of dissolved solids present. Accordingly, appropriate technologies for the treatment of such water are adopted. Cations like calcium and magnesium combined with chlorides and carbonates are responsible for hardness of water, scale formation and bitter taste. Cations like sodium and potassium combined with chloride provide salty or brackish taste to the water and increase the corrosive nature of water. It causes reduced efficiency of water filters and equipments. It is expressed in terms of $CaCO_3$ (Hammer 2009; Eaton et al. 2005; Khandal 2010).

Suspended Solids

Solids that are not dissolved remain suspended in water, and on filtration through a 1.2-μm filter, these are retained on the filter. Suspended solids are expressed in the terms of mg/l. The values of suspended solids can be used for designing treatment system. Drinking water should be free from suspended solids, though no limit has been specified in Indian Standard IS: 10500 (Hammer 2009).

Chemical Parameters

Chloride

Chloride (Cl^-) ions are usually present in the water associated with sodium, potassium, magnesium and calcium. Presence of chloride normally imparts a salty taste to the water depending upon the associated cation. Water would taste salty with as low as 100 mg/l of sodium chloride but with as high as 1,000 mg/l of magnesium or calcium chloride, it may not taste salty. Presence of chloride increases electrical conductivity, corrosivity of metal pipes and levels of metals in water as soluble chlorides. The acceptable limit of chloride ions in drinking water ranges between 200 and 250 mg/l as per majority of the specifications of drinking water but as per the Indian specification on drinking water, that is, IS: 10500, the maximum permissible limit for chloride in the absence of alternate sources is as high as 1,000 mg/l though the acceptable limit is 250 mg/l. Chloride in natural water is due to leaching of chloride from rocks and soil, whereas chloride water from coastal areas is due to the intrusion of salt water (Hammer 2009; Eaton et al. 2005; Khandal 2010).

Fluoride

Fluoride exists naturally in different water sources because of (1) its abundance in earth's crust, (2) use of large number of fluoride compounds in treating municipal water and (3) presence of fluoride in wastewater from electronics industry. Excess presence of fluoride ions in drinking water can cause dental fluorosis, whereas water with no fluoride may cause dental caries. The acceptable limit of fluoride in drinking water as per Indian Standard IS: 10500 is 1.0 mg/l, whereas maximum permissible limit is 1.5 mg/l (Hammer 2009; Patnaik 1997; Eaton et al. 2005; Khandal 2010).

Alkalinity

Alkalinity of water is the measure of its capacity to react with a strong acid (e.g. sulphuric acid). It is expressed as mg/l as $CaCO_3$. Alkalinity in surface water is mainly due to the presence of carbonates, bicarbonates and hydroxides, major source being leaching of calcium carbonate from rocks and soils. Compounds like borates, silicates, phosphates can also contribute to alkalinity of water. Alkalinity is significant in the treatment of wastewater and drinking water. Acceptable limit for alkalinity of drinking water as per IS: 10500 is 200 mg/l, while the permissible limit is 600 mg/l. Water having excess of alkalinity is unsuitable even for irrigation purposes (Hammer 2009; Patnaik 1997; Eaton et al. 2005; Khandal 2010).

Acidity

Acidity of water is the quantitative capacity to react with a strong base (usually NaOH). It is expressed in terms of mg $CaCO_3$/l. Acidic water may lead to corrosion of water pipes. Acidity is also important for regulating biological processes and control of chemical reactions. Water for existence of biological life should have acidity >6 (Hammer 2009; Eaton et al. 2005).

Hardness

Water hardness is the capacity of water to precipitate soap and is mainly due to the presence of carbonates, bicarbonates, sulphates, chlorides and nitrates of calcium and magnesium. It causes scale formation in water distribution systems and in water boilers. The hardness of water is measured in terms of parts per million or mg/l as $CaCO_3$. Water with a concentration of $CaCO_3$ up to 50 ppm is termed as soft water; 50–100 ppm is termed as moderately soft water; 100–150 ppm is slightly hard water; 150–200 ppm is moderately hard water; and with more than 200 ppm, it is known to be very hard water. The acceptable limit of hardness has been prescribed as 200 ppm, whereas the maximum permissible limit is 600 ppm as $CaCO_3$ as per IS-10500 (Hammer 2009; Patnaik 1997; Eaton et al. 2005; Khandal 2010).

Dissolved Oxygen

It is the measure of the amount of the gaseous oxygen (O_2) dissolved, and it is expressed in mg/l or ppm. The dissolved oxygen (DO) is required for the respiration of aerobic microorganisms as well as all other aerobic life forms. Adequate DO is necessary for good water quality, pollution control and strategies for water treatment processes. It prevents the formation of noxious odours. Polluted water with large amount of organic matter would have a low DO content since biological decomposition of organic matter uses the DO. The rate of air supply to aerobic treatment process is monitored based on the measurement of DO. Oxygen levels below 1–2 mg/L even for few hours can be fatal for large fishes. DO in a stream is affected by the following factors: (1) Temperature: DO is high in cold water, (2) Flow: Water flowing with higher velocities is rich in oxygen, whereas stagnant waters have a low DO, (3) Presence of aquatic plants: Green plants and algae in water bodies increase the DO due to the process of photosynthesis, (4) Altitude: DO in water at low altitudes is high, (5) Dissolved or suspended solids: Water having lower levels of dissolved or suspended solids would have higher DO content (Hammer 2009; Patnaik 1997; Eaton et al. 2005; Khandal 2010).

Oil and Grease

Oil and grease include hydrocarbons, fats, oils, waxes and high molecular weight fatty acids; the major sources of contamination being discharges from gas stations, food services facilities, etc. or a myriad of small non-point sources or due to large-scale point sources such as oil spills. The presence of oil and grease in municipal and industrial wastes causes difficulty in water handling and treatment. The grease separates from water and adheres to the interior of pipes and tank walls, thereby reducing biological treatability of wastewater and producing greasy sludge solids, difficult to process (Hammer 2009; Eaton et al. 2005; Khandal 2010).

Chemical Oxygen Demand

Chemical Oxygen Demand (COD) is expressed as mg O_2/l and defined as the amount of oxygen or the specific oxidant required for chemical oxidation of organic and inorganic materials in the water under controlled conditions and is expressed in terms of its oxygen equivalence. COD is often used as a measurement of pollution in wastewater and natural water. High COD means high pollution load (Hammer 2009; Patnaik 1997; Eaton et al. 2005; Khandal 2010).

Biochemical Oxygen Demand

The biological demand is the amount of oxygen required by bacteria for (1) the decomposition of organic matter in water (also known as carbonaceous demand) and (2) oxidation of inorganic materials such as sulphides and ferrous iron and reduced forms of nitrogenous compounds such as ammonia (also known as nitrogenous demand). Biochemical Oxygen Demand (BOD) is expressed in mg/L. Factors contributing towards the presence of BOD in water are either natural sources of organic matter such as plant decay, presence of leaves or could be due to agricultural and urban run-off carrying fertilizers, plant materials, plastics or disposal of untreated industrial water in water bodies (Hammer 2009; Patnaik 1997; Eaton et al. 2005).

Inorganic Chemicals

Nitrogen

Nitrogen occurs in natural waters in various inorganic forms, that is, nitrate (NO_3^-), nitrite (NO_2^-) and ammonia (NH_3) and the organic form. Nitrate is the most common form found in water bodies. It is expressed as mg/l of nitrate–nitrogen (NO_3–N) which means that nitrogen is in the form of nitrate. Nitrite is less stable and usually present in much lower amounts than nitrate. Ammonia is the least stable form of nitrogen and thus difficult to measure accurately. In nature, all these three forms are interrelated through the process of nitrification, where ammonia is biologically oxidized to nitrate producing nitrite as an intermediate product. Discharge of faecal matter in water bodies is the main cause of high content of NO_3^- (Eaton et al. 2005; Patnaik 1997).

Metals

Antimony, arsenic, barium, beryllium, cadmium, copper, zinc, selenium, chromium, lead, tin, mercury, nickel and thallium are affecting the internal organs of the human body due to their toxicity which is dependent upon bioavailability and bioconcentration. Heavy metals may be present in water at the levels of ppm. The presence of these metals arises due to human activities, such as mining and manufacturing, natural process of chemical weathering and soil leaching, discharges from residential dwellings, ground water infiltration, commercial and industrial discharges (Momodu and Anyakora 2010; Hammer 2009; Eaton et al. 2005).

Phosphorus

Phosphorus occurs in natural water and in wastewater as phosphates and is classified as orthophosphates, condensed phosphates (pyrophosphates, meta-phosphates and other polyphosphates) and organically bound phosphates. These forms of phosphates arise from a variety of sources. Small amounts of organophosphates or certain condensed phosphates are added to water supplies during treatment, while large quantities of these compounds are used for laundry and cleaning. Phosphates are also used extensively in treatment of boiler waters. Other non-point sources include soil erosion and water run-off from cropland, lawns and gardens, domestic waste treatment systems, lawn clippings using fertilizers, etc. (Hammer 2009; Eaton et al. 2005).

Sulphur

Sulphur exists in water in the form of sulphates, sulphites and sulphides. Sulphates and sulphites are a part of naturally occurring minerals in some soils and rocks which on dissolution are released into ground water. Sulphate minerals can cause building up of scales in water pipes and may produce bitter taste in water and can have laxative effects on human and livestock. Oxygen-deficient water such as deep wells normally contains sulphur-reducing bacteria which produce large quantities of hydrogen sulphide on reaction with water. Hydrogen sulphide gas also occurs naturally in some ground water due to the decomposition of underground organic matter such as decaying plant materials. It is found in deep or shallow wells and can come into surface water through springs. High concentration of hydrogen sulphide produces rotten egg smell in water and poses several health risks (Hammer 2009, Eaton et al. 2005; Patnaik 1997).

Organic Compounds

Volatile Organic Compounds and Semi-VOCs

Volatile organic compounds (VOCs) are the ones having boiling point less than or equal to 100 °C and/or a vapour pressure greater than

1 mm Hg at 25 °C. Being volatile in nature, their concentration is less in surface water as compared to the ground water. They contribute to an increase in reactive hydrocarbons in the atmosphere, which can lead to the formation of photochemical oxidants. Their presence even at trace levels can cause adverse health effects. The VOCs, being manufactured in large quantities, are one of the major environmental pollutants. They are used for various commercial, agricultural and household activities; the most common are as follows: trichloroethylene, used as a degreasing solvent and an ingredient of household cleaning products; tetrachloroethylene, used for dry-cleaning; carbon tetrachloride, used for manufacturing fluorocarbons for refrigeration; 1,1,1 trichloromethane, a metal cleaner; 1, 2 dichloromethane, an intermediate in the manufacture of vinyl chloride monomer, used for plastics, etc. (Hammer 2009; Eaton et al. 2005; Khandal 2010).

Disinfection By-products

In order to protect the drinking water from various contaminants, such as pathogens and other disease causing organisms, certain disinfectants are added to the water (e.g. chlorine). These may react with naturally occurring materials in water to form by-products, which may pose health problems. Some of these by-products are as follows:

Trihalomethanes and Haloacetic Acids

Trihalomethanes of concern include chloroform, bromodichloromethane, dibromochloromethane and bromoform. Haloacetic acids include monochloroacetic acid, dichloroacetic acid, trichloroacetic acid, bromoacetic acid and dibromoacetic acid. The three most important parameters to be monitored for quality are: (a) *Chloroform:* It is the most commonly found trihalomethane in drinking water. Certain waters may have high concentration of bromated trihalomethanes due to naturally occurring bromide in water. (b) *Bromate:* It occurs when bromide in water reacts with the disinfectant ozone. (c) *Chlorite*: It occurs when chlorine dioxide breaks down. The presence of all the above by-products beyond their MRL values may prove toxic and carcinogenic for human health. As per USEPA guideline, the MRL value for the total four trihalomethanes is 0.080 ng/l or 80 ppb; for the total five haloacetic acids, it is 60 ppb; for bromate, it is 10 ppb; and for chlorite, it is 1 ppm (Hammer 2009; Eaton et al. 2005).

Pesticides

All pesticides are highly toxic to human beings even if ingested at trace levels (as low as ppb) and may cause cancer, disorders of various body organs and physiological systems. The major source of pesticides in water may be due to non-judicious and excess use of pesticides and due to discharge of wastewater from pesticide industry, spillage, infiltration of rainfall and irrigation water. For each of the pesticides, MRL values are specified. Normally, the values are measured as alpha and beta emitters (Hammer 2009; Eaton et al. 2005; Khandal 2010).

Radionuclides

The radioactivity in water originates from leaching of radioactive minerals from rocks and soils into the water bodies. The principal radionuclides are radium and uranium. Exposure to radioactivity through drinking water can be damaging for health (Hammer 2009).

Biological Parameters

Aerobic Microbial Count

It is the enumeration of microorganisms present in drinking water by counting the colonies growing in a solid medium after aerobic incubation at two different temperatures, that is,

22 °C and 37 °C. The total viable colony count should not exceed 100 CFU (colony forming units) per ml at ~22 °C in 72 h and 20 CFU per ml at ~37 °C in 24 h.

Yeast and Mould Count

Yeast and mould count reflects the enumeration of total fungi (yeast and mould) present in drinking water by counting the colonies growing in a solid medium after aerobic incubation at 25 °C. The count should be absent in 250 ml sample.

Coliforms

Total coliforms are gram-negative, oxidase-negative, non-spore forming rods, which ferment lactose with gas production at ~37 °C, after 48 h, in a medium with bile salts and detergents. When the test of coliforms is carried out with environmental waters, several species of four Enterobacteriaceae genera Escherichia, Klebsiella, Enterobacter and Citrobacter give positive results. Faecal coliforms (or thermo-tolerant coliforms) ferment lactose at 44.5 °C in a medium with bile salt and should be absent in 250 ml sample of drinking water.

Escherichia coli

Escherichia, a member of Enterobacteriaceae, are gram-negative, thermo-tolerant, oxidase-negative, catalase-positive, straight rods that ferment lactose. *E. coli* is a natural and essential part of the bacterial flora in the gut of humans and animals. *E. coli* should be absent when 250 ml sample of drinking water is analyzed.

Salmonella

Salmonella is a member of the family Enterobacteriaceae and includes gram-negative motile straight rods. Salmonellae have several endotoxins: antigens O, H and Vi and can cause two types of salmonellosis: typhoid and paratyphoid fever and gastroenteritis. Salmonella water is due to faecal contamination. Municipal sewage, agriculture pollution and storm water run-off are the main sources of these pathogens in natural waters. Salmonella should be absent in 250 ml sample of drinking water.

Shigella

Shigella is a gram-negative, non-spore forming, non-motile, straight rod-like members of the family Enterobacteriaceae. Shigella is found in intestinal tract of humans and other primates and is spread by faecal-contaminated drinking water or food or by direct contact with an infected person. In water, shigella can survive for at least six months at room temperature, and this high survival favours transmission through water. Shigella should be absent in 250 ml sample of drinking water.

Pseudomonas aeruginosa

This is an opportunistic pathogen in humans capable of growing in water even at very low nutrient concentration. Pseudomonas should be absent in 250 ml sample of drinking water.

Vibrio

Vibrio cholerae and *Vibrio parahaemolyticus* are small aquatic bacteria, curved-shaped gram-negative rods, with a single polar flagellum. Vibrios are facultative anaerobes capable of both fermentative and respiratory metabolism. Species distribution depends on sodium concentration and water temperature. Vibrios are very common in marine and estuarine environments, living free or on the surfaces and in the intestines of marine animals. Vibrio should be absent in 250 ml sample of drinking water.

Staphylococcus aureus

Coagulase-positive *Staphylococci* are normal inhabitants of the nose, skin and intestinal tract of humans and are a major component of bacterial flora found in swimming pools. They are highly resistant to halogen disinfectants. Staphylococcus aureus should be absent in 250 ml sample of drinking water.

Faecal Streptococci

Streptococci are gram-positive, non-spore forming, catalase-negative ovoid cells. Cells occur singly, in pairs or short chains. Faecal Streptococci should be absent in 250 ml sample of drinking water.

Clostridium

Clostridia are gram-positive rods, forming endospores and majorities are motile with peritrichous flagellation. Most species are obligate anaerobic, although tolerance to oxygen occurs. They may even be resistant to chlorination at levels, normally used for the treatment of water. They should be absent in 50 ml sample of drinking water.

Quality of Water for Various Purposes

Water has innumerable applications: agriculture consumes 69 %; various industries consume 22 %; household activities, that is, drinking, bathing, cooking, sanitation, gardening, etc., consume about 8 % and water required for recreational activities, such as swimming pools, golf courses and boating, and for environmental purposes, such as creation of wild life habitat, artificial wet lands and lakes, is 1 % of the total water available on earth.

Tables 20.2, 20.3, 20.4 and 20.5 describe the quality specifications of water as prescribed by the Indian Standard for the purpose of industrial use, food and health care, recreation and irrigational, respectively. The specified limits for drinking water have been presented in Tables 20.6 and 20.7. As seen in the Tables 20.2, 20.3, 20.4 and 20.5, there is a difference in the values for certain parameters as per applications.

Water Quality for Industrial Purposes

It may be seen from Table 20.2 that the limits for only the physical and chemical parameters have been specified, whereas no limits have been specified for biological parameters suggesting that biological parameters have little significance for these applications. For major industrial sectors, such as paper, chemical, food and textiles, the required water quality is defined based on the function of water and taking into account the different aspects involved in each sector, that is, product safety, product quality, process stability, workers safety, etc., and is not based upon the origin or the source of water.

In the paper sector, considering the processes and products that are used, most predominant water-related problems are biofouling and microbially induced corrosion. Corrosive waters dissolve iron from parts of the machinery, while algae and mould spores can cause semi-transparent spots. Moreover, processing in pulp and paper industry is greatly affected by certain other impurities present in water. Turbidity and colour affect the brightness and colour of the paper and textile processed. Impurities of iron and manganese can cause staining and loss of brightness to paper and textile. Chemical industry produces a diverse range of products and hence has a broad spectrum of processes. For many of these processes, no references are available regarding required water quality. In such cases, the water

Table 20.2 IS specifications of water for some of the industrial purposes

Quality parameters (physical and chemical)	Limits for various industry (minimum to maximum)				
	Textile (IS 201-2008)	Tanning (IS 4221-1999)	Paper and pulp (IS 2724-1983)	Storage batteries (IS 1069-2003)	Construction purpose (IS 456:2000)
Colour, Hz unit, max.	20	25	20	NS	NS
	Textile = Paper and pulp < Tanning				
Turbidity, NTU, max.	2	20	50	NS	NS
	Textile < Tanning < Paper and pulp				
Odour	Unobjectionable	NS	Unobjectionable	NS	NS
	Textile = Paper and pulp				
pH	6.0–8.5	6.5–8.0	NS	6.5–7.5	Not < 6
	Textile < Storage Batteries < Tanning < Construction purpose				
Specific electrical conductivity at 25 °C (μmhos/cm), max.	NS	NS	NS	5	NS
	Not specified except for storage batteries				
Total dissolved solids, mg/l, max.	NS	NS	500	2	NS
	Storage batteries < Paper and pulp				
Organic solids, mg/L, max.	NS	NS	NS	NS	200
	Not specified except for construction purposes				
Inorganic solids, mg/l, max.	NS	NS	NS	NS	3,000
	Not specified except for construction purposes				
Suspended matter, mg/l, max.	NS	NS	25 (Max)	NS	2,000
	Paper and pulp < Construction, not specified for others				
Non-volatile residue, mg/l, max.	NS	NS	NS	1.0	NS
	Not specified except for storage battery				
Total hardness (as $CaCO_3$), mg/l, max.	50	500	200	Not detectable	NS
	Storage Batteries < Textile < Paper and pulp < Tanning				
Iron (as Fe), mg/l, max.	0.25	NS	1.0	NS	NS
	Textile < Paper and pulp				
Manganese (as Mn), mg/l, max.	0.10	NS	0.50	NS	NS
	Textile < Paper and pulp				
Manganese (as Mn) and Iron (as Fe) added together, mg/l, max.	0.25	1.0	1.0	0.10	NS
	Storage battery < Textile < Tanning = Paper and pulp				
Aluminium (as Al), mg/l, max.	0.10	NS	NS	NS	NS
	Not specified except for textile				
Sulphate (as SO_4), mg/l, max.	100	NS	NS	NS	NS
	Not specified except for textile				
Chloride (as Cl), mg/l, max.	100	NS	NS	1	2,000 for concrete not containing embedded steel and 500 for RCC work
	Storage battery < Textile < Construction purpose				
Total alkalinity (as $CaCO_3$), mg/l, max.	150	150	NS	NS	NS
	Textile = Tanning				

(continued)

Table 20.2 (continued)

Quality parameters (physical and chemical)	Limits for various industry (minimum to maximum)				
	Textile (IS 201-2008)	Tanning (IS 4221-1999)	Paper and pulp (IS 2724-1983)	Storage batteries (IS 1069-2003)	Construction purpose (IS 456:2000)
Corrosivity	Non- corrosive	NS	Non- corrosive	NS	NS
	Textile = Paper and pulp				
Heavy metals (as Pb), mg/l. max.	NS	NS	NS	0.1	NS
	Not specified except for storage batteries				
Oxidizable matter	NS	NS	NS	To pass the test	NS
	Not specified				
Specific electrical conductivity at 25 °C (μmhos per cm), max.	NS	NS	NS	5	NS
	Not specified except for storage batteries				
Total dissolved solids, mg/l, max.	NS	NS	NS	2	NS
	Not specified except for storage batteries				
Volume of 0.02 N H_2SO_4 required to neutralize 100 ml sample using mixed indicator, ml, max.	NS	NS	NS	NS	25.0
	Not specified except for construction purpose				
Volume of 0.02 N NaOH required to neutralize 100 ml sample using phenolphthalein, ml, max.	NS	NS	NS	NS	5
	Not specified except for construction purpose				
Algae and mould spores	Absent	NS	Absent	NS	NS
	Textile = Paper and pulp				
Total residual chlorine, mg/l, max.	NS	NS	2.0	NS	NS
	Not specified except for paper and pulp				

NS Not specified under the respective specifications

with minimum possible interference of contaminants is used.

Water Quality for Food Processing Industries

It may be noted from Table 20.3 that the specifications of water for food processing industries are much more stringent than those for the other industries and even the list of parameters is also elaborate. Emphasis is more on the bacteriological parameters for these industries. Use of contaminated water at any stage of processing would result in a contaminated food product. All possible measures are therefore needed to be taken to ensure the safety of water being used at all stages of processing such as primary, secondary and tertiary. Contaminants of concern include Salmonella, Shigella, Campylobacter and various pathogenic strains of *E. coli*; different types of viral pathogens; protozoan parasites such as Entamoeba histolytica, Giarda lambia and Cyclospora and the various chemical contaminants from environment or chemical or radioactive spill, industrial waste, etc. The water should also be free of any materials or compounds, which could impart discoloration, off-flavour or

Table 20.3 IS specification of water for various food industries

S. No.	Quality parameters	Limits for various food industries		
		Processed food (IS 4251-1992)	Ice manufacturing (IS 3957-1989)	Fermentation (IS 4700-1999)
Physical and chemical parameters				
1	Colour, Hazen units, max.	20	5	20
2	Turbidity, NTU, max.	10	5	10
3	Odour	None	None	None
4	pH	6.5–9.2	6.5–9.2	6.0–9.2
5	Total solids, mg/l, max.	1,000	1,000	1,000
6	Total hardness (as CaCO3), mg/l, max.	600	600	500
7	Sulphate (as SO_4), mg/l, max.	200	200	200
8	Fluoride (as F), mg/l, max.	1.5	1.5	1.5
9	Chloride (as Cl), mg/l, max.	250	250	250
10	Cyanide (as CN), mg/l, max.	0.01	0.01	0.01
11	Selenium (as Se), mg/l, max.	0.05	0.05	0.05
12	Iron (as Fe), mg/l, max.	0.3	0.3	0.3
13	Magnesium (as Mg), mg/l, max.	75.0	125	125
14	Manganese (as Mn), mg/l, max.	0.2	0.2	0.2
15	Copper (as Cu), mg/l, max.	1.0	1.0	1.0
16	Lead (as Pb), mg/l, max.	0.1	0.1	0.1
17	Chromium (as Cr^{+6}), mg/l, max.	0.05	0.05	0.05
18	Zinc (as Zn), mg/l, max.	15.0	15.0	15
19	Arsenic (as As), mg/l, max.	0.2	0.2	0.2
20	Nitrate (as N), mg/l, max.	20	NS	20
21	Phenolic substances (as C_6H_5OH), mg/l, max.	0.001	0.001	0.001
22	Cadmium (as Cd), mg/l, max.	0.01	0.01	0.01
23	Mercury (as Hg), mg/l, max.	0.001	0.001	0.001
24	Total alkalinity (as $CaCO_3$), mg/l, max.	NS	100	NS
Bacteriological parameters				
1	Coliform bacteria, MPN index/100 ml	Less than 1	Less than 1	0
2	Standard plate count/ml, max.	50	100	100
		Processed food < Ice manufacturing = Fermentation		
3	Proteolytic and lipolytic combined organisms, count/ml, max.	5	NS	NS
Radio activity parameters				
1	Alpha emitters, µc/ml, max.	10^{-9}	10^{-9}	10^{-9}
2	Beta emitters, µc/ml, max.	10^{-8}	10^{-8}	10^{-8}
Additional tolerances for individual food industries				

Table 20.3 (continued)

S.No	Name of industry	Parameters	Tolerance (IS 4251-1992)
(I) Bakery			
1	Bread manufacture	pH	Controlled to optimum
		Calcium	Uniformly high
		Magnesium	Low
2	Cracker and cake making	Total hardness (as $CaCO_3$), mg/l	Preferably below 30
3	Cleansing	Total hardness (as $CaCO_3$), mg/l	Preferably below 30
(II) Canning			
1	Legumes (peas, beans, lentils, etc.)	Hardness (as $CaCO_3$), mg/l, max.	75
		Alkalinity (as $CaCO_3$), mg/l, max.	50
		Iron (as Fe), mg/l, max.	0.2
2	Cooker	Carbonate hardness	0
		Alkalinity (as $CaCO_3$), mg/l, max.	50
3	Cleansing	Total hardness (as $CaCO_3$), mg/l	Preferably below 30
4	General	Iron (as Fe), mg/l, max.	0.2
(III) Citrus fruit			
1	Washing citrus fruits	Soft water	NS
2	Pectin, citric acid and syrup making	Total hardness (as $CaCO_3$), mg/l	Preferably below 30
(IV) Confectionery		Total solids, mg/l, max.	100
		Iron (as Fe), mg/l, max.	0.2
		pH	Appropriate adjustment
(V) Dairy			
1	Processing	Thermophilic bacteria	Absent
2	Hot water	Slime forming organisms	Absent
		Total hardness (as $CaCO_3$), mg/l, max.	30
(VI) Edible oil refining			
1	Process water	Iron (as Fe), mg/l, max.	0.2
(VII) Gelatin manufacture			
1	Process water	Demineralized water preferred so that ash content of gelatin is kept low	NS
(VIII) Meat packing			
1	Cleansing and washing of meat for pickling	Soft water preferred to obtain a better colour	NS
		Iron (as Fe), mg/l, max.	0.2
(IX) Starch and corn products manufacture			
1	Process water	Soft water preferred as hard water increases the ash content of starch & high magnesium content leads to cloudiness in corn	
		Iron (as Fe), mg/l, max.	0.2

Table 20.3 (continued)

S.No	Name of industry	Parameters	Tolerance (IS 4251-1992)
(X)	Sugar refining	Iron (as Fe), mg/l, max.	0.2
		Total solids, mg/l	Less than 500 preferred as electrolytes causes inversion of sugar and yield of sucrose is decreased.

Other than the variation in the limits specified for parameters related to colour, turbidity, odour, pH, total solids, total hardness and the standard plate count for various food industries, the values for the rest of the parameters are the same; *NS* Not specified

Table 20.4 IS specifications of water for swimming pool/recreation purpose

S. No.	Quality parameters	Swimming pool/recreation purpose (IS 3328-2003)
Physical and chemical parameters		
1.	Colour, Hazen units, max.	10
2.	Turbidity, NTU, max.	10
3.	Odour	Odourless
4.	Taste	Palatable
5.	pH	7.5–8.5
6.	Total dissolved solids, mg/l, max.	1,500
7.	Chloride (as Cl), mg/l, max.	500
8.	Iron (as Fe), mg/l, max.	0.1
9.	Clearness	Clear
10.	Total alkalinity (as CaCO3), mg/l	50–500
11.	Aluminium (as Al), mg/l, max.	0.1
12.	Total residual chlorine, mg/l.	0.5 (Max. at Inlet); 0.2 (Min. at Outlet)
13.	Oxygen absorbed in 4 h at 27 °C, mg/l, max.	1.0
14.	Heavy metals (as Pb), mg/l, max.	0.1
Bacteriological parameters		
1.	Coliform bacteria, MPN index per 100 ml, max.	10
2.	Standard plate count/ml, max.	100 cfu

Table 20.5 IS specifications of water for irrigation purpose

S. No.	Quality parameters	Irrigation purpose (IS 11624-2001)
1	Sodium adsorption ratio (millimole/litre)	Low (below 10), Medium (10–18), High (18–26), Very high (Above 26)
2	Residual sodium carbonate meq/l	Low (below 1.5), Medium (1.5–3.0), High (3.0–6.0), Very high (Above 6.0)
3	Boron (as B), ppm	Low (below 1.0), Medium (1.0–2.0), High (2.0–4.0), Very high (Above 4.0)
4	Total salt concentration (on the basis of electrical conductivity), at 25 °C μmhos/cm	Low (below 1,500), Medium (1,500–3,000), High (3,000–6,000), Very high (Above 6,000)

Table 20.6 Comparison between quality parameters of drinking water, packaged drinking water and packaged natural mineral water as per IS specifications

S. No.	Quality parameter	Types of drinking water			
		Ordinary drinking water (IS 10500- 2012)		Packaged drinking water (IS 14543: 2004)	Packaged natural mineral water (IS 13428:2005)
	Definition	Water intended for human consumption for drinking and cooking purposes from any potable water source including public drinking water supply systems		Drinking water filled in sealed containers of various materials, shapes and capacities suitable for direct consumption without any further treatment	Water obtained directly from natural sources such as wells, water, spring water which is protected from pollution and contaminants through suitable measures
		Acceptable limit	Permissible limit	Requirements	Requirements
Bacteriological parameters					
1.	E. coli	Shall not be detectable in any 100 ml sample	NS	Absent in 250 ml sample	Absent in 250 ml sample
2.	Coliform bacteria	NS	NS	Absent in 250 ml sample	Absent in 250 ml sample
3.	Faecal streptococci	NS	NS	Absent in 250 ml sample	Absent in 250 ml sample
4.	Sulphite reducing anaerobes	NS	NS	Absent in 50 ml sample	Absent in 50 ml sample
5.	Pseudomonas aeruginosa	NS	NS	Absent in 250 ml sample	Absent in 250 ml sample
6.	Aerobic microbial count	NS	NS		NS
	(i) at 20 °C in 72 h on agar–agar			(i) Shall not exceed 100 per ml	
	(ii) at 37 °C in 24 h on agar–agar			(ii) Shall not exceed 20 per ml	
7.	Yeast and mould	NS	NS	Absent in 250 ml sample	Absent in 250 ml sample
8.	Salmonella and Shigella	NS	NS	Absent in 250 ml sample	Absent in 250 ml sample
9.	Vibrio cholera	NS	NS	Absent in 250 ml sample	Absent in 250 ml sample
Physical Parameters					
10.	Colour, Hazen units, max.	5	15	2	2
11.	Odour	Agreeable	Agreeable	Agreeable	Agreeable
12.	Taste	Agreeable	Agreeable	Agreeable	Agreeable
13.	Turbidity, NTU, max.	1	5	2	2

(continued)

Table 20.6 (continued)

S. No.	Quality parameter	Types of drinking water			
		Ordinary drinking water (IS 10500- 2012)		Packaged drinking water (IS 14543: 2004)	Packaged natural mineral water (IS 13428:2005)
	Definition	Water intended for human consumption for drinking and cooking purposes from any potable water source including public drinking water supply systems		Drinking water filled in sealed containers of various materials, shapes and capacities suitable for direct consumption without any further treatment	Water obtained directly from natural sources such as wells, water, spring water which is protected from pollution and contaminants through suitable measures
		Acceptable limit	Permissible limit	Requirements	Requirements
14.	Total dissolved solids, mg/L	500	2,000	500	150–700
15.	pH value	6.5–8.5	No relaxation	6.5–8.5	6.5–8.5
16.	Total hardness (as $CaCO_3$), mg/L, max.	200	600	Not specified	Not specified
17.	Total alkalinity (as $CaCO_3$) mg/L, max.	200	600	200(as HCO_3)	75–400(as HCO_3)
Chemical parameters, mg/L					
18.	Nitrate (as NO_3)	45	No relaxation	45	50
19.	Nitrite (as NO_2)	NS	NS	0.02	0.02
20.	Sulphide (as H_2S)	0.05	No relaxation	0.05	0.05
21.	Chloramines (as Cl_2)	4.0	No relaxation	NS	NS
22.	Manganese (as Mn)	0.1	0.3	0.1	2.0
23.	Copper (as Cu)	0.05	1.5	0.05	1.0
24.	Zinc (as Zn)	5.0	15.0	5	5
25.	Iron (as Fe)	0.3	No relaxation	0.1	NS
26.	Fluoride (as F)	1.0	1.5	1.0	1.0
27.	Barium (as Ba)	0.7	No relaxation	1.0	1.0
28.	Antimony (as Sb)	NS	NS	0.005	0.005
29.	Borate (as B)	0.5	1.0	5	5
30.	Silver (as Ag)			0.01	0.01
31.	Aluminium (as Al)	0.03	0.2	0.03	NS
32.	Ammonia (as total ammonia-N)	0.5	No relaxation	NS	NS
33.	Silver (as Ag)	0.1	No relaxation	NS	NS
34.	Chloride (as Cl)	250	1,000	200	200

(continued)

Table 20.6 (continued)

S. No.	Quality parameter	Types of drinking water			
		Ordinary drinking water (IS 10500- 2012)		Packaged drinking water (IS 14543: 2004)	Packaged natural mineral water (IS 13428:2005)
	Definition	Water intended for human consumption for drinking and cooking purposes from any potable water source including public drinking water supply systems		Drinking water filled in sealed containers of various materials, shapes and capacities suitable for direct consumption without any further treatment	Water obtained directly from natural sources such as wells, water, spring water which is protected from pollution and contaminants through suitable measures
		Acceptable limit	Permissible limit	Requirements	Requirements
35.	Free residual chlorine	0.2	1.0	0.2	NS
36.	Sulphate (as SO_4)	200	400	200	200
37.	Magnesium (as Mg)	30	100	30	50
38.	Calcium (as Ca)	75	200	75	100
39.	Sodium (as Na)	NS	NS	200	150
40.	Trihalomethanes	0.1	No relaxation	NS	NS
	(a) Bromoform				
	(b) Dibromochloromethane	0.1			
	(c) Bromodichloromethane	0.06			
	(d) Chloroform	0.2			
41.	Selenium (as Se)	0.01	No relaxation	0.01	0.05
42.	Mineral oil	0.5	No relaxation	Absent	Absent
43.	Phenolic compounds (as C_6H_5 OH)	0.001	0.002	Absent	Absent
44.	Anionic detergents (as MBAS)	0.2	1.0	0.2	Not detectable
45.	Arsenic (as As)	0.01	0.05	0.05	0.05
46.	Cadmium (as Cd)	0.003	No relaxation	0.01	0.003
47.	Cyanide (as CN)	0.05	No relaxation	Absent	Absent
48.	Molybdenum (as Mo)	0.07	No relaxation	NS	NS
49.	Chromium (as Cr)	0.05	No relaxation	0.05	0.05
50.	Mercury (as Hg)	0.001	No relaxation	0.001	0.001
51.	Lead (as Pb)	0.01	No relaxation	0.01	0.01

(continued)

Table 20.6 (continued)

S. No.	Quality parameter	Types of drinking water			
		Ordinary drinking water (IS 10500- 2012)		Packaged drinking water (IS 14543: 2004)	Packaged natural mineral water (IS 13428:2005)
	Definition	Water intended for human consumption for drinking and cooking purposes from any potable water source including public drinking water supply systems		Drinking water filled in sealed containers of various materials, shapes and capacities suitable for direct consumption without any further treatment	Water obtained directly from natural sources such as wells, water, spring water which is protected from pollution and contaminants through suitable measures
		Acceptable limit	Permissible limit	Requirements	Requirements
52.	Nickel (as Ni)	0.02	No relaxation	0.02	0.02
53.	Polychlorinated biphenyls (PCB)	0.0005	No relaxation	Not detectable	Not detectable
54.	Polynuclear aromatic hydrocarbons (PAH)	0.0001	No relaxation	Not detectable	Not detectable
55.	Pesticide residues considered individually	See table-20.7a	No relaxation	Not more than 0.0001	BDL
56.	Total pesticide residues	NS	NS	Not more than 0.0005	NS
Radiological, Bq/L					
57.	α-emitters, max.	0.1	No relaxation	0.1	0.1
58.	β-emitters, max.	1.0	No relaxation	1	1

NS Not specified under the respective specifications, *BDL* Below detection limit; Detection limits have not been specified

odours to the product or otherwise adversely affect its quality (Klemes and Chair 2008).

Water Quality for Irrigational Purposes

The standard specifications on water for irrigational purposes help in assessing irrigation water quality in terms of concentration of various soluble salts in water and the degree of harmful effects on soil properties, and crop yield and crop production. Water for irrigation generally refers to the production of commercial crops, gardening and potted plants. The major factors affecting water quality are as follows: (1) *Salinity:* Low-salinity water is desired for most of the crops. High-salinity water requires effective drainage to increase leaching of salts from soil for crop growth. (2) *Sodium content:* It causes dispersion, salinization and damage to plant tissues. (3) *Alkalinity:* It is responsible for high soil pH thereby causing iron deficiency. Highly alkaline soils must be regularly acidified or water treated with the acid agents for agriculture. (Food and Agriculture organization 1985)

Water Quality for Recreational Purposes

Water quality not meeting the desired norms may have potential health hazards due to microbiological pollution, chemical pollution

Table 20.7 Pesticide residue limits as per IS specifications

S. No.	Pesticide	Acceptable limit, µg/L		
		IS 10500-2012	IS 14543:2004	IS 13428:2005
1	Alachlor	20	<0.1	BDL
2	Atrazine	2	<0.1	BDL
3	Aldrin/Dieldrin	0.03	<0.1	BDL
4	α-HCH	0.01	<0.1	BDL
5	β-HCH	0.04	<0.1	BDL
6	Butachlor	125	<0.1	BDL
7	Chlorpyriphos	30	<0.1	BDL
8	Delta-HCH	0.04	<0.1	BDL
9	2,4-dichlorophenoxyacetic acid	30	<0.1	BDL
10	DDT (o,p- and p,p-isomers of DDT, DDE and DDD)	1	<0.1	BDL
11	Endosulphan (α, β and sulphate)	0.4	<0.1	BDL
12	Ethion	3	<0.1	BDL
13	γ-HCH (Lindane)	2	<0.1	BDL
14	Isoproturon	9	<0.1	BDL
15	Malathion	190	<0.1	BDL
16	Methyl parathion	0.3	<0.1	BDL
17	Monocrotophos	1	<0.1	BDL
18	Phorate	2	<0.1	BDL

BDL Below detection limit; Detection limits have not been specified

and toxic algae and their products and can even result in disease outbreaks and illnesses. The hazards would normally depend upon water quality and degree of contact with the water. The different ways in which one can come in contact with water meant for different recreational purposes include the following: (1) Primary contact recreation occurs where the water remains in direct contact with the human body, and there are chances of swallowing water, for example, surfing, water skiing, diving and swimming. Hence, the water quality norms need to be as stringent as that of drinking water; (2) Secondary contact recreation in case of activities such as paddling, wading, boating and fishing which involve contact of the human body with water but the chances of swallowing water are rare; (3) Passive recreation activities have no contact with the water and they include scenic appreciation and walking around the water bodies such as lakes.

Quality of Drinking Water

Assuring safe drinking water for all has always been a priority for policy makers not only for the prevention and control of the waterborne diseases but also for the prosperity of the society (Annan 2003). Drinking water for the common people throughout the world is mostly available in the following forms:

Municipal/Tap Water

It is normally the treated, processed and disinfected water, which is supplied by the municipalities through the water pipes right to the homes of the consumers. It is supposed to be meeting the most stringent specifications of quality.

Ground Water

Since the water from such sources is consumed as such without any treatment, the quality of ground water is a matter of concern. In order to avoid risks, quality of water needs to be checked on the source before consuming.

Purified Water

Water obtained from the source is treated with certain processes such as decantation, filtration, deionization, distillation and reverse osmosis to ensure the removal of all kinds of bacteria, dissolved solids and contaminants and is ready for consumption without giving any further treatment.

Packaged Natural Mineral Water

It is obtained directly from natural or drilled sources from underground for which all possible precautions are taken to avoid any pollution or effect of any external influence on the chemical and physical qualities. The water is packaged in containers of various materials and capacities. It has inherent minerals and micronutrients some of which may be beyond certain specified limits. It should contain not less than 250 ppm of the total dissolved solids. Since it is a commercial product, quality norms are most stringent in this case.

Packaged Drinking Water

It is the treated raw water from surface or ground, packed in containers of various materials and capacities to ensure safety and suitability for drinking purposes. Packaged drinking water again being a commercial product, the quality norms are more stringent than municipal water but a less stringent than mineral water. It may be noted here that the water made available by the municipalities is always treated water meeting the statutory guidelines of quality. The packaged drinking water and packaged mineral water are the products involving industrial processes and their quality should be of desired standards, which are more stringent. In India, there are three different standard specifications as per BIS for drinking water depending upon its type and source: (1) IS: 10500 for Drinking water from any of the sources, (2) IS: 14543 for Packaged drinking water and (3) IS: 13428 for Packaged natural mineral water. The requirements for various parameters are mentioned in Tables 20.6 and 20.7. It may be noted here that for the various quality parameters of ordinary drinking water, Indian Standard specifications have prescribed two different limits, that is, permissible limits and acceptable limits, respectively.

Acceptable Limit and Permissible Limit

Talking about the purity of the water, pure water basically means safe water with no undesired substances but containing all kinds of nutrients such as minerals and salts. Today mineral water has become a leading food product consumed world over.

As can be seen from the Tables 20.6 and 20.7, as per IS: 10500-2012, there are two different categories of acceptability levels, that is, acceptable limit and permissible limit. The limits specified for various parameters as per acceptable levels are stringent as compared to the permissible levels which are considerably relaxed. The difference is because of the fact that large portion of the population, which is not provided with the treated water, should not remain deprived of this essential commodity. Although the acceptable limits are implementable and no values above that are acceptable for drinking water but still in cases such as in certain areas like Rajasthan where no other alternate sources of water are available, the values under the permissible levels are tolerable.

Regulatory Requirements for Drinking Water as per the International Norms

Since drinking water will inevitably contain impurities and contaminants, therefore in order to ensure availability of safe water and protect human life, various international agencies, for example, (WHO 2003, 2008, 2009, 2011) and the regulatory authorities of individual countries, throughout the world have framed policies and regulations and have laid down specifications and guideline values with maximum allowed limits for different quality parameters, for example, Indian Standard (IS) specifications in India, European Union (EU) (European Communities 1998; EPA Handbook 2010) regulations in European countries, USEPA guidelines in USA, (U.S. Environment Protection Agency (U.S.EPA)Handbook 2004; U.S. Deptt. of Agriculture 2003) etc.

Specifications for drinking water as laid down by the regulatory authorities of different countries are given in Table 20.8. Looking at the values in the Table, one finds non-uniformity in the specifications in terms of both quality criteria and the specified limits for the different quality parameters. Although for some of the parameters, the MRL values prescribed are the same in all the specifications, for certain other parameters, there is a wide variation in the values for MRLs, ranging from two to even five times higher, for example, sulphates and total dissolved solids.

In case of concentration of metals such as antimony, arsenic, aluminium, cadmium, chromium, copper, iron, lead, manganese and mercury, MRL values are almost similar as per different specifications except zinc where the MRL value varies from 1.0 mg/l in case of Japanese specification to 15 mg/l as per Indian specification. USEPA drinking water standard allows higher MRL values for barium, that is, 2.0 mg/l as compared to 0.7 mg/l as given in other international standards. Beryllium and thallium are specified as contaminants only in USEPA with a MRL value of 0.004 mg/l and 0.002 mg/l, respectively.

There are a number of parameters for which while some of the regulatory agencies are particular, the others have not specified any limit for those parameters. In case of contaminants of non-metals also, such as boron, cyanide, fluoride, nitrate and nitrite, the different specifications show a significant variation in the MRL values. For chloride in drinking water, limit of 250 mg/l has been prescribed by most of the regulatory authorities except the Japanese specification which specifies 200 mg/l and the Indian specification which allows chloride values, as high as 1,000 mg/l under the permissible limits. Calcium and magnesium ions, which are the major cationic components of water, are mandatory parameters only as per IS specification for drinking water, whereas the other specifications are silent about it.

Anionic detergents and phenolic compounds may also be present as contaminants in drinking water. It is strange that the limits for these two contaminants are mentioned only in the case of Indian and Japanese specifications, whereas all other regulators are silent about it.

Looking at the physical parameters of water, even though the specified values vary in terminology or limits but all the various specifications recommend the use of safe drinking water that does not possess any undesirable odour or taste and is free from any visible colour or turbidity. The acceptable pH range as per the Indian specifications is 6.5–8.5, whereas as per the EU standard, the allowed pH range is 6.5–9.5. The TDS as per the US, Indian and WHO standards vary from 500 mg/l maximum in the case of US specification to 2,000 mg/l maximum in the case of Indian specification under the permissible levels. The values for water hardness have only been specified by the Japanese specification (300 mg/l) and the IS specification (200 mg/l as the acceptable level and 600 mg/l as the permissible level). The limits for this parameter have not been specified in other cases. Similarly, the measurement of alkalinity of water is also mentioned as per the IS specification only with

Table 20.8 Comparison between water quality parameters and their specified limits as per different norms

S. No.	Parameter	EU, 98/83 EC	USEPA, MCL, National primary water regulation	India, IS 10500-2012		WHO 2011	Japan SV 2010	Canada. GV 2010
				AL	PL			
Bacteriological parameters								
1.	E. coli	0 per 250 ml	<5.0 % per month	Shall not be detectable in any 100 ml sample		Must not be detected in any 100 ml sample	Not to be detected	0 per 100 ml
2.	Cyanobacterial toxins Microcystin-LR	NS	NS	NS	NS	0.001	NS	0.0015
3.	Common bacteria	NS	NS	NS	NS	Must not be detected in any 100 ml sample	100 per 1 ml	NS
4.	Total coliform bacteria	0 per 100 ml	<5.0 % per month	NS	NS	Must not be detected in any 100 ml sample	NS	0 per 100 ml
5.	Enterococci	0 per 250 ml	NS	NS	NS	NS	NS	NS
6.	Clostridium perfringens	0 per 100 ml	NS	NS	NS	NS	NS	NS
7.	Colony count, 22 °C	No abnormal change	NS	NS	NS	NS	NS	NS
8.	Pseudomonas aeruginosa	0 per 250 ml	NS	NS	NS	NS	NS	NS

(continued)

Table 20.8 (continued)

S. No.	Parameter	EU, 98/83 EC	USEPA, MCL, National primary water regulation	India, IS 10500-2012		WHO 2011	Japan SV 2010	Canada. GV 2010
				AL	PL			
Physical parameters								
9.	Colour, Hazen units, max	Acceptable to consumers and no abnormal change	NS	5	15	NS	5 degree	≤15
10.	Odour	Acceptable to consumers and no abnormal change	NS	Agreeable	Agreeable	NS	Not abnormal	Inoffensive
11.	Taste	Acceptable to consumers and no abnormal change	NS	Agreeable	Agreeable	NS	Not abnormal	Inoffensive
12.	Turbidity, NTU, max.	Acceptable to consumers and no abnormal change	TT	1	5	NS	2 degree	0.3/1.0/0.1[a]
13.	Total dissolved solids (TDS), mg/L, max.	NS	NS	500	2,000	NS	NS	≤500
14.	pH value	9.5 ≤ 6.5	NS	6.5-8.5	No relaxation	NS	5.8–8.6	6.5–8.5
15.	Total hardness (as $CaCO_3$), mg/L, max.	NS	NS	200	600	NS	300	NS
16.	Alkalinity, mg/l, max.	NS	NS	200	600	NS	NS	NS

(continued)

Table 20.8 (continued)

S. No.	Parameter	EU, 98/83 EC	USEPA, MCL, National primary water regulation	India, IS 10500-2012		WHO 2011	Japan SV 2010	Canada. GV 2010
				AL	PL			
17.	Conductivity, μScm^{-1} at 20 °C	2,500	NS	NS	NS	NS	NS	NS
18.	Oxidisability mg/l O_2	5	NS	NS	NS	NS	NS	NS
Chemical parameters, mg/l, max.								
Metals/cations								
19.	Antimony (as Sb)	0.005	0.006	NS	NS	0.002	0.015^{T}	0.006
20.	Arsenic (as As)	0.01	0.010	0.01	0.05	0.01	0.01	0.01
21.	Aluminium (as Al)	0.2	NS	0.03	0.2	NS	0.2	$0.1/0.2^{b}$
22.	Barium (as Ba)	NS	2	0.7	NR	0.7	NS	1
23.	Beryllium (as Be)	NS	0.004	NS	NS	NS	NS	NS
24.	Boron (as B)	1.00	NS	0.5	1.0	2.4	1.0	5
25.	Cadmium (as Cd)	0.005	0.005	0.003	NR	0.003	0.003	0.005
26.	Chromium (as Cr^{6+})	0.05	0.1	0.05	NR	0.05	0.05	0.05
27.	Copper (as Cu)	2.0	1.3	0.05	1.5	2	1.0	≤1.0
28.	Calcium (as Ca)	NS	NS	75	200	NS	NS	NS
29.	Iron (as Fe	0.2	NS	0.3	NR	NS	0.3	≤0.3
30.	Lead (as Pb)	0.01	TT	0.01	NR	0.01	0.01	0.01
31.	Manganese (as Mn)	0.05	NS	0.1	0.3	NS	0.05	≤0.05
32.	Molybdenum (as Mo)	NS	NS	0.07	NR	NS	0.07^{T}	NS
33.	Magnesium (as Mg)	NS	NS	30	100	NS	NS	NS
34.	Mercury (as Hg)	0.001	0.002	0.001	NR	0.006	0.0005	0.001
35.	Nickel (as Ni)	0.02	NS	0.02	NR	0.07	0.01^{T}	NS
36.	Selenium (as Se)	0.01	0.05	0.01	NR	0.04	0.01	0.01
37.	Silver (as Ag)	NS	NS	0.1	NR	NS	NS	NS
38.	Sodium (as Na)	200	NS	NS	NS	NS	200	≤200
39.	Thallium (as Tl)	NS	0.002	NS	NS	NS	NS	NS
40.	Zinc (as Zn)	NS	NS	5	15	NS	1.0	≤5.0
41.	Uranium	NS	0.03	0.03	NS	0.03	0.002^{T}	0.02

(continued)

Table 20.8 (continued)

S. No.	Parameter	EU, 98/83 EC	USEPA, MCL, National primary water regulation	India, IS 10500-2012		WHO 2011	Japan SV 2010	Canada. GV 2010
				AL	PL			
Non-metals/Anions								
42.	Asbestos, million fibres/litre	NS	7	NS	NS	NS	NS	NS
43.	Ammonium (as NH_4)	0.50	NS	0.5 (ammonia)	NR	NS	NS	NS
44.	Sulphate (as SO_4), max.	250	NS	200	400	NS	NS	≤500
45.	Cyanide (as CN), max.	0.05	0.2	0.05	NR	NS	0.01	0.2
46.	Fluoride (as F), max.	1.5	4.0	1.0	1.5	1.5	0.8	1.5
47.	Nitrate (as NO_3), max	50	NS	45	NR	50	NS	45
48.	Sulphide, max.	NS	NS	0.05	NR	NS	NS	≤0.05
49.	Nitrite (as N), max.	0.50(as NO_2)	1	NS	NS	3 as (NO_2^-)	0.05^T	NS
50.	Nitrate and Nitrite(as N)	NS	10	NS	NS	NS	10	NS
51.	Chloride (as Cl), max.	250	NS	250	1,000	NS	200	≤250
Volatile organic compounds								
52.	Benzene	0.001	0.005	NS	NS	0.01	0.01	0.005
53.	Carbon tetrachloride	NS	0.005	NS	NS	0.002	0.002	0.005
54.	Dichloromethane	NS	0.005	NS	NS	0.02	0.02	0.05
55.	Ethylbenzene	NS	0.7	NS	NS	0.3	NS	≤0.0024
56.	Monochlorobenzene	NS	0.1	NS	NS	NS	NS	0.08
57.	1,2-dichlorobenzene	NS	0.6	NS	NS	1	NS	0.2
58.	1,4-dichlorobenzene	NS	0.075	NS	NS	0.3	NS	0.005
59.	1,2-dichloroethane	0.003	0.005	NS	NS	0.03	0.004^T	0.005
60.	1,1-dichloroethylene	NS	0.007	NS	NS	NS	NS	0.014
61.	cis-1,2-dichloroethylene	NS	0.07	NS	NS	0.05	0.04	NS
62.	trans-1,2-dichloroethylene	NS	0.1	NS	NS		0.04^T	NS
63.	1,2,4-trichlorobenzene	NS	0.07	NS	NS	NS	NS	NS
64.	1,1,1-tri chloroethane	NS	0.2	NS	NS	NS	0.3^T	NS
65.	1,1,2-tri chloroethane	NS	0.005	NS	NS	NS	0.006^T	NS

(continued)

Table 20.8 (continued)

S. No.	Parameter	EU, 98/83 EC	USEPA, MCL, National primary water regulation	India, IS 10500-2012		WHO 2011	Japan SV 2010	Canada. GV 2010
				AL	PL			
66.	1,2-dichloropropane	NS	0.005	NS	NS	0.04	NS	NS
67.	Styrene	NS	0.1	NS	NS	0.02	NS	NS
68.	Tetrachloroethylene	0.01	0.005	NS	NS	0.04	0.01	0.03
69.	Toluene	NS	1	NS	NS	0.7	0.2^{T}	≤0.024
70.	Trichloroethylene	0.01	0.005	NS	NS	0.02	0.01	0.005
71.	Vinyl chloride	0.0005	0.002	NS	NS	0.0003	NS	0.002
72.	Xylenes total	NS	10	NS	NS	0.5	NS	≤0.3
73.	1,2-dibromo-3-chloropropane	NS	0.0002	NS	NS	0.001	NS	NS
74.	1,4-dioxane	NS	NS	NS	NS	0.05	0.05	NS
75.	Methyl tertiary-butyl ether (MTBE)	NS	NS	NS	NS	NS	0.02^{T}	0.015
76.	1,3-dichloropropene	NS	NS	NS	NS	0.02	NS	NS
Pesticides								
77.	Pesticides-total	0.0005	NS	NS	NS	NS	1^{T}	NS
78.	Pesticide-individual	0.0001	–	–	–	–	–	–
79.	Alachlor	NS	0.002	0.02	NR	0.02	NS	NS
80.	Atrazine	NS	0.003	0.002	NR	0.1	NS	0.005
81.	Aldicarb	NS	NS	NS	NS	0.01	NS	0.009
82.	Aldrin/Dieldrin	0.00003	NS	0.00003	NR	0.00003	NS	0.0007
83.	Azinophos-methyl	NS	NS	NS	NS	NS	NS	0.02
84.	Acrylamide	NS	NS	NS	NS	0.0005	0.0005^{T}	NS
85.	α-HCH	NS	NS	0.00001	NR	NS	NS	NS
86.	β-HCH	NS	NS	0.00004	NR	NS	NS	NS
87.	Butachlor	NS	NS	0.125	NR	NS	NS	NS
88.	Bendiocarb	NS	NS	NS	NS	NS	NS	0.04
89.	Bromoxynil	NS	NS	NS	NS	NS	NS	0.005
90.	Chlordane	NS	0.002	NS	NS	0.0002	NS	NS
91.	Carbaryl	NS	NS	NS	NS	NS	NS	0.09

(continued)

Table 20.8 (continued)

S. No.	Parameter	EU, 98/83 EC	USEPA, MCL, National primary water regulation	India, IS 10500-2012		WHO 2011	Japan SV 2010	Canada. GV 2010
				AL	PL			
92.	Carbofuran	NS	0.04	NS	NS	0.007	NS	0.09
93.	Chlorothalonil	NS	NS	NS	NS	NS	NS	NS
94.	Chlorpyriphos	NS	NS	0.03	NR	0.03	NS	0.09
95.	Cyanazine	NS	NS	NS	NS	0.0006	NS	0.01
96.	Dalapon (sodium salt)	NS	0.2	NS	NS	NS	NS	NS
97.	Dicamba	NS	NS	NS	NS	NS	NS	0.12
98.	δ-HCH	NS	NS	0.00004	NR	NS	NS	NS
99.	Diquat	NS	0.02	NS	NS	NS	NS	0.07
100.	Dinoseb	NS	0.007	NS	NS	NS	NS	0.01
101.	DDT and isomers	NS	NS	0.001	NR	0.001	NS	NS
102.	Diazinon	NS	NS	NS	NS	NS	NS	0.02
103.	Diuron	NS	NS	NS	NS	NS	NS	0.15
104.	Diclofop-methyl	NS	NS	NS	NS	NS	NS	0.009
105.	Dimethoate	NS	NS	NS	NS	0.006	NS	0.02
106.	Dichlorprop	NS	NS	NS	NS	0.1	NS	NS
107.	Endosulphan	NS	NS	0.0004	NR	NS	NS	NS
108.	Endothall	NS	0.1	NS	NS	NS	NS	NS
109.	Endrin	NS	0.002	NS	NS	0.0006	NS	NS
110.	Ethion	NS	NS	0.003	NR	NS	NS	NS
111.	Glyphosate	NS	0.7	NS	NS	NS	NS	0.28
112.	Heptachlor	0.00003	0.0004	NS	NS	NS	NS	NS
113.	Heptachlor epoxide	0.00003	0.0002	NS	NS	NS	NS	NS
114.	Hexachlorobenzene	NS	0.001	NS	NS	NS	NS	NS
115.	Hexachlorocyclopentadiene	NS	0.05	NS	NS	NS	NS	NS
116.	Isoproturon	NS	NS	0.009	NR	0.009	NS	NS
117.	Lindane	NS	0.0002	0.002	NR	0.002	NS	NS
118.	Methoxychlor	NS	0.04	NS	NS	0.02	NS	0.9
119.	Malathion	NS	NS	0.19	NR	NS	NS	0.19

(continued)

Table 20.8 (continued)

S. No.	Parameter	EU, 98/83 EC	USEPA, MCL, National primary water regulation	India, IS 10500-2012		WHO 2011	Japan SV 2010	Canada. GV 2010
				AL	PL			
120.	Metolachlor	NS	NS	NS	NS	0.01	NS	0.05
121.	Metribuzin	NS	NS	NS	NS	NS	NS	0.08
122.	Molinate	NS	NS	NS	NS	0.006	NS	NS
123.	Mecoprop	NS	NS	NS	NS	0.01	NS	NS
124.	Monocrotophos	NS	NS	0.001	NR	NS	NS	NS
125.	Oxamyl (Vydate)	NS	0.2	NS	NS	NS	NS	NS
126.	Picloram	NS	0.5	NS	NS	NS	NS	0.19
127.	Paraquat (as dichloride)	NS	NS	NS	NS	NS	NS	0.01
128.	Permethrin	NS	NS	NS	NS	0.02	NS	NS
129.	Pendimethalin	NS	NS	NS	NS	0.02	NS	NS
130.	Parathion	NS	NS	NS	NS	NS	NS	0.05
131.	Phorate	NS	NS	0.0002	NR	NS	NS	0.002
132.	Parathion-methyl	NS	NS	0.0003	NR	NS	NS	NS
133.	Simazine	NS	0.004	NS	NS	0.002	NS	0.01
134.	Toxaphene (camphechlor)	NS	0.003	NS	NS	NS	NS	NS
135.	Terbufos	NS	NS	NS	NS	NS	NS	0.001
136.	Trifluralin	NS	NS	NS	NS	0.02	NS	0.045
137.	Terbuthylazine	NS	NS	NS	NS	0.007	NS	NS
138.	Vernolate	NS	NS	NS	NS	0.007	NS	NS
139.	2,4-D	NS	0.07	0.03	NR	0.03	NS	0.1
140.	2,4,5-TP (Silvex)	NS	0.05	NS	NS	0.009	NS	NS
141.	MCPA	NS	NS	NS	NS	0.002	NS	0.1
Polyaromatic hydrocarbons (PAHs)								
142.	PAHs (total)	0.0001	NS	0.0001	NR	NS	NS	NS
143.	Benzo (a) pyrene	0.00001	0.0002	NS	NS	0.0007	NS	0.00001
Phenols								
144.	Phenolic compounds (as C_6H_5OH), max.	NS	NS	0.001	0.002	NS	0.005	NS

(continued)

Table 20.8 (continued)

S. No.	Parameter	EU, 98/83 EC	USEPA, MCL, National primary water regulation	India, IS 10500-2012		WHO 2011	Japan SV 2010	Canada. GV 2010
				AL	PL			
145.	2,4-dichlorophenol	NS	NS	NS	NS	NS	NS	0.9
146.	2,4,6-trichlorophenol	NS	NS	NS	NS	0.2	NS	0.005
147.	2,3,4,6-tetrachlorophenol	NS	NS	NS	NS	NS	NS	0.1
148.	Pentachlorophenol	NS	0.001	NS	NS	0.009	NS	0.06
Others								
149.	2,3,7,8-TCDD (Dioxin)	NS	3×10^{-8}	NS	NS	NS	NS	NS
150.	PCBs	NS	0.0005	0.0005	NR	NS	NS	NS
151.	Acrylamide	0.0001	NS	NS	NS	0.0005	NS	NS
152.	Mineral oil, max.	NS	NS	0.5	NR	NS	NS	NS
153.	Epichlorohydrin	0.0001	TT	NS	NS	0.0004	0.0004^{T}	NS
154.	Di(2-ethyl hexyl) phthalate	NS	0.006	NS	NS	0.008	0.1^{T}	NS
155.	Di(2-ethyl hexyl) adipate	NS	0.4	NS	NS	NS	NS	NS
156.	Anionic detergents (as MBAS), max.	NS	NS	0.2	1.0	NS	0.2	NS
157.	Nitrilotriacetic acid (NTA)	NS	NS	NS	NS	0.2	NS	0.4
Disinfectants/Disinfection by-products								
158.	Chloramines total	NS	4.0	4.0	NR	NS	NS	3
159.	Monochloramine	NS	NS	NS	NS	3	NS	NS
160.	Residual-free chlorine	NS	4.0	0.2	1	5	1^{T}	NS
161.	Bromate	0.01	0.010	NS	NS	0.01	0.01	0.01
162.	Chlorite	NS	1.0	NS	NS	0.1	0.6^{T}	1
163.	Formaldehyde	NS	NS	NS	NS	NS	0.08	NS
164.	Bromoform	NS	NS	0.1	NR	0.1	0.09	NS
165.	Dibromochloromethane	NS	NS	0.1	NR	0.1	0.1	NS
166.	Bromodichloromethane	NS	NS	0.06	NR	0.06	0.03	NS
167.	Chloroform	NS	NS	0.2	NR	0.3	0.06	NS
168.	Dichloroacetic acid	NS	NS	NS	NS	0.05	0.04	NS
169.	Trichloroacetic acid	NS	NS	NS	NS	NS	0.2	NS

(continued)

Table 20.8 (continued)

S. No.	Parameter	EU, 98/83 EC	USEPA, MCL, National primary water regulation	India, IS 10500-2012		WHO 2011	Japan SV 2010	Canada. GV 2010
				AL	PL			
170.	Dichloroacetonitrile	NS	NS	NS	NS	0.02	0.04^T	NS
171.	Dibromoacetonitrile	NS	NS	NS	NS	0.07	0.06^T	NS
172.	Chlorate	NS	NS	NS	NS	NS	0.6^T	1
173.	Chloroacetic acid	NS	NS	NS	NS	NS	0.02	NS
174.	Trihalomethanes total (THMs)	0.1	0.080	NS	NS	< 1	0.1	0.1
175.	Haloacetic acids total (HAAs)	NS	0.060	NS	NS	NS	NS	0.08
Radiological parameters								
176.	Radioactive materials, Bq/L, max.:	NS					NS	NS
	• α-emitters		15 pCi/L	0.1	NR	0.1		
	• β-emitters		4 millirems per year	1.0	NR	1.0		
177.	Caesium-137, Bq/L	NS	NS	NS	NS	10	NS	10
178.	Iodine-131, Bq/L	NS	NS	NS	NS	10	NS	6
179.	Lead-210, Bq/L	NS	NS	NS	NS	0.1	NS	0.2
180.	Radium-226, Bq/L	NS	5 pCi/L	NS	NS	1	NS	0.5
181.	Strontium-90, Bq/L	NS	NS	NS	NS	10	NS	5
182.	Tritium(^{3}H) Bq/l	100	NS	NS	NS	10,000	NS	7,000
183.	Total indicative dose, mSv/Year	0.10	NS	NS	NS	0.1	NS	NS

Note GV Guideline value—Used by regulatory authorities for surveillance and enforcement purposes, *HV* Health value, *MCL* Maximum contaminant level—It is the maximum permissible level of contaminant in water that is delivered to any user of a public water system. It is an enforceable limit under the safe drinking water act, *AL* Acceptable limit, *PL* Permissible limit, *NS* Not specified, *TT* Treatment technique—A required process intended to reduce the level of a contaminant in drinking water, *T* Target value as given in Japanese Drinking water quality standard

[a] Based on conventional treatment/slow sand or diatomaceous earth filtration/membrane filtration

[b] This is an operational guidance value, designed to apply only to drinking water treatment plants using aluminium-based coagulants. The operational guidance values of 0.1 mg/l applies to conventional treatment plants and 0.2 mg/L applies to other types of treatment systems

maximum limit being 200 mg/l as acceptable and 600 mg/l as permissible.

As per all the specifications, safe water should be free from all microbial contamination. Apart from the above-mentioned quality parameters, it is important that the drinking water should be free from pollutants such as primary volatile and semi-volatile organic compounds like PAHs, phthalates, PCBs, synthetic organic chemicals including different pesticides, phenols, disinfectants, etc. Many of the specifications have specified limits for these parameters but again one finds a lot of disparity between the values specified. Whereas EU specification specifies MRL of 0.5 ppb for the presence of total pesticide residues, as per the WHO specification, it can be seen that for many of the individual pesticides, the MRL values are as high as 300 ppb for bentazone, 100 ppb for dichlorprop, 30 ppb for chlorpyriphos, 2,4-D, etc. (Hamilton et al 2003)

Issues Related to Water Quality

The fact that today the quality of water is gradually deteriorating is one of the major concerns all over the world (Annan 2003). In India, several reports have been published regarding the presence of different contaminants in water from different states. Certain notable ones are listed below: (1) Contaminants like POPs in water from major parts of Delhi, Himachal Pradesh, Jharkhand and West Bengal (Khurana and Sen, 2008); (2) Presence of heavy metals like lead, mercury, cadmium and zinc in the water from the states of Gujarat, Andhra Pradesh, Kerala, Delhi and Haryana (Khurana and Sen, 2008); (3) Presence of arsenic not only in the states of West Bengal, Bihar, Jharkhand, Uttar Pradesh, Gangetic and Brahmaputra plains in India but also in the countries like Bangladesh, Myanmar, Canada, Nepal and Thailand (Kapaj et al. 2006; Smith et al. 2000); (4) Presence of fluoride in different states like Rajasthan, Assam, Gujarat, Karnataka, Madhya Pradesh, Andhra Pradesh, Maharashtra, Tamil Nadu, Tripura, Uttar Pradesh and West Bengal (Khurana and Sen 2008). Pollutants of various types are present in almost all the rivers and major water bodies throughout the country. For most of the studies related to water pollution in India, Shriram Institute for Industrial Research has been involved at both the state level and the national level, and these studies were undertaken mainly for the decisions support systems of the authorities for many of the policy decisions (Agarwal et al. 2010; Thakur et al. 2008).

In 2002, a report published by an NGO on the presence of pesticide residues in packaged drinking water in India led to a lot of controversy in the country resulting in the revision of IS specification for packaged drinking water with the most stringent limits for pesticide residues. Overnight, the MRLs for various pesticide residues were brought down from a level of several ppms to 0.1 ppb for the individual pesticide and 0.5 ppb for total pesticide residues. As a result, the quality norms for packaged drinking water were made at par with the EU specifications.

All these reports create the need for capability building projects at different levels and in different areas. The technical interventions need to be short-listed in order to achieve the desired standards of water quality to ensure health safety of the consumers. In order to appreciate the technical interventions needed in this regard, it is essential to mention about the issues associated with water quality problems in India.

Issue I

Inadequate treatment of human and livestock waste, inappropriate treatment and disposal of industrial waste, unsafe solid waste discharge are the major causes of contamination (Khurana and Sen 2008; Keast and Johnston 2008).

Issue II

Since some pesticides are persistent and have the tendency to accumulate over a period of time, they percolate and thus contaminate both surface and ground water. The excessive uses of pesticides for agricultural as well as non-agricultural uses have become the cause of contamination of pesticide residues in water (Khurana and Sen 2008). Today, the use of pesticides in India per hectare of land may be much lower than even the world average, but the level of residues of pesticides in water is relatively very high because of the poor agricultural practices (Keast and Johnston 2008).

Issue III

Hydrocarbons and solvents in industries, improper handling or disposal of petroleum oils are the sources of water contaminants. Another emerging water quality problem is contamination due to heavy metals. Major heavy metals, which are of concern today, are lead cadmium, arsenic and mercury which are highly toxic for human health (Keast and Johnston 2008; Momodu and Anyakora 2010).

Issue IV

Chemicals and additives used for water treatment cause water contamination. The residues of some of them may be inadvertently retained in water, and some may interact with naturally occurring chemicals leading to the formation of by-products, for example, chlorate, chloramines, chloroform, etc., which in turn leads to contamination. Other sources of water contamination include dumping of garbage and dead bodies into the river, cattle wading, immersion of idols, bathing and washing clothes in river, etc. (Keast and Johnston 2008).

It will be worthwhile describing about the three major contaminants, that is, pesticides, persistent organic pollutants (POPs) and poly chlorinated biphenyls (PCBs), here.

Pesticides

Pesticides are compounds to prevent, control, repel, attract, destroy any pest organism or any undesired plant species which can cause harm during production, processing and storage of food products, etc. in one way or the other. They are also administered to some animals for the control of insects on their bodies. Pesticides are toxic compounds mainly organic in nature that are manufactured and deliberately released into the environment to combat pest organism and disease causing agents (UNDPI 1997; Chung 2000; Zacharia and Tano 2011).

Classification of Pesticides

Pesticides can be classified on the basis of their (1) chemistry, (2) functionality and (3) toxicity (Zacharia and Tano 2011).

Classification on the Basis of Chemistry

Each pesticide has a unique chemical structure, and those having a similar structure exhibit similar properties including mode of their action. Pesticides are mainly of two types: (1) inorganic and (2) organic (Zacharia and Tano 2011). Inorganic pesticides are the different salts of metals and are lesser in number than the organic pesticides. Organic pesticides are the organic molecules with different functional groups in the form of chain, branched or ring. They can be natural (derived directly from plant species) such as pyrethrin, captan, neem and nicotine or synthetic (man-made). Synthetic pesticides are more in number than the naturally occurring pesticides (Zacharia and Tano 2011). Synthetic pesticides are classified into four major categories: organochlorines, organophosphates, carbamates and pyrethroids. Organochlorines are basically the chlorinated compounds which are highly persistent in nature. They were the first synthetic organic pesticides to be used in agriculture and in public health. Organophosphates contain a phosphate group in their basic structure and carbamates contain a carbamate group. Unlike organochlorines, organophosphates and carbamates get easily decomposed in the environment by various chemical and biological reactions. Thus, both these types of

Table 20.9 Classification based on toxicity of technical pesticides

Class	LD_{50} for the rat (mg/kg body weight)				
		Oral		Dermal	
		Solids	Liquids	Solids	Liquids
Ia	Extremely toxic	5 or less	20 or less	10 or less	40 or less
Ib	Highly toxic	5–50	20–200	10–100	40–400
II	Moderately toxic	50–500	200–2,000	100–1,000	400–4,000
III	Slightly toxic	Over 500	Over 2,000	Over 1,000	Over 4,000

Terms 'Solids and liquids' indicate physical state of active ingredient of pesticides

pesticides are not considered as persistent in the environment. Pyrethroids are basically synthetic analogue of naturally occurring pyrethrins (Patnaik 1997; Porto et al. 2011).

Classification on the Basis of Function

Pesticides can also be classified according to the functions they perform such as fungicides, weedicides, herbicides, rodenticides and insecticides (Zacharia and Tano 2011).

On the Basis of Toxicity

Classification of pesticides is also done on the basis of toxicity of the pesticides. Based upon LD_{50} values, pesticides are classified into four major classes: Ia (Extremely toxic), Ib (Highly toxic), II (Moderately toxic) and III (Slightly toxic) (Table 20.9) (WHO 2004).

Physico-Chemical Properties of Pesticides

A pesticide introduced into the environment is distributed in air, water, soil and biota (living organism). The amount of pesticide that goes into each component of environment is governed by its physico-chemical properties which indicate the behaviour of pesticides, their suitability for application, mode of action, dosage and toxicity. Some important properties are mentioned in Table 20.10.

Processes that a Pesticide Undergoes When it is Released into the Environment

Once a pesticide is released into the environment, it undergoes a series of processes also called as chemodynamics of pesticides depending upon various environmental factors such as pH, temperature, moisture, precipitation, salinity and light intensity. The chemodynamic processes determine persistence and the fate of the pesticide in the environment. The pesticides are either adsorbed, transferred or transported or are degraded or transformed (Env. Protection Agency 1994).

Adsorption

Pesticides may get adsorbed onto the soil particles, depending upon the type of soils and chemistry of pesticides. Soils with more organic matter will adsorb the pesticides more firmly than the soil which does not contain organic matter such as sandy and coarse soil. This is due to the fact that organic matter has more sites available to bind. Further, dry soil will absorb more pesticides as compared to the wet soil.

Transfer/Transport

It is important that the pesticide, after its application, is delivered or transferred to the target site instead of getting drifted away which can

Table 20.10 Physico-chemical properties of some pesticides

Pesticides	Physical state	Molecular weight (g/mole)	B.P./M.P (°C)	S.G./Density at 20 °C	Solubility in water (g/l)	Partition coefficient, K_{ow} log P	Toxicity, LD_{50} (mg/kg)	Vapour pressure, mPa, t °C
Organophosphates								
Acephate	S	183.2	88–90	1.35	790	−0.89	1,447	0.226 (24)
Chlorfenvinphos	L	359.6	167–170	1.36	0.121	3.85	10	0.53
Chlorpyriphos	C	350.6	42–43	1.44	Insoluble	4.7	135–163	2.7
Chlorpyriphos methyl	C	322.5	45.5–46.5	1.64	0.0026	4.24	>3,000	3
Diazinon	–	304	–	1.11	0.06	3.30	1,250	1.2×10^7
Dichlorvos	L	221	234	1.42	18	1.9	50	2.1×10^3
Dimethoate	S	229.3	49–52	1131	39.8	0.704	387	0.25
Ethion	L	384	164–165	1.22	0.002	4.28	208	0.20
Fenthion	L	278.3	90	1.25	0.0042	4.84	250	1.4
Malathion	L	330.36	39.8	1.23	0.145	2.75	1,375–5,500	5.3 (30)
Methamidophos	C	141.1	45	1.27	>200	−0.8	15.6	4.7
Monocrotophos	C	223.2	125	1.22	Soluble	−0.22	18	0.29 (20)
Phenthoate	C	320	17–18	1.226	0.01	3.69	270	5.3 (40)
Phorate	L	260	118–120	1.16	0.05	3.92	3.7	85
Phosalone	C	367.8	46.9	1.33	0.0014	4.01	120	0.00777 (20)
Phosphamidon	L	299.7	162	1.21	Miscible	0.79	17.9–30	2.2
Profenophos	L	373.6	100	1.455	0.028	4.44	358	0.124
Temephos	L	466.5	30	1.32	Insoluble	4.91	4,204	0.008
Terbufos	L	288.4	69	1.11	0.0045	2.77	1.6	34.6
Quinalphos	C	298.3	31–32	1.23	0.0178	4.44	71	0.346 (20)
Omethoate	L	213.2	–	1.32	Miscible	−0.74	25	3.3 (20)
Triazophos	L	313.3	>140	1.24	0.039	3.34	57–59	0.39 (30)
Carbamates								
Aldicarb	S	190.3	98–100	1.20	4.93	1.15	0.93	3.84 (24)
Carbaryl	C	201.2	142	1.232	0.12	1.85	264	0.041 (23.5)
Carbofuran	C	221.3	153–154	1.18	320	1.52	8	0.072
Carbosulfan	L	380.6	–	1.054	3 ppm	5.4	250	0.0358

Table 20.10 (continued)

Pesticides	Physical state	Molecular weight (g/mole)	B.P./M.P (°C)	S.G./Density at 20 °C	Solubility in water (g/l)	Partition coefficient, K_{ow} log P	Toxicity, LD_{50} (mg/kg)	Vapour pressure, mPa, t °C
Chlordane	L	409.8	106–107	1.59–1.63	0.1 mg/l	6.0	133–649	1.3
Ferbam	S	416.5	180	0.6	0.13	−1.6	>4,000	Negligible
Methomyl	C	162.2	78–79	1.29	57.9	0.093	34	0.72
Mancozeb	S	271.2	172	1.99	0.0062	0.26	>5,000	<0.0133
Propoxur	C	209.2	90	1.17	1.75	1.56	50	2.8
Thiodicarb	C	354.5	172.6	1.47	Insoluble	1.62	66	0.0027
Thiram	S	240.4	144–146	1.36	0.0165	2.1	3,700	0.02
Zineb	S	275.8	Decompose at 157	NA	0.01	≤1.3	>5,200	<0.01 (20)
Ziram	S	305.8	246	1.66	0.00097–0.0183	1.65	2,068	0.018
Pyrethroids								
Allethrin	L	302.4	281	1.01	Insoluble	4.96	2,150	0.16 (21)
Cyhalothrin	L	449.9	Doesn't boil	1.25	Insoluble	6.9	166	0.0012 (20)
λ-cyhalothrin	S	449.9	49.2	1.33	Insoluble	7	79	0.0002 (20)
Cyfluthrin	C	434.3	Decompose > 220	1.28	Insoluble	6.0	500	0.00096
β-cyfluthrin	C	434.3	81	1.34	Insoluble	5.9	380	1.4×10^{-5}
Cypermethrin	C	416.3	61–83	1.24	Insoluble	6.6	250–4,150	0.0002
α-cypermethrin	C	416	81.5	1.28	Insoluble	6.94	57	0.023 (20)
Cyphenothrin	L	375.5	154	1.08	Insoluble	6.29	318	0.12 (20)
δ-methrin	C	505.5	100–102	0.55	Insoluble	4.6	87–5,000	1.24×10^{-5}
Fenvalerate	L	419.9	39.5–53.7	1.175	Insoluble	5.01	451	0.0192 (20)
Permethrin	L	391.3	34–35	1.29	Insoluble	6.1	430–4,000	0.0029
Prallethrin	L	300.4	313.5	1.03	0.008	4.49	640	<0.013
Organochlorine								
DDT	C	354	108.5–109	1.56	Insoluble	–	113-118	0.025 (20)
Endosulphan	C	406.9	>80	1.8	Insoluble	4.74	70	0.83 (20)
γ-HCH	C	290.8	112.8	1.88	Insoluble	3.5	88–270	4.4 (24)

Note S Solid, *L* Liquid, *C* Crystalline, Melting point for crystalline and solid pesticides is given, whereas boiling point for liquid pesticides is given, *S.G* Specific gravity, *NA* Not available

lead to the contamination of ground and surface water besides being the potential hazard to the non-target species. The pesticides can get transported through the following ways: *Volatilization*: Pesticides with high vapour pressure get volatilized under hot climate, thereby resulting in reduced pest control. Therefore, it is important that pesticides are applied when the environmental conditions are favourable; otherwise, pesticides will not affect target pests rather they will affect non-target species. *Run-off*: After the application of pesticides onto the crops when water moves onto the surface with high speed, it carries away pesticides with it. This process is called as run-off. This leads to surface and ground water contamination and can cause harm to livestock or humans. *Leaching*: This results in pesticide residues reaching the water bodies as they move through the soil along with water. *Absorption*: For the efficacy of the pesticide, it is important that the desired amount of pesticides is delivered to the targeted site. The various phenomena that come into play during the process are driven by the physico-chemical properties of the pesticides. Logically, the pesticide with a high density would have greater tendency to remain adsorbed as well as absorbed in the pores of the soil in comparison with the ones with lower density but the extent of adsorption and absorption would be determined by the hydrophobicity of the pesticides. Pesticides with high water solubility will have less tendency to remain absorbed to the soil surfaces than those with low water solubility. The melting point and boiling point of pesticides would determine the transfer or transport of the pesticide from the site of application due to volatilization. Liquid pesticides would have more tendency to volatilize than the solid ones. Similarly, the pesticides having high vapour pressures would exhibit greater tendency of volatilization (Chung 2000). Looking at the physico-chemical properties (Table 20.10) of the pesticides, one can therefore envisage the life cycle of pesticides applied. All this is important for the analyst to understand while planning to develop a method to estimate the amount of residual pesticides in various matrices including water.

Degradation/Transformation

Degradation of pesticides is the breakdown or chemical transformation of pesticide molecules into other forms or its metabolites that can be either more or less toxic than the parent compounds. For example, degradation of DDT leads to the formation of DDD, which is itself a highly toxic pesticide as compared to DDT. Formation of iso-malathion from malathion is another example of pesticide becoming more toxic on degradation. Some of the metabolites of phorate are known to be more toxic than phorate itself. Degradation can take place by three processes: (1) Chemical, (2) Microbial/Biological and (3) Photodegradation (Vargas 1975; Chung 2000).

Chemical degradation: It occurs in water or atmosphere due to one or more of the following processes: oxidation, reduction, hydrolysis and photolysis.

Microbial/biological degradation: Microbial degradation of pesticides occurs either partially or completely depending upon the presence (aerobic) or absence (anaerobic) of oxygen. Aerobic degradation takes place on the soil surface and anaerobic degradation takes place in waterlogged soils or in the bottom sediments of ponds, lakes, rivers, etc. Thousands of different bacterial species are known to exist in a single gram of fertile soil which cause the degradation to occur. Some microbes also produce enzymes, which also assist in degradation. Most of the time, multiple organisms are involved in the degradation phenomenon. The ability of microbes to degrade a pesticide is related to their metabolic capacity and the complexity of the molecule and to environmental factors (water content, temperature, aeration and nutrient) that regulate microbial activity (Vargas 1975; Porto et al. 2011).

Photodegradation/photolysis: Continuous exposure of pesticide molecules to sunlight leads to the absorption of photons by the molecule causing them to get excited and resulting ultimately in breaking of chemical bonds. There are certain moieties on the pesticide molecules, which are vulnerable to degradation. The photodegradation gets accelerated in the presence of certain metallic compounds as well as under the extreme environmental conditions. Generally, it occurs more at the surface of soil than in the hidden areas. Almost all pesticides undergo photodegradation with varying extent. Pyrethroids, for example, are known for their high tendency of photodegradation. Persistent organic pollutants, however, undergo photodegradation with an extraordinary slow rate and the extent of degradation is very low (Vargas 1975; Katagi 2004).

Photodegradation can also be facilitated by the presence of certain highly reactive compounds like ozone, hydrogen peroxide, etc. In many cases, even the free radicals like H^+ and OH^- due to the absorption of photons by water molecules can also help in enhancing the rate of photodegradation of pesticides (Vargas 1975; Katagi 2004).

Persistence of Pesticides

Persistence of pesticides is the major cause of concern from the point of view of all—environmental safety, health safety and food safety—and is responsible for the safety of a pesticide. Persistence is routinely expressed in terms of 'half-life' of a compound which means the time taken for degradation of a compound to the level of 50 %. The half-life can alternatively be defined also in terms of loss of activity of pesticides. The time taken for bringing down the original activity to 50 % level is referred to as half-life of a pesticide. Typically, a half-life based on degradation of a pesticide is longer than a half-life based on deactivation. The Table 20.11 shows grouping of pesticides based on persistence (based on deactivation and not degradation) in soil (Porto et al. 2011; Tomlin CDS 2009).

Significance of Persistence of Pesticides

Persistent pesticides remain accumulated in the environmental systems as well as in the food chain including water because of the fact that they do not degrade easily even when ingested along with the food. Their tendency to persist in the human body has been the cause of major concern. All organochlorine pesticides like DDT have been established as highly persistent in nature. This is also a fact that highly persistent chemicals are carcinogenic in nature. When evaluating the potential of a particular pesticide

Table 20.11 Persistence levels of various pesticides in soil

Non-persistent (half-life $\leq$ 30 days)		Moderately persistent (half-life > 30 days < 100 days)		Persistent (half-life $\geq$ 100 days)
Herbicides				
Alachlor	Dinoseb	Acifluorfen	Metachlor	Bromacil
Bifenex	EPTC	Atrazine	Metribuzin	Fluchloralin
Butylate	Propachlor	Bentazon	Oryzalin	Paraquat
Chlorpropham	Tridiphane	Chlramben	Pendimethalin	Picloram
Cyanazine	Vernolate	Ethalfluralin	Simazine	Terbacil
Dicamba	2, 4-D	Glyphosate	Triclopyr	
		Linuron	Trifluralin	
Insecticides, fungicides and nematicides		Insecticides, nematicides		Insecticides
Aldicarb	Dizonon	Carbaryl	Endrin	Chlordane
Captan	Malathion	Carbofuran	Fonofos	Dieldrin
	Phorate	Chlorpyriphos	Parathion	Heptachlor
				Lindane

to contaminate the environment, it is essential to consider its partition coefficient (PC) and half-life jointly. For example, a pesticide with small PC value (less than 100) and a long half-life (more than 100 days) poses a considerable threat to ground water through leaching. Pesticides with large PC value (1,000 or more) and a long half-life are likely to remain in soil and increase the chances to be carried away by run-off to lake or streams. Pesticides with short half-life (less than 30 days) pollute the water only when heavy rain or irrigation occurs soon after the application. In terms of ground water quality, pesticides with intermediate to large partition coefficient values and short half-lives may be considered 'safe' because they would not leach into the ground water easily and they would degrade fairly rapidly (Oregon State Univ. 2007).

The Table 20.10 shows ground water contamination potential of pesticide by leaching according to partition coefficient value and half-life. It may be noted here that the solubility of the pesticide in water plays a key role in determining the following: (1) partition coefficient, (2) degradation, (3) halflife, (4) adsorption on the soil surface, (5) absorption by plants and (6) accumulation in the system. It is always advisable to opt for a pesticide, which has high solubility in water. However, highly soluble pesticide would leach out quickly into the ground water, and hence, a balanced approach is always advisable. Soluble pesticides with small partition coefficient values and moderate to long half-lives are the least effective in protecting ground water because they can leach easily and will degrade slowly.

Scenario of Pesticide Residues in India

Pesticides are never applied as such. In fact, they are always formulated so that they can be applied safely for the desired applications. There can be several formulation types for a given pesticide. This means that there can be several pesticide products registered for a given pesticide. As per the latest pesticide manual, total numbers of pesticides developed worldwide are 1,436. In India, the registration of pesticides is regulated by Central Insecticide Board (CIB). As per CIB, 181 pesticides have been registered for use.

India is one of the leading manufacturers of pesticide formulations in the world. In fact, India is also a leading exporter of pesticide formulations. Although the consumption pattern of pesticides is much less as compared to the other countries in the world, the residues of pesticides observed on the different agri-produce are much higher as compared to that found in the other parts of the world. In order to capture the global food trade, there has always been a caution in the use of pesticides on crops such as grapes. The use of banned pesticides is strictly regulated to minimize the residue contamination of pesticides in food including water. The export of marine products demands for strict regulation of water quality from the point of view of contamination of residual pesticides. Since India is a WTO member, it needs to comply with the regulations as per Codex.

Persistent Organic Pollutants

They are organic compounds or mixtures, which include industrial chemicals, pesticides and undesired wastes or by-products. As the name suggests, these compounds are persistent in nature and are highly toxic pollutants. Industrial chemicals in the list of persistent organic pollutants include PCBs and various pesticides like aldrin, dieldrin, DDT, endrin, chlordane, heptachlor, mirex, HCB and toxaphene and the undesired/accidental by-products such as dioxins and furans (Crump 2000; Sanghai 2001; US EPA 2009).

Chemistry of POPs

All POPs are halogenated compounds, the halogen moiety being mostly the chlorine atom. Since chlorine is a heavier atom and C–Cl bond is very strong, it is very difficult to break the bond during hydrolysis. Moreover, as the number of chlorine atom increases in the structure, possibility of degradation becomes even lesser. By virtue of their higher number of chlorine atoms, the POPs are less soluble in water, tend to bioaccumulate more in the fatty tissues, and persist forever. They are semi-volatile compounds, due to which they are transported from one place to another. The different physico-chemical properties of POPs are given in Table 20.11.

Polychlorinated Biphenyls

Polychlorinated biphenyls (PCBs), commonly known as PCBs, are oily liquids or solids, clear to yellow in colour, with no smell or taste. Their presence in environment has been through both use and disposal. For several decades, PCBs were routinely used in the manufacture of wide variety of common products. PCBs find wide range of applications due to their unique and attractive properties such as better conductivity, high dielectric constant, very high breakdown voltage, good resistance to alterations in temperature, non-flammability, high boiling point, chemical stability and lower water solubility. PCBs are highly thermally and chemically resistant, that is, they do not readily break down when exposed to heat or chemical treatment which is one of the desirable requirements for lubricants. However, all the special characteristics of PCBs were later offset by the findings that they are hazardous, less biodegradable and persistent in ecosystem. It was during the 1960s that residues of PCBs were found present in air, water, soil, industrial waste, etc. beyond the permissible limits. Studies were undertaken and it was found that some congeners of PCBs degraded very slowly and have the potential to get into the food chain resulting in the build-up of their concentrations at the levels, well beyond the permissible limits. Though PCBs were banned completely all over the world in 1977, due to their persistent nature, they do exist in different products and their monitoring in different

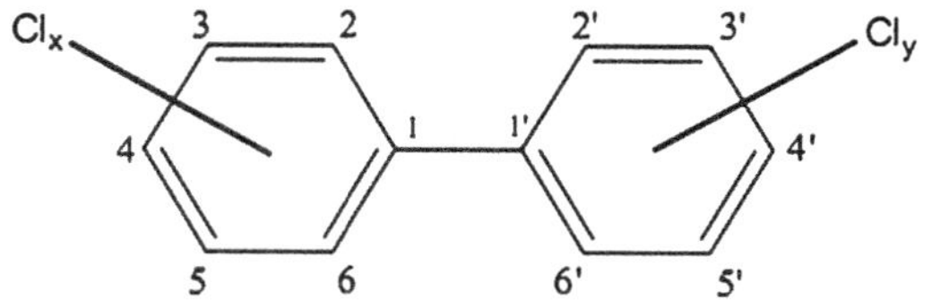

Fig. 20.1 General structure of PCB

consumer products has become a mandatory requirement as per the various regulatory authorities all over the world (Crump 2000; Patnaik 1997).

Structure of PCBs

Basic chemical structure of PCB is shown in Fig. 20.1. The chemical formula of PCBs is $C_{12}H_{10-n}Cl_n$, where n ranges from 1 to 10. They make up a group of 209 individual chlorinated biphenyl rings known as congeners.

Toxic Effects of PCBs

Toxicity of PCBs depends upon the following: (1) Position of chlorine atoms and (2) Number of chlorine atoms. PCBs with chlorine in both the para positions are more toxic than the ones with chlorine in ortho positions. 3,5,3′,5-tetrachloro-biphenyl and 3,3′,4′,4′,5-pentachlorobiphenyl exhibit high toxic effects. Fire, heat or uncontrolled incineration produce highly toxic degradation products, for example, polychlorinated dibenzofurans (PCDFs) and polychlorinated dibenzo-p-dioxins (PCDDs).

Sources of PCB Contamination

PCBs are either produced commercially or formed as by-products in combustion processes as thermodynamically stable compounds. The major sources of PCBs in drinking water are run-off from landfills and discharge of waste chemicals. Some private wells may use old submersible pumps that contain oil with PCBs. If the pump seal fails, PCBs can leak into the well and contaminate the drinking water. They are found principally in water supplies contaminated by transformer oils, in which PCBs have been used as a heat exchanger medium.

Physico-Chemical Properties of PCBs

PCBs are non-flammable and extremely stable compounds having high electrical resistance and good insulating properties. As the number of chlorine atoms in a PCB mixture increases, the flash point rises and the substance becomes less combustible. Also PCBs with large number of chlorine atoms are more stable and thus resistant to biodegradation. Most of the PCBs have poor solubility in water and the solubility decreases with increase in chlorine content. PCB compounds with high aqueous solubility can easily be degraded by micro-organisms than those with low solubility, and highly chlorinated congeners of PCBs are weakly soluble in water and hence resistant to biodegradation. PCBs are highly soluble in fats and lipids. They are insoluble in glycerol and are readily soluble in lower alcohols. The vapour pressure of PCBs depends on degree of chlorination and decreases with increase in chlorination. The persistence of PCBs is a consequence of stable fused aromatic ring structure and low aqueous solubility. Bound or non-extractable residues may arise from either chemical or physical interactions of PCB molecules with soil, hydrophobic sorption, charged transfer complexes or van der Waals forces. They are not subjected to hydrolysis under any condition, nor do they become rancid. They are very resistant to oxidation. Unlike many other synthetic materials, they are not subjected to polymerization.

Degradation of PCBs

PCBs are extremely stable compounds, not reactive chemically under environmental conditions. However, they are inert to hydrolysis except under extreme conditions. They do not undergo oxidation, reduction, nitration, isomerization and nucleophilic reactions. However, PCBs get degraded in the presence of sunlight (photodegradation) and also undergo bio-degradation, loss of chlorine with replacement by hydrogen or hydroxyl groups, and rearrangement, condensation and polar products have been observed. One of the reaction products of sunlight irradiation of PCB mixtures is the replacement of chlorine by hydroxyl groups at the ortho positions. This allows oxygen to bind in a similar position on the other ring and to produce chlorodibenzofurans (CDFs). Both heat and light can accelerate the transformation of PCBs to CDFs. During the biodegradation, bacteria or other micro-organisms act upon PCBs and cause aerobic oxidative dechlorination or hydrolytic dehalogenation. Theoretically, the biological degradation of PCBs results to give CO_2, chlorine and water. The process involves the removal of chlorine from the biphenyl ring followed by cleavage and oxidation of the resulting compound. It is the only process known to degrade PCBs in soil systems or water.

Technological Intervention for Quality of Drinking Water

The regulators fix the parameters and specify the acceptable limits for them before going about ensuring their compliance. The supply of drinking water has to conform to the limits as specified by the regulators without any exception. A number of research studies have to be conducted to provide the necessary justification for the specified limits, toxicity studies to produce the toxicity dossier and life cycle studies for chemical dossier for different contaminants. The basic philosophy that all the consumers must be supplied with safe drinking water will have to prevail across the continents. In India, the task of meeting the desired standards of quality for the water supplied in rural areas across the country is much more complex than fixing the limits of MRLs for various residues.

When it comes to the supply of clean drinking water, the policy makers and the authorities responsible for this purpose will have to seek for technological interventions at various levels and various stages. Technological interventions will also be necessary even for the implementation of mandate (of codex) pertaining to harmonization of standards. The areas where technological interventions would be needed are listed here below: (1) *Development and validation of analytical methods:* The technological capability to be able to develop and validate analytical methods for residue contaminants at such low levels has to be created at local levels across the country. The competency of identifying as well as quantifying the presence of contaminants at trace levels would be the key for this purpose. Further, the expertise to interpret the results obtained from the sophisticated state-of-the-art analytical equipment will have to be acquired to meet the requirements. Today, the necessary facilities as well as the needed expertise are way behind those existing elsewhere in the developed world. The task of matching with the leading countries is certainly going to be a daunting one. (2) *Quality system and accreditation:* In order to be able to stand in league with the leading countries in the world, the first and foremost action will have to do with the establishment of quality systems and accreditations. Unless the quality systems are at par with those needed for the purpose of achieving the higher standards of quality of water, the plans pf provided drinking water to the consumers will remain unfulfilled. The quality systems will have to be accredited according to the best of the international standards. In future, the technical facilities involved with the process of certification of water quality will have to be necessarily GLP compliant. However, the immediate objective to be able to obtain accreditations as per ISO/IEC 17025 which is the minimum desired standard would serve the purpose. During the last five years, there has been an increase in the number of facilities accredited as per ISO/IEC 17025 but the number of laboratories and institutes yet to acquire accreditations is extraordinarily higher. Interventions, therefore, will be necessary not only in the technical front but also on the policy front. (3) *Bio safety studies:* Since the MRL values for the contaminants are jointly decided at the Codex level, all the member countries of WTO must have technical capabilities to be able to carry out biosafety studies on different types of contaminants because the advanced countries have been lowering down the MRL values in the recent times to such low levels that it will become difficult for countries like India to capture the global market of food. If the quality drinking water is the issue, then perhaps this particular aspect of capability building may not be that relevant. However, for the export of marine foods, the stringent specifications of drinking water would become an inhibiting factor for the Indian exporters. At times, the regulators of certain countries, for example, EU, include the contaminants like pesticides, which are not even in use in India. To counter all such matters, the technical interventions pertaining to safety studies by generating detailed toxicity data, in the GLP compliant facilities, would have to be given top priority. (4) *Assuring quality:* Technological interventions would become important when it comes to the assurance of quality of the water supplied by the authorities as well as by the suppliers of packaged drinking water and mineral water is concerned. At the local level, the consumers will have to be provided with the tools such as WFTKs, so that they can assure themselves of the quality of the water before they consume it. Development of state-of-the-art WFTKs would be the first and foremost technological interventions needed for the people depending upon untreated ground water as well as surface water. Assuring the quality of packaged drinking water and mineral water, it will be relatively easier because it will be possible through the capability of method development and validation as described above. The present situation regarding the WFTKs has lot of gaps, when it comes to detection level of contaminants. Thus, this will demand for capability building. On the other hand, assuring quality for packaged and mineral water exists in the country, only need for capacity building in this case.

Development of Sensors

While water quality field test kits can be a good tool for the consumers to differentiate between dirty water and clean water, the presence of many of the residues at trace levels go undetected. For this, it is essential to develop sensors which not only can detect the contaminants but they can also quantify them at levels of their MRL values. Biosensors for heavy metals, pesticides, POPs and microbial contaminants are being developed to help authorities ensure the quality of water. In India, the development of biosensors would be the main area requiring technological intervention.

Water Purifiers

With time, the regulators have been getting stringent about the quality parameters that determine the safety of consumers. At the same time, the consumers have also been becoming more aware about the quality of products (e.g. water) meant for human consumption. The technical interventions are needed to eliminate the impurities from water. The undesirable impurities like heavy metals, pesticides, microbes, etc. must be removed to ensure the availability of safe drinking water. In recent times, several technological breakthroughs have been responsible for exploitation of different types of filtration systems of varying capacities. The use of reverse osmosis systems to remove heavy metals and the microbial pathogens has become popular especially for areas where ground water is being consumed. In this field, satisfactory progress has been made but much more remains to be done, yet superior systems for desalination and for purification of surface water needs to be developed. Moreover, the indigenization of the reverse osmosis systems specially the membranes used for this purpose should be the focus for development (Ahmed et al. 2010; Boussahel et al. 2000; Chian and Fang 1975; Mara and Horan 2003; El-Nashar 2012).

Wastewater Treatment

Efficient and environment-friendly technologies are needed to treat the wastewater from different industries. Industries generating high quantities of wastewater need such technologies more than others. The notable ones include the textile processing, metal plating, agrochemicals, dyes, paints, etc. In textile processing, the wastewater is contaminated with dyes. Often, the dyes which are non-biodegradable percolate down to the ground water. The ground water in many areas around the industry clusters of textile processing has already become coloured. The heavy metals in the ground water as well as in the surface water are attributed to the contamination due to the wastewater from the metal plating industry. In one of the studies conducted by the scientists of SRI, it was observed that heavy metals absorbed in the sediments of the river (Yamuna) basin between Delhi and Agra are responsible for the death of crocodiles in the crocodile park at the confluence of the rivers Yamuna and Chambal. The pesticide residues in the ground water in the area around the clusters of agrochemical industry are a well-known fact.

To resolve all such complex problems of contamination of drinking water, one needs the technological interventions. The treatment of effluents from all these polluting industries should be done by adopting state-of-the-art technologies in an environment-friendly manner. This will not only help recover the good quality water from the wastewater but it will also ensure that the contamination of the water resources is avoided.

Sewage Treatment

One of the most important areas where the technological interventions are urgently required pertains to the sewage treatment. Sewage treatment, if done efficiently and properly, can be a good source of water. Already, the water recovered from the sewage treatment plants is being used for these applications such as toilets and agriculture. The key aspects of recovery of water from sewage treatment plants are as follows: (1) the quality of water and (2) the quality of the sludge. The quality of the water from sewage treatment before it can be disposed off into river has already been specified. Similarly, the sludge from the sewage treatment must also comply with the standards of hygiene. Technological interventions like bioremediation, radiation processing, etc. are to be developed and exploited for achieving the desired standards of sanitation and hygiene.

Path Forward

In the end, it can be said that the sources of water are limited and they need to be managed well to take care of the needs of the society as well as industry. The requirements would be rising but the available sources of water would not. This means that there would be continuous depletion of reservoirs, unless they are replenished on a regular basis. It must be kept in mind that there is sufficient water on this planet to support life on a sustainable basis. However, if the efforts are not made towards the conservation, preservation, waste recovery and avoiding wastage, the situation would get worse!

The policy interventions, technological interventions and the societal interventions would be needed for the sake of availability of quality water for the sustainability of life on this earth.

Acknowledgments The authors are highly thankful to the senior management of Shriram Institute for Industrial Research for their support and valuable technical guidance and permission to publish this chapter. The authors are also thankful to Mr. Vinay Jain who provided all his support in preparation of this chapter.

References

Agarwal A, Pandey RS, Sharma B (2010) Water pollution with special reference to pesticide contamination in india. scientific research. J. Water Resour Prot 2:432–448

Ahmed T, Rafatullah M, Ghazali A, Sulaiman O, Hashim R, Ahmad A (2010) Removal of pesticides from water and wastewater by different adsorbents: a review. J Environ Sci Health C 28:231–271

Annan K (2003) Water for people water for life. In: United Nations educational scientific and cultural organization. UNESCO. The united nations world water development report. ISBN 9231038818. Brussels

Borchardt JA, Walton G (1971) Water quality. In: Water quality and treatment-a handbook of public water supplies. III (ed) The american water works association. Inc. McGraw Hill. USA. pp 1–51

Boussahel R, Bouland S, Moussaoui KM, Montiel A (2000) Removal of pesticide residues in water using the nanofiltration process. Desalination 132:205–209

Bureau of Indian Standards. Paper & pulp specification IS 2724:1964 Reaffirmed 1983. Amendment 1994. New Delhi. India

Bureau of Indian Standards. Ice manufacturing specification IS 3957:1966 Reaffirmed 1983. Amendment 2003. New Delhi. India

Bureau of Indian Standards. Processed food industry specification IS 4251:1967. Reaffirmed 1992. Amendment 2004. New Delhi. India

Bureau of Indian Standards. Tanning specification IS 4221:1967 Reaffirmed 1999. New Delhi. India

Bureau of Indian Standards. Fermentation specification IS 4700:1968. Reaffirmed 1994. New Delhi. India

Bureau of Indian Standards. Irrigation purpose specification IS 11624:1986 Reaffirmed 2001. New Delhi. India

Bureau of Indian Standards. Textile specification. IS 201:1992 Reaffirmed 2008, Amendment 2010. New Delhi. India

Bureau of Indian Standards. Storage batteries specification IS 1069:1993 Reaffirmed 2003. New Delhi. India

Bureau of Indian Standards. Swimming pool: recreation purpose specification IS 3328:1993. Reaffirmed 2003. Amendment 2010. New Delhi. India

Bureau of Indian Standards. Construction Purpose Specification IS 456:2000. Amendment 2005. New Delhi. India

Bureau of Indian Standards. Packaged drinking water specification IS 14543:2004 New Delhi. India

Bureau of Indian Standards. Packaged natural mineral water specification IS 13428:2005. New Delhi. India

Bureau of Indian Standards. Drinking water specification IS 10500: 2012. New Delhi, India

Central Pollution Control Board (2007) Status of groundwater quality in india Part–I. Groundwater quality series GWQS/09/2006-2007. Delhi. India

Central Pollution Control Board (2008a) Guidelines for water quality monitoring Technical Report Series No. MINARS/27/2007-08. Delhi. India

Central Pollution Control Board (2008b) Status of groundwater quality in india part–II. Groundwater quality series GWQS/10/2007-2008. Delhi. India

Central Pollution Control Board (2008c) Status of water treatment plants in india. Delhi, India

Chian ESK, Fang HHP (1975) Removal of pesticides by reverse osmosis. Environ Sci Technol 9(1):52–59

Chung KY (2000) Enhanced degradation of pesticides. Environ Eng Resour 5(1):47–61

Crump J (2000) Persistent organic pollutants: are we close to a solution. 26. Canadian arctic resources committee

Eaton AD, Clesceri LS, Rice EW, Greenberg AE (2005) Standard methods for the examination of water and wastewater, XXI edn. American Public Health Association, New York

El-Nashar AM (2012) Water and wastewater treatment technology—History and current status of membrane desalination process. In: Encyclopedia of life support systems. www.eolss.net. Tokyo. Japan. pp 1–12

Environmental Protection Agency (1994). Physico-Chemical Properties And Environmental Fate Of Pesticides. Technical Report Series No. EH 94-03. Sacramento. California

Environmental Protection Agency Handbook (2010) Standards for drinking water quality In: A handbook on the implementation of the regulations for water service authorities for public water supplies. ISBN 9781840953497. European communities (Drinking Water) (No. 2) Regulations 2007. Co Wexford. Ireland. pp 1–12 (Section2)

European Communities (1998) EU council directive: the quality of water intended for human consumption. Official Journal of the European Communities 98/83/EC No. L 330/32–54. Brussels

European Communities (2008) Ground water protection in Europe. ISBN 9789279098178. Brussels

Facts Green (2009) Facts on water resources: a summary of the united nations world water development report 2. Brussels, Belgium

Food and Agriculture Organization of United Nations. (1985) Water quality for agriculture: FAO irrigation and drainage paper. ISBN 9251022631. Rome, Italy

Hamilton DJ, Ambrus A, Dieterle RM, Felsot AS, Harris CA, Holland PT et al (2003) Regulatory limits for Pesticide residues in water. Pure Applied Chemistry 75(8):1123–1155

Hammer MJ (2009) Water and wastewater technology. VI (ed). ISBN 9788120334991. PHI Learning Pvt. Ltd., New Delhi. India

Health Canada (2010) Guidelines for canadian drinking water quality: summary table. Federal-provincial-territorial committee on drinking water of the federal provincial-territorial committee on health and the environment. Canada

Helmer R, Hespanhol I (1997) Water quality requirements In: Water quality control-a guide to the use of water quality management systems. II (Ed.). E & FN Spon. An imprint of Thomson Professional. Chapter-II. ISBN: 0419229108. London. United Kingdom

Jaiswal RK (2007) Ganga action plan-a critical analysis. http://www.primopdf.com

Kapaj S, Peterson H, Liber K, Bhasttacharya P (2006) Human health effects from chronic arsenic poisoning-a review. J Environ Sci Health Part A 41:2399–2488

Katagi T (2004) Photodegradation of pesticides on plant and soil surfaces. Environ Contam Toxicol 182:1–195

Keast G, Johnston R (2008) UNICEF handbook on water quality. United Nations Childrens Fund, New York, pp 1–191

Khandal RK (2010) A handbook on water quality evaluation and monitoring and mapping. SRI Publication, Delhi

Khurana I, Sen R (2008) Drinking water quality in rural india: Issues and approaches. Background paper. Water aid. India www.wateraid.org

Klemes J, Chair ECMC (2008) Environment technology and management In: Handbook of water and energy management in food processing. ISBN 1845691954. Woodhead publishing series food science. Technology and nutrition No.160. U.K

Mara D, Horen N (2003) Drinking water microbiology In: The handbook of water and wastewater microbiology. ISBN 0124701000. Academic Press, London, p 611–794

Ministry of Health, Labour and Welfare (2010) Water supply in Japan. Revision of drinking water quality standards. Hiroshi Wakayama. Japan

Momodu MA, Anyakora CA (2010) Heavy metal Contamination of Ground water: the surulere case study. Res J Environ Earth Sciences 2(1):39–43

Obreza T, Clark M, Boman B, Borisova T, Cohen M, Dukes M et al (2011) A guide to EPA's numeric nutrient water quality criteria for Florida. SL 316 s. Institute of food and agricultural sciences, Florida, USA

Oregon State University (2007) Understanding pesticide persistence and mobility for ground water and surface water protection. EM 8561-E. Corvallis. http://extension.oregonstate.edu/catalog/

Patnaik P (1997) Handbook of environmental analysis. Chemical pollutants in air. Water. Soil and solid wastes. Lewis Publishers CRC Press, Florida

Porto ALM, Melgar GZ, Kasemodel MC, Nitschke M (2011) Biodegradation of pesticides In: Stoytcheva M (ed) Pesticides in the modern world-pesticides use and management. InTech, New York

Rahaman MM (2006) The Ganges Water Conflict. Aasteriskos 1/2:195–208

Sanghai R (2001) Living in a chemical environment-persistent organic pollutants. resonance. July: 64–72. www.ias.ac.in

Shriram Institute For Industrial Research (2007) Technical Report Series No. Project No. ASD 215. Delhi. India

Smith AH, Lingas EO, Rahman M (2000) Contamination of Drinking Water by Arsenic in Bangladesh: A Public Health Emergency. Bull World Health Organ 78(9):1093–1103

US Geological Survey (2006) Ground water utilization and sustainability. Technical report series No. 060323. USA

Thakur JS, Rao BT, Rajwanshi J, Parwana HK, Kumar R (2008) Epidemiological study of high cancer among rural agricultural community of Punjab in northern India. Int J Environ Res Public Health 5(5):399–407

The International Programme on Chemical Safety (2001) Persistent organic pollutants. An assesment Report prepared by Ritter L, Solomon KR, Forget J, Guelph ON, Canada

Tomlin CDS (2009) The pesticide manual. XV (ed). ISBN: 9781901396188. BCPC. Hampshire. United Kingdom

United Nations Department of Public Information (1997) UN assessment of freshwater resources. Paper presented at the Earth Summit +5. Special session of the general assembly to review & appraise the implementation of agenda 21. DPI/1875/SD. New York. USA

United States Department of Agriculture (2003) National water quality handbook. Technical Report Series No. 450-VI-NWQH. Washington DC. USA

United States Environmental Protection Agency (2004) Water facts. Technical Report Series No. EPA 816-F-04-036. www.epa.gov/safewater

United States Environmental Protection Agency (2009) National recommended water quality criteria for priority pollutants. Technical Report Series No. Office of science & technology (4304T). New York. USA

University of Missouri-Columbia (2010) Pesticides and the environment. Agricultural MU guide. Insects and diseases. Technical Report Series No. G 7520, Columbia

U.S. Department of Agriculture (1989). Pesticides and ground water. Land and Water No 12. Illinois. Urbana

Vargas JM (1975) Pesticide degradation. J Arboric. Presented at international shade tree conference in Detroit, Michigan p 232–233, Dec

World Health Organization (2003) Emerging issues in water and infectious disease. ISBN: 9241590823. Geneva. Switzerland. http://www.who.int/water_sanitation_health

World Health Organization (2004) The WHO recommended classification of pesticide by hazard and guidelines to classification. International programme on chemical society. ISBN: 9241546638, Geneva, Switzerland

World Health Organization (2008) Guidelines for drinking water quality. Volume-I Recommendations. ISBN 9789241547611. Geneva. Switzerland

World Health Organization (2009) Guidelines for drinking water quality. Chapter-8. ISBN 9789241548151. Geneva. Switzerland

World Health Organization (2011) Guidelines for drinking water quality. Geneva. Switzerland

World Health Organization Regional Office for Europe and European Commission (2002) Eutrofication and health. ISBN: 9289444134. Luxembourg

Zacharia JT (2011) Identity, physical and chemical properties of pesticides In: Stoytcheva M (ed) Pesticides in the modern world-trends in pesticides analysis. InTech, New York

Pollution of Aqueous Matrices with Pharmaceuticals

21

Kriti Jindal, Mallikarjun Narayanam, and Saranjit Singh

Abstract

The world is witnessing an increasing contamination of the environment by pharmaceuticals due to their escalating consumption and recalcitrant nature. Water bodies like rivers, lakes and even surface water have been found to be contaminated with drugs. Exposure to these contaminants is already showing detrimental effects in fish, frogs, birds, etc., with the development of antibiotic-resistant pathogens being another repercussion. Chronic exposure to pharmaceuticals, even in trace quantities, may also affect human health adversely in the long term. Although a decade back it was difficult to provide substantial data for such pollution, the recent development of highly sensitive and specific analytical tools has led to the detection of pharmaceuticals in many drinking water sources also. In this chapter, we discuss various aspects of this issue, beginning with the causes of pollution, the enormity of consequences, types of aquatic sources reported to be contaminated, drugs usually found, and finally the variety of techniques that can be used to detect and characterize pharmaceuticals in aqueous matrices.

Keywords

Contamination · Environment · Pharmaceuticals · Pollution · Water

Introduction

Pharmaceuticals play a very important role in health care as they are manufactured and used for specific biological activities in humans and animals. However, in the last two decades, pharmaceuticals have also been recognized as environmental contaminants in various matrices, as corroborated by many studies (Richardson and Bowron 1985). It is obvious that a xenobiotic produced or consumed in large amounts

K. Jindal · M. Narayanam · S. Singh (✉)
Department of Pharmaceutical Analysis, National Institute of Pharmaceutical Education and Research, S. A. S. Nagar, 160 062, Punjab, India
e-mail: ssingh@niper.ac.in

P. P. Singh and V. Sharma (eds.), *Water and Health*,
DOI: 10.1007/978-81-322-1029-0_21, © Springer India 2014

would eventually make its way into the environment and may persist as a contaminant either in an intact form, as a degradation product, or a metabolite of the parent. It is not unreasonable to anticipate that many of the detected pharmaceutical contaminants may have been present in the environment for decades, but have come into notice only recently due to tremendous progress in the analytical techniques for trace analysis (Buchberger 2011). It is also very reasonable to assess that with incessant increase in production and consumption of pharmaceuticals with each passing year, the environmental pollution with pharmaceuticals is rising progressively.

The detection of trace levels of pharmaceuticals in environmental samples, viz. sewage water effluent, surface water, groundwater and even drinking water, was first reported from the USA and Europe in the 1970s (Garrison et al. 1976; Hignite and Azarnoff 1977). Since then, multiple studies have been conducted around the world to check for the presence of pharmaceuticals in various environmental matrices (Waggot 1981; Heberer et al. 1995). However, the extent of distribution is still not very clear, though contamination of aquatic sources is most explored and reported. The main concern regarding pharmaceutical contaminants is the threat they pose to the aquatic flora and fauna, though their long-term eco-toxicological consequences, especially to humans, are still unfathomable. This chapter provides a comprehensive view of the extent of the problem of pollution of aquatic sources with pharmaceuticals, types of drugs found as contaminants, types of matrices found to be contaminated, the reported harmful effects, and the analytical techniques used for the detection as well as the quantitation of pharmaceuticals in aqueous matrices.

The Reasons for Pollution of Aquatic Sources with Pharmaceuticals

As indicated above, the problem of pollution of aquatic sources with pharmaceuticals is most noteworthy as the same is of continuous nature. The major reasons for this are enumerated below:

Consumption on a rise: Consumption of pharmaceuticals is increasing rapidly because of the emergence of new diseases and rise in consumer population due to enhancement of life span (older people consume more drugs). Improvements in living standards and hence better affordability of drugs, complemented with new drug discoveries and the progress of pharmaceutical companies, are also contributing towards increased consumption (Behnke 2012).

Continuous infusion in the environment: Continuous infusion of a pollutant into the aquatic environment is sufficient for effecting incessant, multi-generational life cycle exposures to sensitive aquatic species. Chemical stability is not an issue as long as the pollutant is continually introduced (such as via sewage treatment plant effluent). Current criteria for establishing the importance of pollutants, based on chemical stability as measure of tenacity, may be overlooking entire ranks of potential pollutants, for example, pharmaceutical and the bioactive ingredients in personal care products (Daughton 2000).

Diverse nature of pharmaceuticals: Pharmaceuticals and personal care products (PPCPs) are a very large and diverse suite of pollutants. Many of them may be present together at the same time in an aquatic body, making it difficult to carry out their monitoring.

Inefficient sewage treatment works: Modern sewage treatment works are not constructed to specifically eliminate pharmaceuticals from potable water supplies. Due to their specific chemistry, each pharmaceutical requires an exclusively designed degradation technique for its complete destruction. As waste management plants use generalized degradation techniques, therefore, many pharmaceuticals are not completely destroyed (Vera-Candioti et al. 2008). Also, the presence of antibiotics in wastewater treatment works (WWTWs) influents could disrupt the degradation processes that are driven by bacterial degradation, for example activated sludge, thus hindering the degradation of other drugs as well.

Conversion from inactive to active form: Many metabolites or congeners get transformed in the environment to give higher concentrations of the active metabolite, which may exceed the safe exposure levels and thus cause harm to the aquatic wildlife; for example, pharmacokinetic data indicate that only 1–2 % of carbamazepine is excreted unmetabolized. The drug is transformed to a group of glucuronide conjugates that can be cleaved in the sewage, thus increasing the parent drug concentration. Similar is the case of morphine, where its major urinary metabolite, morphine-3β-D-glucuronide, is deconjugated in the wastewater. Sulphamethoxazole is another drug that enters WWTWs as N4-acetyl-sulphamethoxazole metabolite, but is converted back to the parent compound. Similarly, negative elimination of 3,4-methylenedioxy-amphetamine (MDA), an illicit drug, has been reported, which is suspected to be due to *N*-demethylation of 3,4-methylenedioxy-metamphetamine (MDMA), its conjugation product. Yet another example is that of nitro-aromatic musks, which can be transformed in WWTWs into aniline products, which are even more problematic than parent compounds (Petrovic et al. 2009).

Why It is Such a Big Problem?

Pharmaceuticals are potent in nature as compared to other pollutants. Hence, their coming into water supplies is a serious matter of concern. They can have an impact in the following ways:

Impact on the wildlife: A study was carried out wherein the four most abundantly used pharmaceuticals (namely acetaminophen, carbamazepine, diltiazem and cimetidine) and six sulphonamide-related antibiotics were investigated for their acute toxicity employing three representative model species (1) a marine bacterium (*Vibrio fischeri*), (2) a fresh water invertebrate (*Daphnia magna*) and (3) Japanese medaka fish (*Oryzias latipes*). Endpoints included immobility of test organisms, mortality rate, etc. Utilizing a series of conservative assumptions, the predicted environmental concentrations (PECs) of the pharmaceuticals tested were calculated, and their hazard quotients were derived. The latter were high for acetaminophen and sulphamethoxazole, suggesting potential ecological risks and the need for further investigation (Kim et al. 2007b; Brausch et al. 2012; Liu et al. 2012). Table 21.1 outlines a few more examples of the impact of pharmaceuticals on the wildlife.

Multi-drug resistance as a line of defence for aquatic organisms: Certain pharmaceutical compounds have the ability to inhibit the active transport system required for preventing intracellular exposure of many aquatic organisms (analogous to multi-drug transporters). Once inhibited, the lack of this system could lead to the intracellular accumulation of many toxicants. Since multi-drug transport capacity seems to be better developed in aquatic life as it is routinely exposed to toxicants, organisms in more primeval areas are vulnerable to toxic endpoints when facing first-time exposure (Daughton 2000; Boxall et al. 2012).

Synergistic effects of pharmacologically similar drugs: Isolated pharmaceutical compounds in aquatic environment may be innocuous *per se*, but when a large number of similar drugs are present, their cumulative concentration could exceed safe exposure levels, especially when the drugs act by the same mode of action. In 2006, researchers from the University of Insubria in Italy conducted a study by exposing human embryonic cells HEK 293 to a low-level mixture of 13 drug residues. Even at low doses, the drug residues actually stopped the cells from reproducing (Pomati et al. 2006).

Bioaccumulation: Pharmaceuticals are usually lipophilic and often have a low biodegradability. Still there is a higher probability of increase in the concentration of a chemical in a biological organism over the time, compared to chemical's concentration in the environment (Halling-Sørensen et al. 1998).

Toxicant-induced loss of tolerance: Initial exposure to low doses of drugs sensitizes, and subsequent exposure levels, below those previously tolerated, trigger symptoms (Miller 1997).

Table 21.1 Additional examples of the impact of pharmaceuticals on the wildlife

Contaminant drug	Organism/ animals affected	Adverse effects	References
Diclofenac	Trout fish	Kidney damage	Jones et al. (2004)
Diclofenac given to animals	Vultures	Kidney damage	Green et al. (2004)
Endocrine disrupting compounds like ethinylestradiol	Male smallmouth fish	Feminization	Jones et al. (2004)
Fluoxetine	Goldfish	Anorexigenic effect	Mennigen et al. (2010)
	Japanese medaka	Embryotoxicity	Cheung (2011)
5-Hydroxytryptamine	Crayfish	Release of ovary stimulating hormones	Kulkarni et al. (1992)
Levonorgestrel	Female frogs	Sterility	Kvarnryd et al. (2011)
Oseltamivir	H5N1	Resistance to the drug	Fick et al. (2007)
Tetracycline, penicillin, erythromycin, ampicillin and sulphonamide	Gram-negative bacteria	Resistance to these antibiotics	Ash et al. (2002), Allen et al. (2010), Behnke (2012), Gao et al. (2012)

Allergic responses: Sometimes, very low doses of antibacterial like penicillin can trigger effects in allergic individuals (Wicher and Reisman 1980).

Resistance to antibiotics: The development of antimicrobial resistance in the environment is another eventuality. In a recent study, it was shown that the proportions of multi-resistant bacterial strains sampled downstream from a manufacturing site of penicillin were higher than in the upstream isolates. As these bacteria were multi-resistant, it also indicated towards the cross-resistance prevalent due to penicillin contamination. These resistant bacteria may be directly transmitted to humans via drinking water, if groundwater is also contaminated (Li et al. 2009). Another instance is pandemic of avian flu during which wildfowls ingested oseltamivir at a quantity less than its minimum inhibitory concentration, together with virus in their daily water intake. As both the drug and the pathogen were present together in the gastrointestinal tract of wildfowls, selection of resistant bacteria occurred. These resistant bacteria were perhaps also transmitted to humans, who were living in close proximity to aquatic life. This is corroborated by the fact that 18 % of Japanese children are resistant to oseltamivir due to its widespread use and contamination of surface waters in Japan (Fick et al. 2007; Singer et al. 2007).

Effect on non-target species is less understood: The modes of action of most pharmaceuticals in humans are poorly understood and even less implicit is the gamut of effects on non-target species. Thus, what is visible to us at present may only be the tip of the iceberg (Daughton 2000; Brausch et al. 2012).

Overall impact on human health: Very few reports address the potential effect to human health due to the presence of trace levels of pharmaceuticals in the environment. Christensen and Webb et al. (Christensen 1998; Webb et al. 2003) evaluated potential human health impact for several drug substances found in surface and drinking water and reported no significant impact to human health. A similar study was carried out in the USA for human risk assessment posed by 26 APIs and their metabolites, covering 14 classes of drugs. The predicted no effect concentrations (PNECs) for these drugs were assessed in drinking water and as a consequence of fish ingestion. The PNECs were compared to measured environmental concentrations (MECs) from the published literature and to maximum PECs generated using the pharmaceutical assessment and transport evaluation (PhATE) model. For all the 26 compounds, the study indicated no appreciable human health risks from the presence of trace concentrations of the drugs in surface water and drinking water (Schwab et al. 2005). Yet, there are uncertainties in these assessments as the drugs and their metabolites, although appearing insignificant independently, do positively contribute to the overall toxic load of the environment. It has been insinuated that increasing levels of oestrogen in the environment, via its use for purposes such as menopause symptom relief and birth control pills, could be causing adverse effects on humans, such as reduced male sperm counts and sperm motility and younger ages of puberty in girls (Buhner 2002). Thus, the views regarding potential harmful effects to humans are varied as of now.

Sources of Contamination and Aqueous Matrices Usually Polluted with Pharmaceuticals

As indicated in the above discussion, almost all aquatic matrices have the potential to be contaminated with pharmaceuticals coming from various sources.

Types of sources: The pollution may arise from diffuse sources or point sources. Diffuse sources include contamination from disposal of used and unused drugs by individuals and households. Point sources include contamination due to production facilities, hospitals, etc. Contamination from point sources can be easily monitored, whereas it is hard to estimate for the contamination due to diffuse sources.

Effluents from sewage treatment plants (STPs): Urban STPs receive wastewater and sewage from densely inhabited areas. Some pharmaceuticals are poorly removed in these plants and are consequently detected in surface waters. In a study, an estimate was made for drug contamination in raw and treated sewage of 12 STPs in Finland (Vieno et al. 2007). The study revealed an average concentration in the raw sewage ranging from 100 to 1,060 ng/L, which reduced between <24 and 755 ng/L on treatment. The levels of drugs like carbamazepine were higher in the treated sewage compared to those in the raw sewage. The study concluded that drugs like carbamazepine and beta blockers were not eliminated in the WWTWs and they reached the recipient waters, which could be groundwater or surface water. Another study was conducted in Germany to check for the presence of antiphlogistics, lipid regulators, antipsychotics, antiepileptics, β-blockers and β_2-sympathomimetics in the treated sewage waters of municipal STPs. The concentration levels of carbamazepine were found to be up to 6.3 μg/ml. Above 80 % of the selected drugs were detectable in at least one municipal STP effluent, thus corroborating the role of STPs in defiling recipient waters like rivers and lakes (Ternes 1998).

Groundwater: A major portion of pharmaceuticals in drinking water are those excreted by animals in large concentrated animal feeding operations, where antibiotics and hormones are routinely administered to animals to prevent infection or promote growth. Estimates suggest that nearly two trillion pounds of animal wastes are produced annually in the USA alone, and that between 25 and 75 % antibiotics are excreted unchanged in faeces, wherefrom they persist in the environment. As a result, a large amount of antibiotics (and potentially antibiotic-resistant bacteria) enter waterways and groundwater through overflow of waste lagoons, or by over application of manure as fertilizer in the farm fields (Sarmah et al. 2006). This is supported by studies in China where the presence of widely used veterinary antibiotics of at least four classes was checked in different seasons in the groundwater. The average antibiotic concentration was found to range from 1.6 to 8.6 ng/l in summer and 2.0–7.3 ng/l in winter (Tong et al. 2009).

Surface water: It comprises water in rivers, lakes, seas, etc., where pharmaceuticals gain entry through direct release of effluents from WWTWs handling waste from industries, hospitals and households. In a study carried out by

Ternes (Ternes 1998) in Germany, acidic drugs like the lipid regulators (bezafibrate and gemfibrozil), antiphlogistics (diclofenac, ibuprofen, indomethacin, naproxen and phenazone) and metabolites (clofibric acid, fenofibric acid and salicylic acid) as well as neutral or weak basic drugs like beta blockers (metoprolol and propranolol) and an antiepileptic drug (carbamazepine) were found to be ubiquitously present in the rivers and streams in the ng/l range. In another study, atorvastatin was discovered in the Otonabee River of Canada (Miao and Metcalfe 2003). The widely used veterinary antibiotic, spiramycin, was found to be present in the river of North Italy in the range of ng/l (Calza et al. 2010). Owing to frequent misuse of stanozolol for performance enhancement in athletes as well as for illegal growth promoting purposes in veterinary practices, the Danube River in the UK was tested positive for the presence of this WADA banned synthetic anabolic steroid (Deshmukh et al. 2011). A recent study in Sweden showed higher pharmaceutical input in the river during a large-scale open air festival (Daneshvar et al. 2012).

The drugs that contaminate rivers and seas carry potential to show bioaccumulation in tissues of various aquatic fauna. This was highlighted in a study carried out in a laboratory where >100 times higher concentration of gemfibrozil, a lipid regulator, was found in the blood plasma of goldfilsh after exposure for 14 days to waterborne drug, against 96 h post single intra-peritoneal injection (Mimeault et al. 2005). Schwaiger et al. observed residues >2.8 μg/g wet weight in the livers of fish exposed to diclofenac at a concentration of 1 μg/l (Schwaiger et al. 2004). Similarly, Brooks et al. have reported the presence of antidepressants, viz. fluoxetine and sertraline, in the tissues of fish collected from an effluent dominated stream up to approximately 10 ng/g of wet weight (Brooks et al. 2005).

Soil interstitial water: It accounts for the subterranean water in the pores of rocks, soils and bottom sediments of ocean, seas and lakes. The main route of spoliation of soil interstitial water is via transport of organic waste used as fertilizer in the soil environment. The organic waste comes from animal excreta, thus bringing with itself drug contaminants from those used in veterinary treatment (Montforts et al. 1999). Halling-Sorenson and co-workers carried out a study in Denmark on soil interstitial water to test for the presence of oxytetracyclines or its degradation products. Several degradation products and epimers like epioxytetracycline, *N*-desmethyl-oxytetracycline, *N*-didesmethyl-oxytetracycline and apo-oxytetracyclines were tested positively in the soil water contaminated with oxytetracycline (Halling-Sørensen et al. 2003).

Pharmaceuticals Commonly Found as Contaminants in water

Pharmaceuticals have the potential to exist as pollutants in the environment in the following cases:

Very commonly prescribed and/or consumed: Drugs prescribed for common, short-term ailments like common cold, fever, headaches have a high propensity for contaminating the environment as their high usage facilitates continuous release into the environment, either through dumping of unused drugs or through excretion of consumed drugs.

Less easily degraded in the WWTWs as well as in the environment: Some drugs undergo various degradation processes in the environmental aqueous matrices (like hydrolysis, photodegradation), thus either getting completely destroyed or being converted to less harmful forms. The problem arises when drugs that are resistant to these natural processes are released into the environment. Even if their release is not constant, their persistence in the environment makes them potential harmful contaminants.

Highly potent and thus toxic to non-target species at even low levels: Some highly potent drugs, for example, estrogens, anticancer drugs, even though having low sales volumes and present in very low quantities in the environmental matrices, can still have highly deleterious effects on non-target species. Examples of drugs that have been found in various aqueous matrices are enlisted in Table 21.2.

Table 21.2 Examples of drugs found in various aqueous matrices

Category	Drugs	References
Antibiotics	Tetracycline, oxytetracycline (and their degradation products), ofloxacin, ciprofloxacin, erythromycin, lincomycin, clarithromycin, spiramycin, sulphonamides, amoxicillin, norfloxacin, trimethoprim, triclosan	Kolpin et al. (2002), Andreozzi et al. (2005), Zuccato et al. (2005), Balakrishnan et al. (2006), Díaz-Cruz and Barceló (2006), Godfrey et al. (2007), Kim et al. (2007b), Vieno et al. (2007), Okuda et al. (2009), Fernandez et al. (2010), Hruska and Franek (2012)
Anticonvulsants	Carbamazepine, primidone	Drewes et al. (2003), Kim et al. (2007a, b), Zhang et al. (2008), Fernandez et al. (2010)
Blood lipid regulators	Bezafibrate, clofibric acid, gemfibrozil, fenofibric acid, atorvastatin, simvastatin, lovastatin, pravastatin, pitavastatin	Boyd et al. (2003), Miao and Metcalfe (2003), Radjenovic et al. (2007), Garcia-Ac et al. (2009), Lee et al. (2009), Razavi et al. (2011), Jindal (2012)
Antihypertensives, calcium channel blockers, ACE inhibitors	Metoprolol, propranolol, atenolol, acebutolol, sotalol, enalapril	Paxeus (2004), Hernando et al. (2007), Gallart-Ayala et al. (2011)
Antidepressants	Fluoxetine, norfluoxetine, meprobamate, risperidone	Brooks et al. (2003), Vanderford and Snyder (2006), Chu and Metcalfe (2007), Nentwig (2007)
Analgesics, antiphlogistics	Diclofenac, ibuprofen, indomethacin, naproxen, acetyl salicylic acid, ketoprofen, fenoprofen, propyphenazone, mefenamic acid	Metcalfe et al. (2003), Miao and Metcalfe (2003), Sebok et al. (2008), Zhang et al. (2008), Okuda et al. (2009), Kallio et al. (2010), Baranowska and Kowalski (2012), Jindal (2012)
Reproductive hormones, steroids	Norethisterone, ethinylestradiol, stanozolol	Aherne and Briggs (1989), Boyd et al. (2003), Möder et al. (2007), Deshmukh et al. (2011)
Antidiabetics	Glibenclamide, metformin	Cahill et al. (2004), Radjenovic et al. (2007)
Bronchodilators	Salbutamol	Zuccato et al. (2005)
Proton pump inhibitor/ H_2 receptor antagonists	Omeprazole, ranitidine	Castiglioni et al. (2005), Zuccato et al. (2005), Isidori et al. (2009)
Diuretics	Furosemide, hydrochlorothiazide	Cahill et al. (2004), Castiglioni et al. (2006), Obando et al. (2008), Radjenovic et al. (2008), Okuda et al. (2009), Teijon et al. (2010)
Anticancer drugs	Cyclophosphamide, ifosfamide, methotrexate, bleomycin	Aherne et al. (1990), Buerge et al. (2006), Castiglioni et al. (2006), Garcia-Ac et al. (2009)

Analytical Process for the Detection, Identification, Characterization and Quantitation of Pharmaceuticals in Aquatic Sources

The analysis of aquatic sources for the presence of pharmaceuticals warrants their detection, identification, characterization and quantitation at various concentration levels. The latter are expected to be higher if discharge from industry or hospitals directly reaches the aquatic source. In comparison, if the source itself is diminutive, or the pollutants are labile and/or metabolize to multiple end products, the concentration of each may be much lower. In matrices like drinking water, which is normally sourced from surface canals or underground, pharmaceuticals may only be present at trace levels, unless there is direct contact with an incoming source. Accordingly, the extent of analytical effort, choice of technique and selection of instrumentation may vary based on the levels of pharmaceuticals present. Figure 21.1 summarizes general steps involved in the process of analysis of pollutants in the environmental samples. The steps are discussed individually below:

Defining the purpose: As shown, the first step is to establish the purpose. For example, a purpose can be to look for major pharmaceutical pollutants, while contrastingly, it may be to target those present at minor or trace levels. Similarly, a purpose may be to detect only the targeted pollutants, or it might be a requirement to characterize all those present. The extensiveness of subsequent actions is dependent upon the defined purpose.

Pre-selection of pharmaceuticals to be targeted: Because aqueous matrices may contain a number of dissolved analytes together, therefore, a general prevalent practice is to target only those drug molecules that are of real concern. The pre-selection is done based on a variety of factors, of which important ones are:

1. Drugs that have been detected elsewhere in the environment (for which usually literature reports are screened).
2. Drugs that are likely to be present in discharge from the industry, where criterion usually is the tonnage produced.
3. Those with high prescription and over-the-counter volumes in hospitals or outside like NSAIDs (ibuprofen and acetaminophen), antibacterials, etc.
4. Drugs like diclofenac and oxytocin that are in rampant use in milking animals.
5. Drugs used as a long-term prescription and that are stable in WWTWs, for example, epileptics like carbamazepine and phenobarbital.
6. Antibiotics with concern of bacterial resistance, for example, penicillins and sulphamethoxazole.
7. Drugs with high potency (17α-ethynyl estradiol), toxicity (fluoxetine) and persistence in environment (diazepam).
8. Drugs whose use is hiked in certain season, like bronchodilators are consumed more in winters by patients with asthma.
9. Spurt in use of drug(s) in an event of outbreak of disease.

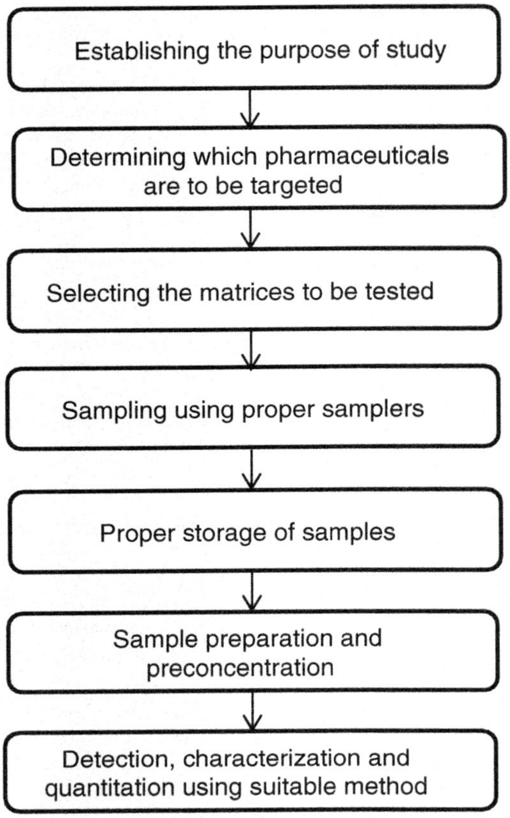

Fig. 21.1 Steps involved in the process of analysis of pollutants in the environmental samples

10. Specific interest in illicit drugs, for example, cocaine, amphetamine and other drugs of abuse.

Sample collection and storage: Usually, three types of sampling modes are prescribed for aquatic sources containing 5 % solids. These are (1) grab sampling wherein a single sample is collected at a particular time and place, which is usually effective for dry weather periods when water quality does not change very rapidly, (2) composite sampling where multiple samples at different times or from different places with same exposure potential are collected, and (3) passive sampling wherein samplers are deployed in the source over days or weeks. The latter uses a special device, for example, polar organic chemical integrative sampler (POCIS), that monitors hydrophilic contaminants, which are

Fig. 21.2 Different types of sampling containers and devices, including pictures of grab sampling from actual sites. The pictures are sourced from different websites and modified suitably, wherever required, to describe the concept

potentially endocrine disrupting or acutely toxic, like pharmaceuticals, hormones (Miege et al. 2012). Figure 21.2 depicts the types of sampling containers and devices used.

The samples may be collected in any thoroughly cleaned container, be a bottle, mug, bucket, plastic jar, etc., but post-collection, immaculate sample handling and storage are the key requirements. The collected samples are transferred similarly in a clean, silanized, amber glass bottles to prevent photodegradation of the analytes of interest. The addition of antimicrobials like sodium azide to prevent microbial degradation and antioxidants, like ascorbic acid to quench any residual oxidant (e.g., chlorine, ozone, chloramine) is also advised (Ferrer et al. 2010). While handling samples anticipated to contain trace pharmaceuticals, avoiding contamination from the laboratory environment is equally important. For the latter, all basic precautions are mandatory, which include thorough cleaning of sample collection containers and also apparatus of all types, which may be used till the analysis is complete. The protocol advised for cleaning of glassware includes sonication with a detergent that is devoid of possible analytes, baking the same for at least 4 h at 450 °C and rinsing with acetone and hexane or similar solvents, prior to the use. Here, a precaution is that the solvents used must also be of high purity. Additionally, method blanks should be run to monitor for contamination from the laboratory environment, contaminated solvents, or other sources (Peck 2006).

The samples are usually stored in dark at low temperatures. These are subjected to extraction within 48 h of sampling. Otherwise, they ought to be frozen, thereby increasing the holding time up to 7 days (Brun et al. 2006; Ferrer et al. 2010).

A detailed treatise on sampling and sample preservation of water and wastewaters is available at the website of US Environment Protection Agency (http://nepis.epa.gov).

Sample preparation: For trace contaminants, a requisite is pre-concentration of the sample, which is also necessary to reduce the matrix effects. For this, most commonly used techniques are solid-phase extraction (SPE), solid-phase microextraction (SPME), single-drop microextraction (SDME), liquid–liquid extraction (LLE), ultrasonic extraction (USE), microwave-assisted extraction (MAE) and pressurized liquid extraction (PLE). For non-polar pharmaceuticals, reversed-phase SPE is preferred, while mixed-mode materials are used for polar pharmaceuticals. Solid-phase microextraction (SPME) and stir bar sorptive extraction (SBSE) are extensively used as they require less organic solvent and show better potential for automation (Balakrishnan et al. 2006; Petrovic et al. 2010; Buchberger 2011)

Detection and characterization: The next step is the detection, identification, characterization and/or quantification of pharmaceuticals in the prepared sample using a method developed employing the most suitable analytical techniques. In recent times, the focus has shifted towards multi-residue analysis against single group detection. However, simultaneous analysis of the compounds from different groups with quite different physicochemical characteristics requires a compromise in the selection of experimental conditions, which in some cases may not be the best for all the analytes under investigation. Due to the wide range of polarities from ionic to non-ionic, it becomes difficult to develop and optimize a single method suitable for the analysis of all the analytes. This problem has been compounded by the current interest in simultaneous analysis of transformation products (e.g., metabolites, degradation and/or interaction products), which are normally more polar than the parent compounds. Therefore, an approach often employed is to develop a method for the targeted components with good resolution and selectivity. One can develop a method initially using standard drugs, along with transformation products as necessary, and then extend the same to the collected/prepared samples. Table 21.3 provides the list of techniques used in literature reports for the detection of major, minor and trace pharmaceuticals in aqueous matrices.

As evident in the table, three common techniques used for the detection are LC, CE and GC, apart from a few others, including immunochemical methods. The former are used as standalones but majority of the studies have involved their use as front end with mass systems. Among the latter, LC–MS systems specifically can be used for all the purposes, viz. detection, identification, characterization and quantification of trace pharmaceuticals in environment. LC-MS instruments are available in many variants and allow multiple modes of testing. The systems available in mid-1980s were direct liquid introduction, thermospray, fast atom bombardment, particle beam interface, etc. These tools had no real applications in environmental analysis, and GC–MS was the technique dominant. Current LC–MS technologies, based on atmospheric pressure ionization (API) interfaces (electrospray, atmospheric pressure chemical ionization and atmospheric pressure photoionization) and with different types of analysers (QqQ, TOF, QTOF, Orbitrap, IT, QqLIT), have proved as versatile and powerful alternatives to GC–MS in environmental analysis. Today, it has become possible to detect pharmaceutical contaminants at levels as low as 0.1 picograms against the levels in the range of nanograms 20 years ago. With the capabilities of modern LC–MS tools (single and multiple reaction monitoring, neutral ion scan, precursor ion scan, product ion scan, etc.), it is possible to detect, characterize and quantify many harmful compounds at the concentrations at which they have a biological effect in the environment.

CE–MS instruments have sensitivity next to LC–MS systems and are useful as an orthogonal technique to confirm results obtained by the latter. GC–MS is an all-time use technique, but with limitation of its applicability to components volatile enough to be transferred directly to gas phase or those that can easily be derivatized to volatile species without any by-products. But the technique has an indisputable advantage of showing less matrix effects when compared to HPLC, due to ionization modes like electrochemical ionization (EI) and chemical ionization (CI). As a consequence, detection limits for GC–MS may be even lower than that obtained with LC–MS. It is obvious that GC–MS methods are robust for certain classes of pharmaceuticals and need not be replaced by LC and LC–MS in all the cases. Figure 21.3 presents a mass spectrometry-based strategy for the determination of pharmaceuticals with and without the use of reference standards.

Table 21.3 Different techniques used for determination of various pharmaceuticals in aqueous matrices

Techniques used	Drugs detected	Matrix tested	References
HPLC/UPLC-UV diode array	Oseltamivir, carbamazepine and 7 NSAIDs	River water	Vera-Candioti et al. (2008), Baranowska and Kowalski (2012)
LC–ESI–MS/MS	Neutral pharmaceuticals including 4-aminoantipyrine, glibenclamide, oseltamivir.	Wastewater and rivers	Ternes et al. (2001), Chenxi et al. (2008)
UHPLC-QqQ-MS/MS	Cocaine, opioids, LSD, cannabinoid and amphetamine like compounds	Raw and treated wastewater, surface water and drinking water	Petrovic et al. (2010)
HPLC/UHPLC–qTOF	β-Blockers (atenolol), lipid regulators (bezafibrate), antiulcer (lansoprazole), antihistamines (loratadine)	Wastewaters	Petrovic et al. (2006), Ferrer and Thurman (2012)
CE–MS	NSAIDs, lipid regulators and antibiotics	Wastewaters	Ahrer and Buchberger (2001), Ahrer et al. (2001), Macia et al. (2004)
GC–MS	Clofibric acid, endocrine disrupting compounds	Lake and sea water	Möder et al. (2007), Cecinato et al. (2009)
LC-LTQ-FT Orbitrap MS	Benzotriazoles and benzothiazoles, β-blockers (atenolol)	River water, sewage treatment plants	Hogenboom et al. (2009), van Leerdam et al. (2009)
HILIC–MS/MS	Estrogens (estrone, estradiol, and estriol) and their conjugates (estrone-3-sulphate, estradiol-3-sulphate, estriol-3-sulphate, etc.)	River water	Qin et al. (2008)
LC with fluorescence detection	Fluoroquinolone antibacterial agents	Urban wastewater	Golet et al. (2001)
Adsorptive voltammetry	Platinum	Hospital effluents	Kümmerer and Helmers (1997)
Square wave voltammetry using antibodies immobilized on magnetic particles	Ethinylestradiol	Water samples	Martinez et al. (2010, 2012)

Extent of Removal of Pharmaceuticals During Conventional and Advanced Wastewater Treatments, Regulatory Efforts, Prevention of Ingress of Pharmaceuticals in Aqueous Sources, and the Role of Consumer Awareness

The developments on the analytical front have helped significantly in understanding the extent of the problem of pollution of water sources with pharmaceuticals. The analytical technologies have been successfully employed to monitor the levels of pharmaceuticals present in influents and effluents of WWTWs, and to the levels, pharmaceuticals are removed with a particular technique used in influent treatment, including tertiary treatment processes. Unfortunately, the list of chemical contaminants in water sources, which may or may not always be of significant human health concern, is ever expanding. Therefore, whether analytical monitoring and up-gradation of treatment facilities are the right approaches, is a matter of concern. Instead, it has been proposed that greater emphasis shall be paid to the prevention of ingress of pharmaceuticals in water matrices and bringing in

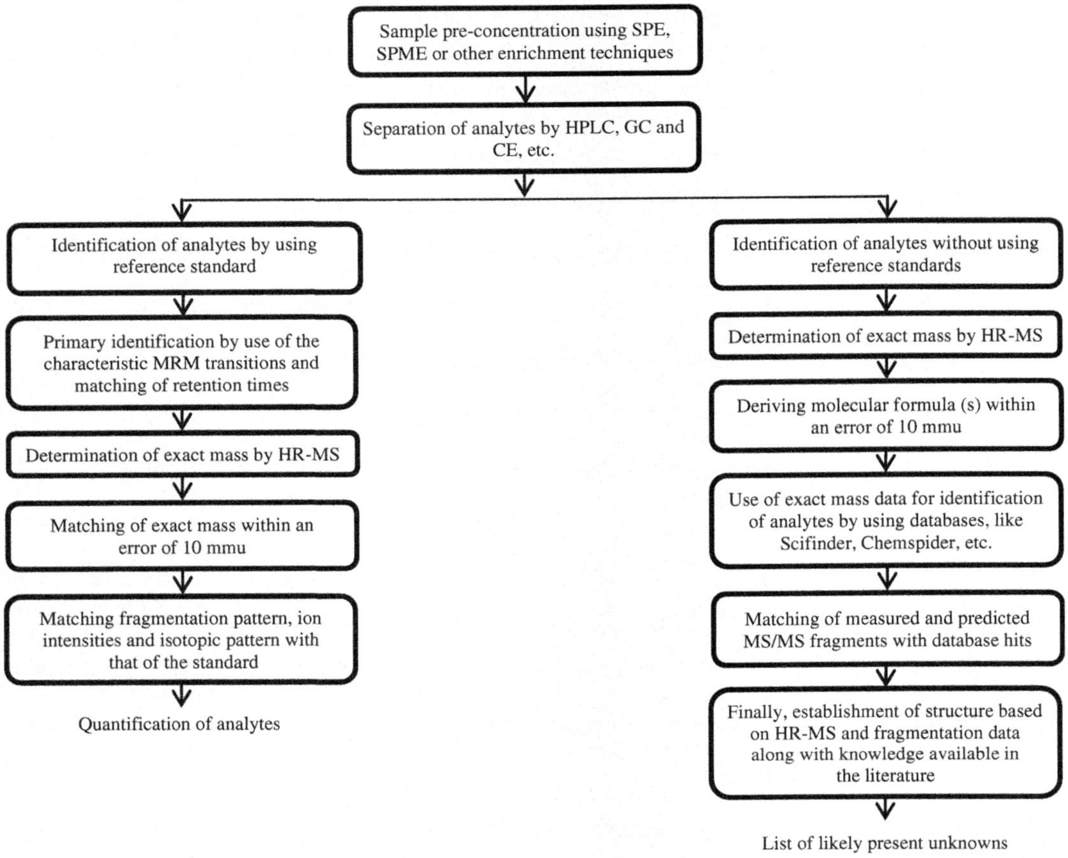

Fig. 21.3 Mass spectrometry-based strategy for the determination of pharmaceuticals in the environment

stronger regulations w.r.t. the same. In addition, it is proposed to increase consumer awareness so as to avert unregulated disposal of drugs in the environment by individuals. These major concerns are strongly highlighted recently by WHO (WHO 2011, 2012).

Removal of pharmaceuticals in WWTWs: As pharmaceuticals have diverse range of physicochemical properties, the same precludes their complete removal by any single method used for treatment of influent in WWTWs. The removal of pharmaceuticals by conventional wastewater treatment plants involves use of activated sludge process or other forms of biological treatment such as biofiltration. These remove pharmaceuticals in the range of <20 % to >90 %, largely dependent upon operational configuration of the WWTWs, age of sludge, activated sludge tank temperature, hydraulic retention time and hydrophobicity of the pharmaceutical. Modern WWTWs have tertiary treatment processes, like reverse osmosis, ozonation and advanced oxidation technologies. The use of oxidation process (chlorine, chlorine dioxide and ozone disinfection) has the ability to remove some pharmaceuticals by transformation. Powder activated carbon (PAC) and granular activated carbon (GAC) were also found to be effective in removal of hydrophobic pharmaceutical compounds from the drinking water. Use of high pressure membranes, like reverse osmosis and microfiltration, is highly efficient in removal of pharmaceuticals, which in some cases goes to 100 % (WHO 2012). But an associated problem is the transformation the pharmaceuticals may undergo during use of these methods. Hence, the minimization of pharmaceutical load in water sources is one solution, but the same may eventually require different combinations of treatment processes, which may not always be practical, considering large volumes of aqueous matrices that may be involved.

Regulatory efforts: Keeping into view of the above, multi-focused regulations have been proposed in various parts of the world, which emphasize on stopping the ingress of pharmaceuticals in aquatic sources by containing them in the first stage itself through incineration and/or landfill disposal of the waste.

1. For categorization of hazardous waste for disposal, pharmaceuticals are also classified based on their toxicity and reactivity. For example, Resource Conservation and Recovery Act (RCRA) of 1976 of US (U.S. EPA 1976) divides hazardous waste for disposal into listed waste and characteristic waste. The former contains four lists (F, K, P and U), of which pharmaceuticals are included in lists P and U. List P contains eight pharmaceuticals that show LD_{50} less than 50 mg/kg. Examples include epinephrine and physostigmine. List U contains twenty-one pharmaceuticals listed for their specific toxicity, examples being chlorambucil and cyclophosphamide. Characteristic waste includes ignitable and highly reactive pharmaceuticals, like erythromycin gel and nitroglycerine. The Bio-Medical Waste (Management and Handling) Rules, 2011, of India categorize bio-medical waste into eight categories, in which discarded medicines and cytotoxic drugs are included in category 5. For this category, the recommended process is incineration, followed by disposal in secured landfill (MoEF 2011).
2. WHO provides for three pharmaceutical disposal techniques to minimize contamination and eventual flow of pharmaceuticals into aquatic sources (WHO 1999). The first is highly engineered sanitary landfill, which is an evacuated pit isolated from watercourses and above the water table. The solid waste is compacted on day to day basis and covered with soil to maintain sanitary conditions. The same is used for disposal of solid, semisolid and powder pharmaceuticals after immobilization. The second is waste immobilization, which is used for solids, semi-solids, powders, liquid forms of antineoplastics and controlled substances. The immobilization is achieved by either encapsulation or inertization. In the former, the pharmaceuticals are immobilized in solid blocks within a plastic or steel drum, wherein they are covered with a mixture of cement, lime and water. The drum is then placed at the base of the landfill and covered with solid waste. Inertization procedure involves making of a paste of ground pharmaceuticals with cement, lime and water, which is then spread on landfill. The third suggested protocol includes high temperature incineration of pharmaceuticals in cement kilns at high temperatures of $\sim 1{,}200\ ^\circ C$. This procedure is preferred for solids, semi-solids, powders and liquid forms of antineoplastics and controlled substance.
3. Requirement of pre-notification of hazard discharge to environmental protection agencies/pollution departments or public water treatment facility and incineration of hazardous waste before land filling (U.S. EPA 2002).
4. Emphasis on environmental risk assessment (ERA) reports on the new drugs by NDA applicants (U.S. FDA 1998), when the expected introduction concentration (EIC) of the active ingredient of the drug in the aquatic environment (EIC-aquatic) exceeds 1 ppb. Also, it is desired that environmental impact assessment phase I should be carried out for all pharmaceuticals meant for non-aquatic, meat-producing animals (VICH 2000, 2005). If the PEC of a pharmaceutical in soil is found to be ≥ 100 ppb, a phase II study must be triggered. The European Agency for the Evaluation of Medicinal Products (EMEA) requires ERAs based on the phase II data before pharmaceuticals are approved for use by the regulatory authorities (EMEA 1996). A similar approach is followed for ERAs in the USA and Japan (Velagaleti et al. 2002).

5. Control on emissions and accountability of raw materials, intermediates and the targeted drug substance during manufacture of bulk or products. Hence, strong emphasis on material balances.
6. Local EPAs issue effluent limitation guidelines and pre-treatment standards for industrial discharges, including restrictions on quantities, rates and concentrations of chemical, physical and biological constituents that are discharged from point source into waters of the USA, the waters of the contiguous zone or the ocean. The Central Pollution Control Board of India provides minimal national standards for WWTWs effluents from pharmaceutical manufacturing units, which give limiting concentrations for BOD, TSS and presence of heavy metals (CPCB 1989).

However, the fact remains that in spite of so many regulations being enacted, pharmaceuticals like hormones, antibiotics, antidepressants, antihypertensives are not regulated as hazardous waste despite their high environmental contamination potential. Also, the regulations deal with individual drug components and overlook the possibility of combined toxicity arising due to additive or supra additive effects of drugs present together in various matrices. Furthermore, many drugs of concern are not regulated as water pollutants, and as such no safe exposure limits have ever been set for such drugs (Velagaleti et al. 2002).

Prevention of ingress of pharmaceuticals in aqueous matrices and consumer education: Other suggestions put forth towards prevention of the ingress of pharmaceuticals in aqueous matrices are summarized below:

1. Use of safe disposal practices like returning medicines to community pharmacies or health care facilities instead of the current practice where tons of unused medicines are flushed down the toilets and sinks or discarded in household waste. Here, there is a vital requirement of increasing the awareness of consumers w.r.t. disposal of medicines. This can be made possible by giving responsibility to pharmacists at retail stores that sell drugs to advise consumers on the proper storage and disposal of unwanted medicines.
2. Initiation of take-back, mail-back or other similar programmes, which have been started by the government and private organizations in several parts of the world. In these programmes, volunteers collect old, unused and expired medicines from the households. These programs are also meant to reduce the misuse and abuse of drugs (WHO 2012). When take-back programme is not available, an alternate suggested method for disposal of medicines is mixing them with coffee seeds or with unpalatable substances and sealing in plastic bag, which may then be thrown into household trash. A very useful publication entitled 'How to Dispose of Unused Medicines' (U.S. FDA 2011) makes an interesting reading in this context.
3. Making it compulsory for all health care facilities to have policies and procedures in place for the correct management of pharmaceutical waste. This may include engaging of licensed waste disposal companies to collect all clinical and pharmaceutical waste for disposal in authorized waste disposal facilities. In this context also, two very useful documents are 'Managing Pharmaceutical Waste: A 10-Step Blueprint for Health Care Facilities in the United States' (Smith 2008) and 'Best Management Practices for Unused Pharmaceuticals at Health Care Facilities' (U.S. EPA 2010).
4. Decrease in the use of pharmaceuticals for non-therapeutic purposes in case of farm animal production (U.S. Senate 2007).
5. Comprehensive risk assessment and risk management approach across the entire water cycle, starting at source (WHO 2012).

Conclusion

The contamination of the environment, particularly aquatic sources, by pharmaceuticals is increasing at an alarming rate. A number of factors are playing a role in this, with inefficient waste disposal system being one of the major ones. Although various regulations and guidelines have been passed in many countries to minimize the impact of pharmaceuticals on the environment, yet in many places, there is lack of compliance with the regulations. The pollution with pharmaceuticals has already started taking toll on the aquatic flora and fauna, and there is not much time left when the adverse effects of this will start showing up in humans. As pharmaceuticals gain entry into the different components of the environment, from a variety of sources and due to the cyclic path they follow in the various environmental matrices, it becomes a difficult task to effectively monitor all of them. Nevertheless, it is essential to obtain some experimental data regarding the levels of pharmaceuticals in the environment, including water sources, in specific areas, so as to be able to warn the local regulatory authorities and public about the seriousness of this issue. Fortunately, newer and improved techniques are available as of now, which can be employed for precise estimation of the types and quantities in which pharmaceuticals exist as pollutants. It is hoped that the situation can be controlled from worsening by greater emphasis on take-back programmes, improving regulations, public guidance, consumer education and by following the risk management approach to the problem.

References

Aherne GW, Briggs R (1989) The relevance of the presence of certain synthetic steroids in the aquatic environment. J Pharm Pharmacol 41:735–736

Aherne GW, Hardcastle A, Nield AH (1990) Cytotoxic drugs and the aquatic environment: estimation of bleomycin in river and water samples. J Pharm Pharmacol 42:741–742

Ahrer W, Buchberger W (2001) Combination of aqueous and non-aqueous capillary electrophoresis with electrospray mass spectrometry for the determination of drug residues in water. Monatshefte für Chemie/Chemical Monthly 132:329–337

Ahrer W, Scherwenk E, Buchberger W (2001) Determination of drug residues in water by the combination of liquid chromatography or capillary electrophoresis with electrospray mass spectrometry. J Chromatogr A 910:69–78

Allen HK, Donato J, Wang HH, Cloud-Hansen KA, Davies J, Handelsman J (2010) Call of the wild: antibiotic resistance genes in natural environments. Nat Rev Microbiol 8:251–259

Andreozzi R, Canterino M, Marotta R, Paxeus N (2005) Antibiotic removal from wastewaters: the ozonation of amoxicillin. J Hazard Mater 122:243–250

Ash RJ, Mauck B, Morgan M (2002) Antibiotic resistance of gram-negative bacteria in rivers, United States. Emerg Infect Dis 8:713–716

Balakrishnan VK, Terry KA, Toito J (2006) Determination of sulfonamide antibiotics in wastewater: a comparison of solid phase microextraction and solid phase extraction methods. J Chromatogr A 1131:1–10

Baranowska I, Kowalski B (2012) A rapid UHPLC method for the simultaneous determination of drugs from different therapeutic groups in surface water and wastewater. Bull Environ Contam Toxicol 89:8–14

Behnke J (2012) Pharmaceuticals in the water: the albatross around Texas's neck. Texas Tech Adm Law J 13:325–391

Boxall ABA, Rudd M, Brooks BW, Caldwell D, Choi K, Hickmann S et al (2012) Pharmaceuticals and personal care products in the environment: what are the big questions. Environ Health Perspect 120:1221–1229

Boyd GR, Reemtsma H, Grimm DA, Mitra S (2003) Pharmaceuticals and personal care products (PPCPs) in surface and treated waters of Louisiana, USA and Ontario, Canada. Sci Total Environ 311:135–149

Brausch JM, Connors KA, Brooks BW, Rand GM (2012) Human pharmaceuticals in the aquatic environment: a review of recent toxicological studies and considerations for toxicity testing. Rev Environ Contam Toxicol 218:1–99

Brooks BW, Foran CM, Richards SM, Weston J, Turner PK, Stanley JK et al (2003) Aquatic ecotoxicology of fluoxetine. Toxicol Lett 142:169–183

Brooks BW, Chambliss CK, Stanley JK, Ramirez A, Banks KE, Johnson RD et al (2005) Determination of select antidepressants in fish from an effluent-dominated stream. Environ Toxicol Chem 24:464–469

Brun GL, Bernier M, Losier R, Doe K, Jackman P, Lee HB (2006) Pharmaceutically active compounds in Atlantic Canadian sewage treatment plant effluents and receiving waters, and potential for environmental effects as measured by acute and chronic aquatic toxicity. Environ Toxicol Chem 25:2163–2176

Buchberger WW (2011) Current approaches to trace analysis of pharmaceuticals and personal care products in the environment. J Chromatogr A 1218:603–618

Buerge IJ, Buser HR, Poiger T, Muller MD (2006) Occurrence and fate of the cytostatic drugs cyclophosphamide and ifosfamide in wastewater and surface waters. Environ Sci Technol 40:7242–7250

Buhner SH (2002) The environmental impacts of technological medicine. In: Buhner SH (ed) The lost language of plants: the ecological importance of plant medicines to life on Earth. Chelsea Green Publishing, Vermont, pp 92–101

Cahill JD, Furlong ET, Burkhardt MR, Kolpin D, Anderson LG (2004) Determination of pharmaceutical compounds in surface- and ground-water samples by solid-phase extraction and high-performance liquid chromatography-electrospray ionization mass spectrometry. J Chromatogr A 1041:171–180

Calza P, Marchisio S, Medana C, Baiocchi C (2010) Fate of antibacterial spiramycin in river waters. Anal Bioanal Chem 396:1539–1550

Castiglioni S, Bagnati R, Calamari D, Fanelli R, Zuccato E (2005) A multiresidue analytical method using solid-phase extraction and high-pressure liquid chromatography tandem mass spectrometry to measure pharmaceuticals of different therapeutic classes in urban wastewaters. J Chromatogr A 1092:206–215

Castiglioni S, Bagnati R, Fanelli R, Pomati F, Calamari D, Zuccato E (2006) Removal of pharmaceuticals in sewage treatment plants in Italy. Environ Sci Technol 40:357–363

Cecinato A, Balducci C, Nervegna G (2009) Occurrence of cocaine in the air of the world's cities: an emerging problem? A new tool to investigate the social incidence of drugs? Sci Total Environ 407:1683–1690

Central Pollution Control Board (1989) Minimal national standards, pharmaceutical manufacturing and formulation industry. Comprehensive industry document series: COINDS/29/1988-89, New Delhi, India

Chenxi W, Spongberg AL, Witter JD (2008) Determination of the persistence of pharmaceuticals in biosolids using liquid-chromatography tandem mass spectrometry. Chemosphere 73:511–518

Cheung M (2011) Embryotoxicity of fluoxetine (Prozac®) to Japanese medaka (Oryzias latipes). Bachelor of Science (Honours) Thesis. School of Environmental Studies, Queen's University, Canada

Christensen FM (1998) Pharmaceuticals in the environment-a human risk? Regul Toxicol Pharmacol 28:212–221

Chu S, Metcalfe CD (2007) Analysis of paroxetine, fluoxetine and norfluoxetine in fish tissues using pressurized liquid extraction, mixed mode solid phase extraction cleanup and liquid chromatography-tandem mass spectrometry. J Chromatogr A 1163:112–118

Daneshvar A, Svanfelt J, Kronberg L, Weyhenmeyer GA (2012) Neglected sources of pharmaceuticals in river water-footprints of a Reggae festival. J Environ Monit 14:596–603

Daughton CG (2000) Pharmaceuticals in the environment: overarching issues and concerns. Pharmaceuticals and Personal Care Products in the Environment. San Francisco, CA, Paper presented at ACS National Meeting

Deshmukh NI, Barker J, Petroczi A, Naughton DP (2011) Detection of stanozolol in environmental waters using liquid chromatography tandem mass spectrometry. Chem Cent J 5:63–70

Díaz-Cruz M, Barceló D (2006) Determination of antimicrobial residues and metabolites in the aquatic environment by liquid chromatography tandem mass spectrometry. Anal Bioanal Chem 386:973–985

Drewes JE, Heberer T, Rauch T, Reddersen K (2003) Fate of pharmaceuticals during ground water recharge. Ground Water Monit Rem 23:64–72

European Medicines Agency (1996) Environmental risk assessment for veterinary medicinal products other than GMO-containing medicinal products. EMEA/CVMP/055/96-FINAL, Brussels, Belgium

Fernandez C, Gonzalez-Doncel M, Pro J, Carbonell G, Tarazona JV (2010) Occurrence of pharmaceutically active compounds in surface waters of the Henares-Jarama-Tajo River system (Madrid, Spain) and a potential risk characterization. Sci Total Environ 408:543–551

Ferrer I, Thurman EM (2012) Analysis of 100 pharmaceuticals and their degradates in water samples by liquid chromatography/quadrupole time-of-flight mass spectrometry. J Chromatogr A 1259:148–157

Ferrer I, Zweigenbaum JA, Thurman EM (2010) Analysis of 70 Environmental Protection Agency priority pharmaceuticals in water by EPA method 1694. J Chromatogr A 1217:5674–5686

Fick J, Lindberg RH, Tysklind M, Haemig PD, Waldenström J, Wallensten A et al (2007) Antiviral oseltamivir is not removed or degraded in normal sewage water treatment: implications for development of resistance by influenza A virus. PLoS ONE 2:986–991

Gallart-Ayala H, Nuenz O, Moyano E, Galceran MT, Martins CPB (2011) Preventing false negatives with high-resolution mass spectrometry: the benzophenone case. Rapid Commun Mass Spectrom 25:3161–3166

Gao P, Mao D, Luo Y, Wang L, Xu B, Xu L (2012) Occurrence of sulfonamide and tetracycline-resistant bacteria and resistance genes in aquaculture environment. Water Res 46:2355–2364

Garcia-Ac A, Segura PA, Gagnon C, Sauve S (2009) Determination of bezafibrate, methotrexate, cyclophosphamide, orlistat and enalapril in waste and surface waters using on-line solid-phase extraction liquid chromatography coupled to polarity-switching electrospray tandem mass spectrometry. J Environ Monit 11:830–838

Garrison AW, Pope JD, Allen FR (1976) GC/MS analysis of organic compounds in domestic wastewater. In: Keith CH (ed) Identification and analysis of organic pollutants in water. Ann Arbor Science, Ann Arbor, pp 517–566

Godfrey E, Woessner WW, Benotti MJ (2007) Pharmaceuticals in on-site sewage effluent and ground water, Western Montana. Ground Water 45:263–271

Golet EM, Alder AC, Hartmann A, Ternes TA, Giger W (2001) Trace determination of fluoroquinolone antibacterial agents in urban wastewater by solid-phase extraction and liquid chromatography with fluorescence detection. Anal Chem 73:3632–3638

Green RE, Newton IAN, Shultz S, Cunningham AA, Gilbert M, Pain DJ et al (2004) Diclofenac poisoning as a cause of vulture population declines across the Indian subcontinent. J Appl Ecol 41:793–800
Halling-Sørensen B, Nors Nielsen S, Lanzky PF, Ingerslev F, Holten Lützhøft HC, Jørgensen SE (1998) Occurrence, fate and effects of pharmaceutical substances in the environment-a review. Chemosphere 36:357–393
Halling-Sørensen B, Lykkeberg A, Ingerslev F, Blackwell P, Tjornelund J (2003) Characterisation of the abiotic degradation pathways of oxytetracyclines in soil interstitial water using LC–MS–MS. Chemosphere 50:1331–1342
Heberer T, Butz S, Stan HJ (1995) Analysis of phenoxycarboxylic acids and other acidic compounds in tap, ground, surface and sewage water at the low ng/l level. Int J Environ Anal Chem 58:43–53
Hernando MD, Gomez MJ, Aguera A, Fernandez-Alba AR (2007) LC-MS analysis of basic pharmaceuticals (beta-blockers and anti-ulcer agents) in wastewater and surface water. TrAC Trends Anal Chem 26:581–594
Hignite C, Azarnoff DL (1977) Drugs and drug metabolites as environmental contaminants: chlorophenoxyisobutyrate and salicyclic acid in sewage water effluent. Life Sci 20:337–341
Hogenboom AC, van Leerdam JA, de Voogt P (2009) Accurate mass screening and identification of emerging contaminants in environmental samples by liquid chromatography-hybrid linear ion trap orbitrap mass spectrometry. J Chromatogr A 1216:510–519
Hruska K, Franek M (2012) Sulfonamides in the environment: a review and a case report. Vet Med 57:1–35
International Cooperation on Harmonisation of Technical Requirements for Registration of Veterinary Medicinal Products (2000) Environmental impact assessment (EIAs) for veterinary medicinal products (VMPs)—phase I. VICH GL 6 (Ecotoxicity phase I), Brussels, Belgium
International Cooperation on Harmonisation of Technical Requirements for Registration of Veterinary Medicinal Products (2005) Environmental impact assessment for veterinary medicinal products phase II guidance. VICH GL 38 (Ecotoxicity phase II), Brussels, Belgium
Isidori M, Parrella A, Pistillo P, Temussi F (2009) Effects of ranitidine and its photoderivatives in the aquatic environment. Environ Int 35:821–825
Jindal K (2012) Determination of trace pharmaceuticals in the environment using advanced analytical modalities. Master of Science (Pharm.) Thesis. Department of Pharmaceutical Analysis, National Institute of Pharmaceutical Education and Research, Punjab, India
Jones OA, Voulvoulis N, Lester JN (2004) Potential ecological and human health risks associated with the presence of pharmaceutically active compounds in the aquatic environment. Crit Rev Toxicol 34:335–350
Kallio JM, Lahti M, Oikari A, Kronberg L (2010) Metabolites of the aquatic pollutant diclofenac in fish bile. Environ Sci Technol 44:7213–7219
Kim SD, Cho J, Kim IS, Vanderford BJ, Snyder SA (2007a) Occurrence and removal of pharmaceuticals and endocrine disruptors in South Korean surface, drinking, and waste waters. Water Res 41:1013–1021
Kim Y, Choi K, Jung J, Park S, Kim PG, Park J (2007b) Aquatic toxicity of acetaminophen, carbamazepine, cimetidine, diltiazem and six major sulfonamides, and their potential ecological risks in Korea. Environ Int 33:370–375
Kolpin DW, Furlong ET, Meyer MT, Thurman EM, Zaugg SD, Barber LB et al (2002) Pharmaceuticals, hormones, and other organic wastewater contaminants in US streams, 1999–2000: a national reconnaissance. Environ Sci Technol 36:1202–1211
Kulkarni GK, Nagabhushanam R, Amaldoss G, Jaiswal RG, Fingerman M (1992) In vivo stimulation of ovarian development in the red swamp crayfish, Procambarus clarkii (Girard), by 5-hydroxytryptamine. Invertebr Reprod Develop 21:231–239
Kümmerer K, Helmers E (1997) Hospital effluents as a source for platinum in the environment. Sci Total Environ 193:179–184
Kvarnryd M, Grabic R, Brandt I, Berg C (2011) Early life progestin exposure causes arrested oocyte development, oviductal agenesis and sterility in adult Xenopus tropicalis frogs. Aquat Toxicol 103:18–24
Lee HB, Peart TE, Svoboda ML, Backus S (2009) Occurrence and fate of rosuvastatin, rosuvastatin lactone, and atorvastatin in Canadian sewage and surface water samples. Chemosphere 77:1285–1291
Li D, Yang M, Hu J, Zhang J, Liu R, Gu X et al (2009) Antibiotic-resistance profile in environmental bacteria isolated from penicillin production wastewater treatment plant and the receiving river. Environ Microbiol 11:1506–1517
Liu X, Lee J, Ji K, Takeda S, Choi K (2012) Potentials and mechanisms of genotoxicity of six pharmaceuticals frequently detected in freshwater environment. Toxicol Lett 211:70–76
Macia A, Borrull F, Calull M, Aguilar C (2004) Determination of some acidic drugs in surface and sewage treatment plant waters by capillary electrophoresis-electrospray ionization-mass spectrometry. Electrophoresis 25:3441–3449
Martinez NA, Schneider RJ, Messina GA, Raba J (2010) Modified paramagnetic beads in a microfluidic system for the determination of ethinylestradiol (EE2) in river water samples. Biosens Bioelectron 25:1376–1381
Martinez NA, Pereira SV, Bertolino FA, Schneider RJ, Messina GA, Raba J (2012) Electrochemical detection of a powerful estrogenic endocrine disruptor: ethinylestradiol in water samples through bioseparation procedure. Anal Chim Acta 723:27–32

Mennigen JA, Sassine J, Trudeau VL, Moon TW (2010) Waterborne fluoxetine disrupts feeding and energy metabolism in the goldfish Carassius auratus. Aquat Toxicol 100:128–137

Metcalfe CD, Koenig BG, Bennie DT, Servos M, Ternes TA, Hirsch R (2003) Occurrence of neutral and acidic drugs in the effluents of Canadian sewage treatment plants. Environ Toxicol Chem 22:2872–2880

Miao X-S, Metcalfe CD (2003) Determination of cholesterol-lowering statin drugs in aqueous samples using liquid chromatography-electrospray ionization tandem mass spectrometry. J Chromatogr A 998:133–141

Miege C, Budzinski H, Jacquet R, Soulier C, Pelte T, Coquery M (2012) Polar organic chemical integrative sampler (POCIS): application for monitoring organic micropollutants in wastewater effluent and surface water. J Environ Monit 14:626–635

Miller CS (1997) Toxicant-induced loss of tolerance-an emerging theory of disease? Environ Health Perspect 105:445–453

Mimeault C, Woodhouse AJ, Miao XS, Metcalfe CD, Moon TW, Trudeau VL (2005) The human lipid regulator, gemfibrozil bioconcentrates and reduces testosterone in the goldfish, Carassius auratus. Aquat Toxicol 73:44–54

Ministry of Environment and Forests (2011) Bio-medical waste (management and handling) rules, India

Möder M, Braun P, Lange F, Schrader S, Lorenz W (2007) Determination of endocrine disrupting compounds and acidic drugs in water by coupling of derivatization, gas chromatography and negative-chemical ionization mass spectrometry. CLEAN—Soil, Air, Water 35:444–451

Montforts MHMM, Kalf DF, van Vlaardingen PLA, Linders JBHJ (1999) The exposure assessment for veterinary medicinal products. Sci Total Environ 225:119–133

Nentwig G (2007) Effects of pharmaceuticals on aquatic invertebrates. Part II: the antidepressant drug fluoxetine. Arch Environ Contam Toxicol 52:163–170

Obando MA, Estela JM, Cerda V (2008) Multi-syringe chromatography (MSC) system for the on-line solid-phase extraction and determination of hydrochlorothiazide and losartan potassium in superficial water, groundwater and wastewater outlet samples. J Pharm Biomed Anal 48:212–217

Okuda T, Yamashita N, Tanaka H, Matsukawa H, Tanabe K (2009) Development of extraction method of pharmaceuticals and their occurrences found in Japanese wastewater treatment plants. Environ Int 35:815–820

Paxeus N (2004) Removal of selected non-steroidal anti-inflammatory drugs (NSAIDs), gemfibrozil, carbamazepine, beta-blockers, trimethoprim and triclosan in conventional wastewater treatment plants in five EU countries and their discharge to the aquatic environment. Water Sci Technol 50:253–260

Peck AM (2006) Analytical methods for the determination of persistent ingredients of personal care products in environmental matrices. Anal Bioanal Chem 386:907–939

Petrovic M, Gros M, Barcelo D (2006) Multi-residue analysis of pharmaceuticals in wastewater by ultra-performance liquid chromatography-quadrupole-time-of-flight mass spectrometry. J Chromatogr A 1124:68–81

Petrovic M, De Alda MJL, Diaz-Cruz S, Postigo C, Radjenovic J, Gros M et al (2009) Fate and removal of pharmaceuticals and illicit drugs in conventional and membrane bioreactor wastewater treatment plants and by riverbank filtration. Philos Trans R Soc A: Math Phys Eng Sci 367:3979–3989

Petrovic M, Farre M, de Alda ML, Perez S, Postigo C, Kock M et al (2010) Recent trends in the liquid chromatography-mass spectrometry analysis of organic contaminants in environmental samples. J Chromatogr A 1217:4004–4017

PharmEcology Associates, LLC (2008) Managing pharmaceutical waste. A 10-Step blueprint for healthcare facilities in the United States, Houston, U.S.A

Pomati F, Castiglioni S, Zuccato E, Fanelli R, Vigetti D, Rossetti C et al (2006) Effects of a complex mixture of therapeutic drugs at environmental levels on human embryonic cells. Environ Sci Technol 40:2442–2447

Qin F, Zhao YY, Sawyer MB, Li XF (2008) Column-switching reversed phase-hydrophilic interaction liquid chromatography/tandem mass spectrometry method for determination of free estrogens and their conjugates in river water. Anal Chim Acta 627:91–98

Radjenovic J, Petrovic M, Barceló D (2007) Analysis of pharmaceuticals in wastewater and removal using a membrane bioreactor. Anal Bioanal Chem 387:1365–1377

Radjenovic J, Petrovic M, Ventura F, Barcelo D (2008) Rejection of pharmaceuticals in nanofiltration and reverse osmosis membrane drinking water treatment. Water Res 42:3601–3610

Razavi B, Ben Abdelmelek S, Song W, O'Shea KE, Cooper WJ (2011) Photochemical fate of atorvastatin (lipitor) in simulated natural waters. Water Res 45:625–631

Richardson ML, Bowron JM (1985) The fate of pharmaceutical chemicals in the aquatic environment. J Pharm Pharmacol 37:1–12

Sarmah AK, Meyer MT, Boxall ABA (2006) A global perspective on the use, sales, exposure pathways, occurrence, fate and effects of veterinary antibiotics (VAs) in the environment. Chemosphere 65:725–759

Schwab BW, Hayes EP, Fiori JM, Mastrocco FJ, Roden NM, Cragin D et al (2005) Human pharmaceuticals in US surface waters: a human health risk assessment. Regul Toxicol Pharmacol 42:296–312

Schwaiger J, Ferling H, Mallow U, Wintermayr H, Negele RD (2004) Toxic effects of the non-steroidal anti-inflammatory drug diclofenac: Part I: histopathological alterations and bioaccumulation in rainbow trout. Aquat Toxicol 68:141–150

Sebok A, Vasanits-Zsigrai A, Palko G, Zaray G, Molnar-Perl I (2008) Identification and quantification of ibuprofen, naproxen, ketoprofen and diclofenac present in waste-waters, as their trimethylsilyl derivatives, by gas chromatography mass spectrometry. Talanta 76:642–650

Singer AC, Nunn MA, Gould EA, Johnson AC (2007) Potential risks associated with the proposed widespread use of tamiflu. Environ Health Perspect 115:102–106

Teijon G, Candela L, Tamoh K, Molina-Diaz A, Fernandez-Alba AR (2010) Occurrence of emerging contaminants, priority substances (2008/105/CE) and heavy metals in treated wastewater and groundwater at Depurbaix facility (Barcelona, Spain). Sci Total Environ 408:3584–3595

Ternes TA (1998) Occurrence of drugs in German sewage treatment plants and rivers. Water Res 32:3245–3260

Ternes T, Bonerz M, Schmidt T (2001) Determination of neutral pharmaceuticals in wastewater and rivers by liquid chromatography-electrospray tandem mass spectrometry. J Chromatogr A 938:175–185

Tong L, Li P, Wang Y, Zhu K (2009) Analysis of veterinary antibiotic residues in swine wastewater and environmental water samples using optimized SPE-LC/MS/MS. Chemosphere 74:1090–1097

United States Department of Health and Human Services, Food and Drug Administration, Center for Drug Evaluation and Research (CDER), Center for Biologics Evaluation and Research (CBER) (1998) Guidance for industry: environmental assessment of human drug and biologics applications, Rockville, U.S.A

United States Environmental Protection Agency (1976) Resources conservation and recovery act (RCRA), Washington, U.S.A

United States Environmental Protection Agency. 2002. Federal water pollution control act. Q:\COMP\WATER2\CLEANWAT.001, Washington, U.S.A

United States Environmental Protection Agency (2010) Best management practices for unused pharmaceuticals at health care facilities. EPA-821-R-10-006, Washington, U.S.A

United States Food and Drug Administration (2011) How to dispose of unused medicines, Rockville, U.S.A

United States Senate (2007) The preservation of antibiotics for medical treatment act. Senate Bill, S.549. U.S.A

van Leerdam JA, Hogenboom AC, van der Kooi MME, de Voogt P (2009) Determination of polar 1H-benzotriazoles and benzothiazoles in water by solid-phase extraction and liquid chromatography LTQ FT orbitrap mass spectrometry. Int J Mass Spectrom 282:99–107

Vanderford BJ, Snyder SA (2006) Analysis of pharmaceuticals in water by isotope dilution liquid chromatography/tandem mass spectrometry. Environ Sci Technol 40:7312–7320

Velagaleti R, Burns PK, Gill M, Prothro J (2002) Impact of current good manufacturing practices and emission regulations and guidances on the discharge of pharmaceutical chemicals into the environment from manufacturing, use, and disposal. Environ Health Perspect 110:213–220

Vera-Candioti L, Gil García MD, Martínez Galera M, Goicoechea HC (2008) Chemometric assisted solid-phase microextraction for the determination of anti-inflammatory and antiepileptic drugs in river water by liquid chromatography-diode array detection. J Chromatogr A 1211:22–32

Vieno N, Tuhkanen T, Kronberg L (2007) Elimination of pharmaceuticals in sewage treatment plants in Finland. Water Res 41:1001–1012

Waggot A (1981) Trace organic substances in the river Lee (Great Britain). In: Cooper WJ (ed) Chemistry in water reuse. Ann Arbor Science, Ann Arbor, pp 55–99

Webb S, Ternes T, Gibert M, Olejniczak K (2003) Indirect human exposure to pharmaceuticals via drinking water. Toxicol Lett 142:157–167

Wicher K, Reisman RE (1980) Anaphylactic reaction to penicillin (or penicillin-like substance) in a soft drink. J Allergy Clin Immunol 66:155–157

World Health Organization (1999) Guidelines for safe disposal of unwanted pharmaceuticals in and after emergencies. WHO/EDM/PAR/99.2, Geneva, Switzerland

World Health Organization (2011) Guidelines for drinking-water quality, 4th edn. Switzerland, Geneva

World Health Organization (2012) Pharmaceuticals in drinking-water. WHO/HSE/WSH/11.05, Geneva, Switzerland

Zhang Y, Geissen SU, Gal C (2008) Carbamazepine and diclofenac: removal in wastewater treatment plants and occurrence in water bodies. Chemosphere 73:1151–1161

Zuccato E, Castiglioni S, Fanelli R (2005) Identification of the pharmaceuticals for human use contaminating the Italian aquatic environment. J Hazard Mater 122:205–209

Chemical Contaminants in Water and Associated Health Hazards

22

Prahlad K. Seth

Abstract

Both surface and groundwater are impacted by chemicals and availability of drinking water meeting the required parameters for health safety continues to remain a challenging task. Chemicals which have led to severe health problems are arsenic, reported to cause skin lesions, cancer, and other ailments and fluoride which lead to dental and skeletal fluorosis when populations are exposed to high levels. Their source is aquifers, containing the mineral salts and health problems are localized to regions. However, their presence in groundwater from new regions is being reported. The other metals of concern are lead, manganese, mercury, cadmium, and chromium which lead to skin, neurological, cardiovascular, reproductive, and immune function disorders and cancer. Source of these metals is mostly industrial, and problem in general is in pockets of heavy industrialization. Persistent organic pollutants (POPs), majority of which are organochlorine pesticides, are also of concern as their presence in variable amounts have been detected in waste- and groundwater. The source of POPs is predominantly discharge of waste and untreated water, sewage, and runoffs from agricultural fields into water bodies. Varying amounts, mostly in small quantities, of pharmaceutical and veterinary drugs residues, nanoparticles, chemicals used for personal hygiene have also been detected in surface water. Some of these have chemical structures similar to hormones and are termed as endocrine disrupters. These are new emerging chemicals of concern. New tools and technologies for their regular monitoring and for treatment of the water are required to safeguard the health from drinking water contaminated by chemicals.

P. K. Seth (✉)
CSIR-Indian Institute of Toxicology Research,
Biotech Park, Sector G, Kursi Road, Jankipuram
226021, Lucknow, India
e-mail: prahladseth@gmail.com

P. P. Singh and V. Sharma (eds.), *Water and Health*,
DOI: 10.1007/978-81-322-1029-0_22, © Springer India 2014

Keywords
Metals · Nitrates · Pesticides · Pharmaceutical residues · Water contaminants

Introduction

Water is one of the most important and precious components of our planet earth, particularly the biosphere, and has been referred as elixir of life. Clean water suitable for drinking is one of the basic needs of life and essential for survival. Yet according to a WHO estimate, around 1,200 million people globally and around 125 million in India lack access to safe water (Kunjappy 2008). The ever increasing population, agricultural activities, industrialization, housing, and requirement of land for man's other activities are severely impacting the water resources. More than 100,000 chemical compounds are being used by the society for comfort and many more are being added every year. Several of these chemicals reach water resources including groundwater and many of these are of health concern. Quality of water is being monitored primarily by the Central Pollution Control Board (CPCB 2005), India (http://www.cpcb.nic.in), with the help of State Pollution Control Boards, and a national monitoring net work has been set up. CSIR National Laboratories like Indian Institute of Toxicology Research, (earlier Industrial Toxicology Research Center), Lucknow, and National Environmental Engineering Institute (NEERI), Nagpur, and other laboratories have made valuable contributions in developing technologies and methods for water quality monitoring and treatment. In major cities, piped water after treatment is available but some parts of the city are solely dependant on groundwater. The large population which lives in villages does not have access to clean or safe water which is a big challenge. The Government is aware of the problem and has taken number of initiatives for providing safe drinking water. In 1986, it launched National Drinking Water Mission which was renamed as Rajiv Gandhi National Drinking Water Mission in 1991 under the Ministry of Drinking Water and Sanitation for providing water to rural areas. While a large number of villages have been covered, yet problem areas exist in rural areas where a large rural population gets exposed to both microbial and chemical pollutants (Rajiv Gandhi National Drinking Water Mission 2010).

The major water resources in India are surface water, groundwater, and 20 river basins. The surface water is affected by runoff from soil which contains agro- and other chemicals and pollutants, municipal waste, and sewage containing chemicals used in household and discharges from uncontrolled minor industries. Sources which pollute the groundwater are geogenic and anthropo-genic contaminants like arsenic, fluoride, nitrate, iron, and sea water intrusion. At point, source chemical contamination can also occur due to leaching of chemicals form pipes and storage containers.

Several studies have shown a surprising connection between the quality of water and healthful longevity. The major sources of morbidity and mortality are the microbial infection but chemical contamination also leads to wide range of health ailments ranging from simple gastrointestinal ailments and liver-, reproductive-,and nervous system disorders to cancer. Chemicals of major concern are metals, for example, arsenic, mercury, lead, minerals such as nitrate, chloride, and industrial chemicals such as pesticides, disinfection agents, residues of pharmaceutical, and veterinary drugs. The common pathogens found in water and their impact on health are covered in detail elsewhere. Here, briefly, the chemical contaminants and their sources and impact on health are described.

Metals and Metalloids

Metals such as calcium, magnesium, potassium, sodium, and copper are needed in trace amount for biological functions of human body and other organisms. However, in higher concentrations, they become injurious to health. Trace metals reach into body through intake of food and water. In addition to useful metals, some others including heavy metals like aluminum (Al) arsenic (As), cadmium (Cd), chromium (Cr), copper (Cu), iron (Fe), manganese (Mn), mercury (Hg), lead (Pb), nickel (Ni), and zinc (Zn) reach into human body through ingestion of water and food.

Among these, arsenic and mercury are volatile, while remaining eight are non-volatile in nature. A survey under the National Drinking Water Mission carried out in Jammu & Kashmir, Himachal Pradesh, Uttar Pradesh, Sikkim, Tripura, Meghalaya, West Bengal, Orissa, Maharashtra, and Rajasthan by Indian Institute of Toxicology Research, Lucknow reported high levels of metals in water. The acceptable levels for metals for drinking water have been prescribed by Bureau of Indian Standards (BIS) for India and relevant agencies of other countries have fixed their permissible limits for effluents and drinking water (Table 22.1).

Table 22.1 Prescribed limits of selected minerals/metals

Metal/minerals	Highest desirable level	Maximum permissible level
Calcium (Ca)	75 mg/L	200 mg/L
Magnesium (Mg)	30 mg/L	100 mg/L
Copper (Cu)	0.5 mg/L	1.5 mg/L
Aluminum (Al)	0.03 mg/L	0.2 mg/L
Iron (Fe)	0.3 mg/L	1.0 mg/L
Nitrate (NO3)	45 mg/L	100 mg/L
Fluoride (F)	1.0 mg/L	1.5 mg/L
Manganese (Mn)	0.1 mg/L	0.3 mg/L
Zinc (Zn)	5.0 mg/L	15.0 mg/L
Chromium (Cr hexavalant)	0.05 mg/L	0.05 mg/L
Lead (Pb)	0.05 mg/L	0.05 mg/L
Cadmium (Cd)	0.01 mg/L	No relaxation
Arsenic (As)	0.01 mg/L	0.05 mg/L ?
Mercury (Hg)	0.001 mg/L	No relaxation

Source BIS 10500-2004

Table 22.2 Health hazards associated with chemical contaminants in water

S. No.	Chemicals	Health hazards
1.	Nitrates	Forms nitrosamines which may cause gastric cancer, methemoglobinemia
2.	Fluorides	Fluorosis, skeletal damage
3.	Arsenic	Skin lesions, nervous system and cardiovascular disorders, carcinogenicity
4.	Cadmium	Itai–itai diseases, kidney dysfunction hypertension, nervous disorders, cancer
5.	Chromium	Skin ulceration
6.	Copper	Hepatic and nervous system disorders
7.	Lead	Abdominal colic, anemia, nervous system disorders, teratogenic, and fetotoxic effects
8.	Manganese	Nervous system disorders
9.	Mercury	Nervous system disorders, kidney damage, mutagenicity, and teratogenicity
10.	Iron	Hemosiderosis, hemochromatosis skin pigmentation, hepatic disorders

According to a 2009 study in Ahmedabad, the range of cadmium (Cd), arsenic (As), and lead (Pb) was above BIS and WHO Standards. High levels of hexavalent chromium in groundwater ranging from 1.05 to 35.34 ppm have been reported from a village of Kanpur Dehat (Personal communication Krishna Gopal, IITR). In Kolkata, presence of four toxic metals, namely lead, chromium, cadmium, mercury, have been reported in the drinking water supply (Chacraverti 2011). Manganese in drinking water is associated with neurological disorders while lead besides its effects on cognitive functions among children also affects other body functions (Table 22.2). Since arsenic has been found to affect large populations, it has been dealt in detail here.

Arsenic

Arsenic poisoning due to consumption of mostly groundwater contaminated by the metal has become a public health threat in many countries, particularly Indian sub-continent (Bangladesh and Bengal), South America, and the Far East. Contamination of water from arsenic has also been reported in other parts of the world, namely Tiwan (Lamm et al. 2006), Argentina, Chile, Hungary, Mexico, Northern China, and many parts of USA (Frost et al. 2003; Peters 2008). The geological composition of the aquifers is the main cause of leaching of arsenic into the groundwater. Due to anthropogenic activities, arsenic is dissolved via oxygenation with the formation of soluble arsenate and arsenite (Kevin 2009; Akter and Ali 2011). According to a study by School of Environmental Studies, Jadavpur University, Kolkatta, West Bengal, 3,417 villages over 111 blocks primarily within 12 districts of the West Bengal namely Koch Bihar, North and South Dingapur, Maldha, Murshidabad, Nadia, Bardhanman, Kolkatta, Hooghly, Howrah, North and South Parganas adjoining river Bharathi are affected (Jadavpur University 2006).

More than 26 million people are exposed to arsenic through drinking water but only 15–20 % exhibit arsenic-induced non-cancerous, precancerous, and cancerous skin lesions (Banerjee et al. 2011). Giri and his associates have provided significant information regarding the polymorphism-based susceptibility of individuals to arsenic poisoning (Banerjee et al. 2011).

Early clinical symptoms of acute arsenic intoxication include abdominal pain, vomiting, diarrhea, muscular pain, and weakness with flushing of the skin. These symptoms are often followed by numbness and tingling of the extremities, muscular cramping, and the appearance of an erythematous rash. Chronic exposure to arsenic contaminated drinking water may result in dermal lesions, peripheral neuropathy, skin, lung and bladder cancer, and peripheral vascular disease. Major dermatological signs are hyperpigmentation, melano-keratosis, hyperkeratosis, melanosis, spotted and diffuse keratosis, leucomelanosis, dorsal keratosis, and peripheral vascular disease (Fawell and Nieuwenhuijsen 2003). Besides cancer and skin problems, cardiovascular ailments, diabetes, and reproductive disorders have been observed among populations exposed to water containing high levels of arsenic. Decrease in immune response in individuals ingesting low levels of arsenic through drinking water has also been reported (Kozul et al. 2009). Detailed studies on the possible mechanism of arsenic poisoning have been described by several investigators in West Bengal (Mazumder and Dasgupta 2011; Singh et al. 2011).

Fluoride

Waterborne fluorosis is a recognized problem and a cause of morbidity in various parts of the world, including the Indian sub-continent, Africa, and the Far East, where concentrations of fluoride can exceed 10 mg/L. In the Southeast Asian region, fluoride problem has been reported from Myanmar, Sri Lanka, and Northern Thailand (WHO 2010).

The adverse effects of ingestion of water containing high levels of fluoride are dental fluorosis with brown mottling of teeth. With increase in concentration of fluoride in water, stiffened and brittle bones and joints, deformities in knee and hip bones and finally crippling fluorosis have been reported (Sharma et al. 2012). According to scientific surveys, skeletal fluorosis in India and China occurs when the fluoride concentration in water exceeds 1 part per million (ppm) and has been found to occur in certain communities with only 0.7 part per million (Susheela et al. 1993; Susheela 1993, 1995; Rao and Puttanna 2000, 2006). Susheela and her colleagues have made significant contribution in monitoring fluoride levels in water and management of fluorosis.

In India, Central Ground Water Board, Public Health Engineering Departments, Water Authority/Board of various states, research institutes, universities, and colleges are engaged in studying fluoride content in drinking- and groundwater for last several years. In surface freshwater, fluoride concentrations are usually low −0.01 to 0.3 ppm. In groundwater, the natural concentration of fluoride depends on the geological and physico-chemical characteristics of the aquifer, the porosity and acidity of the soil and rocks, the temperature, the action of other chemical elements, and the depth of wells. Because of the large number of factors, the fluoride concentrations in groundwater have been found to range between 1 ppm to more than 35 ppm. In Kenya and South Africa, the levels have been reported to exceed 25 ppm. In India, concentrations up to 38.5 ppm have been reported. Some Union territories and 19 out of 35 states have groundwater which is highly contaminated with fluoride, with levels exceeding 1.0 mg/L and going up to 48 mg/L (INSA 2011). The 2011 WHO guidelines suggested that in areas with a warm climate, the optimal fluoride concentration in drinking water should remain below 1 mg/L (1 ppm), while in cooler climates, it could go up to 1.2 mg/L. Level of fluoride concentration in the groundwater in different states of India is shown in Table 22.3.

According to Bureau of Indian Standards for Drinking Water Specification 2004, the maximum permissible limit for fluoride in drinking water is 1.5 ppm and the highest desirable limit is 1.0 ppm. About 20 million people are severely affected by fluorosis, and 40 million are exposed to risk of endemic fluorosis. Presence of fluoride in groundwater poses a serious health problem in most of the states of India. Fluoride endemic states are Andhra Pradesh, Karnataka, Tamil Nadu, Punjab, Haryana, Maharashtra, Gujarat, Rajasthan, Uttar Pradesh, Kerala, Jammu and Kashmir, and Delhi. Severe contamination of fluoride in groundwater of Karbi, Anglong, and Nagaon districts of Assam and its manifestation in the form of fluorosis have been reported (Das et al. 2003). Levels of fluoride in groundwater have been found to be over and above the permissible limits (1.5 mg/L) in some parts of Uttar Pradesh as well. Preliminary investigations indicate severe health disorders in parts of the Kachnarwa region, which is in the upper Panda river basin, Sonbhadra District, Uttar Pradesh (Raju et al. 2009). The concentration of fluoride in the groundwater of the study area varies from 0.483 to 6.7 mg/L. The worst fluoride-affected villages are Rohiniyadamar, Madhuri, Neruiyadamar, Gobardaha, and Kunrwa of Uttar Pradesh. Most people in these villages suffer from dental to skeletal fluorosis such as mottling of teeth, deformation of ligaments, bending of spinal column, and age related problems.

In Tamil Nadu, fluoride levels have been reported to range from 0.1 to 2.8 mg/L and 0.4 to 4.0 mg/L during pre-monsoon and post-monsoon seasons, respectively (Sendesh and Ramasubramanian 2011). High levels of fluoride in groundwater of Mettur region have been reported recently (Srinivasamoorthy et al. 2012).

Table 22.3 Fluoride concentration in the groundwater in different states of India

State	Level (ppm)
Andhra Pradesh	0.56–50.8
Assam	0.4–20.6
Bihar	0.1–2.5
Delhi	0.2–32.5
Gujarat	0.1–40.0
Karnataka	0.33–7.8
Kerala	0.2–5.75
Madhya Pradesh	0.2–6.4
Orissa	0.1–16.4
Rajasthan	0.1–14
Tamil Nadu	0.5–4.0
Uttar Pradesh	0.1–17.5

Nitrate

In the last three to four decades, a gradual decline in the quality of raw water for drinking water supplies has been observed due to increase in nitrate concentrations. Globally, contamination of both surface- and groundwater resources by nitrate has been reported in Europe and United States (Holt 2000; WHO 2010). The main source of nitrates is excessive use of synthetic nitrogenous fertilizers in agriculture. The decayed vegetables, animal matter, domestic effluent, sewage/sludge disposal, industrial discharges, leachates from refuse dumps, atmospheric washout, and precipitation also lead to higher levels of nitrates in water (Makhijani and Mahoharan 1999).

Nitrate compounds are highly soluble in water which enables them to enter water sources easily. Percolation of nitrate through nitrate rich soils is a major factor for higher levels of nitrate in groundwater. Nitrate can reach both surface water and groundwater as a consequence of agricultural activity (including excess application of inorganic nitrogenous fertilizers and manures), from wastewater treatment and from oxidation of nitrogenous waste products in human and animal excreta, including septic tanks. Nitrite can also be formed chemically in distribution pipes by Nitrosomonas bacteria during stagnation of nitrate-containing and oxygen-poor drinking water in galvanized steel pipes or if chloramination is used to provide a residual disinfectant and the process is not sufficiently well controlled (WHO 2011).

High levels of nitrates in groundwater have been reported by National Environmental Engineering Institute, Nagpur in Andhra Pradesh, Bihar, Gujarat, Haryana, Karnataka, Lakshadweep, Madhya Pradesh, Maharashtra, Orissa, Punjab, Rajasthan, Tamil Nadu (NEERI 1991). High nitrate concentration in groundwater in India has been found in almost all hydro-geological formations (Ministry of Water Resources, Government of India). Groundwater levels of nitrate ranging from 0.1 to 870 mg/L with an average of 65 mg/L have been reported (Subramanian 2004). Gujarat, Rajasthan, and West Bengal were the states with high nitrate content in groundwater. A World Bank-assisted study in Karnataka showed that nitrate contamination of groundwater was a problem in a majority of districts (Mahadev and Gholami 2010). Nitrate levels above the desired limits have been reported in more than 33 % of water samples in Punjab and Haryana (Malik 2000). Instances of high concentrations of nitrate in groundwater in most parts of the country, with the exception of Jammu, Kashmir, and the northeastern states, are being reported.

Excessive concentrations of nitrates in drinking water are considered hazardous for infants, because nitrates are reduced to nitrites in their intestinal tract and may cause methemoglobinemia, leading to impairment of proper supply of oxygen to tissues. The WHO has recommended that water supplies containing high levels of nitrate (>100 ppm) should not be used for preparation of infant food. According to the Indian Standards for Safe Drinking Water, the highest desirable limit for nitrate is 45 ppm and maximum permissible limit is 100 ppm (BIS 2004). Besides methemoglobinemia, predominantly in infants, oral cancer, cancer of the colon, rectum and other gastrointestinal cancers, and neural tube defects and cytogenetic effects in children have been reported (Rao and Puttanna 2000, 2006; Lohar et al. 2010). Methemoglobinemia was found to be prevalent in all age groups in areas of Rajasthan, with high nitrate concentrations in drinking water (Gupta et al. 2000). Besides the agricultural resources, there are several non-agricultural sources of groundwater nitrate (Wakida and Lerner 2005).

Polycyclic Aromatic Hydrocarbons

Polycyclic aromatic hydrocarbons (PAHs), also referred as polycyclic nuclear hydrocarbons, are organic compounds composed of two or more benzene rings fused together. A large number of compounds can belong in the PAH category. United States Environmental Protection Agency (USEPA) has categorized that there are more than 100 different PAHs, among which 16 are priority pollutants. Several factors determine the physical, chemical, and biological characteristics of PAHs. The low molecular weight PAHs-those containing two to three benzene rings, for example, naphthalenes, fluorenes, phenanthrenes, and anthracenes are acutely toxic to aquatic organisms. Several high molecular weight PAHs, for example, benzopyrene, are known to be carcinogenic in humans.

PAH can be formed naturally, following agriculture burning or incomplete burning of gas, coal, garbage, or other organic substances and from motor vehicle exhaust. They are also formed through oil spills in sea. PAH has been detected in water sediments and due to their potential toxicity assumes great concern. Both acute and chronic effects following exposure to PAH have been reported. Acute effects lead to eye and skin irritation, inflammation, nausea, vomiting, diarrhea, and confusion while chronic or long-term exposure effects include decreased immune function, cataracts, kidney, and liver damage (e.g. jaundice), breathing problems, like asthma. Only few reports about the PAH levels in water are available from India. According to an UNEP report, the level of PAH in marine water ranged from ND-156.6 μg/L, whereas a mean concentration of 2.91 μg/L (range 0.20–14.47 μg/L) of PAH was found in drinking water samples from Ahmedabad (ITRC 1993; Jani et al. 1991; UNEP 2002). In a study conducted on River Gomti, total concentrations of 16 PAHs in water was found to range between 0.06 and 84.21 μg/L (average ($n = 48$), 10.33 ± 19.94 μg/L) (Malik et al. 2011) and in Coastal water of India, the PAHs ranged between 78 and 1,565 μg/L, respectively (Reddy et al. 2005).

In water, two- and three-ring PAHs and, in sediments, the three- and four-ring PAHs were the dominant species. The ratios of anthracene (An)/An + phenenthrene and fluoranthene (Fla)/Fla + pyrene were calculated to evaluate the possible sources of PAHs (Malik et al. 2011).

Chlorinated Pesticides

Pesticides and insecticides are chemicals which are being widely used in agricultural practices, and their residues both in surface- and groundwater have been detected worldwide (Asi et al. 2008). Presence of pesticides in varying quantities in both surface- and groundwater has been reported from different parts of India and has been reviewed (Seth 2008). In spite of ban and restrictions on use of certain pesticides like DDT, HCH, lindane, etc., for agricultural purposes, their presence in water is still being detected.

The levels of persistent organochlorine pesticide (OCPs), namely aldrin, dieldrin, endrin, HCB, HCH isomers, DDT isomers/metabolites, endosulfan isomers (alpha and beta), endosulfan sulfate, heptachlor and its metabolites, alpha-chlordane, gamma-chlordane, and methoxychlor were studied in the water and bed-sediments of river Gomti at eight different sites over a period of two years by IITR (Malik et al. 2009). The OCPs residues levels in river water and sediments ranged between 2.16 and 567.49 ng/L and 0.92 and 813.59 ng/g, respectively. The results also revealed that bed-sediments are contaminated with lindane, endrin, heptachlor epoxides, and DDT and may contribute to sediment toxicity in the freshwater ecosystem of the river.

In Tamiraparani river basin, South India, the OCP concentrations in surface water and sediments were found to range between 0.1–79.9 ng/L and 0.12–3,938.7 ng/g dry weight (dry weight), respectively (Kumarasamy et al. 2012), and heptachlor, o,p′-DDE, dieldrin, o,p′-DDD, and mirex were the dominant OCPs in water samples indicating heterogenic nature of nonpoint source of pollution.

Among various pesticides, the chlorinated pesticides are of great concern due to their low biodegradability and persistence. Several factors such as solubility of the pesticide in water, environmental factors like soil, weather, season, and distance to water sources, application methods and other practices associated with the pesticide use, influence the pesticide's potential to contaminate water. The pesticide residues enter the water supplies through runoff from pesticides applications areas and leaching through the soil into groundwater (Thodal 2009).

The organo-chemical pesticides are fat-soluble compounds hard to excrete from the body and consequently bio-accumulate in adipose tissue of mammalian species, persist in soil, and also get biomagnified through our food chain (Crinnion 2009). The food material also gets contaminated with pesticides grown on soil contaminated with pesticides or irrigated with pesticide contaminated water. Hence, humans get exposed to pesticides through ingestion of water and food contaminated with pesticides.

Long-term, low-dose exposure to pesticide are increasingly being linked to human health effects such as immune-suppression, hormone

Table 22.4 Chronic toxic effects of pesticides

Metals	Health Hazards
Dieldrin	Disrupts dopamine transport and oxidative damage in the brain (nigrostriatal pathway), Parkinson's disease, increased rates of hypothyroidism, Leydig cells, reducing their production of testosterone, which could contribute to infertility. Increased rates of lung cancer, breast cancer, pancreatic cancer mortality, and non-Hodgkin's lymphoma, superoxide production, and causes neutrophil inflammation
DDT	Nausea, vomiting, headache, neurological changes, facial numbness, partial paralysis, convulsions, loss of perception and vibratory sensation, moderately rapid respiration, slow to normal pulse, kidney and liver dysfunction, general weakening of the immune system (decreased rate of lymphoproliferation), reproductive and carcinogenic effects
HCB	Increased rates of cryptorchidism in male offspring and ability to be an initiator, promoter, and progressor of breast cancer, non-Hodgkin's lymphoma, neurotoxic to the dopaminergic system, may lead to increased risk of parkinsonism and atherosclerosis
Endosulfan	Cancer or birth defects, adverse nervous system effects
Chlordane	Increased risk of diabetes, non-Hodgkin's lymphoma seminoma (testicular germ cell tumor), prostate cancer, and decreased natural-killer cell ability to lyse tumor cells

Source Crinnion (2009)

disruption, diminished intelligence and other neurological problems, reproductive abnormalities, and cancer (Table 22.4).

Chemicals of Emerging Concern

Emerging chemicals include not only those in active production or use in the domestic market, but also those that are a by-product of production and manufacturing, combustion, or a metabolite (i.e. a breakdown product) of a parent compound. The emerging substances of health concern are persistent, bioaccumulative and toxic, global organic contaminants, pharmaceuticals and personal care products (PPCP), nanoparticles, etc.

Municipal wastewater contributes significantly to the micro pollutant load into the aquatic environment (Kolpin et al. 2002). Municipal wastewater include many synthetic substances such as active pharmaceutical ingredients, compounds with biocidal properties (e.g. for material protection or gardening), food additives, cosmetics ingredients or detergents, plasticizers, nanoparticles, etc. Pharmaceutical compounds are highly bioactive and therefore exert undesired effects in organisms. The large-scale production and increasing use of pharmaceuticals is primarily responsible for their residues in waste and natural waters (Rene et al. 2010).

The other chemicals of emerging concern are by-products of drinking water disinfectants (Nieuwenhuijsen et al. 2009). Chlorine and its variants are widely used as a disinfectant in India which results in variety of disinfectant by-products. The levels of chlorine concentrations are directly proportional to the levels of disinfectant by-products particularly in the presence of organic matter. The by-products, released in extremely low concentrations during the disinfection process of water supplies, are potentially toxic chemical compounds. These include chloroform and chloroacetic acids formed during chlorination, and bromate, a by-product of ozone treatment. Different disinfectants produce different types or amounts of disinfection by-products. Disinfection by-products for which regulations have been established for their identification in drinking water, include trihalomethanes, haloacetic acids, bromate, and chlorite.

Chloroform is the disinfectant by-product produced in the largest amount. (At high doses, it is an animal carcinogen.) trihalomethanes (THMs) are the family of disinfection by-products of chlorine and are primary water pollutants, and its analysis and standardization facilities have been developed at CPCB's Trace Organic Laboratory. Recently, it was found that

1H-benzotriazole is a widespread contaminant of wastewater and surface water (Mawhinney et al. 2012). Monitoring of these emerging chemicals is a challenge since they require sophisticated analytical tools and expertise. Several of these chemicals have structural similarities with hormones and therefore have been referred as endocrine disrupters. Although amounts detected are low but because of their toxicogenic nature and potential to affect the biological systems, their monitoring is becoming essential.

Acknowledgments The author is thankful to Ms. Shipra Gupta, Senior Programmer, Bioinformatics Section, Biotech Park, Lucknow and Mr. Shailendra Bakshi, Technical Assistant, Biotech Park for their help in preparation of this manuscript. The support of the Bioinformatics Section of Department of Biotechnology, Government of India is gratefully acknowledged.

References

Akter A, Ali MH (2011) Arsenic contamination in groundwater and its proposed remedial Measures. Int J Environ Sci Technol 8(2):433–443

Asi MR, Hussain A, Muhmood ST (2008) Solid phase extraction of pesticide residues in water samples: DDT and its metabolites. Int J Env Res 2(1):43–48

Banerjee N, Nandy S, Kearns JK, Bandyopadhyay AK, Das JK, Majumder P et al (2011) Polymorphisms in the TNF-alpha and IL 10 gene promoters and risk of arsenic-induced skin lesions and other nondermatological health effects. Toxicol Sci 121:132–139

Bureau of Indian Standards (2004) Specifications for drinking water. Bureau of Indian Standards, New Delhi, p 10500

Central Pollution Control Board (2005) www.cpcb.nic.in/WQ2005/watmain2005.htm

Chacraverti S (2011) The Kolkata metropolitan area: probing for four toxic metals in drinking water. PROJECT REPORT (Project Supported by EEJP) Society for Direct Initiative for Social and Health Action (DISHA)

Crinnion WJ (2009) Chlorinated pesticides: threats to health and importance of detection. Altern Med Rev 14:347–359

Das B, Talukdar J, Sarma S, Gohain B, Dutta RK, Das HB et al (2003) Fluoride and other inorganic constituents in groundwater of Guwahati, Assam, India. Curr Sci 85:657–661

Fawell J, Nieuwenhuijsen MJ (2003) Contaminants in drinking water: environmental pollution and health. Br Med Bull 68:199–208

Frost FJ, Muller T, Petersen HV, Thomson B, Tollestrup K (2003) Identifying US populations for the study of health effects related to drinking water arsenic. J Expo Anal Environ Epidemiol 13:231–239

Gupta SK, Gupta RC, Seth AK, Bassin JR, Gupta DK, Sharma S (2000) Recurrent acute respiratory tract infections in areas with high nitrate concentrations in drinking water. Env Health Prospects 108:363–366

Holt MS (2000) Sources of chemical contaminants and routes into the freshwater environment. Food Chem Toxicol 238:21–27

INSA (2011) Hazardous metals and minerals pollution in India: sources, toxicity and management. A Position Paper. Indian National Science Academy, New Delhi

ITRC (1993) ITRC report on toxicological evaluations of marine and estuarine environment of the northeast coast (India). Department of Ocean Development (DOD), Government of India, New Delhi

Jadavpur University (2006) Report on Ground Water Arsenic Contamination in West Bengal, India—20 years study. (http://www.soesju.org)

Janardhana RN, Sangita D, Das K (2011) Emerging challenge: fluoride contamination in groundwater in Agra District, Uttar Pradesh. Asian J Exp Bio Sci 2:131–134

Jani JP et al (1991) Residues of organochlorine pesticide and polynuclear aromatic hydrocarbons in drinking water of Ahmadabad city. India. Bull Env Contam Toxicol 47:381–385

Kevin H (2009) Arsenic: environmental chemistry, health threats and waste treatment. Wiley, Hoboken, p 588. ISBN: 978-0-470-02758-5

Kolpin DW, Furlong ET, Meyer MT, Thurman EM, Zaugg SD, Barber LB et al (2002) Pharmaceuticals, hormones, and other organic wastewater contaminants in US streams, 1999–2000: a national reconnaissance. Environ Sci Technol 36:1202–1211

Kozul CD, Ely KH, Enelow RI, Hamilton JW (2009) Low-dose arsenic compromises the immune response to influenza A infection in vivo. Env Health Perspects 117:1441–1447

Kumarasamy P, Govindaraj S, Vignesh S, Rajendran RB, James RA (2012) Anthropogenic nexus on organochlorine pesticide pollution: a case study with Tamiraparani river basin, South India. Env Monit Assess 184:3861–3873

Kunjappy R (2008) Progress and prospects on water: for a clean and healthy world with special focus on sanitation. World Water Week in Stockholm-2008

Lamm SH, Engel A, Penn CA, Chen R, Feinleib M (2006) Arsenic cancer risk confounder in southwest Taiwan data set. Environ Health Perspect 114:1077–1082

Lohar V, Arora R, Arora V (2010) Nitrates and associated health hazards. Webmed Central Env Med 1(10):WMC001067

Mahadev J, Gholami S (2010) Heavy metal analysis of Cauvery River water around Krs Dam, Karnataka, India. J Adv Lab Res Biol 1:13–19

Makhijani SD, Mahoharan A (1999) Nitrate pollution problem in drinking water sources: monitoring and surveillance. Paper presented in the workshop Water Quality Field Test Kits for Arsenic, Fluoride and Nitrate held from 8 to 9 Sept 1999 at ITRC, Lucknow

Malik RPS (2000) Agriculture and water quality in India. Towards sustainable management. Water Rep 21:73–85

Malik A, Ojha P, Singh KP (2009) Levels and distribution of persistent organochlorine pesticide residues in water and sediments of Gomti River (India)—a tributary of the Ganges River. Environ Monit Assess 148:421–435

Malik A, Verma P, Singh AK, Singh KP (2011) Distribution of polycyclic aromatic hydrocarbons in water and bed sediments of the Gomti River, India. Env Monit Assess 172:529–545

Mawhinney DB, Vanderford BJ, Snyder SA (2012) Transformation of 1H-Benzotriazole by Ozone in aqueous solution. Environ Sci Technol 46: 7102–7111

Mazumder Guha D, Dasgupta UD (2011) Chronic arsenic toxicity: studies in West Bengal, India. Kaohsiung J Med Sci 27:360–370

Ministry of Water Resources, Government of India. Ground Water Quality Scenario. http://mowr.gov.in

NEERI (1991) Water Mission Report on Participation of NEERI. NEERI (National Environmental Engineering Research Institute), Nagpur, India

Nieuwenhuijsen MJ, Martinez D, Grellier J, Bennett J, Best N, Iszatt N et al (2009) Chlorination disinfection by-products in drinking water and congenital anomalies: review and meta-analyses. Environ Health Perspects 117:1486–1493

Peters SC (2008) Arsenic in groundwaters in the Northern Appalachian Mountain belt: a review of patterns and processes. J Contam Hydrol 99:8–21

Rajiv Gandhi National Drinking Water Mission (2010) Report of an *evaluation study* by the Planning Commission, vol 22. p 17

Raju NJ, Dey S, Das K (2009) Fluoride contamination in groundwaters of Sonbhadra District, Uttar Pradesh, India. Curr Sci 96:979–985

Rao P, Puttanna K (2000) Nitrates, agriculture and environment. Curr Sci 79:1163–1168

Rao P, Puttanna K (2006) Strategies for combating nitrate pollution. Curr Sci 91:1335–1339

Reddy SM, Basha S, Joshi HV, Ramachandraiah G (2005) Seasonal distribution and contamination levels of total PHCs, PAHs and heavy metals in coastal waters of the Alang-Sosiya ship scrapping yard, Gulf of Cambay, India. Chemosphere 61:1587–1593

Rene PS, Thomas E, Thomas BH, von Urs G, Bernhard W (2010) Global water pollution and human health. Ann Rev Environ Res 35:109–136

Sendesh K, Ramasubramanian V (2011) Assessment of fluoride contamination in groundwater using GIS, Dharmapuri District, Tamil Nadu, India. Int J Eng Sci Technol 3:1077–1085

Seth PK (2008) Chemical contamination of water: Indian scenario. In: Nath KJ, Sharma VP (eds) Safe water for community health. The National Academy of Sciences, Allahabad, pp 15–25

Sharma S, Ramani J, Bhalodia J, Thakkar K (2012) Fluoride and fluorosis in context to Gujarat State of India: a review. Res J Pharm Biol Chem Sci 3:85–94

Singh AP, Goel RK, Kaur T (2011) Mechanisms pertaining to arsenic toxicity. Toxicol Int 18:87–93

Srinivasamoorthy K, Vijayaraghavan V, Vasanthavigar M, Sarma S, Chidambaram S, Anandhan P et al (2012) Assessment of groundwater quality with special emphasis on fluoride contamination in crystalline bed rock aquifers of Mettur region, Tamil Nadu, India. Arab J Geosci 5:83–94

Subramanian V (2004) Water quality in South Asia. Asian J Water Env Pollut 1:41–54

Susheela AK (1993) Prevention and control of fluorosis, vol 1. Health Aspects. Published by Ministry of Rural Development (RGNDWMA), Government of India

Susheela AK (1995) Epidemiological studies of health risks from drinking water naturally contaminated with fluoride. Proc Rome Symp 233:123–134

Susheela AK, Kumar A, Bhatnagar M, Bahadur R (1993) Prevalence of endemic fluorosis with gastrointestinal manifestations in people living in some north Indian villages. Fluoride 26:97–104

Thodal CE (2009) Monitoring for pesticides in groundwater and surface water in Nevada, 2008: U.S. Geological Survey Fact Sheet 2009-3093. 4

United Nations Environment Programme (2002) Regionally based assessment of Persistent Toxic Substances. Indian Ocean Regional Report. UNEP Chemicals, Geneva, Switzerland

Wakida FT, Lerner DN (2005) Non-agricultural sources of groundwater nitrate: a review and case study. Water Resour 39:3–16

World Health Organization (2010) Drinking water quality in the South-East Asia Region. SEA-EH-567

World Health Organization (2011) Guidelines for drinking water quality, 4th edn. Geneva, Switzerland

23 Water Quality and Health Issues in South-West Parts of Punjab State, India

K. P. Singh

Abstract

South-west parts of Punjab State comprising Bathinda, Faridkot, Ferozepur, Mansa and Muktsar districts suffer from groundwater quality and associated health problems. Results of hydrogeological and hydrochemical investigations reveal that hot spots of fluoride and uranium having more than permissible concentrations prescribed for drinking water by the World Health Organisation (WHO) occur in most parts of the area. Arsenic hot spots occur locally, but are not widely spread. Other heavy metals (Cr, Se, Hg, Mn, Cu, Zn, etc.) show temporal variations in their concentration and do not occur consistently. Radon concentration in groundwater is within safe limit. In addition, pesticide residues have also been found to occur in groundwater. The cocktail of such toxics and pesticides in groundwater has affected the human health in the area. Fluoride is responsible for causing bone and joint problems, whereas uranium is nephrotoxic and affects the kidneys. Cases of arsenic skin disorders, notably pigmentation changes (melanosis), were not observed in the area. Pesticides are known to be carcinogenic, and their presence in human blood of cancer patients indicates their harmful impact. To alleviate health problems, enhancement of safe canal water supply for drinking purpose is recommended besides installing reverse osmosis plants in the area for removing uranium, fluoride and other toxics from ground water.

Keywords

Arsenic · Fluoride · Health · Uranium · Water

Introduction

The south-west parts of Punjab state comprising Bathinda, Faridkot, Ferozepur, Mansa and Muktsar districts (Fig. 23.1) are popularly known as

K. P. Singh (✉)
Centre of Advanced Study in Geology, Punjab University, Chandigarh, 160014, India
e-mail: kpsingh1948@rediffmail.com

P. P. Singh and V. Sharma (eds.), *Water and Health*,
DOI: 10.1007/978-81-322-1029-0_23,

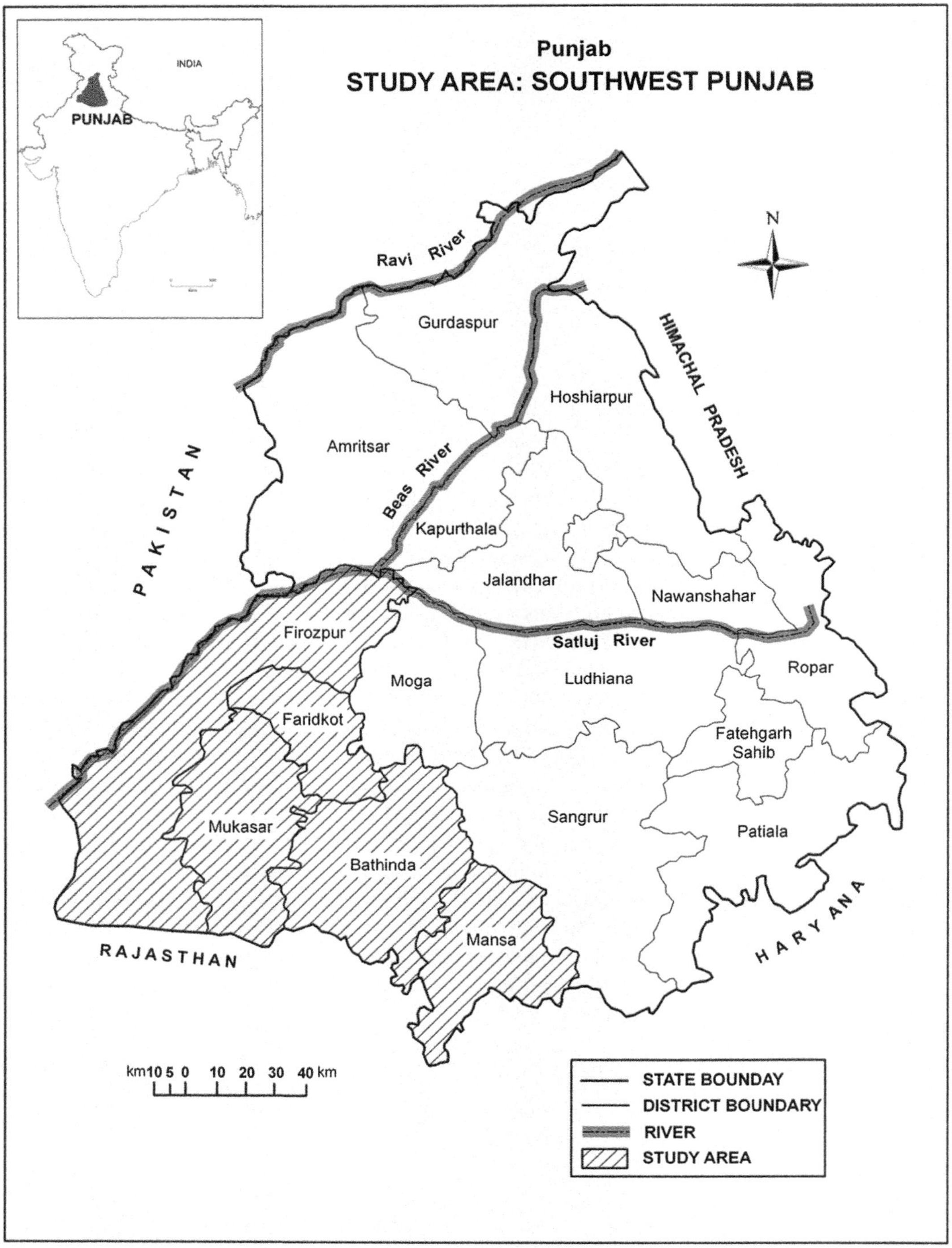

Fig. 23.1 Location of the south-west part of Punjab state

the cotton belt of Malwa region for rich cotton production. Large amounts of pesticides are used to save the cotton crop from various diseases. Of late, large numbers of cancer cases have been reported from this belt by the media and non-government organisations. This led to carrying out detailed epidemiological survey at micro-level in Talwandi Sabo block of Bhatinda district by a team of doctors of P.G.I., Chandigarh (Thakur et al. 2008). The results of survey revealed that prevalence of confirmed cancer cases per 100,000 population was 125 (107/85315) in Talwandi Sabo block as compared to Chamkaur Sahib block 72 (71/97928), a block situated in north-east parts of the state chosen for comparative study (Thakur et al. 2008). They also reported a limited data on concentration of heavy metals such as As, Cr, Se, Hg, Cd and Ni and pesticides such as Heptachlor, Ethion and Chlorpyrifos, which were higher in drinking water, vegetables and blood in comparison with Chamkaur Sahib block (Thakur et al. 2008).

In order to know the effects of water quality on human health, the author carried out detailed hydrogeological and hydrochemical studies in Talwandi Sabo block and Muktsar-Malout belt taking them as representative areas of south-west Punjab. The data reported by other workers in the adjoining parts (Kumar et al. 2010; Sharma et al. 2010) have also been examined and reinterpreted in relation to human health in the present paper. The results of hydrogeological and hydrochemical studies indicate that groundwater in the area has in general higher electrical conductivity and TDS indicating salinity besides presence of fluoride, uranium and arsenic more than permissible limits prescribed by the by World Health Organisation (WHO) (2004), United States Environmental Protection Agency (USEPA) (2002) and Atomic Energy Regulatory Board, Department of Atomic Energy (AERB, DAE) (2004). In addition, pesticide residues appear to be responsible for causing human health problems.

General Features and Hydrogeology

The south-west region of Punjab state comprising districts Bathinda, Faridkot, Ferozepur, Muktsar and Mansa lies between north latitude of 25°56″ and 31°11″ and east latitude of 73°55″ and 75°37″. The slope of the region is from north-east to south-west. The region is bounded on the west side by river Satluj. The annual rainfall varies from 300 to 400 mm though some parts showing slight variation from this range. Most parts of the area are characterised by sand dunes. Other important geomorphological features such as abandoned palaeochannels of rivers are also present as revealed by LANDSAT and IRS data (Singh et al. 1980). The area has a vast network of canal irrigation system.

The area comprises of Indo-Gangetic alluvium of Quaternary age. Groundwater mainly occurs in alluvium comprising sand, silt, kankar and gravel beds. At shallow depths, groundwater occurs under unconfined conditions, whereas at deeper levels confined/semi-confined conditions prevail. Average depth to groundwater ranges from less than 2 m in waterlogged areas in Muktsar-Malout belt to 10 m below ground level. Groundwater flow direction is from NE to SW direction. Thickness of freshwater aquifers below which saline aquifers occur is much less about 20–100 m as compared to north-east parts of state where freshwater sediments exist up to 450 m. Water levels have been rising in most parts of the area since last 100 years (Singh and Kishore 2009). Rise of water levels is responsible for mineral–water interactions giving rise to poor chemical quality of groundwater. The chemical quality of groundwater in general is not good, and the area is characterised by salinity and alkalinity hazards (Sharma et al. 2010). Around canals, freshwater lenses also occur due to seepage of freshwater and are taped by hand pumps (up to 20 m) for domestic water use.

General Chemical Characteristics of Ground Water

Study of results obtained from chemical analysis of water samples collected by the author from Talwandi Sabo block and Muktsar-Malout region during December 2011 and April 2012 (Table 23.1) and results published from adjoining areas by various workers (Rao et al. 2004; Kumar et al. 2007, 2009; Sharma et al. 2010; Kumar et al. 2010; Central Ground Water Board 2005, 2010; Dhanda et al. 2011) revealed the following major chemical characteristics of groundwater of shallow aquifer system up to 50 m.

pH of groundwater ranges from 7.05 to 9 indicating alkaline nature of groundwater. Electrical conductivity values range from 275 to 10,840 micromhos/cm at 25 °C, showing large variations in the total mineralisation of groundwater. Low electrical conductivity values are common around hydrological boundaries, that is, near unlined and lined canals, due to formation of freshwater lenses which float over native saline groundwater (Singh 2002).

Calcium concentration ranges from 6.2 to 740 mg/l. Magnesium concentration ranges from 2.5 to 257 mg/l. Desirable limit of magnesium in drinking water is 30 mg/l, and permissible limit is 100 mg/l according to Bureau Of Indian Standards [Bureau of Indian Standards (BIS) 2003]. Higher concentration of magnesium has laxative effect and can cause gastrointestinal problems. Sodium ranges from 2.9 to 1,450 mg/l. Potassium ranges from 3.7 to 665 mg/l. Excess amount of potassium is cathartic.

Carbonate values range from nil to 12 mg/l. Bicarbonate values range from 132 to 1,463 mg/l. Chloride values range from 7.4 to 1,695 mg/l. The desirable limit of chloride for drinking water is 250 mg/l while permissible limit is 1,000 mg/l [Bureau of Indian Standards (BIS) 2003]. Higher concentration of chloride can prove to be injurious to heart and kidney patients. Sulphate concentration ranges from 5.7 to 3,600 mg/l. The permissible limits for sulphate in drinking water are 400 mg/l [Bureau of Indian Standards (BIS) 2003]. Sulphate ions in association with higher concentration of Mg can act as a laxative and may cause gastrointestinal irritation.

Locally, at places high nitrate values having concentration more than 45 mg/l and high iron values having concentration more than 1 mg/l also occur (Central Ground Water Board 2010). Higher concentration of nitrate more than 45 mg/l can cause methaemoglobinaemia. Iron concentration more than 1 mg/l promotes bacterial growth. Table 23.2 summarises permissible concentration for drinking water by various

Table 23.1 Range of concentration of major parameters in Talwandi Sabo block and Muktsar-Malout belt in south-west Punjab

Parameter	Range in Talwandi Sabo area	Range in Muktsar-Malout belt
pH	8.7–9.14	8.21–9.12
Electrical conductivity (micromhos at 25 °c)	440–6,860	560–10,840
Calcium	10.4–52.1 mg/l	21.9–43.2 mg/l
Magnesium	10.2–167.1 mg/l	21.9–158.8 mg/l
Sodium	15.2–315.9 mg/l	44–1205.4 mg/l
Potassium	10.4–310.5 mg/l	9.5–112.5 mg/l
Carbonate	Nil	Nil
Bicarbonate	125–515 mg/l	230–425 mg/l
Chloride	4.97–440.2 mg/l	106.5–1245.5 mg/l
Sulphate	5.7–18.4 mg/l	8.9–28.0 mg/l
Nitrate	8.71–29.53 mg/l	4.63–29.890 mg/l
Phosphate	0.084–0.622 mg/l	0.12–0.340 mg/l

Table 23.2 Different guidelines for drinking water, health implications and general status in south-west Punjab

Parameter	WHO standards (2004)		ISO—10500: 1991 standard (revised 2003)		Status in south-west Punjab	Health implications
	Desirable limit (DL)	Maximum permissible limit (MPL)	Desirable limit (DL)	Permissible limit (PL)		
pH	7–8.5	9.2	6.5–8.5	No relax	V	Affects taste
EC (μ/cm)	500	1,400	500	1,000	M–H	Gastrointestinal irritation
TH (mg/l)	100	500	300	600	V	Scale formation, calcification of arteries
Na^+ (mg/l)	–	200	–	–	M–H	Scale formation, injurious to heart and kidney problems
K (mg/l)	–	–	–	–	L	Excess amount is cathartic
Ca^{2+} (mg/l)	75	200	75	200	V	Scale formation. Deficiency causes rickets
Mg^{2+} (mg/l)	50	150	30	100	V	Encrustation in water supply structure, salts are cathartics and diuretic
Cl (mg/l)	200	600	250	1,000	H	Salty taste, injurious to heart and kidney patients
SO_4 (mg/l)	200	400	200	400	V	Laxative effect
NO_3 (mg/l)	–	45	45	100	OH	Methaemoglobinaemia
F (mg/l)	–	1.5	1.0	1.5	OH	Fluorosis
Fe (mg/l)	–	0.3	–	1	S	Promotes bacterial growth
Zn (mg/l)	–	5	5	15	L	Causes astringent taste and opalescence in water
Mn (mg/l)	–	0.1	0.1	0.3	L	Adverse effects on domestic users and water supply structure
Cu (mg/l)	–	1	0.5	1.5	L	Discoloration and corrosion of pipe fitting and utensils

V variable, *M* medium, *H* high, *OH* occasionally high, *L* low, *S* sporadically high

agencies, their general status in south-west Punjab and health implications.

Major Health-Related Parameters in Groundwater

The occurrence and impact of major health-related parameters such as fluoride, arsenic, uranium, other heavy metals, radon and pesticides are discussed as under.

Fluoride

South-west parts of Punjab have high fluoride content in groundwater, but its concentration is variable depending upon other geochemical and mineralogical controls. The author determined fluoride content of groundwater samples collected during December 2011 and April 2012 from Talwandi Sabo and Muktsar-Malout belt and found its concentration ranging from 0.11 to 1.88 mg/l. 50 % of water samples contained fluoride more than 1 mg/l. Similarly, Kumar et al. 2010 reported results of 834 samples from 100 villages of Bathinda district and observed a concentration range of 0–9.00 mg/l, and 66 % of samples in the district were found to have more than 1.00 mg/l. In Muktsar district, fluoride concentration range of 0.03–14.5 mg/l was reported by Rao et al. 2004. Central Ground Water Board (2010) also reported fluoride hot spots with concentration of more than 1.5 mg/l in the entire south-west Punjab. The major cause of high concentration of fluoride in ground water is geogenic in this part of the area (Jacks et al. 2005).

Effects of Fluoride on Human Health

World Health Organisation (WHO) (2004) has fixed a upper limit of 1.5 mg/l for fluoride in drinking water as it is known now that exposure to fluoride in drinking water at concentration above 1.5 mg/l can cause dental fluorosis and above 10 mg/l can result in crippling fluorosis (Dissanayake 1991). The health effects of fluoride in drinking water are summarised in Table 23.3.

Table 23.3 Health effects of fluoride concentration in drinking water

Fluoride concentration (mg/l)	Chronic health effects
Nil	Limited growth and fertility
0.0–0.5	Dental caries
0.5–1.5	Promotes dental health, prevents tooth decay
1.5–4.00	Dental fluorosis (mottled teeth)
4.0–10	Dental fluorosis, skeletal fluorosis
>10	Crippling fluorosis

After Dissanayake (1991)

In addition to bone and joint problems, high concentration of fluoride may cause increase in concentration of thyroid-stimulating hormone (TSH) and decrease in concentration of T3 and T4 hormones in the thyroid gland resulting in hypothyroidism (Raja Reddy 1979). It has also been found to affect reproductive potential of humans (Freni 1994). The intelligence quotient and mental work capacity of children who were born and raised in fluorosis-endemic area have been reported to be lower than the normal (Lu et al. 2000).

In south-west parts of Punjab, large population is suffering from bone and joint problems due to consumption of fluoride-rich waters as revealed by field surveys in the area.

Uranium

Uranium concentration in groundwater drawn from hand pumps and shallow tube wells up to 50 m depth ranges between 5 and 98 μg/l in Talwandi Sabo area as revealed by the results of samples collected by the author and Department of Water Supply Sanitation, Punjab, and tested by XRF techniques at Nuclear Research Laboratory of Physics Department of Panjab University, Chandigarh. The author observed large variations in the concentration of uranium in groundwater within a single village.

Table 23.4 Uranium concentration in groundwater from south-west distracts of Punjab

District	No. of samples	Mean concentration of U (μg/L)	Minimum (μg/L)	Maximum (μg/L)
Bathinda	80	80.83 ± 5.3	2	644
Mansa	80	75.83 ± 6.7	<0.2	498
Faridkot	30	68.56 ± 8.3	16.5	473.5
Ferozepur	45	62.12 ± 7.3	2.8	184.4

After Kumar et al. (2011b)

Recent data on uranium content of groundwater from four districts of south-west Punjab were reported by Kumar et al. 2011b by analysing 235 groundwater samples using laser fluorimeter. They reported a concentration range of 2–644 μg/l with a mean value of 73.1 μg/l (Table 23.4). In Bathinda district, 50 % of total subsurface water samples were found to be having uranium concentration greater than 60 μg/l, which is the permissible limit prescribed by Atomic Energy Regulatory Board, Department of Atomic Energy (AERB, DAE) 2004 for drinking water. Groundwater samples from Mansa, Faridkot and Firozpur districts showed 42.5, 36.7 and 31.1 % of the total samples having uranium concentration beyond the permissible limits of 60 μg/l, respectively. They also studied the variation in uranium concentration with depth of tube wells. In Bathinda district, they found bore wells with depth less than 33.3 m (100 ft) showed maximum concentration of U, whereas Mansa and Faridkot districts showed higher concentration of U at depth range of 33–66 m (100–200 ft). In case of Firozpur district, maximum concentration was observed to be at depth of more than 100 m (300 ft). No distinct relationship of depth of tube well and concentration of U was noticed. However, Kumar et al. 2011a observed a strong correlation of uranium with high salinity and TDS. In view of the above observations, there is a need to collect aquifer-wise data on U concentration rather than bore well depth wise. A deep tube well may be tapping single aquifer or it could be tapping the confined or semi-confined aquifer only or both unconfined and confined aquifers depending upon the local hydrogeology and design of the well.

The causes of high concentration of U in groundwater can be because of geogenic (natural) or anthropogenic (man's activity) reasons (Singh 2010). The geogenic sources are basement granite and aquifer mineralogy, whereas anthropogenic causes include fly ash from thermal plants, air blown from Afghanistan (nuclear reactors) and use of phosphatic fertilisers (Singh 2010). At present, there are several data gaps and it is difficult to comment on the exact cause of occurrence of U hot spots. Systematic data on water chemistry, soils and aquifer mineralogy should be collected to know the exact cause.

Effects of Uranium on Human Health and Cancer Risk

The World Health Organisation (WHO) (2004) strictly recommends a reference limit of 15 μg/l of U in drinking water considering it to be nephrotoxic. There are enough scientific evidence about the association of nephritis due to excess amount of U exposure [Hursh and Spoor 1973, World Health Organisation (WHO) 2006]. In 1991, the US Environmental Protection Agency (USEPA) classified U as a confirmed human carcinogen (group A) and suggested zero tolerance only as a safe acceptable limit for carcinogenic risk. Health effects of U can be divided into two types of group: Carcinogenic (radiological risk) and non-carcinogenic (chemical risk) [Health Canada 1999; World Health Organisation (WHO) 1998].

Kumar et al. 2011b calculated cancer risk on the basis of uranium concentration in groundwater determined by them. According to their findings, over a lifetime consumption of groundwater at an average 4.05 l/day and present U levels, the mean cancer risk in all the four districts, that is, Mansa, Bathinda, Faridkot and Ferozepur, in south-west Punjab was found to be 2.03×10^{-4} (about 2 per 10,000 people), whereas at the 95th percentile (worst case), it rose to 5×10^{-4} (about 5 per 10,000). At the 50th percentile (medium), they observed that both radiological risk (carcinogenicity) and chemical risk (non-carcinogenic toxicity) for kidney toxicity were observed to be well within acceptable limits. However, the author is of the opinion that more data on U concentration in groundwater as well intake of U through crops, milk, etc. along with epidemiological data, specially related to kidney diseases, need to be collected to know the effects on human health.

Arsenic

In India, the effects of arsenic contamination in groundwater were first noticed in July 1983 in West Bengal where a total of 8 million people from 61 blocks were found to be living at a risk of arsenic poisoning. Later, some more areas of Bihar, Chhattisgarh and UP states have been found to be suffering from arsenic contamination. In Punjab state, Hundal et al. 2007 reported large scale of contamination of arsenic in groundwater. In the south-western districts (Bathinda, Mansa, Ferozepur and Faridkot), they reported concentration range of 11.4–688 μg/l with an average concentration of about 76.8 μg/l. The safe limit of arsenic in groundwater for drinking water is 10 μg/l as per WHO standards 2004.

The author also investigated the arsenic problem in groundwater, especially in some parts of south-western Punjab where very high concentration of arsenic was reported by Hundal et al. 2007. According to the results of the studies, the arsenic hazard is not as severe as reported by Hundal et al. 2007. However, local arsenic hot spots having concentration more than 10 μg/l do occur sparingly, and occurrence of such hot spots is controlled by geomorphic, hydrological and mineralogical factors. On the basis of data reported by Hundal et al. 2007; Jain and Kumar 2007 and Central Ground Water Board 2005, a map (Fig. 23.2) has been prepared showing distribution of arsenic hot spots in the state. The map shows that high values of arsenic hot spots do occur along hydrological boundaries, that is, near flood plains or along palaeo-flood plains. The Water Supply and Sanitation Department, Punjab (WSSP), also found high concentration of arsenic around areas of Ravi River in Gurdaspur district (personal communication). Further, it has been found that the relationship of occurrence of arsenic in Punjab state follows similar relationship observed in Uttar Pradesh, Bangladesh and elsewhere where mobile arsenic is associated with younger Pleistocene-Holocene alluvial sediments, which contain more organic matter and thus have a more reducing geochemical environment facilitating release of arsenic in groundwater. (Nickson et al. 2000; Srivastava et al. 2008).

In view of the above observations, it is suggested to monitor arsenic content of groundwater in villages located close to rivers and flood plains on priority to formulate an action plan to tackle the problem.

Health Effects

Health problems resulting from chronic exposure to arsenic in drinking water consist mainly of skin disorders, notably pigmentation changes (melanosis) and keratosis, although skin cancers have also been identified (Smedley and Kinnisburgh 2005). In addition, it also causes decrease in heart contractility, blood vessels damage and injuries to nervous system, kidney and liver. Cases of arsenic skin disorders were not observed in the area during field surveys.

Radon

Radon is natural radioactive gas that cannot be seen, smelt or tasted, but can be detected by

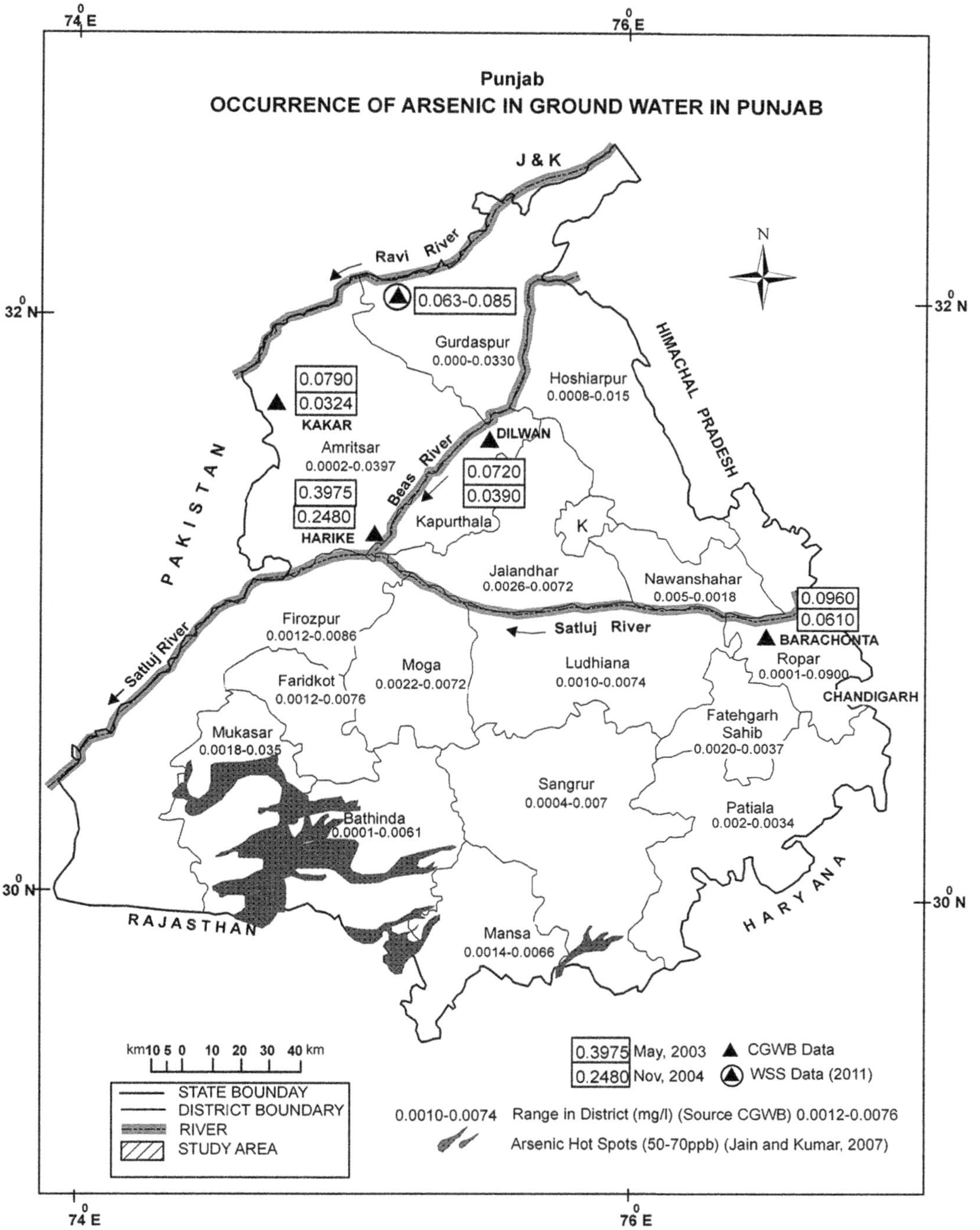

Fig. 23.2 Occurrence of arsenic hot spots in groundwater in Punjab state based on the data collected by the author and other researchers

special geophysical equipment. It is produced by the radioactive decay of radium, which in turn is derived from the radioactive decay of U. Kochhar and Dadwal (2004) reported radon concentration in groundwater from adjoining areas of Talwandi Sabo and found its concentration ranging between 2.28 and 7.79 Bq/l against the safe limit of 400 Bq/l prescribed by International Council for Radiation Protection.

Health Effects

Radon decays to form radioactive particles that can enter the human body by inhalation. Inhalation of radon has been linked to an increase in risk of developing cancers of respiratory tracts, especially of lungs (Appleton 2005). In the south-west Punjab, very low concentration of radon has been reported within safe limits, indicating that it has not affected the health of people.

Other Heavy Metals

Due to absence of major industrial units in south-west Punjab, groundwater contamination by heavy metals such as chromium, cadmium, nickel and lead observed in Ludhiana area (Singh 1994) is not prevalent in south-west parts of Punjab. However, localised point occurrence of such heavy metals in low concentration is often observed (Kochar et al. 2007; Thakur et al. 2008). Kochar et al. (2007) attributed such occurrence due to use of pesticides. Kumar et al. (2009) reported the results of Cu, Mn and Zn analysis from groundwater in Muktsar district. Cu and Zn were not detected in most of the samples, whereas Mn showed a highest concentration of 6.6 mg/l in post-monsoon data, which is more than the permissible limit of 0.3 mg/l prescribed by WHO (2004). Singh et al. 2010 reported high concentration of strontium in association of U in Talwandi Sabo. Alrakabi et al. 2012 also reported high concentration of strontium and bromide, besides U in Jajal, Giana and Malkana villages of Talwandi Sabo block in Bathinda district. They observed Mo concentration below EPA limit of 0.07 mg/l in samples taken from Bathinda city. Bromide, strontium and uranium concentrations showed an increase in concentration with increase in salinity.

Thakur et al. (2008) detected As, Cd, Cr and Se, in nine groundwater samples of Talwandi Sabo area. As and Se were detected more than maximum permissible limit (MPL) during March 2004. However, in April 2004, only As, Cr and Se were detected, but none of these exceeded MPL for drinking water. In May 2004, only Hg exceeded the MPL. Such temporal behaviour of heavy metals indicates that they are not present consistently. Temporal data of nine surface water samples collected in March–April and May 2004 indicate that none of the elements such as Cr, Se, Hg, Cd and As exceeded MPL prescribed for drinking water by WHO 2004.

Health Effects

Heavy metals (Cr, Cd, Se, Hg, Zn, Cu, etc.) discussed above are known to have adverse effect on human health when present more than the prescribed concentration by WHO 2004, but it has been observed by the author that such localised high values are often seasonal and not consistent as revealed by the data reported by Thakur et al. 2008 and Kumar et al. 2009. Therefore, with very limited data available, it is not possible to comment on their adverse impact on human health in the area. However, monitoring of heavy metals concentration in surface as well as groundwater needs to be taken up to keep a watch on heavy metal contamination.

Pesticides

Farmers in south-west Punjab use pesticides such as Lindan, Heptachlor and Endosulfan, Ethion and Chlorpyrisfos to save the cotton crops from pests. It has been observed that they use 30–35 sprays against 8–10 recommended by agriculture experts. Depending upon the nature of unsaturated zone sediments and depth to water table, they get leached out into the groundwater. Thakur et al. 2008 reported concentration of Heptachlor and Malathion γ-HCH more than the MPL, that is, 0.00003 and 0.0005 mg/l for Heptachlor and Malathion, respectively, in Talwandi Sabo area. They observed that Heptachlor was consistently present in groundwater having concentration more than MPL in March, April and May 2004. Studies carried out by Rao et al. (2004) and Rao and Suriniadu (2010) in Muktsar area revealed the persistence of Organo-Chlorine pesticides of BHC (Lindan), Heptachlor, Epoxide, Endosulphane 1,4,4—DDT, Methoxychlor and Organo-Phosphorus pesticides of Methyl Parathion, Dimethoate and Malathion during July and November 2003 and May 2004. The authors attributed higher concentration of pesticides in Muktsar area, which was found to be 6–8 times higher than in Ludhiana district, to cropping pattern (cotton), higher water table and fast leaching rate.

Health Effects

Pesticides enter into human body from consumption of pesticide-laden water, through vegetables and milk and also from direct inhalation during sprays which farmers carry out without taking necessary precautions as revealed during field observations and affect human health. In such a scenario, likelihood occurrence of cancer gets increased in vulnerable population. Further presence of pesticides in the blood samples reported by Narain and Chandra (2005) support the above statement. The long-term exposure to pesticides such as Organo-Chlorine, Creosote and Sulfallate have been correlated with higher cancer rates, whereas Organo-Chlorines, DDT, Chlordane and Lindan are known to be tumour promoters in animals (Bhalla 2010). A comprehensive study carried out by Thakur et al. 2008 has underlined the direct relationship between in discriminate use of these pesticides and the increased incidence of the cancer in the south-west parts of Punjab.

Conclusions

South-west parts of Punjab state suffer from severe water quality problems such as salinity, alkalinity and sodicity. Groundwater of top aquifer, which is tapped by hand pumps up to 30 m and shallow tube wells up to 50–70 m depth, has fluoride as well as uranium hazard.

High fluoride content in groundwater is responsible for large-scale bone and joint problems and fluorosis noticed during health surveys. Uranium hot spots having U content more than 60 μ/l (AERB, DAE 2004) are widespread especially in highly saline zones. High uranium content in drinking water is nephrotoxic. Therefore, there is a need to take up detailed health survey in the area to examine nephritis and related problems. Arsenic hot spots occur locally depending upon favourable geomorphological, hydrochemical and mineralogical regime, but are not widespread. No patient with arsenical skin lesions in Talwandi Sabo and Muktsar-Malout belt was noticed during field surveys.

Heavy metals such as Cr, Cd, Hg and Se show temporal variations and are not consistently present. Radon in groundwater is present within the safe limits. Pesticides are used in excess to save the cotton crops from pests and insects without taking any precaution resulting also in inhalation of pesticides. Pesticide residues are consistently present in groundwater and have affected the human health.

In order to alleviate health problems in the area, alternate surface water–based drinking supply needs to be enhanced in the area to meet the drinking water shortage. In addition, reverse osmosis plants need to be installed to remove uranium, fluoride and other toxics. Awareness and education programme regarding restricted use of pesticides and water and health issues need also to be taken up.

Acknowledgments The author is thankful to University Grants Commission (UGC) for awarding Emeritus Fellowship and sanctioning a major research project which helped in carrying out more research on these aspects. Thanks also to Chairman Geology Department, Panjab University, Chandigarh, for providing facilities.

References

Alrakabi M, Singh G, Bhalla A, Kumar S, Srivastava A, Rai B et al (2012) Study of uranium contamination of ground water in Punjab state in India using x-ray fluorescence technique. J Radioanal Nucl Chem. doi: 10.1007/s10967-011-1585-x

Appleton JD (2005) Radon in air and water. In: Olle S, Brian A, Centeno JA, Finkelman RB, Fuge R, Lindh U, Smedley P (eds) Essentials of medical geology: impacts of the natural environment on public health. Elsevier Academic Press, London, pp 227–262

Atomic Energy Regulatory Board, Department of Atomic Energy (AERB, DAE) (2004) Drinking water specifications in India. Mumbai, India

Bhalla A (2010) The green revolution, organochlorine compounds and cancer. In: Gill SS (ed) Impact of uranium and other heavy metals on health. Saurabh Medical Publishers, Chandigarh, pp 209–211

Bureau of Indian Standards (BIS) (2003) Indian standard-drinking water-specification (first revision) IS10500:1991, 2.2 edn (2003–2009)

Central Ground Water Board (2005) Occurrence of arsenic in shallow groundwater of Haryana, Punjab and U.T of Chandigarh. Technical report, North–Western region, Chandigarh

Central Ground Water Board (2010) Groundwater quality in shallow aquifers of India. Published technical report, central ground water board, ministry of water resources, Govt. of India, Faridabad

Dhanda DS, Singh KP, Chand H, Virk P, Brar KK (2011) Assessment of groundwater quality for drinking purpose in Talwandi Sabo block of southwest part of Punjab. Int J Environ Sci 1:159–166

Dissanayake CB (1991) The fluoride problem in the groundwater of Sri Lanka: environmental management and health. Int J Environ Stud 38:137–156

Freni SC (1994) Exposure to high fluoride concentration in drinking water associated with decreased birth rates. J Toxicol Environ Health 42(1):109–121

Health Canada (1999) Uranium in drinking water. Document for public comment prepared by federal provincial sub committee on drinking water, Ottawa, ON, Canada

Hundal HS, Kumar R, Singh K, Singh D (2007) Occurrence and geochemistry of arsenic in groundwater of Punjab, northwest India. Commun Soil Plant Anal 38:2257–2277

Hursh JB, Spoor NL (1973) Data on man. In: Hodge HC, Stannard JN, Hursh JB (eds) Handbook of experimental pharmacology, 36. Uranium, Plutonium Transplutonium elements. Springer, Berlin, pp 197–240

Jacks G, Bhattacharya P, Chaudhary V, Singh KP (2005) Controls on the genesis of some high fluoride groundwaters in India. Appl Geochem 20:221–228

Jain AK, Kumar R (2007) Water management issues: Punjab, North–West India. Paper presented at the Indo-US workshop on innovative e-technologies for distance education and extension/outreach for effective water management. ICRISAT, Hyderabad, India

Kochhar N, Dadwal V (2004) Radon and chemical quality of ground water in parts of SW Punjab in relation to the buried Aravalli Delhi ridge. In: Proceedings workshop on medical geology, IGCP—454, special publication no. 83, geological survey of India, Calcutta, pp 293–298

Kochhar N, Gill GS, Tuli N, Dadwal V, Balram V (2007) Chemical quality of ground water in relation to cancer in parts of SW Punjab, India. Asian J Water Environ Pollut 4:107–112

Kumar M, Kumari K, Ramanathan AL, Saxena R (2007) A comparative evaluation of groundwater suitability for irrigation and drinking purposes in two intensively cultivated districts of Punjab, India. Environ Geol 53:553–574

Kumar M, Kumari K, Singh UK, Ramanathan AL (2009) Hydrogeochemical processes in the groundwater environment of Muktsar, Punjab: conventional graphical and multivariate statistical approach. Environ Geol 57:873–884

Kumar R, Singh B, Hundal HS, Singh K, Singh Y (2010) Geospatial quality of under groundwater of Punjab. Published report, department of soils, Punjab Agricultural University, Ludhiana, pp 1–72

Kumar A, Rout S, Naranayanan U, Mishra MK, Tripathi RM, Singh J et al (2011a) Geochemical modelling of uranium speciation in subsurface aquatic environment of Punjab State in India. J Geol Min Res 3(5):137–146

Kumar A, Usha N, Sawant PD, Tripathi RM, Sanu S, Mishra M et al (2011b) Risk assessment for natural uranium in subsurface water of Punjab State India. Int J 17(2):381–393

Lu Y, Sun ZR, Wu LN, Wang X, Lu W, Liu SS (2000) Effect of high fluoride water on intelligence in children. Flouride 33(2):74–78

Narain S, Chandra B (2005) Evidence of damage. Down to Earth 28:27–29

Nickson RT, McArthur JM, Ravenscroff P, Burges WG, Ahmed KM (2000) Mechanism of arsenic release to groundwater in Bangladesh and West Bengal. Appl Geochem 15:403–413

Raja Reddy D (1979) In: Vinken PJ and Bruyn GW (eds) Hand book of neurology. North Holland Publishing co., Amsterdam, pp 4–65

Rao VVSG, Suriniadu (2010) Simulation of persistence residues in groundwater of Ludhiana and Muktsar districts, Punjab. In: Rao MS, Khobragade S, Kumar B, and Singh RD (eds) Water availability and management in Punjab. National institute of hydrology, Roorkee, pp 131–138

Rao VVSG, Sankaran S, Prakash BA, Kumar KM, Yadiah P, Chandrashekar SVN (2004) Assessment of groundwater quality in Ludhiana and Muktsar districts, Punjab. In: Dilawari VK, Brar LS, Jalota SK (eds) Sustaining agriculture problems and prospects. Punjab Agricultural University, Ludhiana and Technology Information and Forecasting Assessment Council, Department of Science and Technology, New Delhi, pp 47–55

Sharma PK, Sood A, Setia RK, Singh N, Menakshi, Verma VK, et al (2010) Groundwater quality for irrigation in Punjab. Published report, Punjab remote sensing center, Ludhiana, pp 1–163

Singh KP (1994) Temporal changes in the chemical quality of groundwater in Ludhiana area. Curr Sci 66:375–377

Singh KP (2002) Development of water logging and salinity in South West Punjab. In: Pathan AM, Thingale SS (eds) Contribution to environmental geosciences. Commemorative volume in honour of Prof. K.B. Powar, Aravali Books International Pvt. Ltd., New Delhi, pp 97–106

Singh KP (2010) Uranium in groundwater in Punjab state, India: status paper. Punjab Envis Cent 8:1–13

Singh KP, Kishore N (2009) Hydrogeology and hydrochemistry of parts of Talwandi Sabo block, district Bhatinda, Punjab with special reference to cancer cases. Indian J Environ Prot 2:97–103

Singh KP, Tewari BS, Williams THL (1980) Application of Landsat imagery to groundwater studies in parts of Punjab and Haryana states. In: The contribution of space observations to water resources management, 22nd COSPAR, Pergaman Press, Oxford and New York, pp 107–111

Singh KP, Kishore N, Mehta D (2010) Uranium in groundwater in Punjab state India: some observations. In: Gill SS (ed) Impact of uranium and other heavy metals on health. Saurabh Medical Publishers, Chandigarh, pp 39–52

Smedley P, Kinnisburgh DG (2005) Arsenic in ground water and the environment. In: Olle S, Brian A, Centeno JA, Finkelman RB, Fuge R, Lindh U, Smedley P (eds) Essentials of medical geology: impacts of the natural environment on public health. Elsevier Academic Press, London, pp 263–299

Srivastava AK, Govil PC, Tripathi RS, Shukla RS, Srivastava RS, Vaish DP, Nickson RT (2008) Initial data on arsenic in groundwater and development of a state action plan, Uttar Pradesh, India. In: Bhattacharya P, Ramanathan AL, Mukherjee AB, Bundschuh J, Chandrasekharam D, Keshari AK (eds) Groundwater for sustainable development: problems perspectives and challenges. Taylor and Francis, London, pp 271–281

Thakur JS, Rao BT, Rajwanshi A, Parwana HK, Kumar R (2008) Epidemiological study of high cancer among rural agricultural community of Punjab in Northern India. Int J Environ Res Public Health 5(5):399–407

United States Environmental Protection Agency (USEPA) (1991) EPA standard default exposure factors. Washington DC, USA

United States Environmental Protection Agency (USEPA) (2002) National primary drinking water regulation, radionuclides. Final Rule, Washington DC, USA

World Health Organisation (WHO) (1998) Guidelines for drinking water quality, vol. 2, 2nd edn. Addendum to: health criteria and other supporting information, WHO/EOS/98.1, Geneva, Switzerland, p 283

World Health Organisation (WHO) (2004) Guidelines for drinking water quality, 3rd edn. Geneva, Switzerland

World Health Organisation (WHO) (2006) Guidelines for drinking water quality. First addendum to, vol 1, 3rd edn. Recommendations, p 455

Groundwater Chemical Contamination: The Implication for Human Health

24

Anita Rana and D. M. Craig

Abstract

Daurala is a small agricultural town located in western Uttar Pradesh. The majority of inhabitants rely on groundwater as their primary source of drinking water. In 2005, research revealed high concentrations of toxic metals in drinking water samples. Specifically, lead exceeding the maximum contaminant level (U.S. EPA) of 0.015 mg/l was found to be present in each sample, with the highest level recorded at 5.51 mg/l. Being an agricultural hub, waste water in Daurala is regularly utilized for crop irrigation. With concentrations of arsenic, chromium, copper and cyanide also found in waste water samples, soil was also analysed to acquire an understanding of the scale and spread of contamination. Concentrations of lead were again found in each sampled location, in addition to toxic metals such as arsenic and cyanide. To determine the impact of this on the health of Daurala's inhabitants, a survey covering 14,888 of the population was conducted, with results indicating that approximately 23 % of residents were in a state of ill health. Gastro-intestinal diseases were some of the common conditions experienced in Daurala, with 1,007 people suffering from similar ailments. Exposure to copper is a known cause of these health problems, and this is one of many links between the chemical contamination of water and human health which have been identified through this study. This research also provides an insight into how effective communication and partnership between local stakeholders can both address and resolve water pollution issues for the benefit of a community as a whole.

Keywords

Contamination · Groundwater · Industrialization · Health · Heavy metals

A. Rana (✉) · D. M. Craig
Janhit Foundation, 180/7, Shastri Nagar, Meerut
250002, India
e-mail: janhitfoundation@gmail.com

P. P. Singh and V. Sharma (eds.), *Water and Health*,
DOI: 10.1007/978-81-322-1029-0_24,

Introduction

With various estimates suggesting that groundwater accounts for nearly 90 % of rural and 30 % of urban domestic water needs in India (Srikanth 2009), maintaining a clean and plentiful groundwater supply is critical for the ever increasing demand for drinking water in India to be met. The population of India currently stands at approximately 1.21 billion, over one sixth of the entire global population, and this, combined with rapid economic growth, places a significant strain on environmental resources, of which groundwater is one (Greenstone and Hanna 2011).

A complex framework of pollution control legislation exists in India, and in 1986, the Environment Protection Act was enshrined into statute, giving control to State Pollution Boards to prosecute those who fail to comply with regulations. This provides the legal basis for what is widely known as the 'Polluter Pays Principle' to be implemented.

However, despite these changes, implementation of this legislation has proved difficult. Tort law is applied by law courts in civil prosecution cases, where environmental polluters are penalized financially, with the proceeds being used to remedy the damage done to the environment. In India, Tort law is in the early stages of development, and, therefore, the judiciary does not have sufficient access to past precedents on which they can base new legal interpretations (Ramanathan 2001).

This chapter focuses on an area which has been badly affected by severe groundwater contamination. The case in point is the town of Daurala, located in Meerut district in western Uttar Pradesh. A study evaluating groundwater quality and the health of the local population was undertaken in 2005. This research revealed that the groundwater of Daurala was not suitable for drinking, with the presence of various heavy metals, including arsenic, lead and cyanide. This evidence was corroborated by the results of an area wide health survey which indicated that approximately 23 % of the surveyed population of Daurala were showing symptoms of illnesses which were directly attributable to the toxic substances present in the groundwater. Furthermore, the results of this study identified a clear link between the contamination of drinking water and the discharge of industrial effluent in Daurala.

This paper not only demonstrates the devastating impact of contaminated drinking water on human health, but also highlights how positive communication and dialogue between stakeholders in a given area can resolve existing problems and provide a clean, pure water supply for all.

Context

The small town of Daurala is situated within the district of Meerut, in western Uttar Pradesh. It is located 14 km from district headquarters and has a population of approximately 20,000. It lies within Daurala block and is bordered by Muzaffarnagar district to the north. Traditionally, Daurala has been a rural agricultural community, largely dependent on the growth of sugarcane and the sugar processing industries to which the crop is supplied. Throughout the latter half of the twentieth century, as India began to experience rapid economic development, urbanization came to Daurala and various industrial firms began to locate their premises within the locality.

In 2003, local media coverage reported a rise in severe health problems and deaths in Daurala, with reports suggesting that the problems were being caused by groundwater contamination. Janhit Foundation was approached by local sources to verify these claims and subsequently undertook an initial basic study of groundwater quality. The preliminary results showed levels of dissolved solids (TDS) to be between 1,610 and 2,590 mg/l, far exceeding the acceptable limit of 500 mg/l, as set out by the U.S. Government's Environmental Protection Agency (EPA) Drinking Water Standards (EPA 2012).

Furthermore, concentrations of heavy metals were found to be present in the drinking water sampled, including aluminium and chromium. These results provided the necessary scope to undertake a full and detailed study of groundwater quality in Daurala, in addition to a survey to establish the health of the residents of Daurala.

Water Quality Analysis

At the time of the study in 2005, the residents of Daurala obtained their water from two sources. The vast majority of surveyed households (75 %) sourced their drinking water solely from handpumps, whilst only a quarter of the surveyed population had access to overhead tanks, which were filled with groundwater extracted by electrically operated submersible pumps. To establish the quality of water in Daurala, various parameters were measured, with samples of drinking water, waste water and agricultural soil collected.

Drinking water samples were taken from handpumps and tubewells at seven separate sites, with high concentrations of lead found in each of the samples analysed. The U.S. EPA Drinking Water Standards (U.S. Environmental Protection Agency (EPA) 2012) stipulate that lead should be completely absent from drinking water, and if levels are found to be over 0.015 mg/l (maximum contaminant level (MCL)), action must be taken to remediate the situation. The highest recorded level of 5.51 mg/l of lead was found in water sourced from the handpump located on the Main Sardhana Road. The remaining samples were found to have concentrations of lead ranging from 0.13 mg/l to 0.48 mg/l. Each drinking water sample therefore exceeded the MCL for lead. Concentrations of iron, nickel and zinc were also found to be present in many of the drinking water samples collected.

Waste water samples were collected from three different sites, two of which included a road drain and an industrial collection pond. Levels of cyanide in the industrial collection pond were recorded to be 0.8 mg/l, 0.6 mg/l above the maximum contaminant level (MCL) of 0.2 mg/l, as set by the U.S. EPA Drinking Water Standards. The MCL for chromium is 0.1 mg/l, and 0.19 mg/l of chromium was recorded in the groundwater close to the site of the industrial collection pond. Levels of 0.5 mg/l of arsenic were found in the water of the collection pond, which exceed the MCL of 0.010 mg/l. In addition to this, lead concentrations above the MCL were found at all three sampling sites, with concentrations peaking at 77.32 mg/l in the collection pond, whilst 6.36 mg/l of lead was found to be present in the wastewater of the Daurala Sardhana road drain and 0.63 mg/l in the groundwater close to the collection pond.

Soil samples were taken from four separate sites in Daurala, with concentrations of lead found in all samples, each exceeding the permissible U.S. EPA MCL. The soil sample taken from Bablu's field had the highest concentration of lead, at 310 parts per million (ppm), which is more than double the normal permissible limit of 15 ppm. Furthermore, cyanide was also detected in this soil sample. In terms of other contaminants, all the soil samples analysed were found to contain zinc and copper, whilst in Satendra's field, arsenic was found to be present.

The aforementioned water quality analysis revealed significantly high concentrations of heavy metals present in both drinking water and waste water in Daurala. With the practice of utilizing waste water for the irrigation of agricultural land commonplace in Daurala, and being a largely agricultural area, it is significant that the soil samples collected indicated the presence of the same contaminants which were found in drinking water and waste water.

Human Health in Daurala

To determine the health of the local population, a door-to-door questionnaire was conducted, surveying 2,291 households and 14,888 people in May 2005. The findings of this study conclude that many of the diseases and illnesses suffered by the people of Daurala were directly attributable to the contamination of water by heavy metals.

Of the total surveyed population of Daurala, 3,488 are suffering from various illnesses, approximately 23 %. In the five years leading up to the year of the survey (2000 to 2005), 192 deaths were recorded, of which 54 were due to cancer, primarily thyroid cancer. Ailments affecting the nervous system were the cause of 15 deaths, whilst 33 people died of heart disease, 6 from kidney failure and 42 from varying gastrointestinal disorders.

Figure 24.1 provides a synoptic overview of those people among the surveyed population of Daurala who were in a state of ill health at the time of the survey in 2005. It is evident that approximately 1,338 patients were suffering from epidermal complaints, 89 from neurological disorders and 14 from cancer. Furthermore, 1,007 were suffering from gastrointestinal ailments and 77 from heart disease. The remaining 963 people were reported to be suffering from respiratory diseases like asthma, arthritis and abnormal blood pressure.

The most prominent heavy metal found in both water and soil in Daurala is lead, concentrations of which were detected in all drinking water samples above the maximum contaminant level. Lead is one of the number of toxic metals which have established health effects. Chronic exposure to lead is known to delay physical and mental development in children and cause kidney problems and high blood pressure in adults (U.S. Environmental Protection Agency (EPA) 2012). Furthermore, lead has no biological utility and once present in the human body it is extremely hard to eradicate.

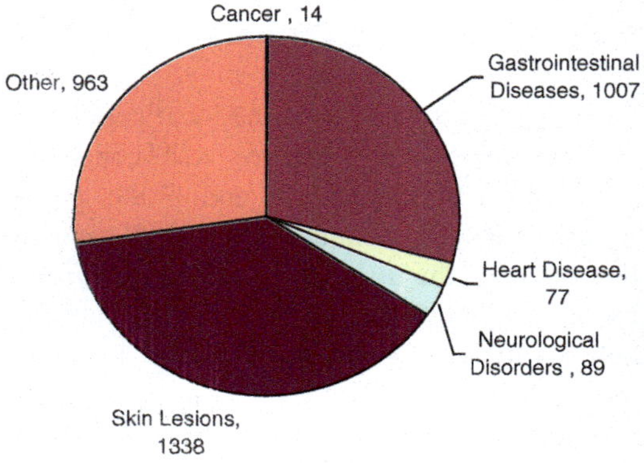

Fig. 24.1 Health patients in Daurala 2005

Two of the most toxic heavy metals are arsenic and cyanide, both found within water and soil in Daurala. Prolonged exposure to arsenic can cause skin damage, problems with the circulatory system and most significantly increase the risk of contracting cancer, whilst cyanide is widely documented to cause thyroid problems and nerve damage (U.S. Environmental Protection Agency (EPA) 2012).

Copper is also present in the majority of drinking water and soil samples collected from Daurala. Short-term exposure to this heavy metal can cause gastrointestinal distress, which is consistent with the symptoms which have been experienced by the population of Daurala, 1007 of whom had gastrointestinal problems at the time of the survey. Long-term exposure to copper can also cause kidney and liver damage, both of which have again been experienced by the inhabitants of Daurala.

Daurala: Post-Study Recommendations

Upon completion of this primary research study, a series of detailed recommendations were set out to remediate the situation in Daurala. Principally, this focused on preventing the discharge of untreated industrial effluent into open drains, effectively eliminating the problem of pollution at source. Local farmers were encouraged to change their practices, notably by stopping the use of waste water for irrigation purposes. Furthermore, to improve the overall drinking water quality, it was advised that a piped water supply system from filtered overhead tanks be installed to supply all households in Daurala.

These recommendations formed the basis for the initiation of a positive dialogue between Janhit Foundation and local industries. This provided a forum for the discussion of appropriate remedial action. In line with the Polluter Pays Principle, local industries acknowledged their moral and financial obligations to the community and environment, and subsequently implemented substantial measures to improve the social, economic and welfare conditions of

Table 24.1 Heavy metals analysis of drinking water samples

S. No.	Parameters (mg/l)	Railway crossing H/P	Daurala organics gate H/P	Daurala sugar works colony	Main Sardhana road (H/P)	Main Sardhana road (T/W)	Daurala village (H/P)	Jhal hotel (H/P)
1	Cyanide	LDL	LDL	LDL	LDL	LDL	LDL	LDL
2	Chromium	LDL	LDL	LDL	–	–	–	–
3	Iron	0.11	0.23	0.55	–	–	–	–
4	Nickel	LDL	0.01	0.02	LDL	LDL	LDL	LDL
5	Arsenic	LDL	LDL	LDL	LDL	LDL	LDL	LDL
6	Lead	0.46	0.22	0.48	5.51	0.29	0.13	0.18
7	Copper	–	–	–	0.04	0.01	0.05	0.01
8	Zinc	–	–	–	0.27	0.35	0.13	0.38

Key LDL Less than the detectable limit. *H/P* Handpump. *T/W* Tube well

Table 24.2 Heavy metals analysis of waste water samples

S.No.	Parameters (mg/l)	Daurala Sardhana road drain	Daurala organics collection pond	Near collection pond
1	Cyanide	0.05	0.80	LDL
2	Chromium	0.02	0.20	LDL
3	Iron	0.83	15.40	0.33
4	Nickel	0.08	0.17	0.01
5	Arsenic	0.10	0.50	LDL
6	Lead	6.36	77.32	0.63

Table 24.3 Heavy metals analysis of soil samples

S.No.	Parameters (ppm)	Satendra's field	Yogendra's field	Bablu's field	Main road field
1	Nickel	300	17,800	1,100	LDL
2	Copper	3,100	4,000	4,700	3,700
3	Arsenic	2,00	LDL	LDL	LDL
4	Cyanide	LDL	LDL	800	LDL
5	Zinc	7,600	7,200	7,600	7,900
6	Lead	220	270	310	260

Table 24.4 Human health in Daurala

Type of illness/disease	Number of people suffering within surveyed population of Daurala
Gastrointestinal diseases	1,007
Cancer	14
Heart disease	77
Skin lesions	1,338
Neurological disorders	89
Others	963

the local population of Daurala. This is testament to what can be achieved when all stakeholders in a specific location come together to address a common objective. (Tables 24.1, 24.2, 24.3, 24.4).

Acknowledgments We would like to acknowledge the Department of Civil Engineering at the Indian Institute of Technology (IIT), Roorkee and the Centre of Science and Environment's Pollution Monitoring Laboratory, New Delhi, for carrying out the relevant scientific water and soil analysis. We are also extremely grateful to A. Nair,

P. Bharadwaj, N. Sharma, R. Tyagi, B. Singh and A. Kumar for their administrative and technical assistance in undertaking the study in 2005. We would also like to recognize and pay tribute to the community of Daurala and DCM Shriram Industries Limited for their cooperation and assistance in remediating the situation in Daurala. Our final thanks are reserved for The Blacksmiths Institute, for without their financial support, this study would not have been realizable.

References

Greenstone M, Hanna R (2011) Environmental regulations, air and water pollution, and infant mortality in India. HKS Faculty Research Working Paper Series RWP11-034, Sept 2011. http://web.hks.harvard.edu/publications/citation.aspx?PubId=7964

Ramanathan U (2001) Tort Law in India. Annual survey of Indian law , pp 615–628 (2002). http://www.ielrc.org/content/a0206.pdf

Srikanth R (2009) Challenges of sustainable water quality management in rural India. Curr Sci 97(3): 317–325

U.S. environmental protection agency (EPA) (2012) National primary drinking water regulations. Available at: http://water.epa.gov/drink/contaminants/index.cfm#List (accessed Feb 24, 2012)

The manufacturer's authorised representative in the EU is Springer Nature Customer Service Centre GmbH, Europaplatz 3, 69115 Heidelberg, Germany. If you have any concerns regarding our products, please contact ProductSafety@springernature.com

Printed and bound by CPI Group (UK) Ltd, Croydon, CR0 4YY
30/09/2026
02225445-0020